Truss Optimization

Vimal Savsani • Ghanshyam Tejani • Vivek Patel

Truss Optimization

A Metaheuristic Optimization Approach

Springer

Vimal Savsani
Trades and Technology
Canadore College
North Bay, ON, Canada

Ghanshyam Tejani
School of Engineering and Technology
LJ University
Ahmedabad, Gujarat, India

Vivek Patel
Department of Mechanical Engineering
Pandit Deendayal Energy University
Gandhinagar, Gujarat, India

ISBN 978-3-031-49294-5 ISBN 978-3-031-49295-2 (eBook)
https://doi.org/10.1007/978-3-031-49295-2

This Springer imprint is published by the registered company Springer Nature Switzerland AG
The registered company address is: Gewerbestrasse 11, 6330 Cham, Switzerland

Paper in this product is recyclable.

Preface

There are three categories of truss optimization: size optimization, shape optimization, and topology optimization. Size optimization involves finding the optimal cross-sectional areas for truss elements, while shape optimization involves determining the optimal nodal positions for certain nodes. Topology optimization involves adding or removing elements, which makes it more challenging since it requires consideration of all possible topologies rather than just a specific one. The effect of simultaneous topology, shape, and size (TSS) variables on both the objective function and constraints is fairly unlike. Therefore, simultaneous TSS optimization is a challenging problem for optimization algorithms. The TSS variables have a different effect on the objective function and constraints, making simultaneous optimization a difficult task.

Designing an optimal truss that can withstand dynamic behavior is a challenging area of research, which has been actively pursued for many years. The fundamental natural frequencies of a truss are an important parameter for improving its dynamic behavior. Certain restrictions on the natural frequencies can prevent resonance with external excitations. Additionally, trusses should be lightweight, but weight reduction conflicts with frequency constraints and adds complexity to the optimization process. Thus, simultaneous TSS optimization with multiple natural frequency constraints is even more complex and often leads to divergence. Therefore, an efficient metaheuristic optimization approach is needed to design trusses with fundamental frequency constraints, and various metaheuristics have been employed for this purpose. However, this area has not been thoroughly addressed.

In this book, 10 fundamental benchmark trusses (i.e., 10-bar, 14-bar, 15-bar, 24-bar, 20-bar, 72-bar (3D), 39-bar, 45-bar truss, 25-bar (3D), and 39-bar (3D) trusses) subjected to static (i.e., stress, displacement, and buckling) and dynamic (i.e., natural frequency) constraints with multiload conditions are proposed. There are 30 benchmark problems in total, with 10 for size optimization, 10 for topology and size optimization, and 10 for topology, shape, and size optimization. The goal of the book is to study the search performance of metaheuristic algorithms developed after 2011 for truss optimization problems.

The book examines various metaheuristic methods, including the dragonfly algorithm (DA), multi-verse optimizer (MVO), sine cosine algorithm (SCA), whale optimization algorithm (WOA), ant lion optimizer (ALO), heat transfer search (HTS), passing vehicle search (PVS), symbiotic organisms search (SOS), gray wolf optimizer (GWO), and teaching-learning based optimization (TLBO) algorithms. These algorithms are applied successfully to the 30 benchmark trusses proposed in this study to evaluate their search performance. Metaheuristics are problem-independent methodologies that provide a set of rules for solving optimization problems. The proposed algorithms are tested on 30 problems, and many publications based on the application of these algorithms can be found in various international journals published by reputable publishers such as Springer, Elsevier, IEEE, ASME, ASCE, IGI Global, and Taylor and Francis.

This book offers an in-depth understanding of truss structure modeling, metaheuristic algorithms, and their application for optimizing various truss structures. Specifically, the book covers 10 metaheuristic algorithms and 30 truss structures that have continuous and discrete sections. The optimization of all 30 truss structures is implemented using each algorithm, and this book provides a detailed comparison of the performance of each algorithm. Each metaheuristic algorithm simulates different natural phenomena, and their search characteristics are unique, which affects their performance on different types of problems. Therefore, a thorough performance comparison of each algorithm for the optimization of each truss structure is included in the book. Additionally, the book includes statistical analysis of the results, providing a ranking of the various metaheuristic algorithms for optimizing each truss structure.

The book is anticipated to be beneficial for undergraduate and postgraduate students, as well as research scholars in mechanical, civil, and architectural engineering, as it offers a comprehensive understanding of finite element analysis and optimization aspects of various truss structures. It is expected that this book will serve as a valuable reference for individuals interested in conducting research in the field of metaheuristic algorithms and their application to optimize real-world problems. Additionally, engineers, managers, and institutes engaged in applied optimization work, as well as graduate students from various engineering disciplines intending to pursue a career in applied optimization, will find this book useful.

Our project has received continuous support and interest from our publisher, Springer, and we express our gratitude to them. Additionally, we would like to thank several researchers and international journal publishers for granting us permission to reuse certain portions of their research work. Lastly, we would like to convey our special thanks to our family members for their unwavering support throughout the entire book preparation process.

Overall, we put forth wholehearted efforts to ensure that the book is free from any errors, whether it be printing or otherwise. Nevertheless, the possibility of errors in the book cannot be completely ruled out. Therefore, we would appreciate it if the readers could kindly bring to our attention any errors that may have been overlooked in the book.

North Bay, ON, Canada — Vimal Savsani
Ahmedabad, Gujarat, India — Ghanshyam Tejani
Gandhinagar, Gujarat, India — Vivek Patel

Contents

1 Introduction ... 1
References ... 6

2 Methodology ... 9
2.1 Introduction ... 9
2.2 Finite Element Method ... 11
2.2.1 FEM Analysis of 2D Trusses ... 14
2.3 Problem Formulation ... 20
2.4 Proposed Methodology ... 26
2.5 Design Problems ... 29
2.5.1 Problem Complexity of the TSS Optimization Problems ... 29
2.5.2 Analysis of Feasible Search Space in TSS Optimization ... 31
References ... 33

3 Metaheuristics Methods ... 37
3.1 Introduction ... 37
3.2 Metaheuristics ... 39
3.2.1 The DA Algorithm ... 39
3.2.2 The MVO Algorithm ... 40
3.2.3 The SCA Algorithm ... 42
3.2.4 The WOA Algorithm ... 43
3.2.5 The ALO Algorithm ... 44
3.2.6 The HTS Algorithm ... 45
3.2.7 The PVS Algorithm ... 47
3.2.8 The SOS Algorithm ... 48
3.2.9 The GWO Algorithm ... 50
3.2.10 The TLBO Algorithm ... 51
References ... 53

4 Size Optimization 57
4.1 Introduction 57
4.2 Problem Formulation 58
4.3 Results and Discussion on Size Optimization with Continuous Cross Sections 60
4.3.1 Size Optimization of the 10-Bar Truss with Continuous Cross Sections 60
4.3.2 Size Optimization of the 14-Bar Truss with Continuous Cross Sections 73
4.3.3 Size Optimization of the 15-Bar Truss with Continuous Cross Sections 74
4.3.4 Size Optimization of the 24-Bar Truss with Continuous Cross Sections 78
4.3.5 Size Optimization of the 20-Bar Truss with Continuous Cross Sections 82
4.3.6 Size Optimization of the 72-Bar 3D Truss with Continuous Cross Sections 85
4.3.7 Size Optimization of the 39-Bar Truss with Continuous Cross Sections 88
4.3.8 Size Optimization of the 45-Bar Truss with Continuous Cross Sections 91
4.3.9 Size Optimization of the 25-Bar 3D Truss with Continuous Cross Sections 95
4.3.10 Size Optimization of the 39-Bar 3D Truss with Continuous Cross Sections 99
4.3.11 A Comprehensive Analysis 102
4.3.12 The Friedman Rank Test 105
4.4 Results and Discussion on Size Optimization with Discrete Cross Sections 106
4.4.1 Size Optimization of the 10-Bar Truss with Discrete Cross Sections 106
4.4.2 Size Optimization of the 14-Bar Truss with Discrete Cross Sections 109
4.4.3 Size Optimization of the 15-Bar Truss with Discrete Cross Sections 112
4.4.4 Size Optimization of the 24-Bar Truss with Discrete Cross Sections 114
4.4.5 Size Optimization of the 20-Bar Truss with Discrete Cross Sections 118
4.4.6 Size Optimization of the 72-Bar 3D Truss with Discrete Cross Sections 121
4.4.7 Size Optimization of the 39-Bar Truss with Discrete Cross Sections 124

4.4.8 Size Optimization of the 45-Bar Truss with Discrete Cross Sections 127
4.4.9 Size Optimization of the 25-Bar 3D Truss with Discrete Cross Sections 131
4.4.10 Size Optimization of the 39-Bar 3D Truss with Discrete Cross Sections 132
4.4.11 A Comprehensive Analysis 136
4.4.12 The Friedman Rank Test 139
4.5 Multi-objective Optimization for Structure Design 142
4.5.1 Mathematical Formulation of Multi-objective Structure Optimization Problem 142
4.6 The CEC2014 Benchmark Functions 147
References 150

5 Topology and Size Optimization 155
5.1 Introduction 155
5.2 Truss Topology Optimization 156
5.3 Problem Formulation 157
5.4 Results and Discussion on Truss Topology Optimization with Continuous Cross-Sections 159
5.4.1 Topology Optimization of the 10-Bar Truss with Continuous Cross-Sections 159
5.4.2 Topology Optimization of the 14-Bar Truss with Continuous Cross-Sections 173
5.4.3 Topology Optimization of the 15-Bar Truss with Continuous Cross-Sections 176
5.4.4 Topology Optimization of the 24-Bar Truss with Continuous Cross-Sections 177
5.4.5 Topology Optimization of the 20-Bar Truss with Continuous Cross-Sections 180
5.4.6 Topology Optimization of the 72-Bar 3D Truss with Continuous Cross-Sections 183
5.4.7 Topology Optimization of the 39-Bar Truss with Continuous Cross-Sections 186
5.4.8 Topology Optimization of the 45-Bar Truss with Continuous Cross-Sections 190
5.4.9 Topology Optimization of the 25-Bar 3D Truss with Continuous Cross-Sections 193
5.4.10 Topology Optimization of the 39-Bar 3D Truss with Continuous Cross-Sections 197
5.4.11 A Comprehensive Analysis 200
5.5 Results and Discussion on Truss Topology Optimization with Discrete Cross-Sections 203
5.5.1 Topology Optimization of the 10-Bar Truss with Discrete Cross-Sections 203

5.5.2 Topology Optimization of the 14-Bar Truss with Discrete Cross-Sections 206
5.5.3 Topology Optimization of the 15-Bar Truss with Discrete Cross-Sections 209
5.5.4 Topology Optimization of the 24-Bar Truss with Discrete Cross-Sections 212
5.5.5 Topology Optimization of the 20-Bar Truss with Discrete Cross-Sections 215
5.5.6 Topology Optimization of the 72-Bar 3D Truss with Discrete Cross-Sections 216
5.5.7 Topology Optimization of the 39-Bar Truss with Discrete Cross-Sections 219
5.5.8 Topology Optimization of the 45-Bar Truss with Discrete Cross-Sections 224
5.5.9 Topology Optimization of the 25-Bar 3D Truss with Discrete Cross-Sections 227
5.5.10 Topology Optimization of the 39-Bar 3D Truss with Discrete Cross-Sections 231
5.5.11 A Comprehensive Analysis 234
References 236

6 Topology, Shape, and Size Optimization 241
6.1 Introduction 241
6.2 Problem Formulation 242
6.3 Results and Discussion on TSS Optimization with Continuous Sections 245
6.3.1 TSS Optimization of the 10-Bar Truss with Continuous Cross-Sections 245
6.3.2 TSS Optimization of the 14-Bar Truss with Continuous Cross-Sections 266
6.3.3 TSS Optimization of the 15-Bar Truss with Continuous Cross-Sections 268
6.3.4 TSS Optimization of the 24-Bar Truss with Continuous Cross-Sections 270
6.3.5 TSS Optimization of the 20-Bar Truss with Continuous Cross-Sections 272
6.3.6 TSS Optimization of the 72-Bar 3D Truss with Continuous Cross-Sections 274
6.3.7 TSS Optimization of the 39-Bar Truss with Continuous Cross-Sections 278
6.3.8 TSS Optimization of the 45-Bar Truss with Continuous Cross-Sections 283
6.3.9 TSS Optimization of the 25-Bar 3D Truss with Continuous Cross-Sections 286

6.3.10 TSS Optimization of the 39-Bar 3D Truss with Continuous Cross-Sections 290
6.3.11 A Comprehensive Analysis 292
6.3.12 The Friedman Rank Test 303
6.4 Results and Discussion on TSS Optimization with Discrete Sections 305
6.4.1 TSS Optimization of the 10-Bar Truss with Discrete Cross-Sections 305
6.4.2 TSS Optimization of the 14-Bar Truss with Discrete Cross-Sections 309
6.4.3 TSS Optimization of the 15-Bar Truss with Discrete Cross-Sections 313
6.4.4 TSS Optimization of the 24-Bar Truss with Discrete Cross-Sections 316
6.4.5 TSS Optimization of the 20-Bar Truss with Discrete Cross-Sections 320
6.4.6 TSS Optimization of the 72-Bar 3D Truss with Discrete Cross-Sections 324
6.4.7 TSS Optimization of the 39-Bar Truss with Discrete Cross-Sections 329
6.4.8 TSS Optimization of the 45-Bar Truss with Discrete Cross-Sections 334
6.4.9 TSS Optimization of the 25-Bar 3D Truss with Discrete Cross-Sections 338
6.4.10 TSS Optimization of the 39-Bar 3D Truss with Discrete Cross-Sections 342
6.4.11 A Comprehensive Analysis of the Best Results Obtained for TSS with Discrete Sections 346
6.4.12 The Friedman Rank Test 355
References 357

7 Validation 361
7.1 The Finite Element Analysis Process Using ANSYS 361
7.2 Validation of the 10-Bar Truss 363
7.3 Validation of the 14-Bar Truss 364
7.4 Validation of the 15-Bar Truss 367
7.5 Validation of the 24-Bar Truss 369
7.6 Validation of the 20-Bar Truss 371
7.7 Validation of the 72-Bar 3D Truss 373
7.8 Validation of the 39-Bar Truss 375
7.9 Validation of the 45-Bar Truss 378
7.10 Validation of the 25-Bar 3D Truss 380
7.11 Validation of the 39-Bar 3D Truss 382
Reference 384

8 MATLAB Codes of Metaheuristics Methods 385
8.1 The Dragonfly Algorithm 385
8.2 The Multiverse Optimizer 389
8.3 The Sine Cosine Algorithm 393
8.4 The Whale Optimization Algorithm 395
8.5 The Ant Lion Optimizer 398
8.6 The Heat Transfer Search 400
8.7 The Passing Vehicle Search 406
8.8 The Symbiotic Organisms Search 411
8.9 The Grey Wolf Optimizer 417
8.10 The Teaching-Learning-Based Optimization 420

Index 427

List of Figures

Fig. 1.1	The 10-bar truss: (**a**) ground structure and (**b**) size optimization	2
Fig. 1.2	The 52-bar truss: (**a**) ground structure and (**b**) shape and size optimization	2
Fig. 1.3	The 20-bar truss: (**a**) ground structure and (**b**) topology and size optimization	3
Fig. 1.4	The 39-bar truss: (**a**) ground structure and (**b**) topology, shape, and size optimization	4
Fig. 2.1	Ground structures	10
Fig. 2.2	1D bar element	12
Fig. 2.3	2D truss element	12
Fig. 2.4	A three-bar truss structure	14
Fig. 2.5	Free-body diagram	14
Fig. 2.6	Global and local numbering for the three-bar truss	15
Fig. 2.7	Truss topology based on ground structures	21
Fig. 2.8	Flowchart of the TTO problem	27
Fig. 2.9	Schematic representation of proposed methodology	29
Fig. 4.1	The 10-bar truss	61
Fig. 4.2	The 14-bar truss	73
Fig. 4.3	The 15-bar truss	76
Fig. 4.4	The 24-bar truss	79
Fig. 4.5	The 20-bar truss	82
Fig. 4.6	The 72-bar 3D truss	85
Fig. 4.7	The 39-bar truss	88
Fig. 4.8	The 45-bar truss	91
Fig. 4.9	The 25-bar 3D truss: (**a**) ground structure and (**b**) optimal structure	96
Fig. 4.10	The 39-bar 3D truss	99
Fig. 4.11	The 10-bar truss	107
Fig. 4.12	The 14-bar truss	109

Fig. 4.13 The 15-bar truss ... 112
Fig. 4.14 The 24-bar truss ... 114
Fig. 4.15 The 20-bar truss ... 118
Fig. 4.16 The 72-bar 3D truss ... 122
Fig. 4.17 The 39-bar truss ... 125
Fig. 4.18 The 45-bar truss ... 127
Fig. 4.19 The 25-bar 3D truss: (**a**) ground structure and (**b**) optimal structure ... 131
Fig. 4.20 The 39-bar 3D truss ... 135
Fig. 4.21 The 942-bar truss ... 145
Fig. 4.22 Best Pareto fronts of the 942-bar truss ... 146
Fig. 4.23 The Friedman rank test of mean fitness values obtained the algorithms ... 150

Fig. 5.1 The 10-bar truss ... 160
Fig. 5.2 The 14-bar truss ... 174
Fig. 5.3 The 15-bar truss ... 176
Fig. 5.4 The 24-bar truss ... 179
Fig. 5.5 The 20-bar truss ... 182
Fig. 5.6 The 72-bar 3D truss ... 185
Fig. 5.7 The 39-bar truss ... 188
Fig. 5.8 The 45-bar truss ... 191
Fig. 5.9 The 25-bar 3D truss: (**a**) ground structure and (**b**) optimal structure ... 194
Fig. 5.10 The 39-bar 3D truss ... 197
Fig. 5.11 The 10-bar truss ... 203
Fig. 5.12 The 14-bar truss ... 206
Fig. 5.13 The 15-bar truss ... 209
Fig. 5.14 The 24-bar truss ... 212
Fig. 5.15 The 20-bar truss ... 215
Fig. 5.16 The 72-bar 3D truss ... 218
Fig. 5.17 The 39-bar truss ... 221
Fig. 5.18 The 45-bar truss ... 224
Fig. 5.19 The 25-bar 3D truss: (**a**) ground structure and (**b**) optimal structure ... 228
Fig. 5.20 The 39-bar 3D truss ... 231

Fig. 6.1 The 10-bar truss ... 255
Fig. 6.2 Optimal structure of the 10-bar truss ... 265
Fig. 6.3 The 14-bar truss ... 266
Fig. 6.4 Optimal structure of the 14-bar truss ... 268
Fig. 6.5 The 15-bar truss ... 269
Fig. 6.6 Optimal structure of the 15-bar truss ... 270
Fig. 6.7 The 24-bar truss ... 271
Fig. 6.8 Optimal structure of the 24-bar truss ... 272
Fig. 6.9 The 20-bar truss ... 273

Fig. 6.10 Optimal structure of the 20-bar truss 274
Fig. 6.11 The 72-bar 3D truss 275
Fig. 6.12 Optimal structure of the 72-bar 3D truss 277
Fig. 6.13 The 39-bar truss 278
Fig. 6.14 Optimal structure of the 39-bar truss 282
Fig. 6.15 The 45-bar truss 283
Fig. 6.16 Optimal structure of the 45-bar truss 286
Fig. 6.17 The 25-bar 3D truss 287
Fig. 6.18 Optimal structure of the 25-bar 3D truss 290
Fig. 6.19 The 39-bar 3D truss 291
Fig. 6.20 Optimal structure of the 39-bar 3D truss 294
Fig. 6.21 The 10-bar truss: TSS with continuous sections 299
Fig. 6.22 The 14-bar truss: TSS with continuous sections 300
Fig. 6.23 The 15-bar truss: TSS with continuous sections 300
Fig. 6.24 The 24-bar truss: TSS with continuous sections 300
Fig. 6.25 The 20-bar truss: TSS with continuous sections 301
Fig. 6.26 The 72-bar 3D truss: TSS with continuous sections 301
Fig. 6.27 The 39-bar truss: TSS with continuous sections 301
Fig. 6.28 The 45-bar truss: TSS with continuous sections 302
Fig. 6.29 The 25-bar 3D truss: TSS with continuous sections 302
Fig. 6.30 The 39-bar 3D truss: TSS with continuous sections 302
Fig. 6.31 The Friedman rank test of mean mass obtained using TSS for continuous sections 304
Fig. 6.32 The Friedman rank test of success rate of TSS for continuous sections 304
Fig. 6.33 The 10-bar truss 305
Fig. 6.34 Optimal structure of the 10-bar truss 308
Fig. 6.35 The 14-bar truss 309
Fig. 6.36 Optimal structure of the 14-bar truss 312
Fig. 6.37 The 15-bar truss 313
Fig. 6.38 Optimal structure of the 15-bar truss 316
Fig. 6.39 The 24-bar truss 317
Fig. 6.40 Optimal structure of the 24-bar truss 319
Fig. 6.41 The 20-bar truss 320
Fig. 6.42 Optimal structure of the 20-bar truss 324
Fig. 6.43 The 72-bar 3D truss 325
Fig. 6.44 Optimal structure of the 72-bar 3D truss 328
Fig. 6.45 The 39-bar truss 329
Fig. 6.46 Optimal structure of the 39-bar truss 333
Fig. 6.47 The 45-bar truss 335
Fig. 6.48 Optimal structure of the 45-bar truss 337
Fig. 6.49 The 25-bar 3D truss 338
Fig. 6.50 Optimal structure of the 25-bar 3D truss 342
Fig. 6.51 The 39-bar 3D truss 343

Fig. 6.52 Optimal structure of the 39-bar 3D truss 346
Fig. 6.53 The 10-bar truss: TSS with discrete sections 351
Fig. 6.54 The 14-bar truss: TSS with discrete sections 351
Fig. 6.55 The 15-bar truss: TSS with discrete sections 351
Fig. 6.56 The 24-bar truss: TSS with discrete sections 352
Fig. 6.57 The 20-bar truss: TSS with discrete sections 352
Fig. 6.58 The 72-bar 3D truss: TSS with discrete sections 352
Fig. 6.59 The 39-bar truss: TSS with discrete sections 353
Fig. 6.60 The 45-bar truss: TSS with discrete sections 353
Fig. 6.61 The 25-bar 3D truss: TSS with discrete sections 353
Fig. 6.62 The 39-bar 3D truss: TSS with discrete sections 354
Fig. 6.63 The Friedman rank test of mean mass obtained using TSS with discrete sections 356
Fig. 6.64 The Friedman rank test of success rate of TSS with discrete sections 356

Fig. 7.1 Displacement diagrams of the 10-bar truss under multiple load conditions 365
Fig. 7.2 Displacement diagrams of the 14-bar truss under multiple load conditions 366
Fig. 7.3 Displacement diagrams of the15-bar truss under multiple load conditions 368
Fig. 7.4 Displacement diagrams of the 24-bar truss under multiple load conditions 370
Fig. 7.5 Displacement diagrams of the 20-bar truss under multiple load conditions 372
Fig. 7.6 Displacement diagrams of the 72-bar 3D truss under multiple load conditions 376
Fig. 7.7 Displacement diagrams of the 39-bar truss under multiple load conditions 377
Fig. 7.8 Displacement diagrams of the 45-bar truss under multiple load conditions 379
Fig. 7.9 Displacement diagrams of 45-bar truss under multiple load conditions 381
Fig. 7.10 Displacement of the 39-bar 3D truss 384

List of Tables

Table 2.1 The element connectivity 15
Table 2.2 Problem complexity 30
Table 2.3 Proportion of infeasible trusses in the search space (continuous sections) 32
Table 2.4 Proportion of infeasible trusses in the search space (discrete sections) 33

Table 4.1 Design parameters of the 10-bar truss 61
Table 4.2 Size optimization of the 10-bar truss (continuous section) 72
Table 4.3 Design parameters of the 14-bar truss 74
Table 4.4 Size optimization of the 14-bar truss (continuous section) 75
Table 4.5 Design parameters of the 15-bar truss 76
Table 4.6 Size optimization of the 15-bar truss (continuous section) 77
Table 4.7 Design parameters of the 24-bar truss 79
Table 4.8 Size optimization of the 24-bar truss (continuous section) 80
Table 4.9 Design parameters of the 20-bar truss 83
Table 4.10 Size optimization if the 20-bar truss (continuous section) 84
Table 4.11 Design parameters of the 72-bar truss 86
Table 4.12 Size optimization of the 72-bar truss (continuous section) 87
Table 4.13 Design parameters of the 39-bar truss 89
Table 4.14 Size optimization of the 39-bar truss (continuous section) 90
Table 4.15 Design parameters of the 45-bar truss 92
Table 4.16 Size optimization of the 45-bar truss (continuous section) 93
Table 4.17 Design parameters of the 25-bar 3D truss 97
Table 4.18 The 25-bar 3D truss size optimization (continuous section) 98
Table 4.19 Design parameters of the 39-bar 3D truss 100
Table 4.20 The 39-bar 3D truss size optimization (continuous section) 101
Table 4.21 Minimum mass obtained using size optimization with continuous sections 103
Table 4.22 Mean mass obtained using size optimization with continuous sections 104
Table 4.23 The Friedman test 105

Table 4.24 Design parameters of the 10-bar truss 107
Table 4.25 Size optimization of the 10-bar truss (discrete section) 108
Table 4.26 Design parameters of the 14-bar truss 110
Table 4.27 Size optimization of the 14-bar truss (discrete section) 111
Table 4.28 Design parameters of the 15-bar truss 112
Table 4.29 Size optimization of the 15-bar truss (discrete section) 113
Table 4.30 Design parameters of the 24-bar truss 115
Table 4.31 Size optimization of the 24-bar truss (discrete section) 116
Table 4.32 Design parameters of the 20-bar truss 118
Table 4.33 Size optimization of the 20-bar truss (discrete section) 119
Table 4.34 Design parameters of the 72-bar truss 122
Table 4.35 Size optimization of the 72-bar truss (discrete section) 123
Table 4.36 Design parameters of the 39-bar truss 125
Table 4.37 Size optimization of the 39-bar truss (discrete section) 126
Table 4.38 Design parameters of the 45-bar truss 128
Table 4.39 Size optimization of the 45-bar truss (discrete section) 129
Table 4.40 Design parameters of the 25-bar 3D truss 133
Table 4.41 Size optimization of the 25-bar 3D truss (discrete section) 134
Table 4.42 Design parameters of the 39-bar 3D truss 136
Table 4.43 Size optimization of the 39-bar 3D truss (discrete section) 137
Table 4.44 Minimum mass obtained using size optimization with discrete sections 138
Table 4.45 Mean mass obtained using size optimization with discrete sections 140
Table 4.46 The Friedman rank test 141
Table 4.47 The hypervolume of the 942-bar truss 146
Table 4.48 The CEC2014 benchmark functions (Liang et al., 2014) 147
Table 4.49 Results of the CEC2014 benchmark functions 148

Table 5.1 Design parameters of the 10-bar truss 160
Table 5.2 Truss topology optimization of the 10-bar truss (continuous section) 172
Table 5.3 Design parameters of the 14-bar truss 174
Table 5.4 Truss topology optimization of the 14-bar truss (continuous section) 175
Table 5.5 Design parameters of the 15-bar truss 177
Table 5.6 Truss topology optimization of the 15-bar truss (continuous section) 178
Table 5.7 Design parameters of the 24-bar truss 179
Table 5.8 Truss topology optimization of the 24-bar truss (continuous section) 181
Table 5.9 Design parameters of the 20-bar truss 183
Table 5.10 Truss topology optimization of the 20-bar truss (continuous section) 184
Table 5.11 Design parameters of the 72-bar truss 185

Table 5.12 Truss topology optimization of the 72-bar 3D truss (continuous section) 187
Table 5.13 Design parameters of the 39-bar truss 188
Table 5.14 Truss topology optimization of the 39-bar truss (continuous section) 189
Table 5.15 Design parameters of the 45-bar truss 191
Table 5.16 Truss topology optimization of the 45-bar truss (continuous section) 192
Table 5.17 Design parameters of the 25-bar 3D truss 195
Table 5.18 Truss topology optimization of the 25-bar 3D truss (continuous section) 196
Table 5.19 Design parameters of the 39-bar 3D truss 198
Table 5.20 Truss topology optimization of the 39-bar 3D truss (continuous section) 199
Table 5.21 Minimum mass obtained using TTO with continuous sections 201
Table 5.22 Mean mass obtained using TTO with continuous sections 202
Table 5.23 Design parameters of the 10-bar truss 204
Table 5.24 Truss topology optimization of the 10-bar truss (discrete section) 205
Table 5.25 Design parameters of the 14-bar truss 207
Table 5.26 Truss topology optimization of the 14-bar truss (discrete section) 208
Table 5.27 Design parameters of the 15-bar truss 210
Table 5.28 Truss topology optimization of the 15-bar truss (discrete section) 211
Table 5.29 Design parameters of the 24-bar truss 213
Table 5.30 Truss topology optimization of the 24-bar truss (discrete section) 214
Table 5.31 Design parameters of the 20-bar truss 216
Table 5.32 Truss topology optimization of the 20-bar truss (discrete section) 217
Table 5.33 Design parameters of the 72-bar truss 218
Table 5.34 Truss topology optimization of the 72-bar 3D truss (discrete section) 220
Table 5.35 Design parameters of the 39-bar truss 222
Table 5.36 Truss topology optimization of the 39-bar truss (discrete section) 223
Table 5.37 Design parameters of the 45-bar truss 225
Table 5.38 Truss topology optimization of the 45-bar truss (discrete section) 226
Table 5.39 Design parameters of the 25-bar 3D truss 229
Table 5.40 Truss topology optimization of the 25-bar 3D truss (discrete section) 230
Table 5.41 Design parameters of the 39-bar 3D truss 232

Table 5.42 Truss topology optimization of the 39-bar 3D truss (discrete section) 233
Table 5.43 Minimum mass obtained using TTO with discrete sections 235
Table 5.44 Mean mass obtained using TTO with discrete sections 236

Table 6.1 Design parameters of the 10-bar truss 245
Table 6.2 Topology, shape, and size optimization of the 10-bar truss (continuous section) 246
Table 6.3 Design parameters of the 14-bar truss 247
Table 6.4 Topology, shape, and size optimization of the 14-bar truss (continuous section) 248
Table 6.5 Design parameters of the 15-bar truss 249
Table 6.6 Topology, shape, and size optimization of the 15-bar truss (continuous section) 250
Table 6.7 Design parameters of the 24-bar truss 251
Table 6.8 Topology, shape, and size optimization of the 24-bar truss (continuous section) 252
Table 6.9 Design parameters of the 20-bar truss 253
Table 6.10 Topology, shape, and size optimization of the 20-bar truss (continuous section) 254
Table 6.11 Design parameters of the 72-bar truss 275
Table 6.12 Topology, shape, and size optimization of the 72-bar 3D truss (continuous section) 276
Table 6.13 Design parameters of the 39-bar truss 279
Table 6.14 Topology, shape, and size optimization of the 39-bar truss (continuous section) 280
Table 6.15 Design parameters of the 45-bar truss 284
Table 6.16 Topology, shape, and size optimization of the 45-bar truss (continuous section) 285
Table 6.17 Design parameters of the 25-bar 3D truss 288
Table 6.18 Topology, shape, and size optimization of the 25-bar 3D truss (continuous section) 289
Table 6.19 Design parameters of the 39-bar 3D truss 291
Table 6.20 Topology, shape, and size optimization of the 39-bar 3D truss (continuous section) 293
Table 6.21 Minimum mass obtained using TSS optimization with continuous sections 295
Table 6.22 Mean mass obtained using TSS optimization with continuous sections 296
Table 6.23 The success rate of TSS optimization with continuous sections 297
Table 6.24 The mean number of FE required to reach the best mean optimum value with 25% error in TSS with continuous sections 299

Table 6.25 The Friedman rank test of mean mass and success rate obtained using TTO for continuous sections 303
Table 6.26 Design parameters of the 10-bar truss 306
Table 6.27 Topology, shape, and size optimization of the 10-bar truss (discrete section) 307
Table 6.28 Design parameters of the 14-bar truss 310
Table 6.29 Topology, shape, and size optimization of the 14-bar truss (discrete section) 311
Table 6.30 Design parameters of the 15-bar truss 313
Table 6.31 Topology, shape, and size optimization of the 15-bar truss (discrete section) 315
Table 6.32 Design parameters of the 24-bar truss 317
Table 6.33 Topology, shape, and size optimization of the 24-bar truss (discrete section) 318
Table 6.34 Design parameters of the 20-bar truss 321
Table 6.35 Topology, shape, and size optimization of the 20-bar truss (discrete section) 322
Table 6.36 Design parameters of the 72-bar truss 326
Table 6.37 Topology, shape, and size optimization of the 72-bar 3D truss (discrete section) 327
Table 6.38 Design parameters of the 39-bar truss 330
Table 6.39 Topology, shape, and size optimization of the 39-bar truss (discrete section) 331
Table 6.40 Design parameters of the 45-bar truss 334
Table 6.41 Topology, shape, and size optimization of the 45-bar truss (discrete section) 336
Table 6.42 Design parameters of the 25-bar 3D truss 339
Table 6.43 Topology, shape, and size optimization of the 25-bar 3D truss (discrete section) 341
Table 6.44 Design parameters of the 39-bar 3D truss 344
Table 6.45 Topology, shape, and size optimization of the 39-bar 3D truss (discrete section) 345
Table 6.46 Minimum mass obtained using TSS optimization with discrete sections 347
Table 6.47 Mean mass obtained using TSS optimization with discrete sections 348
Table 6.48 The success rate of the TSS optimization with discrete sections 349
Table 6.49 The mean number of FE required to reach the best mean optimum value in the algorithms with discrete sections 350
Table 6.50 The Friedman rank test of mean mass and success rate obtained using the algorithms with discrete sections 356

Table 7.1 The 10-bar truss 364
Table 7.2 The 14-bar truss 367
Table 7.3 The 15-bar truss 369
Table 7.4 The 24-bar truss 371
Table 7.5 The 20-bar truss 373
Table 7.6 The 72-bar 3D truss 374
Table 7.7 The 39-bar truss 378
Table 7.8 The 45-bar truss 380
Table 7.9 The 25-bar 3D truss 382
Table 7.10 The 39-bar 3D truss 383

Chapter 1
Introduction

Nature has been a great source of inspiration for scholars to practice optimization methods in structural engineering. Take the natural topology of a coconut tree, for instance. Its wood contains porous material at the center and dense material at the circumference, gradually distributed from the center axis to the circumference. Likewise, the topology of bones increases structural strength and optimally supports body weight (Sigmund, 1994). As a result, these structures are lightweight and provide significant resistance to failures like buckling and bending.

According to Christensen and Klarbring (2009), Gordon defined a structure as "any assemblage of material, which is intended to sustain loads." Galileo envisioned the shape optimization problem of a beam in his novel sketch in 1638 (Arora, 2007), while Maxwell (1890) made pioneering contributions to the field of structural optimization. He presented fundamental principles for designing minimum-weight structures for specified loads and materials. Additionally, Michell (1904) published significant literature on topology optimization, proposing the least weight layout trusses based on his optimal criteria method, which is governed by stress and displacement. However, the optimal criteria method is an analytical approach with limited applications. In 1964, Dorn, Gomory, and Greenberg introduced the ground structure method and demonstrated all possible node connections in a truss structure, opening up a broad path for truss topology optimization (TTO). Sheu and Schmit (1972) used an extensive search method for topology optimization, while Kirsch (1989) presented optimal truss structure topologies using optimal criteria (Tejani, 2017).

A truss is a 2D or 3D structure consisting of linear elements connected at nodes to withstand load and subjected to tension or compression. Truss optimization has emerged as a rapidly growing field of structural optimization over the past three decades, which can be categorized into three types: size optimization, shape optimization, and topology optimization (Tejani et al., 2016a, b, 2017a, b).

Size optimization aims to determine the optimal cross-sectional areas of truss elements, as depicted in Fig. 1.1. Shape optimization involves the relocation of nodal coordinates of the truss, as shown in Fig. 1.2; nodal coordinates are employed for

V. Savsani et al., *Truss Optimization*, https://doi.org/10.1007/978-3-031-49295-2_1

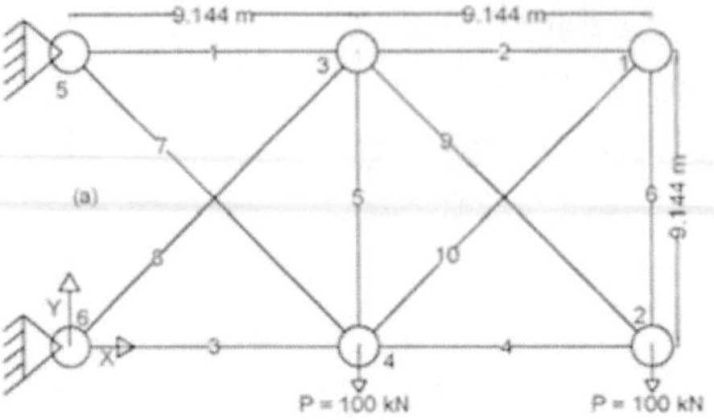

Fig. 1.1 The 10-bar truss: (**a**) ground structure and (**b**) size optimization

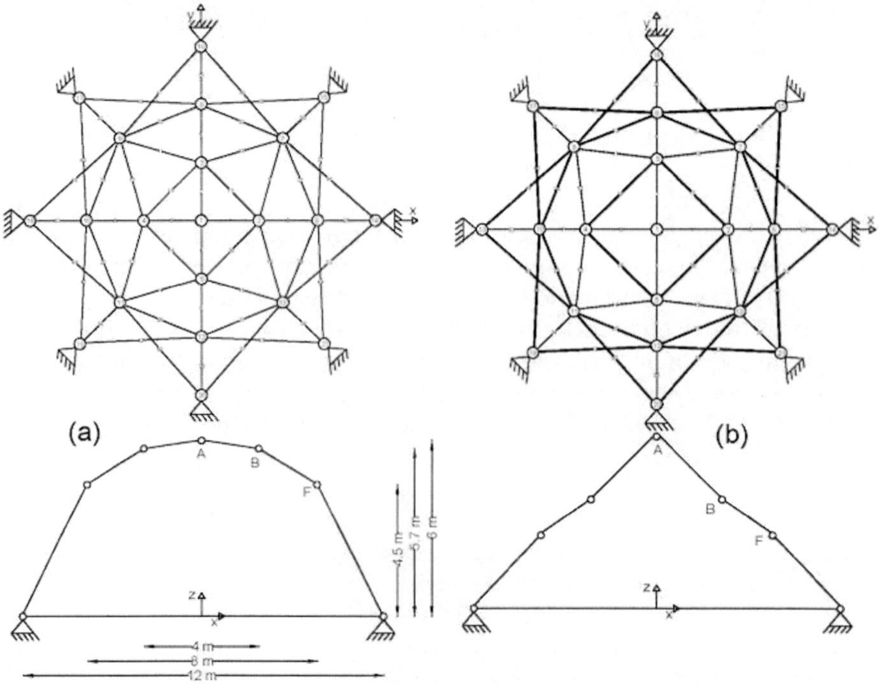

Fig. 1.2 The 52-bar truss: (**a**) ground structure and (**b**) shape and size optimization

shape optimization of the truss. Topology optimization, on the other hand, deals with the addition and removal of nodes and elements, as presented in Fig. 1.3; several elements are removed from the ground structure for topology optimization of the truss. Topology optimization is the most challenging problem in this field since it involves exploring all possible topologies instead of a particular one, leading to significant weight savings. The simultaneous consideration of topology, shape, and size (TSS) optimization yields the lightest truss (Tejani et al., 2017a, b), as illustrated in Fig. 1.4.

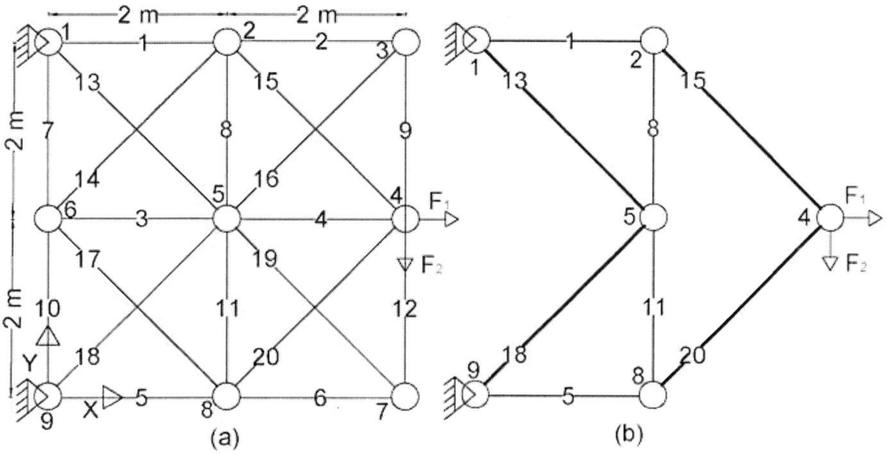

Fig. 1.3 The 20-bar truss: (**a**) ground structure and (**b**) topology and size optimization

Most TTO problems in the literature only consider stress and displacement constraints, with few studies incorporating frequency constraints in addition to these. The natural frequency of a truss is an essential parameter, particularly when it is subject to dynamic excitations. Many engineering trusses experience dynamic excitations due to operational conditions or unexpected events, which can lead to unwanted vibration and noise. Such conditions can become dangerous if the dynamic response produces resonance, and therefore, restrictions must be enforced on natural frequencies to protect the truss (Savsani et al., 2016; Tejani et al., 2018; Kumar et al., 2022).

Frequency constraints increase the complexity of TTO problems, and buckling can also have an adverse effect, adding further complexity. The simultaneous consideration of natural frequencies and buckling constraints increases the complexity even more. However, these constraints cannot be neglected to ensure practicality. Kinematic instability and invalid trusses are major obstacles in the TTO process, and they must be identified and handled efficiently to avoid a large number of unwanted analyses (Savsani et al., 2017).

For a long time, analytical or numerical methods have been used to optimize engineering problems by calculating the optimum values of a function. However, these methods may not be sufficient for more complex design situations, even if they work well in simpler cases. In the case of TTO problems, the number of design variables can be large, and their effect can be difficult to assess. Furthermore, the presence of multiple loading conditions, along with stress, displacement, buckling, frequency, and kinematic stability constraints, makes the truss optimization problems even more challenging for metaheuristic optimization methods. To address these difficulties, an efficient method is required to solve such problems. Classical optimization techniques may not be able to handle these complex problems or may only compute local extrema. Therefore, advanced metaheuristic algorithms are

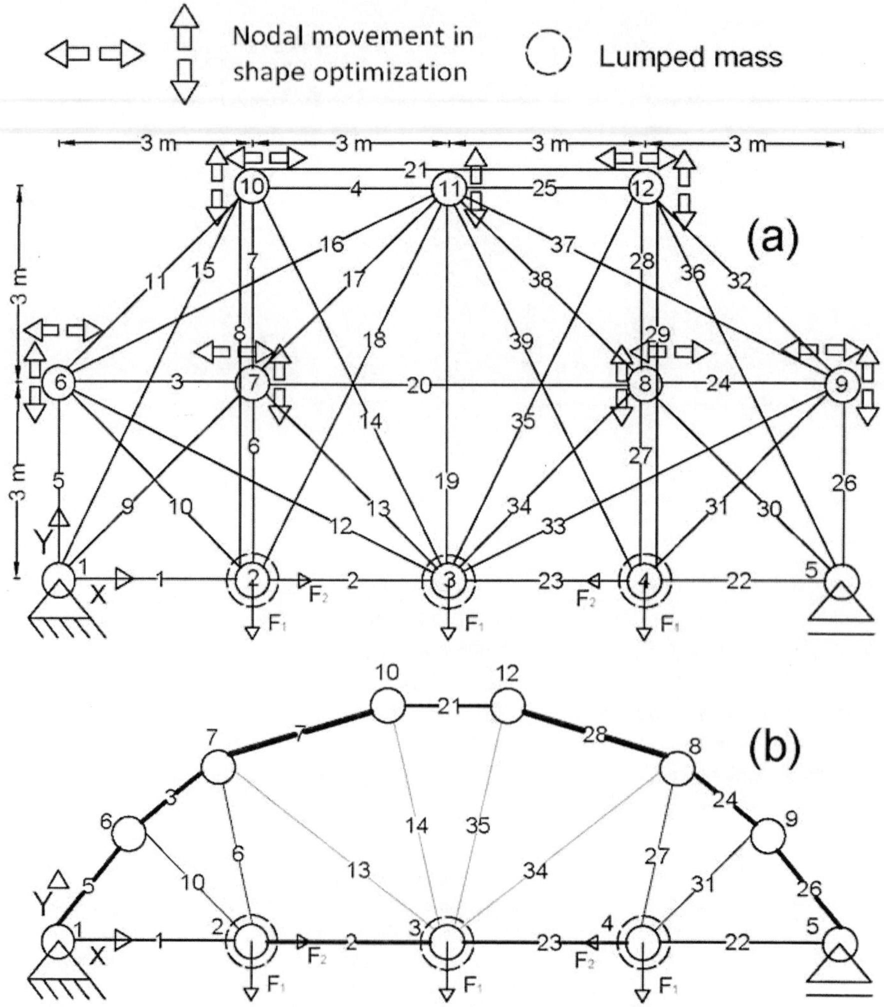

Fig. 1.4 The 39-bar truss: (**a**) ground structure and (**b**) topology, shape, and size optimization

needed to provide solutions to these problems. These algorithms search for a solution close to the global optimum with less computational effort.

Optimization techniques can be classified into two distinct categories as given below:

(a) Classical metaheuristic techniques: These are analytical algorithms that utilize methods of differential calculus to find optimal solutions. These techniques have been employed in engineering design problems for a considerable amount of time and have demonstrated success. Examples of these techniques include linear/nonlinear programming, geometric programming, quadratic programming, dynamic programming, and integer programming. Despite their success, classical optimization techniques have limited applicability in real-world scenarios.

(b) Advanced metaheuristic techniques: These techniques are characterized by their heuristic nature, which involves probabilistic transition rules based on a range of factors such as biological behaviors, physical laws, social hierarchies, and neurobiological systems, among others. These techniques are relatively new and have gained popularity as effective tools for optimizing complex engineering problems. Some of the popular advanced metaheuristic techniques include genetic algorithm (GA), simulated annealing, differential evolution, particle swarm optimization, biogeography-based optimization, artificial bee colony, and artificial immune algorithm.

Despite the use of classical optimization techniques to address optimization problems in TSS optimization, they have certain limitations:

- Firstly, these techniques are not effective for a wide range of problem domains, as cited by Yang (2010) and Pholdee and Bureerat (2014).
- Secondly, classical optimization techniques struggle with solving multimodal problems, as they only provide locally optimal solutions, as highlighted by Arora (2007) and Miguel et al. (2013).
- Thirdly, classical optimization techniques are not suitable for solving multi-objective optimization problems, as noted by Su et al. (2011) and Tejani et al. (2019).
- Finally, classical optimization techniques are not suitable for problems with a large number of constraints. Given these shortcomings, researchers are exploring advanced metaheuristic techniques to optimize TTO problems.
- Furthermore, the no free lunch theorem (Wolpert & Macready, 1997) states that there is no metaheuristic that can effectively optimize all problem types. An algorithm that performs well on one set of challenges may not perform as well on another set of problems. Therefore, it is crucial to continue developing more powerful and robust optimization techniques and to modify existing algorithms to enhance their accuracy in providing solutions.

This book aims to achieve the following objectives:

- To recommend the use of ten advanced metaheuristic techniques developed after 2011 to solve truss design problems. These techniques include the dragonfly algorithm (DA) (Mirjalili, 2016a), multiverse optimizer (MVO) (Mirjalili et al., 2016), sine cosine algorithm (SCA) (Mirjalili, 2016b), whale optimization algorithm (WOA) (Mirjalili & Lewis, 2016), ant lion optimizer (ALO) (Mirjalili, 2015), heat transfer search (HTS) (Patel & Savsani, 2015), passing vehicle search (PVS) (Savsani & Savsani, 2015), symbiotic organisms search (SOS) (Cheng & Prayogo, 2014), grey wolf optimizer (GWO) (Mirjalili et al., 2014), and the teaching-learning-based optimization (TLBO) (Rao et al., 2011, 2012) algorithm, which is an effective technique with a positive impact on various engineering optimization problems. As these metaheuristics have distinct search mechanisms, it is challenging to predict their effectiveness over different engineering applications.

- To introduce new benchmark problems for (1) size optimization, (2) topology and size optimization, and (3) topology, shape, and size optimization, considering static and dynamic constraints, and continuous and discrete element cross sections with multiple load conditions.

The book is organized as follows: In Chap. 2, a methodology and problem formulation of truss design problems are presented. The design problems aim to minimize structural mass while satisfying static and dynamic constraints. A ground structure method is employed to address truss topology, and a finite element approach is used for structural analysis to determine the objective and constraint functions. Chapter 3 provides detailed information on the advanced metaheuristic algorithms (the DA, MVO, SCA, WOA, ALO, HTS, PVS, SOS, GWO, and TLBO algorithms) utilized in this book, including the modifications made to these algorithms. In Chap. 4, the results and discussions of the applications of the advanced metaheuristic algorithms to size optimization problems are presented. Chapter 5 covers topology and size optimization, while Chap. 6 focuses on topology, shape, and size optimization. In Chap. 7, the validation of proposed methodologies is simulated using ANSYS software.

The following chapter presents a detailed and comprehensive methodology for truss optimization.

References

Arora, J. S. (2007). *Optimization of structural and mechanical systems*. World Scientific.

Cheng, M. Y., & Prayogo, D. (2014). Symbiotic organisms search: A new metaheuristic optimization algorithm. *Computers and Structures, 139*, 98–112. https://doi.org/10.1016/j.compstruc.2014.03.007

Christensen, P. W., & Klarbring, A. (2009). In G. M. L. Glaswell (Ed.), *An introduction to structural optimization* (p. 153). Springer Science + BusinessMedia B.V.

Dorn, W., Gomory, R., & Greenberg, H. (1964). Automatic design of optimal structures. *Journal de Mecanique, 3*, 25–52.

Kirsch, U. (1989). Optimal topologies of truss structures. *Computer Methods in Applied Mechanics and Engineering, 72*(1), 15–28. https://doi.org/10.1016/0045-7825(89)90119-9

Kumar, S., Tejani, G. G., Pholdee, N., & Bureerat, S. (2022). Performance enhancement of metaheuristics through random mutation and simulated annealing-based selection for concurrent topology and sizing optimization of truss structures. *Soft Computing, 26*, 5661–5683. https://doi.org/10.1007/s00500-022-06930-2

Maxwell, J. C. (1890). On reciprocal figures, frames and diagrams of forces. *Scientific Papers, Cambridge University Press, 2*, 175–177.

Michell, A. G. M. (1904). The limits of economy of material in frame structures. *Philosophical Magazine, 8*, 589–597.

Miguel, L. F. F., Lopez, R. H., & Miguel, L. F. F. F. (2013). Multimodal size, shape, and topology optimisation of truss structures using the firefly algorithm. *Advances in Engineering Software, 56*, 23–37. https://doi.org/10.1016/j.advengsoft.2012.11.006

Mirjalili, S. (2015). The ant lion optimizer. *Advances in Engineering Software, 83*, 80–98. https://doi.org/10.1016/j.advengsoft.2015.01.010

Mirjalili, S. (2016a). Dragonfly algorithm: A new meta-heuristic optimization technique for solving single-objective, discrete, and multi-objective problems. *Neural Computing and Applications, 27*(4), 1053–1073. https://doi.org/10.1007/s00521-015-1920-1

Mirjalili, S. (2016b). SCA: A sine cosine algorithm for solving optimization problems. *Knowledge-Based Systems, 96*, 120. https://doi.org/10.1016/j.knosys.2015.12.022

Mirjalili, S., & Lewis, A. (2016). The whale optimization algorithm. *Advances in Engineering Software, 95*, 51–67. https://doi.org/10.1016/j.advengsoft.2016.01.008

Mirjalili, S., Mirjalili, S. M., & Lewis, A. (2014). Grey wolf optimizer. *Advances in Engineering Software, 69*, 46–61. https://doi.org/10.1016/j.advengsoft.2013.12.007

Mirjalili, S., Mirjalili, S. M., & Hatamlou, A. (2016). Multi-verse optimizer: A nature-inspired algorithm for global optimization. *Neural Computing and Applications, 27*(2), 495–513. https://doi.org/10.1007/s00521-015-1870-7

Patel, V. K., & Savsani, V. J. (2015). Heat transfer search (HTS): A novel optimization algorithm. *Information Sciences, 324*, 217–246. https://doi.org/10.1016/j.ins.2015.06.044

Pholdee, N., & Bureerat, S. (2014). Comparative performance of meta-heuristic algorithms for mass minimisation of trusses with dynamic constraints. *Advances in Engineering Software, 75*, 1–13. https://doi.org/10.1016/j.advengsoft.2014.04.005

Rao, R. V., Savsani, V. J., & Vakharia, D. P. (2011). Teaching–learning-based optimization: A novel method for constrained mechanical design optimization problems. *Computer-Aided Design, 43*(3), 303–315. https://doi.org/10.1016/j.cad.2010.12.015

Rao, R. V., Savsani, V. J., & Vakharia, D. P. (2012). Teaching–learning-based optimization: An optimization method for continuous non-linear large scale problems. *Information Sciences, 183*(1), 1–15. https://doi.org/10.1016/j.ins.2011.08.006

Savsani, P. V., & Savsani, V. J. (2015). Passing vehicle search (PVS): A novel metaheuristic algorithm. *Applied Mathematical Modelling, 40*(5–6), 3951–3978. https://doi.org/10.1016/j.apm.2015.10.040

Savsani, V. J., Tejani, G. G., & Patel, V. K. (2016). Truss topology optimization with static and dynamic constraints using modified subpopulation teaching–learning-based optimization. *Engineering Optimization, 48*, 1–17. https://doi.org/10.1080/0305215X.2016.1150468

Savsani, V. J., Tejani, G. G., Patel, V. K., & Savsani, P. (2017). Modified meta-heuristics using random mutation for truss topology optimization with static and dynamic constraints. *Journal of Computational Design and Engineering, 4*, 106–130. https://doi.org/10.1080/0305215X.2016.1150468

Sheu, C. Y., & Schmit, L. A., Jr. (1972). Minimum weight Design of Elastic Redundant Trusses under multiple static loading conditions. *AIAA Journal, 10*(2), 155–162.

Sigmund, O. (1994). *Design of material structures using topology optimization*. Technical University of Denmark.

Su, R., Wang, X., Gui, L., & Fan, Z. (2011). Multi-objective topology and sizing optimization of truss structures based on adaptive Multi-Island search strategy. *Structural and Multidisciplinary Optimization, 43*, 275–286. https://doi.org/10.1007/s00158-010-0544-4

Tejani G. G. (2017). *Investigation of advanced metaheuristic techniques for simultaneous size, shape, and topology optimization of truss structures*. PhD thesis, Pandit Deendayal Petroleum University, India.

Tejani, G. G., Savsani, V. J., & Patel, V. K. (2016a). Modified sub-population teaching-learning-based optimization for Design of Truss Structures with natural frequency constraints. *Mechanics Based Design of Structures and Machines, 44*, 495–513. https://doi.org/10.1080/15397734.2015.1124023

Tejani, G. G., Savsani, V. J., & Patel, V. K. (2016b). Adaptive symbiotic organisms search (SOS) algorithm for structural design optimization. *Journal of Computational Design and Engineering, 3*, 226–249. https://doi.org/10.1016/j.jcde.2016.02.003

Tejani, G. G., Savsani, V. J., & Patel, V. K. (2017a). Modified sub-population based heat transfer search algorithm for structural optimization. *The International Journal of Applied Metaheuristic Computing, 8*, 1–23. https://doi.org/10.4018/IJAMC.2017070101

Tejani, G. G., Savsani, V. J., Patel, V. K., & Bureerat, S. (2017b). Topology, shape, and size optimization of truss structures using modified teaching-learning based optimization. *Advances in Computational Design, 2*, 313–331. https://doi.org/10.12989/acd.2017.2.4.313

Tejani, G. G., Savsani, V. J., Patel, V. K., & Mirjalili, S. (2018). An improved heat transfer search algorithm for unconstrained optimization problems. *Journal of Computational Design and Engineering, 6*, 13. https://doi.org/10.1016/j.jcde.2018.04.003

Tejani, G. G., Kumar, S., & Gandomi, A. H. (2019). Multi-objective heat transfer search algorithm for truss optimization. *Engineering Computations, 37*, 641. https://doi.org/10.1007/s00366-019-00846-6

Wolpert, D. H., & Macready, W. G. (1997). No free lunch theorems for optimization. *IEEE Transactions on Evolutionary Computation, 1*(1), 67–82.

Yang, X. S. (2010). A new metaheuristic bat-inspired algorithm. In *Nature inspired cooperative strategies for optimization (NICSO 2010)* (pp. 65–74). Springer.

Chapter 2
Methodology

2.1 Introduction

Truss optimization is generally classified into three categories: topology optimization, shape optimization, and size optimization (Kirsch, 1989; Souza et al., 2016). In topology optimization, the structural initial configuration or layout is determined by topological design variables (Kumar et al., 2022; Savsani et al., 2016, 2017). This is a challenging task that requires significant computational effort, as the goal is to find the optimal initial configuration. As a result, the design problem is typically large scale (Ahrari et al., 2015; Li, 2015). Furthermore, topology optimization aims to minimize the structural mass while searching for the best topology and element cross-sectional areas for the final design (Tejani, 2017). Due to these reasons, many researchers and engineers have investigated truss topology optimization (TTO) problems.

The ground structure method is a widely used technique in TTO due to its simplicity and flexibility. This approach involves defining a ground structure composed of user-defined truss elements or connections between nodes, based on designer preference or construction feasibility. The optimal structure is then determined as a subset of the members in the selected ground structure. Kaveh and Kalatjari (2003) proposed a graph theory to select a suitable ground structure, where a truss with minimum members is known as a simple graph-based ground structure (Fig. 2.1a), while a truss with all pairs of nodes being connected by single members is known as a complete graph-based ground structure, Fig. 2.1d. In topology optimization, a star graph or modified star graph-based ground structure is typically used (Fig. 2.1b), where every node is connected to neighboring nodes, with modified star graphs adding a few members to the next neighboring nodes (Fig. 2.1c).

It is important to limit the number of ground elements to avoid increasing the complexity of the optimization problem and generating impractical structures. Therefore, a star graph or modified star graph-based ground structure is suitable

V. Savsani et al., *Truss Optimization*, https://doi.org/10.1007/978-3-031-49295-2_2

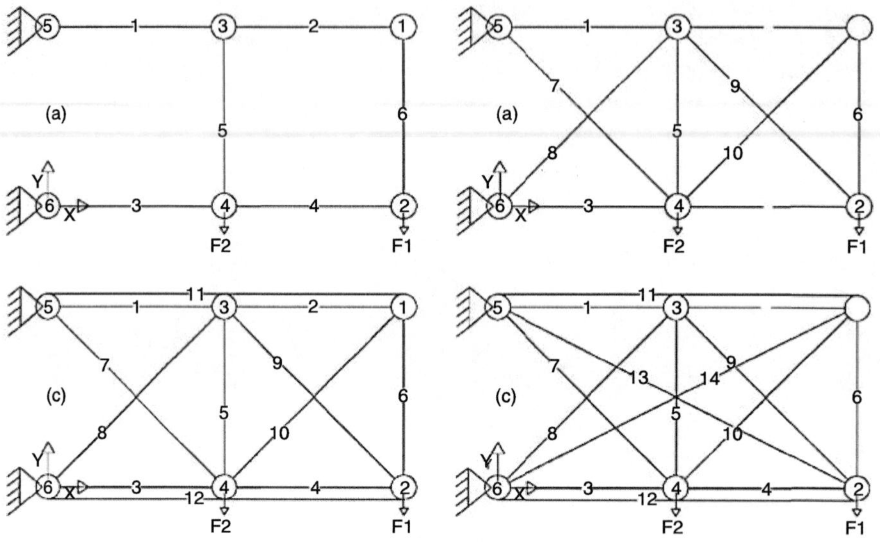

Fig. 2.1 Ground structures

for TTO problems. During the TTO process, ground elements are removed or kept until the best element connection is found, which can lead to a singular structural global stiffness matrix and optimization failure. To avoid this, assigning significantly small values to the cross-sectional areas of deleted elements may be used, but this increases computing time and may affect the accuracy of natural frequency values in truss optimization with natural frequency constraints. Instead, this study uses a restructuring approach where the finite element model is restructured based on topological design variables that control the existence of elements. A binary string with 0/1 bits represents removed or existing elements, and additional design variables are used to determine element cross-sectional areas, avoiding unnecessary analysis of removed elements. Concentrated masses at nodes, which can affect overall structural mass, are often overlooked in previous studies and should be considered (Ohsaki, 1995).

Topological truss design with natural frequency constraints is a challenging problem due to the complex nature of the optimization process (Kumar et al., 2020). Natural frequency constraints require consideration of the dynamic behavior of the structure, which adds an additional layer of complexity to the problem. Furthermore, the combinatorial nature of topology optimization makes it even more difficult to solve. Despite these challenges, researchers have made significant progress in recent years, using various methods such as genetic algorithms, particle swarm optimization, and simulated annealing.

A usual constrained optimization problem of trusses is assigned to optimize structural mass and performances while fulfilling various design constraints. Some researchers and engineers have investigated truss sizing using metaheuristic methods by considering various constraints such as natural frequency, displacement, stress, and buckling. Nevertheless, size optimization with static loading constraints, i.e., stress and displacement, has been studied by many researchers. The combined shape and sizing design of trusses with only natural frequency constraints have been proposed by some researchers (Gholizadeh & Barzegar, 2013; Kaveh & Zolghadr, 2014; Pholdee & Bureerat, 2014; Tejani et al., 2016a, b, 2017a; Kumar et al., 2019). Much work related to shape and sizing optimization with natural frequency constraints has been reported, but studies on truss topology optimization with the natural frequency constraints have been somewhat limited (Ohsaki et al., 1999; Xu et al., 2003; Jin & De-yu, 2006; Bai et al., 2009; Noilublao & Bureerat, 2011; Kaveh & Zolghadr, 2013; Gonçalves et al., 2015; Kaveh & Mahdavi, 2015; Tejani et al., 2019; Singh et al., 2022). The development of efficient and accurate optimization algorithms will allow engineers to design more lightweight and efficient structures, leading to significant improvements in the performance of the structure while reducing material and manufacturing costs.

So far, two design strategies have been utilized to handle truss topology and sizing optimization, as documented by Zhong et al. (2015). These are the two-stage approach, which was introduced by Xu et al. (2003) and Kaveh and Zolghadr (2013), and the single-stage (simultaneous) approach, which was introduced by Deb and Gulati (2001) and Tejani et al. (2018c). In the two-stage approach, ground elements with the constant cross-sectional area are first created, and then topological design is performed to either retain or delete some ground elements while keeping the cross-sectional areas and/or nodal coordinates of the truss elements constant. Once the optimal topology is determined, sizing and/or shape optimization is then conducted to search for the optimal element cross-sectional areas. However, the two-stage approach may not yield a global optimum solution; thus, its single-stage counterpart or an automated design was developed. The single-stage approach involves simultaneously searching for topology, shape, and size (TSS) variables in a single optimization run. Although the single-stage approach requires more computational effort as it deals with TSS optimization simultaneously, it is capable of designing a lightweight structure, as noted by Deb and Gulati (2001); Ahrari et al. (2015); and Tejani et al. (2017b).

2.2 Finite Element Method

The finite element method is currently the most commonly used computational technique for truss analysis to determine elemental stresses and nodal displacements (Ferreira, 2009). Initially developed as a matrix method for structural analysis of trusses and frames, the finite element method considers a bar element with two nodes and two degrees of freedom associated with two axial displacements, δ_1 and δ_2, as illustrated in Fig. 2.2. The element is subjected to applied loads or specified displacements and is constrained by boundary conditions. Assuming an element of length L, cross-sectional area A, modulus of elasticity E, and stiffness k, the equations are conveniently expressed in matrix form.

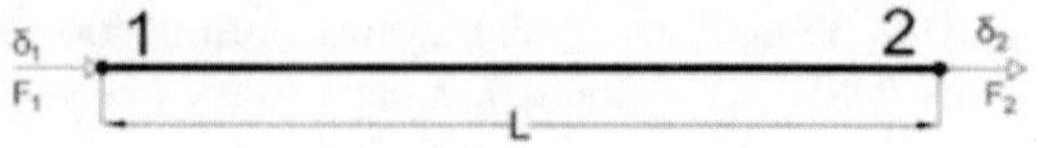

Fig. 2.2 1D bar element

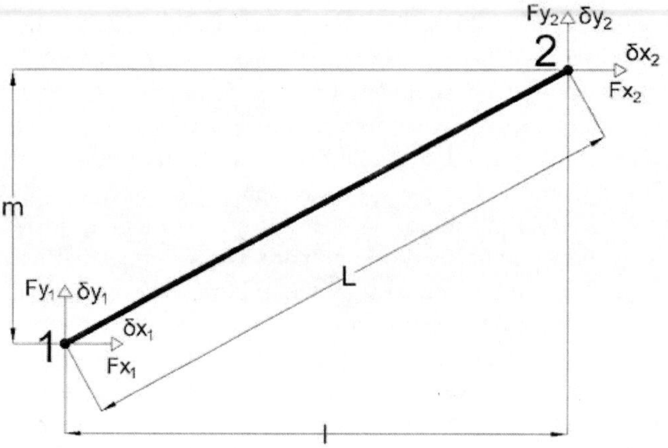

Fig. 2.3 2D truss element

The element is subjected to applied loads (or specified displacements) and boundary conditions:

$$\begin{Bmatrix} F_1 \\ F_2 \end{Bmatrix} = \frac{AE}{L} \begin{bmatrix} 1 & -1 \\ -1 & 1 \end{bmatrix} \begin{Bmatrix} \delta_1 \\ \delta_2 \end{Bmatrix} \tag{2.1}$$

The stress of the bar is represented by

$$\sigma = \frac{E(\delta_2 - \delta_1)}{L} \tag{2.2}$$

The truss bar can be viewed as a line element with simply supported ends, as depicted in Fig. 2.3. Assuming an element of a 2D truss with length L, cross-sectional area A, modulus of elasticity E, and mass density ρ, the bar element is subjected to forces and displacements. These equations can be conveniently expressed in matrix form as:

$$\begin{Bmatrix} F_{x_1} \\ F_{y_1} \\ F_{x_2} \\ F_{y_2} \end{Bmatrix} = \frac{AE}{L} \begin{bmatrix} l^2 & lm & -l^2 & -lm \\ lm & m^2 & -lm & -m^2 \\ -l^2 & -lm & l^2 & lm \\ -lm & -m^2 & lm & m^2 \end{bmatrix} \begin{Bmatrix} \delta_{x_1} \\ \delta_{y_1} \\ \delta_{x_2} \\ \delta_{y_2} \end{Bmatrix} \tag{2.3}$$

in compact form, it is represented by

$$\{F\} = [K]\{\delta\} \tag{2.4}$$

Here, K represents the stiffness matrix; δ represents the displacement vector, and F represents the force vector.

The stress of the bar is represented by

$$\sigma = \frac{E}{L}[-l \ \ -m \ \ l \ \ m]\begin{Bmatrix} \delta_{x_1} \\ \delta_{y_1} \\ \delta_{x_2} \\ \delta_{y_2} \end{Bmatrix} \tag{2.5}$$

Mass matrix is represented by

$$M = \frac{\rho AL}{6}\begin{bmatrix} 2 & 0 & 1 & 0 \\ 0 & 2 & 0 & 1 \\ 1 & 0 & 2 & 0 \\ 0 & 1 & 0 & 2 \end{bmatrix} \tag{2.6}$$

Eigenvalues describe the system's natural frequencies and are critical in understanding its dynamic behavior. Eigenvalues are the roots of the characteristic equation of the dynamic matrix or Eigenvector:

$$\{D\} = [K]^{-1}[M] \tag{2.7}$$

where

[D] is the dynamic matrix
[K] is the stiffness matrix of the system
[M] is the mass matrix of the system

To find the eigenvalues of the dynamic matrix [D], we first construct the characteristic equation. The characteristic equation for [D] is given by:

$$\det([D]–\lambda I) = 0 \tag{2.8}$$

Here, λ (lambda) represents an eigenvalue, I is the identity matrix, and det() denotes the determinant of the matrix. To find the eigenvalues of [D], you need to solve the characteristic equation for λ. This involves substituting [D] – λI into the equation and setting the determinant equal to zero:

Natural angular frequencies are written by

$$\{\omega\}^2 = \{\lambda\} \tag{2.9}$$

Natural frequencies are represented by

$$\{f\} = \frac{\{\omega\}}{2\pi} \tag{2.10}$$

The finite element method solves for displacements as primary unknowns by inverting the stiffness matrix, while stresses, strains, and natural frequencies are obtained as derived unknowns, as noted by Ferreira (2009). In this book, a combination of the single-stage approach design strategy, the ground structure method, and restructuring of a finite element model will be utilized.

2.2.1 *FEM Analysis of 2D Trusses*

We can employ the methodologies we have studied to calculate the stresses within a truss presented in Fig. 2.4.

In the numerical solution of this problem, an important step is the discretization process, which involves assigning numbers to both the finite elements and nodes. Hence, Fig. 2.5 illustrates the numerical representation of the truss structure. The truss consists of three nodes and three elements. In our two-dimensional analysis, each node is limited to moving either horizontally (X direction) or vertically (Y direction). We refer to these motion directions as degrees of freedom, abbreviated as DOF. With two degrees of freedom assigned to each node, there are a total of 6 degrees of freedom for the three nodes (see Fig. 2.6).

The element connectivity table is presented below (Table 2.1):

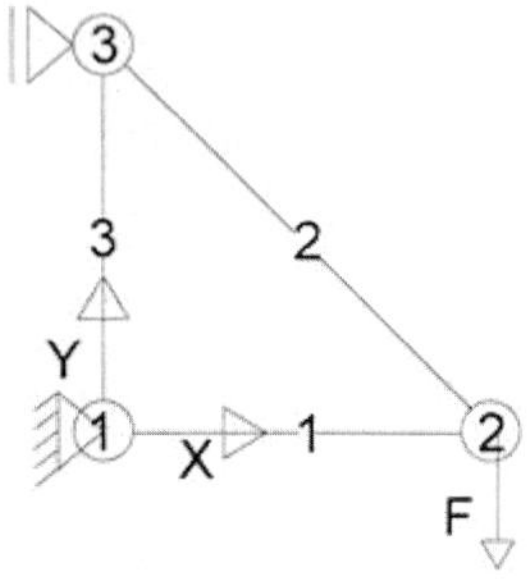

Fig. 2.4 A three-bar truss structure

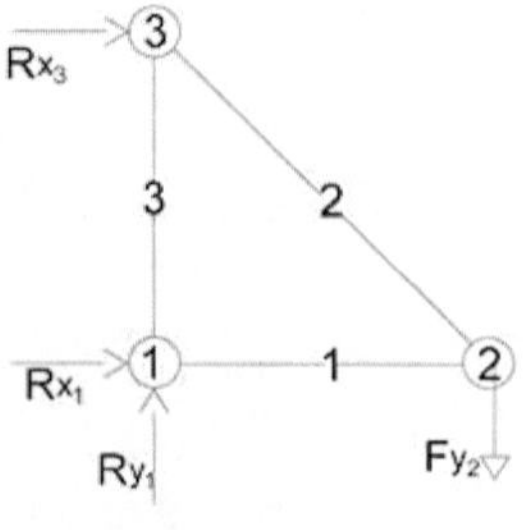

Fig. 2.5 Free-body diagram

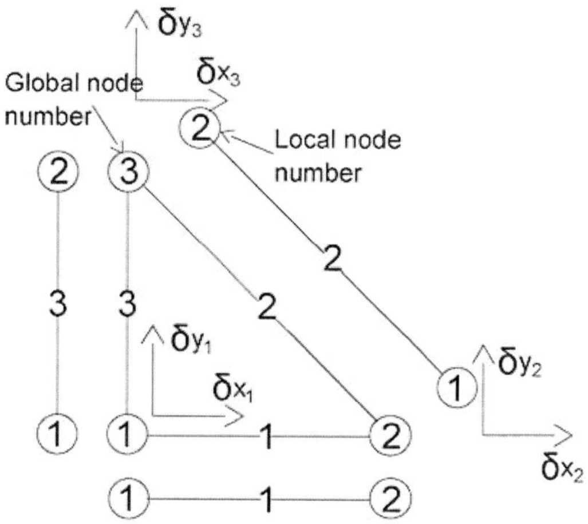

Fig. 2.6 Global and local numbering for the three-bar truss

Table 2.1 The element connectivity

	Local DOF	Local DOF	Global DOF	Global DOF
Number of element	First element node (i)	Second element node (j)	First element node (i)	Second element node (j)
1	1	2	1	2
2	1	2	2	3
3	1	2	1	3

This stiffness matrix is specific to an element. The element is connected to two nodes, and each of these nodes has two degrees of freedom. The arrangement of rows and columns in the stiffness matrix corresponds to these degrees of freedom.

The local stiffness matrices can be defined using Eq. 2.3.

$$K^{(1)} = \begin{bmatrix} \text{Global DoF} & 1 & 2 & 3 & 4 \\ 1 & k_{11}^{(1)} & k_{12}^{(1)} & k_{13}^{(1)} & k_{14}^{(1)} \\ 2 & k_{21}^{(1)} & k_{22}^{(1)} & k_{23}^{(1)} & k_{24}^{(1)} \\ 3 & k_{31}^{(1)} & k_{32}^{(1)} & k_{33}^{(1)} & k_{34}^{(1)} \\ 4 & k_{41}^{(1)} & k_{42}^{(1)} & k_{43}^{(1)} & k_{44}^{(1)} \end{bmatrix} \tag{2.11}$$

$$K^{(2)} = \begin{bmatrix} \text{Global DoF} & 3 & 4 & 5 & 6 \\ 3 & k_{11}^{(2)} & k_{12}^{(2)} & k_{13}^{(2)} & k_{14}^{(2)} \\ 4 & k_{21}^{(2)} & k_{22}^{(2)} & k_{23}^{(2)} & k_{24}^{(2)} \\ 5 & k_{31}^{(2)} & k_{32}^{(2)} & k_{33}^{(2)} & k_{34}^{(2)} \\ 6 & k_{41}^{(2)} & k_{42}^{(2)} & k_{43}^{(2)} & k_{44}^{(2)} \end{bmatrix} \tag{2.12}$$

$$K^{(3)} = \begin{bmatrix} \text{Global DoF} & 1 & 2 & 5 & 6 \\ 1 & k_{11}^{(3)} & k_{12}^{(3)} & k_{13}^{(3)} & k_{14}^{(3)} \\ 2 & k_{21}^{(3)} & k_{22}^{(3)} & k_{23}^{(3)} & k_{24}^{(3)} \\ 5 & k_{31}^{(3)} & k_{32}^{(3)} & k_{33}^{(3)} & k_{34}^{(3)} \\ 6 & k_{41}^{(3)} & k_{42}^{(3)} & k_{43}^{(3)} & k_{44}^{(3)} \end{bmatrix} \tag{2.13}$$

After obtaining the element matrices, the subsequent step involves assembling these matrices into a global finite element matrix. Given that the total degrees of freedom in the structure amount to six, the resulting global stiffness matrix will be a 6×6 matrix.

Transforming local stiffness matrices to global stiffness matrices:

$$K^{(1)} = \begin{bmatrix} k_{11}^{(1)} & k_{12}^{(1)} & k_{13}^{(1)} & k_{14}^{(1)} & 0 & 0 \\ k_{21}^{(1)} & k_{22}^{(1)} & k_{23}^{(1)} & k_{24}^{(1)} & 0 & 0 \\ k_{31}^{(1)} & k_{32}^{(1)} & k_{33}^{(1)} & k_{34}^{(1)} & 0 & 0 \\ k_{41}^{(1)} & k_{42}^{(1)} & k_{43}^{(1)} & k_{44}^{(1)} & 0 & 0 \\ 0 & 0 & 0 & 0 & 0 & 0 \\ 0 & 0 & 0 & 0 & 0 & 0 \end{bmatrix} \tag{2.14}$$

$$K^{(2)} = \begin{bmatrix} 0 & 0 & 0 & 0 & 0 & 0 \\ 0 & 0 & 0 & 0 & 0 & 0 \\ 0 & 0 & k_{11}^{(2)} & k_{12}^{(2)} & k_{13}^{(2)} & k_{14}^{(2)} \\ 0 & 0 & k_{21}^{(2)} & k_{22}^{(2)} & k_{23}^{(2)} & k_{24}^{(2)} \\ 0 & 0 & k_{31}^{(2)} & k_{32}^{(2)} & k_{33}^{(2)} & k_{34}^{(2)} \\ 0 & 0 & k_{41}^{(2)} & k_{42}^{(2)} & k_{43}^{(2)} & k_{44}^{(2)} \end{bmatrix} \tag{2.15}$$

$$K^{(3)} = \begin{bmatrix} k_{11}^{(3)} & k_{12}^{(3)} & 0 & 0 & k_{13}^{(3)} & k_{14}^{(3)} \\ k_{21}^{(3)} & k_{22}^{(3)} & 0 & 0 & k_{23}^{(3)} & k_{24}^{(3)} \\ 0 & 0 & 0 & 0 & 0 & 0 \\ 0 & 0 & 0 & 0 & 0 & 0 \\ k_{31}^{(3)} & k_{32}^{(3)} & 0 & 0 & k_{33}^{(3)} & k_{34}^{(3)} \\ k_{41}^{(3)} & k_{42}^{(3)} & 0 & 0 & k_{43}^{(3)} & k_{44}^{(3)} \end{bmatrix} \tag{2.16}$$

Since the coefficient for each stiffness matrix is uniform, combining them is straightforward. We simply aggregate the degrees of freedom for each element stiffness matrix into corresponding degrees of freedom in the structural matrix. The resulting structural stiffness matrix is depicted below.

Global stiffness matrix,

$$K = K^{(1)} + K^{(2)} + K^{(3)} \tag{2.17}$$

$$K=\begin{bmatrix} k_{11}^{(1)}+k_{11}^{(3)} & k_{12}^{(1)}+k_{12}^{(3)} & k_{13}^{(1)} & k_{14}^{(1)} & k_{13}^{(3)} & k_{14}^{(3)} \\ k_{21}^{(1)}+k_{21}^{(3)} & k_{22}^{(1)}+k_{22}^{(3)} & k_{23}^{(1)} & k_{24}^{(1)} & k_{23}^{(3)} & k_{24}^{(3)} \\ k_{31}^{(1)} & k_{32}^{(1)} & k_{33}^{(1)}+k_{11}^{(2)} & k_{34}^{(1)}+k_{12}^{(2)} & k_{13}^{(2)} & k_{14}^{(2)} \\ k_{41}^{(1)} & k_{42}^{(1)} & k_{43}^{(1)}+k_{21}^{(2)} & k_{44}^{(1)}+k_{22}^{(2)} & k_{23}^{(2)} & k_{24}^{(2)} \\ k_{31}^{(3)} & k_{32}^{(3)} & k_{31}^{(2)} & k_{32}^{(2)} & k_{33}^{(2)}+k_{33}^{(3)} & k_{34}^{(2)}+k_{34}^{(3)} \\ k_{41}^{(3)} & k_{42}^{(3)} & k_{41}^{(2)} & k_{42}^{(2)} & k_{43}^{(2)}+k_{43}^{(3)} & k_{44}^{(2)}+k_{44}^{(3)} \end{bmatrix} \tag{2.18}$$

$$\{F\}=[K]\{\delta\}$$

$$\begin{Bmatrix} F_{x_1} \\ F_{y_1} \\ F_{x_2} \\ F_{y_2} \\ F_{x_3} \\ F_{y_3} \end{Bmatrix}=\begin{bmatrix} k_{11}^{(1)}+k_{11}^{(3)} & k_{12}^{(1)}+k_{12}^{(3)} & k_{13}^{(1)} & k_{14}^{(1)} & k_{13}^{(3)} & k_{14}^{(3)} \\ k_{21}^{(1)}+k_{21}^{(3)} & k_{22}^{(1)}+k_{22}^{(3)} & k_{23}^{(1)} & k_{24}^{(1)} & k_{23}^{(3)} & k_{24}^{(3)} \\ k_{31}^{(1)} & k_{32}^{(1)} & k_{33}^{(1)}+k_{11}^{(2)} & k_{34}^{(1)}+k_{12}^{(2)} & k_{13}^{(2)} & k_{14}^{(2)} \\ k_{41}^{(1)} & k_{42}^{(1)} & k_{43}^{(1)}+k_{21}^{(2)} & k_{44}^{(1)}+k_{22}^{(2)} & k_{23}^{(2)} & k_{24}^{(2)} \\ k_{31}^{(3)} & k_{32}^{(3)} & k_{31}^{(2)} & k_{32}^{(2)} & k_{33}^{(2)}+k_{33}^{(3)} & k_{34}^{(2)}+k_{34}^{(3)} \\ k_{41}^{(3)} & k_{42}^{(3)} & k_{41}^{(2)} & k_{42}^{(2)} & k_{43}^{(2)}+k_{43}^{(3)} & k_{44}^{(2)}+k_{44}^{(3)} \end{bmatrix}\begin{Bmatrix} \delta_{x_1} \\ \delta_{y_1} \\ \delta_{x_2} \\ \delta_{y_2} \\ \delta_{x_3} \\ \delta_{y_3} \end{Bmatrix} \tag{2.19}$$

The force vector of this problem:

$$\{F\}=\begin{Bmatrix} F_{x_1} \\ F_{y_1} \\ F_{x_2} \\ F_{y_2} \\ F_{x_3} \\ F_{y_3} \end{Bmatrix} = > \{F\}=\begin{Bmatrix} R_{x_1} \\ R_{y_1} \\ 0 \\ F_{y_2} \\ R_{x_3} \\ 0 \end{Bmatrix} \tag{2.20}$$

The boundary condition of this problem:

$$\{\delta\}=\begin{Bmatrix} \delta_{x_1} \\ \delta_{y_1} \\ \delta_{x_2} \\ \delta_{y_2} \\ \delta_{x_3} \\ \delta_{y_3} \end{Bmatrix} = > \{\delta\}=\begin{Bmatrix} 0 \\ 0 \\ \delta_{x_2} \\ \delta_{y_2} \\ 0 \\ \delta_{y_3} \end{Bmatrix} \tag{2.21}$$

Remembering our basic equation:

$$\{F\}=[K]\{\delta\}$$

$$\begin{Bmatrix} F_{x_1} \\ F_{y_1} \\ F_{x_2} \\ F_{y_2} \\ F_{x_3} \\ F_{y_3} \end{Bmatrix}=\begin{bmatrix} k_{11}^{(1)}+k_{11}^{(3)} & k_{12}^{(1)}+k_{12}^{(3)} & k_{13}^{(1)} & k_{14}^{(1)} & k_{13}^{(3)} & k_{14}^{(3)} \\ k_{21}^{(1)}+k_{21}^{(3)} & k_{22}^{(1)}+k_{22}^{(3)} & k_{23}^{(1)} & k_{24}^{(1)} & k_{23}^{(3)} & k_{24}^{(3)} \\ k_{31}^{(1)} & k_{32}^{(1)} & k_{33}^{(1)}+k_{11}^{(2)} & k_{34}^{(1)}+k_{12}^{(2)} & k_{13}^{(2)} & k_{14}^{(2)} \\ k_{41}^{(1)} & k_{42}^{(1)} & k_{43}^{(1)}+k_{21}^{(2)} & k_{44}^{(1)}+k_{22}^{(2)} & k_{23}^{(2)} & k_{24}^{(2)} \\ k_{31}^{(3)} & k_{32}^{(3)} & k_{31}^{(2)} & k_{32}^{(2)} & k_{33}^{(2)}+k_{33}^{(3)} & k_{34}^{(2)}+k_{34}^{(3)} \\ k_{41}^{(3)} & k_{42}^{(3)} & k_{41}^{(2)} & k_{42}^{(2)} & k_{43}^{(2)}+k_{43}^{(3)} & k_{44}^{(2)}+k_{44}^{(3)} \end{bmatrix}\begin{Bmatrix} \delta_{x_1} \\ \delta_{y_1} \\ \delta_{x_2} \\ \delta_{y_2} \\ \delta_{x_3} \\ \delta_{y_3} \end{Bmatrix} \tag{2.22}$$

Boundary conditions are applied at the fixed supports, assuming that these joints remain stationary in the constrained direction. Consequently, we eliminate these joints from our matrix, specifically the constrained displacements at degrees of freedom (DOF) 1, 2, and 5. This suggests that the first, second, and fifth rows, along with their corresponding columns, have no impact on the solution of the matrix equation. Consequently, these particular rows and columns can be deleted.

$$\begin{Bmatrix} R_{x_1} \\ R_{y_1} \\ 0 \\ F_{y_2} \\ R_{x_3} \\ 0 \end{Bmatrix} = \begin{bmatrix} - & - & - & - & - & - \\ - & - & - & - & - & - \\ - & - & k_{33}^{(1)} + k_{11}^{(2)} & k_{34}^{(1)} + k_{12}^{(2)} & - & k_{14}^{(2)} \\ - & - & k_{43}^{(1)} + k_{21}^{(2)} & k_{44}^{(1)} + k_{22}^{(2)} & - & k_{24}^{(2)} \\ - & - & - & - & - & - \\ - & - & k_{41}^{(2)} & k_{42}^{(2)} & - & k_{44}^{(2)} + k_{44}^{(3)} \end{bmatrix} \begin{Bmatrix} 0 \\ 0 \\ \delta_{x_2} \\ \delta_{y_2} \\ 0 \\ \delta_{y_3} \end{Bmatrix} \tag{2.23}$$

The resulting matrix is:

$$\begin{Bmatrix} 0 \\ F_{y_2} \\ 0 \end{Bmatrix} = \begin{bmatrix} k_{33}^{(1)} + k_{11}^{(2)} & k_{34}^{(1)} + k_{12}^{(2)} & k_{14}^{(2)} \\ k_{43}^{(1)} + k_{21}^{(2)} & k_{44}^{(1)} + k_{22}^{(2)} & k_{24}^{(2)} \\ k_{41}^{(2)} & k_{42}^{(2)} & k_{44}^{(2)} + k_{44}^{(3)} \end{bmatrix} \begin{Bmatrix} \delta_{x_2} \\ \delta_{y_2} \\ \delta_{y_3} \end{Bmatrix} \tag{2.24}$$

Utilizing Gaussian elimination or other appropriate solution techniques to address the system of equations presented above, this process yields solutions for the displacement vector.

Solve δ_{x_2}, δ_{y_2}, and δ_{y_3} using the above matrix.

The stress of the bar is represented by

$$\sigma = \frac{E}{L}[-l \quad -m \quad l \quad m] \begin{Bmatrix} \delta_{x_1} \\ \delta_{y_1} \\ \delta_{x_2} \\ \delta_{y_2} \end{Bmatrix} \tag{2.25}$$

Solve σ_1, σ_3, and σ_3.

$$\text{Element 1}, \sigma_1 = \frac{E_1}{L_1}[-l \quad -m \quad l \quad m] \begin{Bmatrix} \delta_{x_1} \\ \delta_{y_1} \\ \delta_{x_2} \\ \delta_{y_2} \end{Bmatrix} = > \sigma_1 = \frac{E_1}{L_1}[-l \quad -m \quad l \quad m] \begin{Bmatrix} 0 \\ 0 \\ \delta_{x_2} \\ \delta_{y_2} \end{Bmatrix} \tag{2.26}$$

$$\text{Element 2}, \sigma_2 = \frac{E_2}{L_2}[-l \quad -m \quad l \quad m]\begin{Bmatrix} \delta_{x_2} \\ \delta_{y_2} \\ \delta_{x_3} \\ \delta_{y_3} \end{Bmatrix} = > \sigma_2 = \frac{E_2}{L_2}[-l \quad -m \quad l \quad m]\begin{Bmatrix} \delta_{x_2} \\ \delta_{y_2} \\ 0 \\ \delta_{y_3} \end{Bmatrix} \tag{2.27}$$

$$\text{Element 3}, \sigma_3 = \frac{E_3}{L_3}[-l \quad -m \quad l \quad m]\begin{Bmatrix} \delta_{x_1} \\ \delta_{y_1} \\ \delta_{x_3} \\ \delta_{y_3} \end{Bmatrix} = > \sigma_3 = \frac{E_3}{L_3}[-l \quad -m \quad l \quad m]\begin{Bmatrix} 0 \\ 0 \\ 0 \\ \delta_{y_3} \end{Bmatrix} \tag{2.28}$$

Finite Element Method (FEM) Analysis Process

The FEM process involves a systematic series of steps aimed at approximating solutions to differential equations governing physical phenomena. Here is a comprehensive breakdown of the steps involved in executing a successful FEM analysis:

Step 1: Problem definition (pre-processing).

- Clearly define the physical problem and objectives of analysis.
- Identify the domain and boundaries for the analysis.

Step 2. Enter nodal points.

- Establish nodal points within the domain.
- Define the coordinates and locations of these nodes.

Step 3: Define element type.

- Choose suitable elements for discretizing the domain.
- Determine the type and shape of elements for approximation.

Step 4: Define element cross-sectional areas.

- Assign cross-sectional areas or properties to the selected elements.
- Specify the geometric characteristics of elements.

Step 5: Define material property.

- Define material properties such as elasticity, density, etc.
- Assign material properties to elements or nodes.

Step 6: Enter element connectivity and assign their element cross-sectional areas.

- Establish connections between nodes to form elements.
- Assign previously defined cross-sectional areas to corresponding elements.

Step 7: Assign nodal elements and their mass.

- Assign mass properties to nodal points.
- Define the distribution of mass within the domain.

Step 8: Define boundary conditions.

- Prescribe constraints or fixed conditions at specific nodes or boundaries.
- Define immovable or restricted regions within the domain.

Step 9: Define loads.

- Apply external loads and forces to designated areas or nodes.
- Specify the magnitude, direction, and distribution of loads.

Step 10: Define the material behavior.

- Detail material behavior under the defined loading and environmental conditions.

Step 11: Problem solution (analysis).

- Perform mathematical computations or simulations to solve the problem based on defined inputs.
- Apply numerical techniques or solvers to obtain results.

Step 12: Post-processing: examine the results.

- Analyze and interpret the obtained results comprehensively.
- Visualize and evaluate the response of the system.
- Validate against theoretical expectations or experimental data if available.

2.3 Problem Formulation

The ground structure approach is a widely used method for truss optimization problems, which enables fair comparisons of results with benchmark problems. In this approach, truss ground elements representing all possible combinations of available truss elements are initially generated (as shown in Fig. 2.7). The optimization algorithm is then applied to determine whether to remove or maintain the elements forming a structural topology (as illustrated in Fig. 2.7a–i). A typical truss problem involves finding element sizes, nodal positions, and topology that minimize truss mass subject to design constraints, such as displacement, stress, buckling instability, natural frequencies, and kinematic stability. Trusses are subjected to multiple static load cases, and nodal and element masses are added to the objective function where applicable. The optimization problem can be expressed as follows:

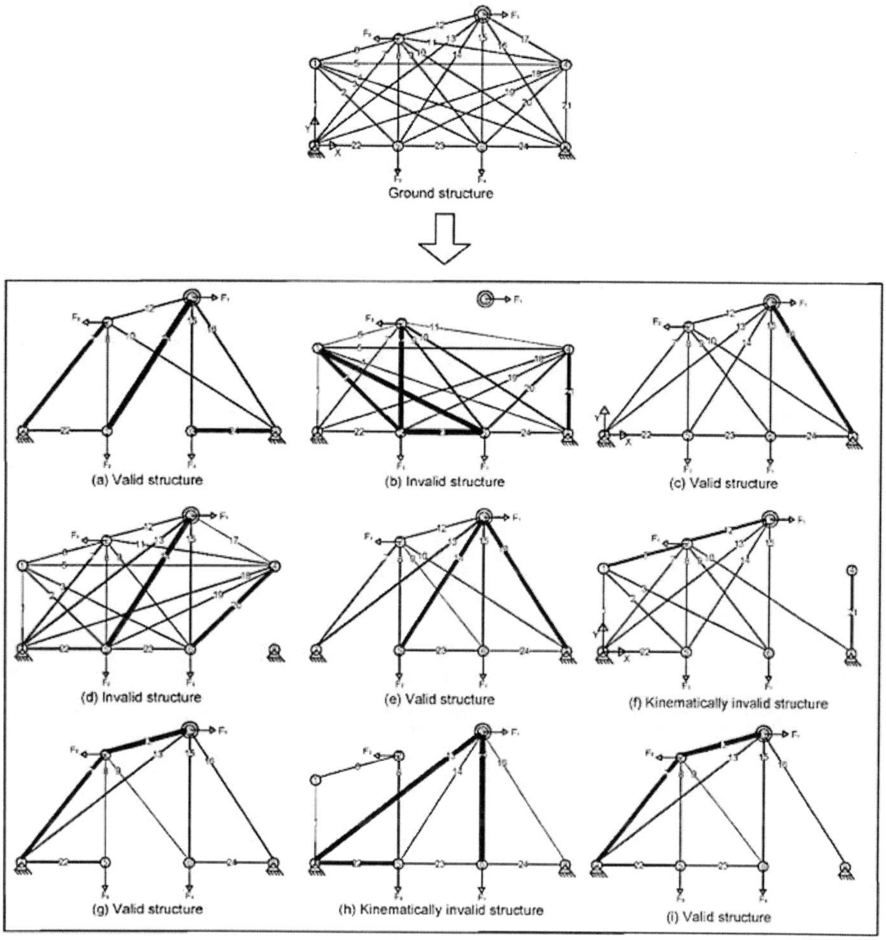

Fig. 2.7 Truss topology based on ground structures

1. Problem definition for size optimization

$$\text{Find} \quad X = \{A_1, A_2, .., A_m\} \tag{2.29}$$

to minimize mass of truss,

$$F(X) = \sum_{i=1}^{m} A_i \rho_i L_i + \sum_{j=1}^{n} b_j$$

Subject to:

Behavior constraints:

$$g_1(X) : \text{stress constraints}, \;\; |\sigma_i| - \sigma_i^{\max} \leq 0$$

$$g_2(X) : \text{displacement constraints}, \;\; \left|\delta_{xj/yj/zj}\right| - \delta_{xj/yj/zj}^{\max} \leq 0$$

$$g_3(X) : \text{Euler buckling constraints}, \;\; \left|\sigma_i^{\text{comp}}\right| - \sigma_i^{cr} \leq 0, \;\; \text{where } \sigma_i^{cr} = \frac{k_i A_i E_i}{L_i^2}$$

$g_4(X) : f_r - f_r^{\min} \geq 0$ for some natural frequencies

Side constraints:

$$\text{Cross} - \text{sectional area constraints}, \;\; A_i^{\min} \leq A_i \leq A_i^{\max}$$

$$\text{where } \; i = 1, 2, \ldots, m; j = 1, 2, \ldots, n$$

2. Problem definition for topology and size optimization

$$\text{Find } \; X = \{A_1, A_2, .., A_m\} \tag{2.30}$$

to minimize mass of truss,

$$F(X) = \sum_{i=1}^{m} B_i A_i \rho_i \, L_i + \sum_{j=1}^{n} b_j; \text{if TSS optimization or TS optimization}$$

Subject to:

Behavior constraints:

$$g_1(X) : \text{stress constraints}, \;\; |B_i \sigma_i| - \sigma_i^{\max} \leq 0$$

$$g_2(X) : \text{displacement constraints}, \;\; \left|\delta_{xj/yj/zj}\right| - \delta_{xj/yj/zj}^{\max} \leq 0$$

$$g_3(X): \text{Euler buckling constraints, } \left|B_i\sigma_i^{\text{comp}}\right| - \sigma_i^{cr} \le 0, \text{ where } \sigma_i^{cr} = \frac{k_iA_iE_i}{L_i^2}$$

$$g_4(X): f_r - f_r^{\min} \ge 0 \text{ for some natural frequencies}$$

Side constraints:

$$\text{Cross} - \text{sectional area constraints, } A_i^{\min} \le A_i \le A_i^{\max}$$

Check on kinematic stability

Check on validity of structure

$$\text{where } i = 1, 2, \ldots, m; j = 1, 2, \ldots, n$$

3. Problem definition for topology, shape, and size optimization

$$\text{Find } X = \{A_1, A_2, \ldots, A_m, \xi_1, \xi_2, \ldots, \xi_n\} \tag{2.31}$$

to minimize mass of truss,

$$F(X) = \sum_{i=1}^{m} B_iA_i\rho_i L_i + \sum_{j=1}^{n} b_j$$

Subject to:

Behavior constraints:

$$g_1(X): \text{stress constraints, } |B_i\sigma_i| - \sigma_i^{\max} \le 0$$

$$g_2(X): \text{displacement constraints, } \left|\delta_{x_j/y_j/z_j}\right| - \delta_{x_j/y_j/z_j}^{\max} \le 0$$

$$g_3(X): \text{Euler buckling constraints, } \left|B_i\sigma_i^{\text{comp}}\right| - \sigma_i^{cr} \le 0, \text{ where } \sigma_i^{cr} = \frac{k_iA_iE_i}{L_i^2}$$

$$g_4(X): f_r - f_r^{\min} \ge 0 \text{ for some natural frequencies}$$

Side constraints:

$$\text{Cross}-\text{sectional area constraints,}\ \ A_i^{\min} \le A_i \le A_i^{\max}$$

$$\text{Shape constraints,}\ \ \xi_{x_j/y_j/z_j}^{\min} \le \xi_{x_j/y_j/z_j} \le \xi_{x_j/y_j/z_j}^{\max}$$

Check on kinematic stability

Check on validity of structure

$$\text{where}\ \ i = 1, 2, \ldots, m; j = 1, 2, \ldots, n$$

where B_i is a binary element, $B_i = 0$ (if $A_i <$ critical area) and $B_i = 1$ (if $A_i \ge$ critical area), for deleting and retaining the ith element, respectively. $A_i, \rho_i, L_i, E_i, \sigma_i$, and σ_i^{cr} stand for cross-sectional area, mass density, element length, Young modules, stress, and critical buckling stress on the element "i," respectively. $\delta_{x_j/y_j/z_j}$, $\xi_{x_j/y_j/z_j}$, and b_j are values of nodal displacement, nodal position, and mass values of node "j," respectively, where x, y, and z present the x, y, and z axes, respectively. f_r is the structural natural frequent obtained from solving structural free vibration analysis of the rth mode. Superscripts "max," "min," and "comp" denote the maximum allowable limit, minimum allowable limit, and compressive stress, respectively. k_i is the Euler buckling coefficient calculated from elements' cross sections. The detailed graphical representation of the formulation of the generalize TTO problem is illustrated in Fig. 2.7.

It is worth noting that kinematic stability is included as a design constraint because it can potentially prevent an ill-conditioned structural analysis. For instance, as shown in Fig. 2.7f, element number 21 can rotate, and elements 1, 6, 8, and 22 may result in a mechanism, as illustrated in Fig. 2.7h. Based on the work by Deb and Gulati (2001), two criteria for kinematic instability are considered:

Step (I). Evaluate Grubler's criterion (Ghosh & Mallik, 1994) to check the degree of freedom (DOF) of the truss:

$$\text{Degree of freedom} = d \times m - n - m_r \tag{2.32}$$

In these criteria, d is equal to 2 for planar trusses and 3 for space trusses, while n and m denote the numbers of elements and nodes of the truss, respectively, and m_r is the restricted number of degrees of freedom at support nodes. It is essential that the generated truss is kinematically stable and does not turn into a mechanism. If the degree of freedom is non-positive, the truss is not a mechanism, as shown in Fig. 2.4f, where element number 21 can rotate about node 8. On the other hand, if the truss is a mechanism, we penalize the solution by assigning a large value. Subsequently, the truss is not sent for the finite element analysis (FEA) model for further analysis.

Step (II). To check the singularity of the truss, we evaluate the positive definiteness of the global stiffness matrix (K) using the method described by Rao (2009). Grubler's criterion is a necessary, but not sufficient criterion for kinematic stability. Therefore, we need to check the positive definiteness of the global stiffness matrix, as stated by Richardson et al. (2012). Analytically, a truss is considered kinematically unstable if the determinant of the stiffness matrix is zero. However, since FEA is a numerical method, the determinant of the stiffness matrix may not be exactly zero (Ahrari & Deb, 2016). To determine the positive definiteness of the stiffness matrix, we use the "eig(K)" command in MATLAB. If the first value returned by "eig(K)" is greater than 10^{-5} (a small number near zero), we assume that the truss is kinematically stable (positive definite). If the truss is nonpositive-definite, we penalize the solution by assigning a large value. In such a case, we do not send the truss for further analysis to evaluate stresses, displacements, Euler buckling, and natural frequencies.

This is because a truss element can only transmit forces between its two nodes, and if none of its nodes are connected to the external environment or to other elements, the element cannot carry any load and is essentially redundant. Therefore, an unusable truss topology is not structurally meaningful and should be avoided in the design process. Any truss topology is said to be unusable if none of the truss elements are connected to nodes with applied loads (i.e., node 3 is free as shown in Fig. 2.7b), nodes with boundary conditions (i.e., support node 8 is having no connection with the truss as shown in Figure 2.7d), and unchangeable nodes predefined by a user.

The penalty function technique is a method to handle constraints in optimization problems. It involves adding a penalty term to the objective function to penalize solutions that violate the constraints. The penalty term is a function of the constraint violation, and its value increases as the violation becomes larger.

In the case of truss optimization, the design constraints can be related to stress, displacement, buckling instability, natural frequency, and kinematic stability. If the solution violates any of these constraints, the penalty function is activated to penalize the solution.

The penalty function is formulated based on the severity of the violation and the user-specified penalty parameters. The severity of the violation is measured by the amount of constraint violation, and the penalty parameters are determined by the user to balance the trade-off between the objective function and the constraints.

The penalty function technique is effective in handling constraints in optimization problems, but it can also result in a large search space and numerical instability if the penalty parameters are not carefully chosen. To handle constraints, a penalty function technique is used. For no violation of the constraints, the penalty becomes zero; otherwise, the penalty is intended by the following criteria (Deb & Gulati, 2001; Kaveh & Zolghadr, 2013):

For size optimization:

$$\text{Penalized } F(X) = F(X) * F_{\text{penalty}} \tag{2.33}$$

For TTO and TSS optimization:

$$\text{Penalized } F(X) = \begin{cases} 10^9 \text{ if invalid structure} \\ 10^8 \text{ if violation in DOF} \\ 10^7 \text{ if violation in positive definiteness} \\ F(X) * F_{\text{penalty}} \quad \text{otherwise} \end{cases} \tag{2.34}$$

$$\text{where, } F_{\text{penalty}} = (1 + \varepsilon_1 * C)^{\varepsilon_2}, \; C = \sum_{i=1}^{q} C_i, \; C_i = \left|1 - \frac{p_i}{p_i^*}\right| \tag{2.35}$$

where p_i is the level of constraint violation having the bound as p_i^*. The parameter q is a number of active constraints. The variables ε_1 and ε_2 are predetermined by a user. In this study, the values of both ε_1 and ε_2 are set as 1.5, which were obtained from experimenting their effect on the balance of the exploitation-exploration balance. The detailed graphical representation of the formulation of the TTO problem is illustrated in Fig. 2.8.

2.4 Proposed Methodology

1. The proposed methodology for size optimization can be outlined in the following steps:

   Step 1: Begin by defining the ground structure of the truss. This involves generating nodes over a predefined design domain and assigning all possible element connections, as well as material properties, loading, and boundary conditions.

   Step 2: Proceed to the optimization algorithm, where the objective function, population size, design variables, bounds, algorithm controlling parameters, and termination criterion are defined.

   Step 3: Initiate a randomly generated set of size variables (i.e., population) within its upper and lower bounds.

   Step 4: Go to truss configurations: Generate trusses according to size variables.

   Step 5: Proceed to finite element analysis.

   Step 6: Compute the mass matrix, force vector, and displacement vector using finite element analysis.

   Step 7: Compute natural frequencies, element stresses, nodal displacements, and Euler buckling of the truss using finite element analysis.

   Step 8: Check for constraint violations in the penalty function. If there is a violation, assign a penalty as per Eq. 2.35; otherwise, compute the total mass of the truss.

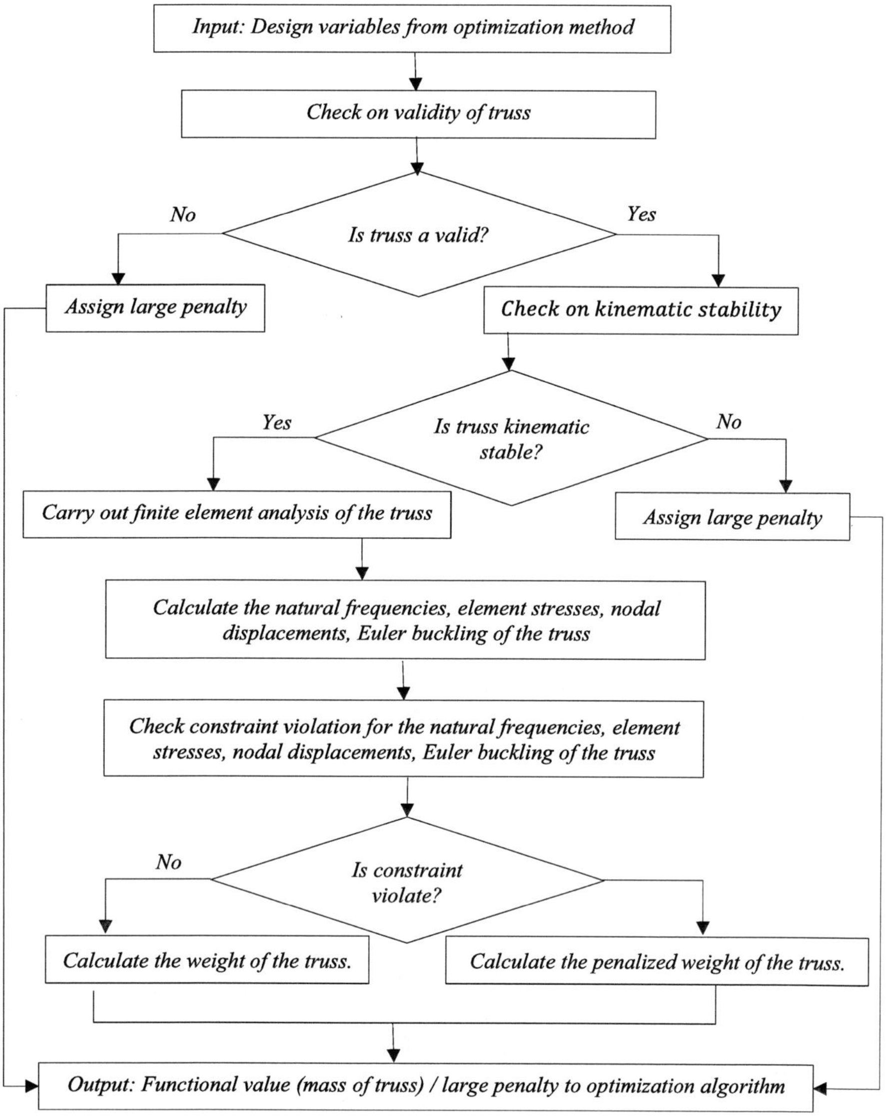

Fig. 2.8 Flowchart of the TTO problem

Step 9: Assign functional values and proceed to the optimization algorithm.
Step 10: Check the termination criteria. If not satisfied, generate new trusses (i.e., solutions) according to the algorithm and proceed to Step 4.
Step 11: Output: The best solution with natural frequencies, natural frequencies, element stresses, nodal displacements, and Euler buckling values.

2. The proposed methodology for TTO and TSS optimization (Tejani, 2017) can be outlined in the following steps:

   Step 1: Begin by defining the ground structure of the truss. This involves generating nodes over a predefined design domain and assigning all possible element connections, as well as material properties, loading, and boundary conditions.

   Step 2: Proceed to the optimization algorithm, where the objective function, population size, design variables, bounds, algorithm controlling parameters, and termination criterion are defined.

   Step 3: Initiate a randomly generated set of trusses (i.e., population) within its upper and lower bounds.

   Step 4: Go to truss configurations: Generate trusses according to the ground structure.

   Step 5: Identify invalid trusses, which are those that lack connections to unchangeable nodes (i.e., nodes subjected to load or support), as discussed in Sect. 2.2. Evaluate the degree of freedom of valid trusses as per Eq. 2.32. If a truss is invalid or its degree of freedom is a positive number, assign a large penalty to the objective function and proceed to Step 12.

   Step 6: If the truss is a valid structure, proceed to finite element analysis.

   Step 7: Compute the global stiffness matrix of each truss.

   Step 8: Check the positive definiteness of the global stiffness matrix to assess the singularity of trusses. If a truss is nonpositive-definite, assign a large penalty to the objective function and proceed to Step 12.

   Step 9: Compute the mass matrix, force vector, and displacement vector using finite element analysis.

   Step 10: Compute natural frequencies, element stresses, nodal displacements, and Euler buckling of the truss using finite element analysis.

   Step 11: Check for constraint violations in the penalty function. If there is a violation, assign a penalty as per Eq. 2.34; otherwise, compute the total mass of the truss.

   Step 12: Assign functional values and proceed to the optimization algorithm.

   Step 13: Check the termination criteria. If not satisfied, generate new trusses (i.e., solutions) according to the algorithm and proceed to Step 4.

   Step 14: Output: The best solution with natural frequencies, natural frequencies, element stresses, nodal displacements, and Euler buckling values.

A graphical representation of the proposed methodology is presented in Fig. 2.9.

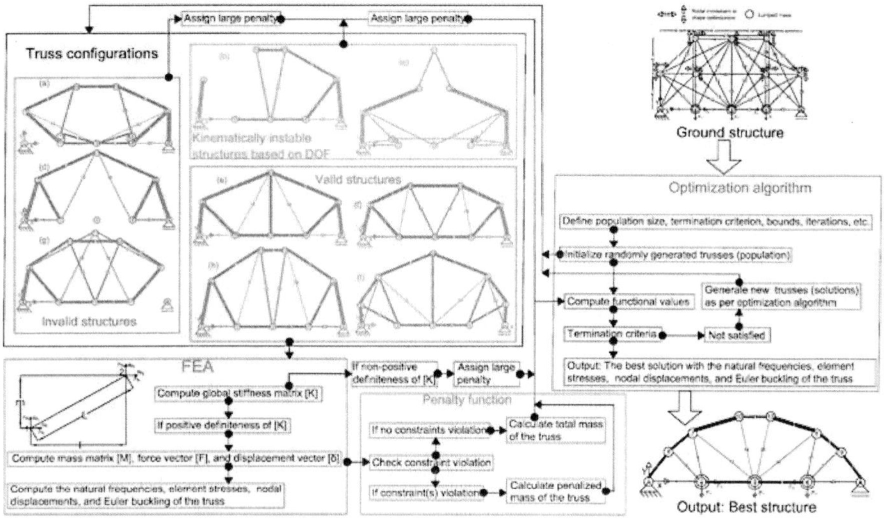

Fig. 2.9 Schematic representation of proposed methodology

2.5 Design Problems

Three benchmark trusses of simultaneous topology and size optimization with continuous cross-sectional areas from Xu et al. (2003), Kaveh and Zolghadr (2013), Tejani et al. (2018b), and seven new benchmark problems (Hajela & Lee, 1995; Deb & Gulati, 2001; Richardson et al., 2012; Miguel et al., 2013; Tejani et al., 2018a) are modified for size optimization, Truss topology optimization, and simultaneous TSS optimization to investigate the performance of proposed algorithms. Trusses have continuous and discrete design variables. The Euler buckling coefficient (k_i, $i = 1, 2, \ldots, m$) and mass of nodes (b_j, $j = 1, 2, \ldots, n$) are set as 4.0 and 5 kg, respectively, for all the problems. The lower and upper limits of design variables are $X^{\text{lower}} = -A^{\max}$ and $X^{\text{upper}} = A^{\max}$, respectively. The critical area is $A^{\min}$, which is applied to remove elements from the ground structure. The ten truss optimization problems are detailed in Appendix Chap. 6. Truss constraints consist of displacement, stress, natural frequency, and buckling while trusses are acted upon by multiple load cases. The search space is approximately converted into two times of the design variable limits to deal with the topology optimization.

2.5.1 Problem Complexity of the TSS Optimization Problems

Table 2.2 in this study provides a comprehensive overview of various complexity-measuring parameters that are associated with the optimization problems related to TSS (time synchronization service). These parameters encompass a wide range of

Table 2.2 Problem complexity

Problem	Objective function	Stress constraints		Displacement constraints	Buckling constraints		Frequency constraints	Total no. of constraints	Number of design variables	Feasible space (%)	
		L	NL	NL	L	NL	L			Continuous	Discrete
10-bar truss	NL	4	16	4	4	16	1	45	13	0.19	0.18
14-bar truss	NL	6	22	4	6	22	1	61	17	1.02	0.98
15-bar truss	NL	0	30	4	0	30	1	65	23	0.12	0.12
24-bar truss	NL	28	20	4	28	20	1	101	26	0.30	0.29
20-bar truss	NL	4	36	4	4	36	2	86	26	0.04	0.04
72-bar 3D truss	NL	0	144	16	0	144	2	306	19	0.36	0.35
39-bar truss	NL	8	70	12	8	70	1	169	28	0.11	0.08
45-bar truss	NL	56	34	12	56	34	1	193	33	4.34	3.99
25-bar 3D truss	NL	2	48	36	2	48	1	137	13	0.51	0.50
39-bar 3D truss	NL	2	76	36	2	76	1	193	16	3.93	3.83

L and NL signify linear and nonlinear functions, respectively factors, including the nature of the objective function, the details of design variables and constraints, and the feasible space for optimization.

Additionally, this study goes beyond examining the complexity-measuring parameters and delves into the impact of both feasible and infeasible search spaces on the optimization process. To conduct this analysis, the study randomly selects 100,000 points for 10 independent runs, using a random number technique. This allows for a thorough investigation of the effects of different search space conditions on the optimization results.

The objective functions of the optimization problems proposed in this study are characterized as nonlinear, meaning that they do not adhere to a linear relationship between input and output variables. On the other hand, the stress and buckling constraints involve a combination of linear and nonlinear components, while the displacement constraints are linear in nature.

As a result, the optimization problems in this study are classified as mixed linear and nonlinear types, due to the presence of both linear and nonlinear constraints. Moreover, the feasible space for these problems is narrow, meaning that the region where feasible solutions exist is limited in size. This narrow feasible space poses significant challenges for optimization methods, as finding optimal solutions within such constraints becomes highly complex and demanding.

In summary, the optimization problems analyzed in this study exhibit a high level of complexity, as evidenced by the nonlinear nature of the objective functions, the mixed linear and nonlinear constraints, and the narrow feasible space. These factors collectively make these problems highly challenging for optimization methods and underscore the need for advanced and robust optimization techniques to achieve effective solutions.

2.5.2 *Analysis of Feasible Search Space in TSS Optimization*

Tables 2.3 and 2.4 in this study provide a detailed analysis of the feasible and infeasible solutions for the TSS optimization problems of structurally constrained optimization. Specifically, the validity of a truss is determined based on whether its elements are connected to nodes with applied loads, nodes with boundary conditions, or unchangeable nodes predefined by a user, as illustrated in Fig. 2.9b, d. Trusses that do not meet these conditions are considered invalid and are rejected as solutions.

The analysis reveals that the 15-bar truss presents the most challenging problem, with 56.26% and 41.58% of its trusses being invalid and kinematically unstable, respectively. In contrast, the 45-bar truss proves to be the least difficult problem, with only 2.57% and 4.87% of its trusses being invalid and kinematically unstable for continuous sections, as shown in Table 2.3. These results indicate that the 15-bar truss poses more difficulties in optimization due to the higher percentage of invalid and kinematically unstable trusses, while the 45-bar truss is relatively easier to optimize.

Table 2.3 Proportion of infeasible trusses in the search space (continuous sections)

Problem	Infeasible solutions				Feasible solutions (%)
	Invalid trusses (%)	Kinematic instable trusses in DOF (%)	Kinematic instable trusses in the global stiffness matrix (%)	Constraint violation (%)	
10-bar truss	52.82	41.44	1.48	4.06	0.19
14-bar truss	17.42	43.85	9.34	28.37	1.02
15-bar truss	56.26	41.58	0.79	1.25	0.12
24-bar truss	12.32	35.47	16.59	35.32	0.30
20-bar truss	26.93	66.10	4.39	2.54	0.04
72-bar 3D truss	34.14	53.45	10.67	1.37	0.36
39-bar truss	11.38	51.11	12.71	24.69	0.11
45-bar truss	2.57	4.87	29.71	58.51	4.34
25-bar 3D truss	24.79	58.66	12.26	3.79	0.51
39-bar 3D truss	4.92	30.53	25.84	34.78	3.93

Invalid and kinematically unstable trusses present additional challenges in optimization problems, as they restrict the feasible solutions. The feasible space for the optimization problems ranges from 0.11% to 4.34% for continuous sections and 0.08% to 3.99% for discrete sections, as reported in Tables 2.3 and 2.4, respectively. This indicates that the feasible region is very small for all the problems, regardless of whether continuous or discrete sections are considered. Furthermore, the feasible space for discrete problems is even smaller than that of continuous sections, indicating that discrete problems are more challenging in terms of feasibility.

In conclusion, the analysis of feasible and infeasible solutions in this study reveals that the TSS optimization problems of structurally constrained optimization pose significant challenges, particularly due to the presence of invalid and kinematically unstable trusses and the small feasible space. Additionally, discrete problems are more challenging than continuous problems, as the feasible space is even more limited. These findings emphasize the need for robust and advanced optimization techniques to address the complexities associated with these problems.

Table 2.4 Proportion of infeasible trusses in the search space (discrete sections)

Problem	Infeasible solutions: Invalid trusses (%)	Kinematic instable trusses in DOF (%)	Kinematic instable trusses in the global stiffness matrix (%)	Constraint violation (%)	Feasible solutions (%)
10-bar truss	52.39	41.76	1.50	4.18	0.18
14-bar truss	17.09	43.49	9.41	29.03	0.98
15-bar truss	55.38	42.91	0.88	0.71	0.12
24-bar truss	11.79	33.83	16.71	37.38	0.29
20-bar truss	26.46	66.39	4.52	2.59	0.04
72-bar 3D truss	33.82	53.25	11.17	1.42	0.35
39-bar truss	10.97	49.71	13.37	25.87	0.08
45-bar truss	2.43	4.41	28.93	60.23	3.99
25-bar 3D truss	24.17	58.77	12.67	3.89	0.50
39-bar 3D truss	4.66	29.72	25.72	36.07	3.83

References

Ahrari, A., & Deb, K. (2016). An improved fully stressed design evolution strategy for layout optimization of truss structures. *Computers & Structures, 164*(February), 127–144. https://doi.org/10.1016/j.compstruc.2015.11.009

Ahrari, A., Atai, A. A., & Deb, K. (2015). Simultaneous topology, shape and size optimization of truss structures by fully stressed design based on evolution strategy. *Engineering Optimization, 47*(8), 37–41. https://doi.org/10.1080/0305215X.2014.947972

Bai, Y., De Klerk, E., Pasechnik, D., & Sotirov, R. (2009). Exploiting group symmetry in truss topology optimization. *Optimization and Engineering, 10*(3), 331–349. https://doi.org/10.1007/s11081-008-9050-6

Deb, K., & Gulati, S. (2001). Design of truss-structures for minimum weight using genetic algorithms. *Finite Elements in Analysis and Design, 37*(5), 447–465. https://doi.org/10.1016/S0168-874X(00)00057-3

Ferreira, A. J. (2009). MATLAB codes for finite element analysis. *Solids and Structures*, 157. Netherlands: Springer. https://doi.org/10.1007/978-1-4020-9200-8

Gholizadeh, S., & Barzegar, A. (2013). Shape optimization of structures for frequency constraints by sequential harmony search algorithm. *Engineering Optimization, 45*(6), 627–646. https://doi.org/10.1080/0305215X.2012.704028

Ghosh, A., & Mallik, A. K. (1994). *Theory of mechanisms and machines*. East-West Press.

Gonçalves, M. S., Lopez, R. H., & Miguel, L. F. F. (2015). Search group algorithm: A new metaheuristic method for the optimization of truss structures. *Computers & Structures, 153*, 165–184. https://doi.org/10.1016/j.compstruc.2015.03.003

Hajela, P., & Lee, E. (1995). Genetic algorithms in truss topological optimization. *International Journal of Solids and Structures, 32*(22), 3341–3357.

Jin, P., & De-yu, W. (2006). Topology optimization of truss structure with fundamental frequency and frequency domain dynamic response constraints. *Acta Mechanica Solida Sinica, 19*(3), 231–240. https://doi.org/10.1007/s10338-006-0628-2. The Chinese Society of Theoretical and Applied Mechanics.

Kaveh, A., & Kalatjari, V. (2003). Topology optimization of trusses using genetic algorithm, force method and graph theory. *International Journal for Numerical Methods in Engineering, 791-*(October 2002), 771–791. https://doi.org/10.1002/nme.800

Kaveh, A., & Mahdavi, V. R. (2015). Colliding bodies optimization for size and topology optimization of truss structures. *Structural Engineering and Mechanics, 53*(5), 847–865. https://doi.org/10.12989/sem.2015.53.5.847

Kaveh, A., & Zolghadr, A. (2013). Topology optimization of trusses considering static and dynamic constraints using the CSS. *Applied Soft Computing Journal, 13*(5), 2727–2734. https://doi.org/10.1016/j.asoc.2012.11.014. Elsevier B.V.

Kaveh, A., & Zolghadr, A. (2014). Democratic PSO for truss layout and size optimization with frequency constraints. *Computers and Structures, 130*, 10–21. https://doi.org/10.1016/j.compstruc.2013.09.002. Elsevier Ltd.

Kirsch, U. (1989). Optimal topologies of truss structures. *Computer Methods in Applied Mechanics and Engineering, 72*(1), 15–28. https://doi.org/10.1016/0045-7825(89)90119-9

Kumar, S., Tejani, G. G., & Mirjalili, S. (2019). Modified symbiotic organisms search for structural optimization. *Engineering with Computers, 35*, 1269–1296. https://doi.org/10.1007/s00366-018-0662-y

Kumar, S., Tejani, G. G., Pholdee, N., & Bureerat, S. (2020). Improved metaheuristics through migration based search and an acceptance probability for truss optimization. *Asian Journal of Civil Engineering*. https://doi.org/10.1007/s42107-020-00271-x

Kumar, S., Tejani, G. G., Pholdee, N., & Bureerat, S. (2022). Performance enhancement of meta-heuristics through random mutation and simulated annealing-based selection for concurrent topology and sizing optimization of truss structures. *Soft Computing, 26*, 5661–5683. https://doi.org/10.1007/s00500-022-06930-2

Li, J. (2015). Truss topology optimization using an improved species-conserving genetic algorithm. *Engineering Optimization, 47*(1), 107–128. https://doi.org/10.1080/0305215X.2013.875165

Miguel, L. F. F., Lopez, R. H., & Miguel, L. F. F. F. (2013). Multimodal size, shape, and topology optimisation of truss structures using the firefly algorithm. *Advances in Engineering Software, 56*, 23–37. https://doi.org/10.1016/j.advengsoft.2012.11.006. Elsevier Ltd.

Noilublao, N., & Bureerat, S. (2011). Simultaneous topology, shape and sizing optimisation of a three-dimensional slender truss tower using multiobjective evolutionary algorithms. *Computers and Structures, 89*(23–24), 2531–2538. https://doi.org/10.1016/j.compstruc.2011.08.010. Elsevier Ltd.

Ohsaki, M. (1995). Genetic algorithm for topology optimization of trusses. *Computers & Structurers, 57*(2), 219–225. https://doi.org/10.1016/0045-7949(94)00617-C

Ohsaki, M., Fujisawa, K., Katoh, N., & Kanno, Y. (1999). Semi-definite programming for topology optimization of trusses under multiple eigenvalue constraints. *Computer Methods in Applied Mechanics and Engineering, 180*(1–2), 203–217. https://doi.org/10.1016/S0045-7825(99)00056-0

Pholdee, N., & Bureerat, S. (2014). Comparative performance of meta-heuristic algorithms for mass minimisation of trusses with dynamic constraints. *Advances in Engineering Software, 75*, 1–13. https://doi.org/10.1016/j.advengsoft.2014.04.005. Elsevier Ltd.

Rao, S. S. (2009). *Engineering optimization theory and practice* (4th ed.). Wiley.

Richardson, J. N., Bouillard, P., Adriaenssens, S., Bouillard, P., & Filomeno, R. (2012). Multiobjective topology optimization of truss structures with kinematic stability repair. *Structural and Multidisciplinary Optimization, 46*, 513–532.

Savsani, V. J., Tejani, G. G., & Patel, V. K. (2016). Truss topology optimization with static and dynamic constraints using modified subpopulation teaching–learning-based optimization. *Engineering Optimization, 48*, 1–17. https://doi.org/10.1080/0305215X.2016.1150468

Savsani, V. J., Tejani, G. G., Patel, V. K., & Savsani, P. (2017). Modified meta-heuristics using random mutation for truss topology optimization with static and dynamic constraints. *Journal of Computational Design and Engineering, 4*, 106–130. https://doi.org/10.1080/0305215X.2016.1150468

Singh, P., Kottath, R., & Tejani, G. G. (2022). Ameliorated follow the leader: Algorithm and application to truss design problem. *Structures, 42*, 181–204. https://doi.org/10.1016/j.istruc.2022.05.105

Souza, R. R., Miguel, L. F. F., Lopez, R. H., Miguel, L. F. F., & Torii, A. J. (2016). A procedure for the size, shape and topology optimization of transmission line tower structures. *Engineering Structures, 111*, 162–184. https://doi.org/10.1016/j.engstruct.2015.12.005. Elsevier Ltd.

Tejani, G. G. (2017). *Investigation of advanced metaheuristic techniques for simultaneous size, shape, and topology optimization of truss structures*. Ph.D. thesis, Pandit Deendayal Petroleum University, India.

Tejani, G. G., Savsani, V. J., & Patel, V. K. (2016a). Modified sub-population teaching-learning-based optimization for design of truss structures with natural frequency constraints. *Mechanics Based Design of Structures and Machines, 44*, 495–513. https://doi.org/10.1080/15397734.2015.1124023

Tejani, G. G., Savsani, V. J., & Patel, V. K. (2016b). Adaptive symbiotic organisms search (SOS) algorithm for structural design optimization. *Journal of Computational Design and Engineering, 3*, 226–249. https://doi.org/10.1016/j.jcde.2016.02.003

Tejani, G. G., Savsani, V. J., & Patel, V. K. (2017a). modified sub-population based heat transfer search algorithm for structural Optimization. *International Journal of Applied Metaheuristic Computing, 8*, 1–23. https://doi.org/10.4018/IJAMC.2017070101

Tejani, G. G., Savsani, V. J., Patel, V. K., & Bureerat, S. (2017b). Topology, shape, and size optimization of truss structures using modified teaching-learning based optimization. *Advances in Computational Design, 2*, 313–331. https://doi.org/10.12989/acd.2017.2.4.313

Tejani, G. G., Savsani, V. J., Patel, V. K., & Mirjalili, S. (2018c). An improved heat transfer search algorithm for unconstrained optimization problems. *Journal of Computational Design and Engineering*. https://doi.org/10.1016/j.jcde.2018.04.003

Tejani, G. G., Savsani, V. J., Patel, V. K., & Mirjalili, S. (2018d). Truss optimization with natural frequency bounds using improved symbiotic organisms search. *Knowledge-Based Systems, 143*, 162–178. https://doi.org/10.1016/j.knosys.2017.12.012

Tejani, G. G., Savsani, V. J., Patel, V. K., & Savsani, P. V. (2018a). Size, shape, and topology optimization of planar and space trusses using mutation-based improved metaheuristics. *Journal of Computational Design and Engineering, 5*, 198–214. https://doi.org/10.1016/j.jcde.2017.10.001

Tejani, G. G., Savsani, V. J., Bureerat, S., & Patel, V. K. (2018b). Topology and size optimization of trusses with static and dynamic bounds by modified symbiotic organisms search. *Journal of Computing in Civil Engineering, 32*, 1–11. https://doi.org/10.1061/(ASCE)CP.1943-5487.0000741

Tejani, G. G., Savsani, V. J., Bureerat, S., et al. (2019). Topology optimization of truss subjected to static and dynamic constraints by integrating simulated annealing into passing vehicle search algorithms. *Engineering with Computers, 35*, 1–19.

Xu, B., Jiang, J., Tong, W., & Wu, K. (2003). Topology group concept for truss topology optimization with frequency constraints. *Journal of Sound and Vibration, 261*(5), 911–925. https://doi.org/10.1016/S0022-460X(02)01021-0

Zhong, W., Su, R., Gui, L., & Fan, Z. (2015). Topology and sizing optimization of discrete structures using a cooperative coevolutionary genetic algorithm with independent ground structures. *Engineering Optimization*. https://doi.org/10.1080/0305215X.2015.1064119

Chapter 3
Metaheuristics Methods

3.1 Introduction

Engineering optimization functions are complex and challenging due to their large search space, multidimensional, multimodal, nonlinear, nonconvex, and implicit nature. Gradient-based information may not be available, rendering analytical methods insignificant and often leading to local optima. To address these challenges, numerous metaheuristics have been developed, modified, and widely used over the past four decades (Tejani, 2017).

The most commonly used metaheuristic is the genetic algorithm (GA), which is based on the Darwinian evolutionary theory (Holland, 1975). Subsequently, many algorithms have been developed, including simulated annealing (SA) which is based on the concept of annealing in metallurgy (Kirkpatrick et al., 1983; Cerny, 1985), the artificial immune system algorithm inspired by the biological immune system (Farmer et al., 1986), and the ant colony optimization (ACO) algorithm which works on the foraging behavior of ants towards food (Dorigo, 1992).

Other algorithms include the particle swarm optimization (PSO) which operates based on the social behavior of a swarm towards food (Kennedy & Eberhart, 1995), the differential evolution (DE) algorithm that is similar to GA but utilizes specialized crossover and selection methods (Storn & Price, 1997), and the bacterial foraging optimization algorithm which simulates the behavior of *Escherichia coli* bacteria (Passino, 2002). Additionally, the invasive weed optimization algorithm was inspired by colonizing weeds (Mehrabian & Lucas, 2006), the artificial bee colony algorithm mimics the cooperative behavior of honeybee colonies (Karaboga & Basturk, 2007, 2008), and the biogeography-based optimization algorithm simulates the principle of species migration (Simon, 2008).

Other metaheuristics include the gravitational search algorithm (GSA), inspired by the law of gravity and mass interactions (Rashedi et al., 2009), the firefly algorithm inspired by the mating behavior of firefly insects (Yang, 2009), the cuckoo search algorithm based on the reproductive strategy of cuckoo birds (Yang & Deb,

V. Savsani et al., *Truss Optimization*, https://doi.org/10.1007/978-3-031-49295-2_3

2010), the hunting search algorithm that operates based on a model of group hunting of animals (Oftadeh et al., 2010), and the bat algorithm inspired by the echolocation behavior of bats (Yang, 2010). Other algorithms include the grenade explosion method that operates based on grenade explosion behavior (Ahrari & Atai, 2010), the charged system search algorithm that utilizes the electric laws of physics and the Newtonian laws of mechanics (Kaveh & Talatahari, 2010), the teaching-learning based optimization (TLBO) algorithm that works based on teaching-learning in a classroom (Rao et al., 2011, 2012), the animal migration optimization algorithm inspired by animal migration behavior (Li et al., 2014), and the water wave optimization that simulates the shallow water wave theory (Zheng, 2015).

Although metaheuristics can effectively solve numerous real and complex problems, they are not universally suitable for optimizing all types of problems, as the no free lunch (NFL) theorem (Wolpert & Macready, 1997) indicates. Consequently, it is likely that one algorithm will perform better than another in solving one particular set of challenges, while it may perform poorly on a different set of problems. This theorem has been a fundamental concept in this field, providing a strong rationale for the application of metaheuristics in various domains.

The dragonfly algorithm (DA) (Mirjalili, 2016a), multiverse optimizer (MVO) (Mirjalili et al., 2016), sine cosine algorithm (SCA) (Mirjalili, 2016b), whale optimization algorithm (WOA) (Mirjalili & Lewis, 2016), antlion optimizer (ALO) (Mirjalili, 2015), heat transfer search (HTS) (Patel & Savsani, 2015), passing vehicle search (PVS) (Savsani & Savsani, 2015), symbiotic organisms search (SOS) (Cheng & Prayogo, 2014), grey wolf optimizer (GWO) (Mirjalili et al., 2014), and TLBO (Rao et al., 2011, 2012) algorithms are a set of recently developed techniques that can be explored for solving challenging optimization problems. Among these algorithms, the TLBO method is a highly effective approach that has demonstrated success in various engineering optimization problems. However, each of these metaheuristics has unique search mechanisms, and the impact of modifying them on different applications is difficult to predict. Therefore, it is essential to study and evaluate these algorithms in various domains to determine their efficacy for specific problems.

Numerous metaheuristics have been examined by researchers to address structural optimization (SO) problems. While the genetic algorithm (GA) and its variations have been widely used for SO (Rajan, 1995; Deb & Gulati, 2001; Tang et al., 2005; Rahami et al., 2008; Richardson et al., 2012), other modern and promising metaheuristics have been utilized across a broad range of SO problems. These include the optimality criteria method (Canfield et al., 1989), simulated annealing (SA) and ant colony optimization (ACO) (Luh & Lin, 2008), group search optimizer (GSO) and improved GSO (Li & Liu, 2011), adaptive multi-population differential evolution (AMPDE) (Wu & Tseng, 2010), firefly algorithm (FA) (Miguel et al., 2013), improved species-conserving GA (Li, 2015), fully stressed design based on evolution strategy (Ahrari et al., 2014), and FA with proportional topology optimization method (Fu et al., 2016).

In this book, ten basic metaheuristics are employed to investigate three distinct trusses for truss optimization problems. The proposed algorithms are evaluated on size optimization, TTO, and TSS optimization problems, considering overall mass as an objective function and stress, displacement, buckling, frequency, and kinematic stability as constraints, along with multiload conditions. The study also incorporates a single-stage approach, ground structure method, and restructuring of finite element analysis model for TTO.

3.2 Metaheuristics

In this comparative study, ten population-based metaheuristics are analyzed for their ability to solve structural optimization problems. These algorithms start with a random population and use mathematical formulas inspired by nature to update and improve the solutions. A greedy selection scheme is employed by some of the metaheuristics to decide whether to retain or replace a solution. It should be noted that duplicate solutions are not removed in the DA, MVO, SCA, WOA, ALO, HTS, PVS, SOS, GWO, and TLBO algorithms, as this may require more function evaluations (FE) (Liu & Mernik, 2012; Waghmare, 2013). Details about the algorithms and their proposed improvement strategies are summarized in the following sections.

3.2.1 The DA Algorithm

The DA algorithm, proposed by Mirjalili et al. (2016), initiates with a randomly generated population within its upper and lower bounds, where "n" is the population size (i.e., dragonflies) and "m" the design variable1s. Next, the objective values of the population are calculated. In the following stage, the population is updated by separation, alignment, and cohesion phenomena in each generation. The separation phase updates the position vector of each dragonfly based on the distance between the dragonfly and its neighbors to avoid collision. The alignment phase updates the velocity vector of each dragonfly by matching its velocity with that of its neighbors. The cohesion phase updates the position vector of each dragonfly by attracting it towards the center of the population.

Moreover, the progression of the DA algorithm is controlled by various parameters such as separation weight, alignment weight, cohesion weight, food attraction weight, and enemy distraction weight. The food attraction weight and enemy distraction weight mimic the dragonflies' behavior of searching for food and avoiding enemies, respectively. The neighboring radius, velocity vector, and position vector of each dragonfly are updated based on the weighted parameters, food attraction weight, and enemy distraction weight. New solutions are then searched towards the velocity vector or position vector, and the bounds of the new solution are checked. This process is repeated until the termination criterion is satisfied.

Overall, the DA algorithm is a promising metaheuristic algorithm that has shown good performance in various optimization problems. It has the advantage of being easy to implement and requiring only a few control parameters. However, the performance of the DA algorithm can be affected by the choice of parameter values, and it may converge to local optima in complex optimization problems.

The DA algorithm can be detailed as below:

```
START
Define objective function F(X), population size (n), set number of design variables (m), limits on design variables (LB, UB),
and set termination criterion ('FEmax', or 'gmax'); where F(X) is the objective function and 'X' is the design vector. Dragonflies
are assumed to be population size (i=1,2,...,n).                    /* Initialization /*
Initialize the randomly generated population (X) of dragonflies and velocity vector (ΔX) within its upper and lower bounds
and evaluate the population.                                        /* Initialize population /*
FE = 0
for g = 1 to gmax do                                                /* Initialize the optimization loop /*
Identify the best solution (X_best) and worse solution (X_worse) of the population.
    r = (UB-LB)/4 + (UB - LB) × (g/gmax) × 2                        /* Algorithm specific parameters /*
    w = 0.9 - g × ((0.9-0.4)/gmax) /* 'w' is the inertia weight /*
    my_c = 0.1 - (0.1 × g)/(gmax/2)
    if my_c < 0 then
        my_c = 0
    end if
    s = 2 × rand(0,1) × my_c /* 's' is separation weight /*
    a = 2 × rand(0,1) × my_c /* 'a' is alignment weight /*
    c = 2 × rand(0,1) × my_c /* 'c' is cohesion weight /*
    f = 2 × (0,1) /* 'f' is food attraction weight /*
    e = my_c /* 'e' is enemy distraction weight /*
    for i = 1 to n do                                               /* Update the population /*
Find the neighbouring solutions (X_j; j=1,2,...,M). M is the number of neighbouring individuals, which stand within 'r'
distance of i^th population.
        S_i = -Σ_j^M (X_i - X_j)                                    /* Separation /*
        A_i = (Σ_j^M X_j)/M                                         /* Alignment /*
        C_i = (Σ_j^M X_j)/M - X_j                                   /* Cohesion /*
        if distance(X_i, X_best) ≤ r then                           /* Attraction to food /*
            F_i = X_best - X_i
        else
            F_i = 0
        end if
        if distance(X_i, X_worse) ≤ r then                          /* Distraction outwards an enemy /*
            F_i = X_worse + X_i
        else
            F_i = 0
        end if
        if a dragonfly has at least one neighboring dragonfly then  /* Update solution /*
            ΔX_i = (s × S_i + a × A_i + c × C_i + f × F_i + e × E_i) + w × ΔX_i /*ΔX_i is velocity vector /*
            X'_i = X_i + ΔX_i /* Where, X'_i is updated solution /*
        else if
            X'_i = X_i + Lévy × X_i                                 /* position vector /*
        end if
/* Where, Lévy = 0.01 × (rand_1(0,1)×σ)/(|rand_2(0,1)|^(1/β)); 'β' is constant (equal to 1.5 in this study);
σ = ( Γ(1+β)×sin(πβ/2) / (Γ((1+β)/2)×β×2^((β-1)/2)) )^(1/β); Γ(x) = (x - 1)! /*
        FE = FE + 1                                                 /* Count Function evaluation /*
        if FE ≥ FEmax then                                          /* Termination criterion /*
            break optimization loop
        end if
    end for                                                         /* Population loop ends /*
end for                                                             /* Optimization loop ends /*
STOP
```

3.2.2 The MVO Algorithm

The MVO algorithm is a novel metaheuristic inspired by the theories of cosmology proposed by Mirjalili et al. (2016). The term "multiverse" refers to the existence of multiple universes, including our own, that may interact and collide with each other. The MVO algorithm incorporates three theories of cosmology: white hole, black hole, and wormhole. The white hole theory is presumed to be the big bang, where the collisions among the universes result in the birth of a new universe. The black hole absorbs everything with its strong gravitational force, while wormholes connect different sections of the universe and function as time/space travel channels. In the MVO algorithm, the white hole and black hole theories assist in exploring the search space, while the wormholes govern the exploitation of the search space.

The MVO algorithm starts with a randomly generated set of universes within the specified upper and lower bounds, where "n" represents the population size and "m" is the number of design variables. The objective values of the population are then calculated. Next, the solution is updated by the white hole, black hole, and wormhole operators at an indefinite time interval (i.e., generation). The selection of the white hole of the population is made by the roulette wheel in each generation. The MVO algorithm's progression is controlled by various parameters such as inflation rates, wormhole existence probability, and traveling distance rate. In this method, new solutions search near the best solution. Additionally, the updated solution in the MVO algorithm is accepted if it has a better function value compared to the previous solutions. The process is repeated until it satisfies the termination criterion. The detailed stepwise process of the MVO algorithm is presented below:

```
START
Define objective function F(X), population size (n), set number of design variables (m), limits on design variables (LB, UB),
and set termination criterion ('FEmax', or 'gmax'); where F(X) is the objective function and 'X' is the design vector. Universes
is assumed to be population size (i=1,2,...,n).                    /* Initialization /*
Initialize the randomly generated population within its upper and lower bounds and evaluate it.    /* Initialize population
/*
FE = 0
WEP_max = 1                                                    /* Wormhole existence probability (WEP) /*
WEP_min = 0.2
for g = 1 to gmax do /* generation is considered to be time interval /*     /* Initialize the optimization loop /*
    WEP = WEP_min + g × (WEP_max − WEP_min) / g_max
    TDR = 1 − (g / g_max)^(1/6)                                /* Travelling distance rate (TDR)/*
Arrange the population in ascending order of F(X_i) values Identify the best universe (X_best) of the population.
Find normalized inflation rates (NI)                      /* NI is normalized value of evaluated population /*
    for i = 1 to n do                                          /* Update the population /*
        for j = 1 to m do /* update design variable /*
            if rand(0,1)< NI_i then    /* Select White hole (X_k) based on roulette wheel selection criteria /*
                X_i,j = X_k,j /* j indicates design variable number /*
            end if
            if rand(0,1) < WEP then
                if rand(0,1) < 0.5 then
                    X_i,j = X_best,j + TDR × ((UB_j − LB_j) × rand(0,1) + LB_j)
                else
                    X_i,j = X_best,j − TDR × ((UB_j − LB_j) × rand(0,1) + LB_j)
            end if
        end for
        FE = FE + 1                                            /* Count Function evaluation /*
        if F(X_i) < F(X_best) then    /* Update the best solution /*    /* Greedy selection /*
            X_best = X_i
        end if
        if FE ≥ FEmax then                                     /* Termination criterion /*
            break optimization loop
        end if
    end for                                                    /* Population loop ends /*
end for                                                        /* Optimization loop ends /*
STOP
```

3.2.3 The SCA Algorithm

The SCA algorithm, which was proposed by Mirjalili (2015), is a population-based metaheuristic. Its theory is based on a mathematical model that involves sine and cosine functions. These functions are arranged in a cyclic pattern, which enables the algorithm to explore and exploit the search space. During the early stages, the sine and cosine functions are used to explore new regions, while the algorithm gradually shifts towards exploiting the search space as it progresses.

To begin, the SCA algorithm generates a set of random solutions within the upper and lower bounds of the search space, with a population size of "n" and "m" design variables. Next, it calculates the objective values of each solution in the population. In each generation, the population is updated based on a set of rules that involve the sine and cosine functions. Additionally, the SCA algorithm uses a combination of random and adaptive variables to gradually shift from exploration to exploitation. The algorithm directs new solutions towards the best solution found so far. Furthermore, updated solutions are accepted only if they have a better function value than previous solutions. The process is repeated until the termination criterion is met. The detailed step-by-step process of the SCA algorithm is provided below according to Mirjalili (2015):

```
START
Define objective function F(X), population size (n), set number of design variables (m), limits on design variables (LB,UB),
and set termination criterion ('FEmax', or 'gmax'); where, F(X) is the objective function and 'X' is the design vector. The set of
chromosomes is considered to be population size (i=1,2,...,n).          /* Initialization /*
Initialize the randomly generated population within its upper and lower bounds and evaluate it.     /* Initialize population
/*
FE = 0;
Identify the best solution (X_best) of the population.
for g = 1 to gmax do                                         /* Initialize the optimization loop /*
    r1 = a - (a×g)/gmax ; a = 2 /* r1 is SCA parameter /*
    for i = 1 to n do                                        /* Update the population /*
        for j = 1 to m do                                    /* Update design variable /*
            r2 = 2 × π × rand(0,1)                           /* r2, r3, and r4 are SCA parameters /*
            r3 = 2 × rand(0,1)
            r4 = rand(0,1)
            if r4 < 0.5 then
                X'i,j = Xi,j + r1 × sin r2 × |r3 × Xbest,j − Xi,j|    /* j is design variable number /*
            else
                X'i,j = Xi,j + r1 × cos r2 × |r3 × Xbest,j − Xi,j|
            end if
        end for
        FE = FE + 1                                          /* Count Function evaluation /*
        if F(Xi) < F(Xbest) then    /* Update the best solution /*    /* Greedy selection /*
            Xbest = Xi
        end if
        if FE ≥ FEmax then                                   /* Termination criterion /*
            break optimization loop
        end if
    end for                                                  /* Population loop ends /*
end for                                                      /* Optimization loop ends /*
STOP
```

3.2.4 The WOA Algorithm

The WOA algorithm is a novel metaheuristic proposed by Mirjalili and Lewis (2016) that takes inspiration from the social behavior of humpback whales during bubble-net hunting. This hunting strategy involves the whales creating a circular or spiral-shaped bubble net to trap and catch krill or small fishes. The bubble-net hunting strategy consists of three stages: the coral loop, lobtail, and capture loop.

To begin the WOA algorithm, a set of population (referred to as ants and antlions) is randomly generated within upper and lower bounds. This population includes "n" size and "m" design variables. Next, the objective values of the population are calculated. In the subsequent stage, the population is updated through various mechanisms such as encircling prey, shrinking the encircling mechanism, spiral updating position, and searching for prey in each generation. The WOA algorithm's progression is controlled by various parameters. New solutions are searched for near the best solution and a randomly selected solution. The updated solution in the WOA algorithm is accepted if it has a better function value than the previous solutions. The process is repeated until the termination criterion is met. The stepwise process of the WOA algorithm is presented in detail below:

```
START
Define objective function F(X), population size (n), set number of design variables (m), limits on design variables (LB, UB),
and set termination criterion ('FEmax', or 'gmax'); where, F(X) is the objective function and 'X' is the design vector. The whale's
population are considered to be population size (i=1,2,...,n).          /* Initialization /*
Initialize the randomly generated population within its upper and lower bounds and evaluate it.
                                                                        /* Initialize population /*
FE = 0
Identify the leader, the best solution (X_best), of the population.
for g = 1 to gmax do                                                    /* Initialize the optimization loop /*
    a = 2 − g × 2/gmax                                  /* 'a' and 'a2' are WOA parameters /*
    a2 = −1 + −g/gmax
    for i = 1 to n do                                                   /* Update the population /*
        A = 2 × a × rand1(0,1) − a                      /* 'A' and 'C' are coefficient parameters /*
        C = 2 × rand2(0,1)
        b = 1                   /* 'b' is a constant for defining the shape of the logarithmic spiral /*
        l = (a2 − 1) × rand(0,1) + 1                    /* 'l' and 'p' are random numbers /*
        p = rand(0,1) /*
        if p < 0.5 then                                 /* Update position of the current population /*
            if |A| ≥ 1 then
                Di = |C × Xk − Xi|                      /* 'k' is a randomly selected population /*
                X'i = Xk − A × Di                       /* Generate new population in Search for prey phase /*
            else
                Di = |C × Xbest − Xi|
                X'i = Xbest − A × Di                    /* Generate new population in Encircling prey phase /*
            end if
        else
            D'i = |Xbest − Xi|          /* D'i is distance of the i^th population to the best /*
            X'i = D'i × e^(b×l) × cos(2 × π × l) + Xbest    /* Generate new population in Bubble-net attacking method /*
        end if
        FE = FE + 1                                     /* Count Function evaluation /*
        if F(Xi) < F(Xbest) then                        /* Update the best solution /* Greedy selection /*
            Xbest = Xi
        end if
        if FE ≥ FEmax then                              /* Termination criterion /*
            break optimization loop
        end if
    end for                                             /* Population loop ends /*
end for                                                 /* Optimization loop ends /*
STOP
```

3.2.5 The ALO Algorithm

Mirjalili (2015) proposed the ALO algorithm, which is a newly developed algorithm inspired by nature. This algorithm imitates the hunting behavior of antlions, which includes the random walk of ants inside the antlion's trap, the building of traps, trapping ants, catching prey, and reconstructing traps. The algorithm assumes that antlions and ants represent a population, where the antlion's life span involves two primary stages: larvae and adulthood. The antlion larva creates a cone-shaped ditch trap in sand and patiently waits for insects, such as ants, to become trapped. Antlions primarily hunt during their larval stage, while the adulthood period is for reproduction. Therefore, the ALO algorithm integrates the hunting strategies of antlions.

The ALO algorithm is a novel nature-inspired algorithm, suggested by Mirjalili (2015). The ALO algorithm works on the hunting mechanism (viz., the random walk of ants inside an antlion's trap, building traps, entrapment of ants in traps, catching preys, and rebuilding traps) of antlions in nature. In this algorithm, antlions and ants are assumed to be a population. The life span of antlions comprises mainly two stages: larvae and adult. An antlion larva digs a cone-shaped ditch trap in sand and waits for insects (ants) to be stuck in the ditch. Antlions typically hunt in larvae and the adulthood period is for reproduction. Thus, the ALO incorporates hunting strategies of antlions.

The ALO algorithm begins by generating a set of populations randomly, constrained within upper and lower bounds. The population size is "n," and the design variables are "m." The objective values of the population are then calculated, and the best antlions are identified and chosen as the elite. In the next stage, the population is updated using predefined states in each generation. In each generation, the selection of antlions is done using the roulette wheel selection process. Furthermore, various parameters are used to control the ALO algorithm's progress and decrease bounds towards the antlion after each generation.

The ALO algorithm employs two steps to search for new solutions. The first step is the random walk around the elite, which searches for solutions around the mean of the best solution. The second step is the random walk of ants, which searches for solutions around the previous solution. The updated solution is accepted only if it has a better function value. The process is repeated until it satisfies the termination criterion.

The ALO algorithm comprises several steps, including the random walks of ants, trapping in antlion's pits, building a trap, sliding ants towards the antlion, catching prey, and rebuilding the pit, along with elitism. These steps are illustrated in detail below.

START
Define objective function F(X), population size (n), set number of design variables (m), limits on design variables (LB, UB), and set termination criterion ('FE_{max}', or 'g_{max}'); where F(X) is the objective function and 'X' is the design vector. Ants and antlions are assumed to be population size (i=1,2,...,n). ***/* Initialization /****
Initialize the randomly generated population of ants and antlions within its upper and lower bounds and evaluate it. Sort the population. ***/* Initialize population /****
FE = 0; c = L; d = U / c is current lower bound, and d is upper current bound /** ***/* Set current bounds /****
for** g = 1 to g_{max} **do ***/* Initialize the optimization loop /****
Identify the best solution (X_{elite}) of the population and calculate mean of each design variables (X_{mean}).
 for** i = 1 to n **do ***/* Update the population /****
Select antlions ($Antlion_i$) based on roulette wheel selection criteria. ***/* Selection of antlions /****
 $c_i = \frac{c_i}{I}$ & $d_i = \frac{d_i}{I}$; $I = \frac{10^w \times g}{g_{max}}$ ***/* Decrease bounds to converge towards antlion /****
/ Where, w = 2 when $g > 0.1 \times g_{max}$, w = 3 when $g > 0.5 \times g_{max}$, w = 4 when $g > 0.75 \times g_{max}$, w = 5 when $g > 0.9 \times g_{max}$, and w = 6 when $g > 0.95 \times g_{max}$ /**
 $c_i = Antlion_i + c_i$ ***/* Trapping in antlion's pits /****
 $d_i = Antlion_i + d_i$ ***/* Move the current bounds around the antlion /****
 $RA_i = \frac{(X_i - a_i) \times (d_i - c_i)}{(b_i - a_i)} + c_i$ ***/* Random walks of ants /****
/ Where, X_i' is updated solution, 'a_i' is the minimum value of variables, 'b_i' is the maximum of variable. /**
 $RE_i = \frac{(X_{elite} - a_i) \times (d_i - c_i)}{(b_i - a_i)} + c_i$ ***/* Random walk around the elite /****
 $X_i' = \frac{RA_i + RE_i}{2}$ ***/* Generate new population /****
 FE = FE + 1 ***/* Count Function evaluation /****
 if** $F(X_i') < F(X_i)$ **then ***/* Greedy selection/****
 $X_i = X_i'$
 end if
Update the position of elite antlions (best solution)
 if** $FE \geq FE_{max}$ **then ***/* Termination criterion /****
 break optimization loop
 end if
 end for ***/* Population loop ends /****
end for ***/* Optimization loop ends /****
STOP

3.2.6 The HTS Algorithm

Patel and Savsani (2015) proposed the HTS algorithm, which aims to achieve thermal equilibrium by considering heat transfer between system molecules and the surrounding. The law of thermodynamics dictates that any system will always strive to reach equilibrium with its surroundings. Consequently, a system that is thermodynamically imbalanced will transfer heat with its surroundings to achieve thermal equilibrium. Conduction, convection, and radiation are the primary modes of heat transfer that play a crucial role in setting thermal equilibrium. Therefore, the HTS algorithm incorporates "the conduction phase," "the convection phase," and "the radiation phase" to achieve equilibrium (Tejani et al., 2017a, 2018a). The algorithm gives equal opportunities to all three modes of heat transfer, and the heat transfer mode for each generation is chosen randomly.

The HTS algorithm starts with the generation of a randomly populated system containing "n" molecules, with the temperature level as the design variable. Objective values are then calculated for each member of the population. In each subsequent generation, the population is updated using one of the three randomly selected heat transfer modes. If the updated solution has a better functional value than the current one, it is accepted.

During the conduction phase, the new solution searches for nearby random solutions. In the convection phase, the new solution searches for nearby solutions from the best solution. In the radiation phase, the new solution searches for nearby solutions calculated as the difference between two random solutions or the mean value of the population. The search for a better solution involves computing the difference between the current solution and either the best solution, a random solution, or the mean value of the population. This process is repeated until the termination criterion is met.

The HTS algorithm assumes a conduction factor (CDF) of 2, a convection factor (COF) of 10, and a radiation factor (RDF) of 2, as suggested by Patel and Savsani (2015). The detailed stepwise process of the algorithm is outlined below.

```
START
Define objective function F(X), population size (n), set number of design variables (m), limits on design variables (LB, UB),
set controlling parameters (i.e. CDPF, COPF, RAPF, IDFE, p_f, and p_r. Where, CDPF, COPF, RAPF, p_f, and p_r ∈ [0,1] and
CDPF + COPF + RAPF = 1), and set termination criterion ('FE_max', or 'g_max'); where F(X) is the objective function and 'X'
is the design vector. Molecules of the system are considered to be a part of the population (i=1,2,...,n) and the temperature
levels of the molecule are considered as the design variables (j=1,2,...,m).                /* Initialization /*
Initialize the randomly generated set of the population within its upper and lower bounds and evaluate it.
                                                                        /* Initialize population /*
  FE = 0
  for g = 1 to g_max do                                         /* Initialize the optimization loop /*
  Arrange the population in ascending order of F(X_i) values and select the best solution (X_1) as the surrounding (X_s).
  Generate values for R randomly to decide the heat transfer phase with equal probability; where, R ∈ [0, 1].
      for i = 1 to n do                                              /* Update the population /*
  Generate new population in the conduction phase.
          if R < 0.3333 then                    /* Generate new population in the conduction phase */
              if FE ≤ FE_max/CDF then /* Take CDF = 2 /*
                  if F(X_i) > F(X_k) then /* 'k' is a randomly selected population /*
                      X'_i = X_k + (−R^2 * X_k)
                  else
                      X'_i = X_i + (−R^2 * X_i)
                  end if
              else
                  if F(X_i) > F(X_k) then
                      X'_i = X_k + (−rand(0,1)_i * X_k)
                  else
                      X'_i = X_i + (−rand(0,1)_i * X_i)
                  end if
              end if
          else if R < 0.6666 then               /* Generate new population in the radiation phase /*
              if FE ≤ FE_max/RDF then  /* Where, RDF = 2 /*
                  if F(X_i) > F(X_k) then
                      X'_i = X_i + R * (X_k − X_i)
                  else
                      X'_i = X_i + R * (X_i−X_k)
                  end if
              else
                  if F(X_i) > F(X_k) then
                      X'_i = X_i + rand(0,1)_i * (X_k − X_i)
                  else
                      X'_i = X_i + rand(0,1)_i * (X_i−X_k)
                  end if
              end if
          end if
          else
              if FE ≤ FE_max/COF then /* Where, COF = 10; /*Generate new population in the convection phase */
                  X'_i = X_i + R * (X_s − X_ms * abs(R − rand(0,1)_i)) /* X_ms is populations' mean value /*
              else
                  X'_i = X_i + R * (X_s − X_ms * round(1 + rand(0,1)_i))
              end if
          end if
          FE = FE + 1                                              /* Count Function evaluation /*
          if F(X'_i) < F(X_i) then                                 /* Greedy selection/*
              X_i = X'_i
          end if
          if FE ≥ FE_max then                                      /* Termination criterion /*
              break optimization loop
          end if
      end for                                                      /* Population loop ends /*
  end for                                                          /* Optimization loop ends /*
STOP
```

3.2.7 The PVS Algorithm

Savsani and Savsani (2015) proposed the PVS algorithm, a novel metaheuristic algorithm that imitates the mechanism of vehicle passing on a two-lane highway. The primary objective of the PVS algorithm is to identify safe overtaking opportunities on a two-lane vehicle passing mechanism. This mechanism depends on various complex and interdependent parameters, such as the availability of gaps in the opposing traffic stream, individual vehicle speed and acceleration, driver skills, traffic, and road and weather conditions (Tejani et al., 2019).

The PVS algorithm considers three types of vehicles, namely, the back vehicle (BV), front vehicle (FV), and oncoming vehicle (OV), responsible for the passing mechanism on two-lane highways. The BV intends to pass the FV, but this is only possible if the FV's speed is slower than the BV's speed. If the FV's speed is higher than the BV's speed, then no passing is possible. Furthermore, the passing mechanism depends on the position and speed of the OV, as well as the distance between the vehicles and their velocities. Therefore, Savsani and Savsani (2015) considered various conditions, as follows:

Let us assume three types of vehicles (BV, FV, and OV) on a two-lane highway with different velocities (V_1, V_2, and V_3), where x represents the distance between BV and FV, and y represents the distance between FV and OV at a particular time. Based on the velocity of FV and BV, two primary conditions arise: when FV is slower than BV ($V_1 > V_3$) and vice versa. If FV is faster than BV, then no passing is possible, and BV can continue with its desired velocity. Passing is only possible when FV is slower than BV. Even in this situation, overtaking is only possible if the distance x_1 from the FV at which overtaking occurs is less than the distance traveled by OV ($y-y_1$). Thus, different conditions arise for the selected vehicles.

The PVS algorithm begins by generating a set of random solutions within the upper and lower bounds, where the population size is "n" and the number of design variables is "m." Next, the objective values of the population are calculated. Then, in each generation, the population is updated using one of the vehicle passing mechanisms. The updated solution is accepted if it has a better functional value. This process continues until the termination criterion is met. The detailed stepwise process of the PVS algorithm is presented below:

```
START
Define objective function F(X), population size (n), set number of design variables (m), limits on design variables (LB, UB),
and set termination criterion ('FEmax', or 'gmax'); where F(X) is the objective function and 'X' is the design vector. The
population of vehicles on a two-lane highway are considered to be a part of the population size (i=1,2,...,n).
                                                                        /* Initialization /*
Initialize the randomly generated the set of the population within its upper and lower bounds and evaluate it. Sort the
population.                                                             /* Initialize population /*
FE = 0
for g = 1 to gmax do                                                    /* Initialize the optimization loop /*
    for i = 1 to n do                                                   /* update the population /*
        r1 = ri  /* r1 is current solution; BV = r1) /*        /* Select three vehicles (BV, FV, and OV). /*
        Select two random vehicles r2 and r3                   /* r1≠r2≠r3; FV = r2; OV = r3 /*
        Dk = rk/n /* k = 1, 2, 3        /* Calculate the distances (D) and velocities (V) of the vehicles. /*
        Vk = Rk*(1-Dk) /* R ∈ [0,1] /*
        y = r3 − r2                                            /* Calculate the distance x, y, x1, and y2. /*
        x = r3 − r1
        x1 = (x*V3)/(V1-V3)
        y1 = (x*V2)/(V1-V3)
        if V1 > V3 then                                        /* Generate new population. /*
            if (y-y1) > x1 then
                Vco = V1/(V1-V3)
                X'r1 = Xr1 + Vco * rand(0,1) * (Xr1 − Xr3) /* rand ∈ [0,1] /*
            else
                X'r1 = Xr1 + rand(0,1) * (Xr1 − Xr2)
            end if
        else
            X'r1 = Xr1 + rand(0,1) * (Xr3 − Xr1)
        end if
        FE = FE + 1                                            /* Count function evaluation /*
        if F(X'r1) < F(Xr1) then                               /* Greedy selection/*
            Xr1 = X'r1
        end if
        if FE ≥ FEmax then                                     /* Termination criterion /*
            break optimization loop
        end if
    end for                                                    /* Population loop ends /*
end for                                                        /* Optimization loop ends /*
STOP
```

3.2.8 The SOS Algorithm

Cheng and Prayogo (2014) proposed the SOS algorithm, which is a powerful and simple metaheuristic algorithm inspired by the cooperative behavior observed in nature among organisms for survival and food. The relationship between two different species that results in mutual benefits is called mutualism, and it is one of the most common symbiotic relations found in the ecosystem. Commensalism is a type of cooperative behavior between two different species that provides benefits to one of them without affecting the other. Parasitism is another type of relationship between two different species that benefits one and harms the other. The interdependency between two species in nature can be used to develop algorithms for optimization.

The SOS algorithm commences with a population, which is referred to as an ecosystem, generated randomly within the search space. The ecosystem comprises "n" number of organisms, equivalent to the population size. In the subsequent phase, every organism achieves an advantage through the symbiotic relationship during the mutualism, commensalism, and parasitism phases, respectively. Additionally, in each phase, the updated population is accepted only if it demonstrates a superior fitness level compared to the pre-interaction fitness. The process is repeated until the termination criteria are met. This optimization algorithm leverages the symbiotic relationship between the current solution vector and either another randomly selected solution or the best alternative from the current population to obtain the optimal solution (Tejani et al., 2016b, 2018b, 2018c; Kumar et al., 2019). A detailed explanation of the SOS algorithm is presented below.

START
Define objective function F(X), population size (n), set number of design variables (m), limits on design variables (LB, UB), and set termination criterion ('FE_{max}', or 'g_{max}'); where, F(X) is the objective function and 'X' is the design vector. Organisms of the ecosystem are considered to be a part of the population (i=1,2,...,n).
/* Initialization /*
Initialize the randomly generated population within its upper and lower bounds and evaluate it.
/* Initialize population /*
FE = 0
for** g = 1 to g_{max} **do ***/* Initialize the optimization loop /****
Arrange the population in ascending order of $F(X_i)$ values Identify the best solution (X_{best}) of the population.
  for** i = 1 to n **do ***/* Update the population /****
    BF_1 = 1 + round [rand(0,1)] ***/* The mutualism phase /****
    BF_2 = 1 + round [rand(0,1)]
    $MV = \frac{X_i+X_k}{2}$ */* 'k' is a randomly selected population of the ecosystem, $k \neq i$ /**
    $X_i' = X_i + rand(0,1) * (X_{best} - MV * BF_1)$ ***/* Generate new population in mutualism phase /****
    $X_k' = X_k + rand(0,1) * (X_{best} - MV * BF_2)$
    FE = FE + 2 ***/* Count Function evaluation /****
    if** $F(X_i') < F(X_i)$ **then ***/* Greedy selection/****
      $X_i = X_i'$
    end if
    if** $F(X_k') < F(X_k)$ **then ***/* Greedy selection/****
      $X_k = X_k'$
    end if
    $X_i' = X_i + rand(-1,1) * (X_{best} - X_k)$ */* 'k' is a randomly selected population of the ecosystem, $k \neq i$ /**
    /* The commensalism phase /*
    FE = FE + 1 ***/* Count Function evaluation /****
    if** $F(X_i') < F(X_i)$ **then ***/* Greedy selection/****
      $X_i = X_i'$
    end if
Select Parasite_Vector / Parasite vector is a fusion of design variables of the organism 'i' and randomly generated design variables within it's bound /** ***/* The parasitism phase /****
/ If parasite vector is fitter than the organism 'k,' the parasite will kill organism 'k' and acquire its position in the ecosystem. /**
    FE = FE + 1 ***/* Count Function evaluation /****
    if** $F(Parasite\_Vector) < F(X_k)$ **then */* 'k' is a randomly selected population of the ecosystem, $k \neq i$ /**
      $X_k = Parasite\_Vector$ ***/* Greedy selection/****
    end if
    if** FE ≥ FE_{max} **then ***/* Termination criterion /****
      break optimization loop
    end if
  end for ***/* Population loop ends /****
end for ***/* Optimization loop ends /****
STOP

3.2.9 The GWO Algorithm

The GWO algorithm, proposed by Mirjalili, Mirjalili, and Lewis in 2014, is a metaheuristic algorithm that mimics the social behavior and hunting activity of grey wolves in a group. Grey wolves live in packs, where they have a well-defined social hierarchy consisting of four levels, namely, alpha (α), beta (β), delta (δ), and omega (ω). The alpha wolf is the leader of the pack, responsible for making critical decisions regarding hunting, sleeping, and other activities. The beta and delta wolves are the second and third levels in the hierarchy, while the remaining wolves are considered omega.

The hunting activity of grey wolves is a critical aspect of their survival in the wild. It involves three primary stages, starting with tracking, chasing, and approaching the prey, followed by pursuing, encircling, and harassing the target until it becomes steady. Finally, the pack attacks the prey in a coordinated manner, leading to a successful hunt. The GWO algorithm mimics this hunting activity by searching for the optimal solution to a given problem.

The GWO algorithm's progression is controlled by various parameters such as the number of search agents (i.e., the number of grey wolves in the pack), the maximum number of iterations, and the search space. In each iteration, the GWO algorithm searches for the optimal solution by updating the position of each search agent (i.e., grey wolf) based on its position, the best position found so far, and the position of the alpha wolf.

The GWO algorithm's update process is similar to how grey wolves behave in a group. The alpha wolf has the highest priority in decision-making and takes the lead in the search process. The beta and delta wolves follow the alpha wolf, while the omega wolves are responsible for exploring new areas of the search space. The GWO algorithm uses the concept of social learning and hunting behavior to update the position of each search agent, leading to the discovery of better solutions.

In summary, the GWO algorithm is an effective metaheuristic algorithm that mimics the social behavior and hunting activity of grey wolves in a group. By using this algorithm, researchers can solve complex optimization problems in a faster and more efficient way. The detailed step-by-step process of the GWO algorithm is outlined below.

```
START
Define objective function F(X), population size (n), set number of design variables (m), limits on design variables (LB, UB),
and set termination criterion ('FEmax', or 'gmax'); where F(X) is the objective function and 'X' is the design vector. Grey wolves
are considered to be population size (i=1,2,...,n). /* Initialization /*
Initialize the randomly generated population within its upper and lower bounds and evaluate it.    /* Initialize population
/*
FE = 0
Identify the best solution (Xα) of the population.                      /* Alpha wolf /*
Identify the second best solution (Xβ) of the population.               /* Beta wolf /*
Identify the third best solution (Xδ) of the population.                /* Delta wolf /*
/* rest of the population is assumed to be Xω /*                         /* Omega wolf /*
for g = 1 to gmax do                                                    /* Initialize the optimization loop /*
    a = 2 − (2/gmax) /* 'a' decreases linearly from 2 to 0 /*
    for i = 1 to n do                                                   /* Update population /*
        A1 = 2 × a × rand(0,1) − a
        C1 = 2 × rand(0,1) /* 'A' and 'C' are coefficient vectors /*
        Dα = C1 × Xα − Xi
        X1 = Xα − A1 × Dα                                               /* Positions of search agent alpha /*
        A2 = 2 × a × rand(0,1) − a
        C2 = 2 × rand(0,1)
        Dβ = C2 × Xβ − Xi
        X2 = Xβ − A2 × Dβ                                               /* Positions of search agent beta /*
        A3 = 2 × a × rand(0,1) − a
        C3 = 2 × rand(0,1)
        Dδ = C3 × Xδ − Xi
        X3 = Xδ − A3 × Dδ                                               /* Positions of search agent delta /*
        X'i = (X1+X2+X3)/3                                              /* Update the positions of search agent /*
        FE = FE + 1                                                     /* Count Function evaluation /*
        if F(X'i) < F(Xα) then                                          /* Update the best solution (Xα) /*
            Xα = X'i                                                    /* Greedy selection/*
        end if
        if F(X'i) ≥ F(Xα) & F(X'i) < F(Xβ) then                         /* Update second best solution (Xβ) /*
            Xβ = X'i                                                    /* Greedy selection/*
        end if
Step XIV: Update the second best solution (Xδ).
        if F(X'i) ≥ F(Xα) & F(X'i) ≥ F(Xβ) & F(X'i) < F(Xδ) then /* Update third best solution (Xδ) /*
            Xδ = X'i                                                    /* Greedy selection /*
        end if
        if FE ≥ FEmax then                                              /* Termination criterion /*
            break optimization loop
        end if
    end for                                                             /* Population loop ends /*
end for                                                                 /* Optimization loop ends /*
STOP
```

3.2.10 The TLBO Algorithm

In the TLBO algorithm (Rao et al., 2011, 2012), the teacher represents the best solution, which guides the learners to achieve better results. The algorithm mimics the process of how a teacher trains the learners in a classroom setting. The learners (population) interact with each other and with the teacher to improve their grades (objective function values).

In the TLBO algorithm, the teacher influences the population through two stages: the teaching phase and the learning phase. In the teaching phase, the teacher provides guidance to the learners based on their current performance, and the learners update their solutions accordingly. In the learning phase, the learners interact with each other to improve their solutions through information sharing and collaboration (Savsani et al., 2016; Tejani et al., 2016a, 2017b).

The TLBO algorithm is a simple yet effective optimization algorithm that has been successfully applied to various optimization problems. Its effectiveness lies in its ability to strike a balance between exploration and exploitation, which allows it to converge quickly to the global optimum. Furthermore, the algorithm is easy to implement, and its population-based approach allows for parallel processing, making it suitable for high-dimensional problems.

Overall, the TLBO algorithm provides a novel approach to solving optimization problems by drawing inspiration from the classroom setting. Its effectiveness and ease of implementation make it a popular choice among researchers and practitioners alike.

The TLBO algorithm commences with a population of "n" number of students (i.e., population size) studying various subjects (i.e., design variables) randomly generated for the classroom. In the subsequent stage, the solution is updated by the teacher and learner phases in each generation. The TLBO algorithm accepts an updated solution only if it has a better function value.

During the teacher phase, the current solution interacts with the teacher ($X_{teacher}$) and a portion of the mean solution (X_{mean}) of the learners. This interaction allows the new solution to search in the vicinity of the teacher. In the learner phase, the current solution interacts with randomly selected solutions (X_k and X_l), moving the new solution closer to other randomly selected solutions. Finally, the worst solution in the population is replaced by the elite solution. These steps are repeated until the termination criterion is satisfied.

In summary, the TLBO algorithm uses the teaching and learning process in a classroom as an analogy for updating the population. The teacher phase and learner phase update the solutions in each generation, followed by the replacement of the worst solution with the elite solution. The stepwise process of the TLBO algorithm is outlined below for a better understanding.

```
START
Define objective function F(X), population size (n), set number of design variables (m), limits on design variables (LB, UB),
and set termination criterion ('FEmax', or 'gmax'); where, F(X) is the objective function and 'X' is the design vector. The class
has 'n' number of learners (i.e. population size, i=1,2,...,n) studying 'm' number of the subjects (i.e. design variables,
j=1,2,...,m).                                          /* Initialization /*
Initialize the randomly generated population within its upper and lower bounds and evaluate it./* Initialize population /*
FE = 0
for g = 1 to gmax do                                   /* Initialize the optimization loop /*
Arrange the population in ascending order of F(Xi) values, Identify the best solution (Xteacher) of the population and
calculate mean of each design variables (Xmean).
    for i = 1 to n do                                  /* The teacher phase /*
    TF = 1 + round [rand(0,1)]        /* TF is teaching factor /*
    X'i = Xi + rand(0,1) * (Xteacher - TF * Xmean)    /* Generate new population in teacher phase /*
    FE = FE + 1                                        /* Count Function evaluation /*
    if F(X'i) < F(Xi) then                             /* Greedy selection/*
        Xi = X'i
    end if
    if F(X'k) < F(Xl) then                             /* The learner phase /*
        X'i = Xi + rand(0,1) * (Xk - Xl) /*'k' and 'l' are randomly selected populations, k ≠ l ≠ i /*
    else                                               /* Generate new population in learner phase /*
        X'i = Xi + rand(0,1) * (Xl - Xk)
    end if
    FE = FE + 1                                        /* Count Function evaluation /*
    if F(X'i) < F(Xi) then                             /* Greedy selection /*
        Xi = X'i
    end if
    if FE ≥ FEmax then                                 /* Termination criterion /*
        break optimization loop
    end if
    end for                                            /* Population loop ends /*
end for                                                /* Optimization loop ends /*
STOP
```

References

Ahrari, A., & Atai, A. A. (2010). Grenade explosion method – A novel tool for optimization of multimodal functions. *Applied Soft Computing, 10*, 1132–1140.

Ahrari, A., Atai, A. A., & Deb, K. (2014). Simultaneous topology, shape and size optimization of truss structures by fully stressed design based on evolution strategy. *Engineering Optimization, 47*(8), 37–41. https://doi.org/10.1080/0305215X.2014.947972

Canfield, R. A., Venkayya, V. B., & Grandhi, R. V. (1989). Structural optimization with stiffness and frequency constraints. *Mechanics of Structures and Machines, 17*(1), 95–110. https://doi.org/10.1080/089054508915631

Cerny, V. (1985). A thermodynamical approach to the travelling salesman problem: An efficient simulation algorithm. *Journal of Optimization Theory and Applications, 45*, 41–51.

Cheng, M. Y., & Prayogo, D. (2014). Symbiotic organisms search: A new metaheuristic optimization algorithm. *Computers and Structures, 139*, 98–112. https://doi.org/10.1016/j.compstruc.2014.03.007

Deb, K., & Gulati, S. (2001). Design of truss-structures for minimum weight using genetic algorithms. *Finite Elements in Analysis and Design, 37*(5), 447–465. https://doi.org/10.1016/S0168-874X(00)00057-3

Dorigo, M. (1992). *Optimization, learning and natural algorithms*. Ph.D. thesis, Politecnico di Milano, Italy.

Farmer, J. D., Packard, N., & Perelson, A. (1986). The immune system, adaptation and machine learning. *Physica D, 22*, 187–204.

Fu, Z., Wang, C., & Zhao, J. (2016). Truss topology optimization under uncertain nodal locations with proportional topology optimization method. *Mechanics Based Design of Structures and Machines, 7734*(April), 15397734.2016.1163640. https://doi.org/10.1080/15397734.2016.1163640

Holland, J. H. (1975). *Adaptation in natural and artificial systems: An introductory analysis with applications to biology, control, and artificial intelligence*. University of Michigan Press.

Karaboga, D., & Basturk, B. (2007). A powerful and efficient algorithm for numerical function optimization: Artificial bee colony (ABC) algorithm. *Journal of Global Optimization, 39*(3), 459–471.

Karaboga, D., & Basturk, B. (2008). On the performance of artificial bee colony (ABC) algorithm. *Applied Soft Computing, 8*(1), 687–697.

Kaveh, A., & Talatahari, S. (2010). A novel heuristic optimization method: Charged system search. *Acta Mechanica, 213*, 267–286.

Kennedy, J., & Eberhart, R. (1995). Particle swarm optimization, in neural networks. In *Proceedings of the IEEE international conference, 1942–1948*.

Kirkpatrick, S., Gelatt, C. D., & Vecchi, M. P. (1983). Optimization by simulated annealing. *Science, 220*(4598), 671–680. https://doi.org/10.1126/science.220.4598.671

Kumar, S., Tejani, G. G., & Mirjalili, S. (2019). Modified symbiotic organisms search for structural optimization. *Engineering Computations, 35*, 1269–1296. https://doi.org/10.1007/s00366-018-0662-y

Li, J. (2015). Truss topology optimization using an improved species-conserving genetic algorithm. *Engineering Optimization, 47*(1), 107–128. https://doi.org/10.1080/0305215X.2013.875165

Li, L., & Liu, F. (2011). Group search optimization for applications in structural design. *Adaptation, Learning, and Optimization (BD), 9*. https://doi.org/10.1007/978-3-642-20536-1

Li, X., Zhang, J., & Yin, M. (2014). Animal migration optimization: An optimization algorithm inspired by animal migration behavior. *Neural Computing and Applications, 24*(7–8), 1867–1877. https://doi.org/10.1007/s00521-013-1433-8

Liu, S., & Mernik, L. (2012). A note on teaching – Learning-based optimization algorithm. *Information Sciences, 212*, 79–93. https://doi.org/10.1016/j.ins.2012.05.009

Luh, G., & Lin, C. (2008). Optimal design of truss structures using ant algorithm. *Structural and Multidisciplinary Optimization, 36*(4), 365–379. https://doi.org/10.1007/s00158-007-0175-6

Mehrabian, A. R., & Lucas, C. (2006). A novel numerical optimization algorithm inspired from weed colonization. *Ecological Informatics, 1*(4), 355–366.

Miguel, L. F. F., Lopez, R. H., & Miguel, L. F. F. F. (2013). Multimodal size, shape, and topology optimisation of truss structures using the firefly algorithm. *Advances in Engineering Software, 56*, 23–37. https://doi.org/10.1016/j.advengsoft.2012.11.006. Elsevier Ltd.

Mirjalili, S. (2015). The ant lion optimizer. *Advances in Engineering Software, 83*, 80–98. https://doi.org/10.1016/j.advengsoft.2015.01.010. Elsevier Ltd.

Mirjalili, S. (2016a). Dragonfly algorithm: A new meta-heuristic optimization technique for solving single-objective, discrete, and multi-objective problems. *Neural Computing and Applications, 27*(4), 1053–1073. https://doi.org/10.1007/s00521-015-1920-1. Springer London.

Mirjalili, S. (2016b). SCA: A sine cosine algorithm for solving optimization problems. *Knowledge-Based Systems*. https://doi.org/10.1016/j.knosys.2015.12.022

Mirjalili, S., & Lewis, A. (2016). The whale optimization algorithm. *Advances in Engineering Software, 95*, 51–67. https://doi.org/10.1016/j.advengsoft.2016.01.008. Elsevier Ltd.

Mirjalili, S., Mirjalili, S. M., & Lewis, A. (2014). Grey wolf optimizer. *Advances in Engineering Software, 69*, 46–61. https://doi.org/10.1016/j.advengsoft.2013.12.007. Elsevier Ltd.

Mirjalili, S., Mirjalili, S. M., & Hatamlou, A. (2016). Multi-verse optimizer: A nature-inspired algorithm for global optimization. *Neural Computing and Applications, 27*(2), 495–513. https://doi.org/10.1007/s00521-015-1870-7

Oftadeh, R., Mahjoob, M. J., & Shariatpanahi, M. (2010). A novel meta-heuristic optimization algorithm inspired by group hunting of animals: Hunting search. *Computers and Mathematics with Applications, 60*(7), 2087–2098. https://doi.org/10.1016/j.camwa.2010.07.049. Elsevier Ltd.

Passino, K. M. (2002). Biomimicry of bacterial foraging for distributed optimization and control. *Control Systems, 22*(3), 52–67. IEEE.

Patel, V. K., & Savsani, V. J. (2015). Heat Transfer Search (HTS): A novel optimization algorithm. *Information Sciences, 324*, 217–246. https://doi.org/10.1016/j.ins.2015.06.044. Elsevier Ltd.

Rahami, H., Kaveh, A., & Gholipour, Y. (2008). Sizing, geometry and topology optimization of trusses via force method and genetic algorithm. *Engineering Structures, 30*, 2360–2369. https://doi.org/10.1016/j.engstruct.2008.01.012

Rajan, S. D. (1995). Sizing, shape, and topology design optimization of trusses using genetic algorithm. *Journal of Structural Engineering, 121*(10), 1480–1487.

Rao, R. V., Savsani, V. J., & Vakharia, D. P. (2011). Teaching–learning-based optimization: A novel method for constrained mechanical design optimization problems. *Computer-Aided Design, 43*(3), 303–315. https://doi.org/10.1016/j.cad.2010.12.015. Elsevier Ltd.

Rao, R. V., Savsani, V. J., & Vakharia, D. P. (2012). Teaching–learning-based optimization: An optimization method for continuous non-linear large scale problems. *Information Sciences, 183*(1), 1–15. https://doi.org/10.1016/j.ins.2011.08.006. Elsevier Inc.

Rashedi, E., Nezamabadi-pour, H., & Saryazdi, S. (2009). GSA: A gravitational search algorithm. *Information Sciences, 179*(13), 2232–2248.

Richardson, J. N., Bouillard, P., Adriaenssens, S., Bouillard, P., & Filomeno, R. (2012). Multiobjective topology optimization of truss structures with kinematic stability repair. *Structural and Multidisciplinary Optimization, 46*, 513–532. https://doi.org/10.1007/s00158-012-0777-5

Savsani, P. V., & Savsani, V. J. (2015). Passing vehicle search (PVS): A novel metaheuristic algorithm. *Applied Mathematical Modelling, 40*(5–6), 3951–3978. https://doi.org/10.1016/j.apm.2015.10.040

Savsani, V. J., Tejani, G. G., & Patel, V. K. (2016). Truss topology optimization with static and dynamic constraints using modified subpopulation teaching–learning-based optimization. *Engineering Optimization, 48*, 1–17. https://doi.org/10.1080/0305215X.2016.1150468

Simon, D. (2008). Biogeography-based optimization. *IEEE Transactions on Evolutionary Computation, 12*(6), 702–713.

Storn, R., & Price, K. (1997). Differential evolution–a simple and efficient heuristic for global optimization over continuous spaces. *Journal of Global Optimization, 11*(4), 341–359.

Tang, W., Tong, L., & Gu, Y. (2005). Improved genetic algorithm for design optimization of truss structures with sizing, shape and topology variables. *International Journal for Numerical Methods in Engineering, 62*, 1737–1762. https://doi.org/10.1002/nme.1244

Tejani, G. G. (2017). *Investigation of advanced metaheuristic techniques for simultaneous size, shape, and topology optimization of truss structures*. Ph.D. thesis, Pandit Deendayal Petroleum University, India.

Tejani, G. G., Savsani, V. J., & Patel, V. K. (2016a). Modified sub-population teaching-learning-based optimization for design of truss structures with natural frequency constraints. *Mechanics Based Design of Structures and Machines, 44*, 495–513. https://doi.org/10.1080/15397734.2015.1124023

Tejani, G. G., Savsani, V. J., & Patel, V. K. (2016b). Adaptive symbiotic organisms search (SOS) algorithm for structural design optimization. *Journal of Computational Design and Engineering, 3*, 226–249. https://doi.org/10.1016/j.jcde.2016.02.003

Tejani, G. G., Savsani, V. J., & Patel, V. K. (2017a). Modified sub-population based heat transfer search algorithm for structural optimization. *International Journal of Applied Metaheuristic Computing, 8*, 1–23. https://doi.org/10.4018/IJAMC.2017070101

Tejani, G. G., Savsani, V. J., Patel, V. K., & Bureerat, S. (2017b). Topology, shape, and size optimization of truss structures using modified teaching-learning based optimization. *Advances in Computational Design, 2*, 313–331. https://doi.org/10.12989/acd.2017.2.4.313

Tejani, G. G., Savsani, V. J., Patel, V. K., & Mirjalili, S. (2018a). An improved heat transfer search algorithm for unconstrained optimization problems. *Journal of Computational Design and Engineering*. https://doi.org/10.1016/j.jcde.2018.04.003

Tejani, G. G., Savsani, V. J., Patel, V. K., & Mirjalili, S. (2018b). Truss optimization with natural frequency bounds using improved symbiotic organisms search. *Knowledge-Based Systems, 143*, 162–178. https://doi.org/10.1016/j.knosys.2017.12.012

Tejani, G. G., Savsani, V. J., Bureerat, S., & Patel, V. K. (2018c). Topology and size optimization of trusses with static and dynamic bounds by modified symbiotic organisms search. *Journal of Computing in Civil Engineering, 32*, 1–11. https://doi.org/10.1061/(ASCE)CP.1943-5487.0000741

Tejani, G. G., Savsani, V. J., Bureerat, S., et al. (2019). Topology optimization of truss subjected to static and dynamic constraints by integrating simulated annealing into passing vehicle search algorithms. *Engineering Computations, 35*, 1–19.

Waghmare, G. (2013). Comments on: A note on teaching–learning-based optimization algorithm. *Information Sciences, 229*, 159–169. https://doi.org/10.1016/j.ins.2012.11.009

Wolpert, D. H., & Macready, W. G. (1997). No free lunch theorems for optimization. *IEEE Transactions on Evolutionary Computation, 1*(1), 67–82.

Wu, C., & Tseng, K. (2010). Truss structure optimization using adaptive multi-population differential evolution. *Structural and Multidisciplinary Optimization, 42*(4), 575–590. https://doi.org/10.1007/s00158-010-0507-9

Yang, X. S. (2009). Firefly algorithms for multimodal optimization. In *Stochastic algorithms: Foundations and applications* (pp. 169–178). Springer.

Yang, X. S. (2010). A new metaheuristic bat-inspired algorithm. In *Nature inspired cooperative strategies for optimization (NICSO 2010)* (pp. 65–74). Springer.

Yang, X. S., & Deb, S. (2010). Engineering optimisation by cuckoo search. *International Journal of Mathematical Modelling and Numerical Optimisation, 1*(4), 330–343.

Zheng, Y. (2015). Water wave optimization: A new nature-inspired metaheuristic. *Computers & Operations Research, 55*, 1–11. Elsevier. https://doi.org/10.1016/j.cor.2014.10.008

Chapter 4
Size Optimization

4.1 Introduction

The goal of size optimization is to determine the optimal cross-sectional areas of the truss elements, as depicted in Fig. 1.1. Rajeev and Krishnamoorthy (1997) optimized both 10-bar and 18-bar trusses for this purpose. Tang et al. (2005) presented a 15-bar truss that included stress constraints. Fenton et al. (2014) introduced a 10-bar truss with multiple loading conditions and a 17-bar truss with a single loading condition. Gonçalves et al. (2015) utilized a 10-bar truss in three different cases. Souza et al. (2016) compared size optimization to both size and topology optimization and TSS optimization in the design of a Conseil International des Grands Réseaux Electriques tower and a 115-kV transmission line tower. Mortazavi and Toğan (2016) designed a 582-bar transmission tower truss with a focus on size variables.

Studies by Ringertz (1986), Deb and Gulati (2001), and Ahrari, Atai, and Deb have placed significant emphasis on design criteria such as structural strength and rigidity in the investigation of structural optimization, particularly in truss sizing. However, limited research has been conducted to incorporate dynamic structural parameters such as natural frequency into the design problem, as demonstrated by studies conducted by Bennage and Dhingra (1995), Xu et al. (2003), Jin and De-yu (2006), Noilublao and Bureerat (2011), Kaveh and Zolghadr (2013), and Savsani et al. (2016). Natural frequency is a crucial design criterion for structures under dynamic loadings because it measures dynamic structural stiffness. Most engineering structures are exposed to various sources of dynamic excitations, which can lead to unwanted vibration and noise (Hajirasouliha et al., 2011; Tejani et al., 2016a, b; 2017a, b). To prevent undesirable vibration phenomena and structural resonance, natural frequency constraints should be included in structural design problems (Nakamura & Ohsaki, 1992; Xu et al., 2003; Kaveh & Zolghadr, 2013; Savsani et al., 2017; Tejani et al., 2018a, b, c, d, e; Kumar et al., 2019). However, incorporating this constraint often results in non-convex feasible regions in structural optimization problems. Furthermore, buckling can complicate truss optimization,

V. Savsani et al., *Truss Optimization*, https://doi.org/10.1007/978-3-031-49295-2_4

creating non-convex feasible regions for the optimization problem (Achtziger, 1999a, b; Mela, 2014; Rozvany, 1996; Smith, 1996; Stolpe & Svanberg, 2003; Zhou, 1996). Simultaneously including natural frequencies and buckling constraints in structural optimization problems can make it difficult to solve the problem and require significant computational effort (Xu et al., 2003; Kumar et al., 2022). However, these constraints cannot be ignored to ensure the usability of structures. To avoid ill-conditioning in structural finite element analysis, kinematic instability and illegitimate trusses should be identified before conducting structural analyses (Tejani et al., 2017a, b).

Metaheuristic methods have become increasingly popular as optimizers for a range of engineering optimization problems. This is due to their simplicity, robustness, and gradient-free nature. In recent decades, thousands of these methods have been introduced. However, despite this, the dragonfly algorithm (DA) (Mirjalili, 2016a), multiverse optimizer (MVO) (Mirjalili et al., 2016), sine cosine algorithm (SCA) (Mirjalili, 2016b), whale optimization algorithm (WOA) (Mirjalili & Lewis, 2016), antlion optimizer (ALO) (Mirjalili, 2015), heat transfer search (HTS) (Patel & Savsani, 2015), passing vehicle search (PVS) (Savsani & Savsani, 2015), symbiotic organisms search (SOS) (Cheng & Prayogo, 2014), grey wolf optimizer (GWO) (Mirjalili et al., 2014), and teaching-learning-based optimization (TLBO) (Rao et al., 2011, 2012) algorithms have not yet been examined for truss optimization problems subject to static and dynamic constraints with multi-load conditions. This lack of examination has motivated us to incorporate these basic algorithms and evaluate their effectiveness in solving size optimization problems. In this book, these algorithms are tested on ten distinct trusses, with the objective function being structural mass and design constraints including stress, displacement, buckling, and frequency as constraints, along with multi-load conditions.

4.2 Problem Formulation

The typical goal of a size optimization problem is to minimize the weight or mass of a truss structure while meeting certain design requirements and constraints. These constraints can include displacement limits, stress limits, buckling instability limits, natural frequency requirements, and kinematic stability criteria.

Additionally, if the truss includes nodal and element masses, they are also considered in the optimization objective function. The objective function represents the mathematical formulation of the optimization problem, and its aim is to minimize the total mass of the truss structure.

In summary, the problem definition for a typical size optimization problem involves finding the optimal design of a truss structure that meets specified constraints while minimizing the overall weight or mass of the structure, including both nodal and element masses, if applicable. The size optimization problem can be expressed as follows:

$$\text{Find} \quad X = \{A_1, A_2, .., A_m\} \tag{4.1}$$

to minimize mass of truss,

$$F(X) = \sum\nolimits_{i=1}^{m} A_i \rho_i L_i + \sum\nolimits_{j=1}^{n} b_j$$

Subject to:
Behavior constraints:

$$g_1(X) : \text{Stress constraints,} \quad |\sigma_i| - \sigma_i^{\max} \leq 0$$

$$g_2(X) : \text{Displacement constraints,} \quad \left|\delta_{x_j/y_j/z_j}\right| - \delta_{x_j/y_j/z_j}^{\max} \leq 0$$

$$g_3(X) : \text{Euler buckling constraints,} \quad \left|\sigma_i^{\text{comp}}\right| - \sigma_i^{cr} \leq 0, \quad \text{where } \sigma_i^{cr} = \frac{k_i A_i E_i}{L_i^2}$$

$$g_4(X) : f_r - f_r^{\min} \geq 0 \quad \text{for some natural frequencies}$$

Side constraints:

$$\text{Cross} - \text{sectional area constraints,} \quad A_i^{\min} \leq A_i \leq A_i^{\max}$$

where

$$i = 1, 2, .., m; j = 1, 2, .., n$$

where X_i, $\rho_i, L_i, E_i, \sigma_i$, and σ_i^{cr} stand for design variable, mass density, element length, Young modules, stress, and critical buckling stress, respectively, on the element "*i*" in that order. δ_j and b_j are values of nodal displacement and mass values of node "*j*," respectively. f_r is structural natural frequent of the *rth* mode. The superscripts "max," "min," and "comp" denote maximum allowable stress, minimum allowable stress, and compressive stress, respectively. k_i is the Euler buckling coefficient calculated from elements' cross sections.

The penalty function technique is employed to manage the constraints. This technique penalizes violated solutions by adding positive penalty terms to their objective function values, with the aim of minimizing them. The computation of the penalty value can be done by following the methodology described by Deb and Gulati (2001) and Kaveh and Zolghadr (2013):

$$\text{Penalized } F(X) = F(X) * F_{\text{penalty}} \tag{4.2}$$

$$\text{where } F_{\text{penalty}} = (1 + \varepsilon_1 * C)^{\varepsilon_2}, \quad C = \sum\nolimits_{i=1}^{q} C_i, \quad C_i = \left|1 - \frac{p_i}{p_i^*}\right| \tag{4.3}$$

where p_i is the level of constraint violation having the bound as p_i^*. The parameter q is a number of active constraints. The variables ε_1 and ε_2 are predetermined by a user. In this study, the values of both ε_1 and $\varepsilon_2$are set as 1.5, which were obtained from experimenting their effect on the balance of the exploitation-exploration balance.

4.3 Results and Discussion on Size Optimization with Continuous Cross Sections

In this section, the effectiveness of algorithms for continuous cross sections is evaluated using ten benchmark problems. These benchmark problems comprise different truss structures, including the 10-bar, 14-bar, 15-bar, 24-bar, 20-bar, 72-bar 3D, 39-bar, 45-bar, 25-bar 3D, and 39-bar 3D trusses. The design-proposed problems consider the overall mass as an objective function and stress, displacement, buckling, and frequency as constraints along with multi-load conditions. Euler buckling coefficient ($k_i, i = 1, 2, \ldots, m$) and mass of the nodes ($b_j, j = 1, 2, .., n$) are assumed to be 4.0 and 5 kg, respectively, for all problems. The suggested algorithms are employed to address every truss problem through 100 independent runs, taking into account the stochastic nature of metaheuristics. The population size remains constant at 50 for all problems, and each run involves a total of 20,000 function evaluations (FEs). The ten truss optimization problems are detailed in the subsequent sections.

The main objective of this evaluation is to assess how well the chosen algorithms can solve the truss size optimization problems that have continuous characteristics. By using these benchmark problems, we can compare the performance of different algorithms and determine which ones are most effective in solving this type of problem.

4.3.1 Size Optimization of the 10-Bar Truss with Continuous Cross Sections

In this particular section of our research, we focused on optimizing a 10-bar truss for size using 10 continuous size variables. The ground structure of the first benchmark truss with load and boundary condition is presented in Fig. 4.1. This problem has been investigated by many researchers (Deb & Gulati, 2001) for static constraints such as elemental stress and nodal displacement. This truss has been also examined for size and shape optimization by several researchers, including Wei et al., Gomes, Wei et al., Kaveh and Zolghadr, Miguel and Miguel, Kaveh and Zolghadr, and Farshchin et al. for dynamic constraints (i.e., natural frequencies).

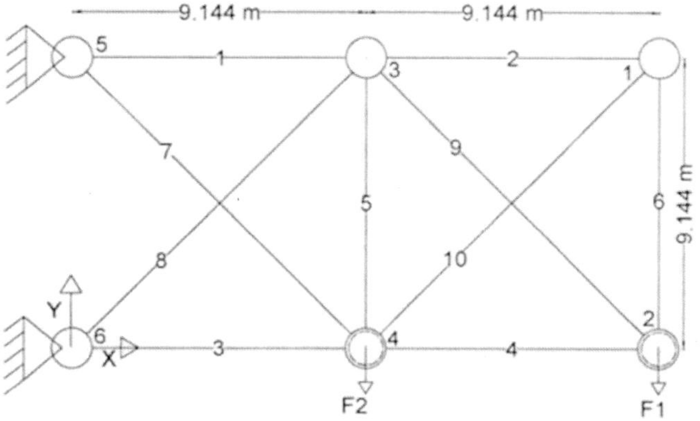

Fig. 4.1 The 10-bar truss

Table 4.1 Design parameters of the 10-bar truss

Size variables: $X_i = A_i$, $i = 1, 2, \ldots, 10$	
Multiple load conditions	
Condition 1 : $F_1 = 44.537\ KN$, $F_2 = 44.537\ KN$	Condition 2 : $F_1 = 44.537\ KN$, $F_2 = 0\ KN$
Lumped masses on node 2 and 4 are 500 kg	
Stress and displacement constraints: $\sigma_i^{max} = 172.43$ MPa and $\delta_{y_5 \& y_6}^{max} = 100$ mm (0.1 m)	
Natural frequency constraint: $f_1 \geq 4$ Hz	
The continuous cross − sectional areas : $[A^{min}, A^{max}] = [1, 100]\ cm^2 = ([1, 100] * 10^{-4}\ m^2)$	
Material properties: $E = 69\ GPa$ and $\rho = 2740 kg/m^3$	

Previously, the researchers had investigated this problem with static constraints (i.e., stress, buckling, and displacement) or dynamic constraints (i.e., natural frequencies) but not examined with simultaneous static and dynamic constraints. In this study, this problem is modified to introduce the effect of multi-load conditions, lumped masses, natural frequency constraints, and continuous cross-sectional area. In this problem, the continuous cross-sectional areas (A_i, where $i = 1, 2, \ldots, 10$) are considered over [1, 100] cm^2. Table 4.1 shows all details of material properties, design variables, loading conditions, lumped mass, and other important data of the test problem.

- The 10-bar truss is considered as a demonstration of the truss optimization procedure. The step-by-step procedure for implementing the size optimization with a sample calculation is outlined below:

Step 1: Begin by defining the ground structure of the truss. This involves generating nodes over a predefined design domain and assigning all possible element connections, as well as material properties, loading, and boundary conditions.

- Figure 4.1 illustrates the ground structure of the 10-bar truss, showcasing the loading and boundary conditions, while Table 4.1 provides the material properties.

Step 2: Define the objective function, population size, design variables, constraints, algorithm controlling parameters, and termination criterion.

- The objective function and constraints of the 10-bar truss are:

$$\text{Find}\quad X = \{A_1, A_2, .., A_{10}\} \tag{4.4}$$

to minimize mass of truss,

$$F(X) = \sum\nolimits_{i=1}^{10} A_i \rho_i L_i + \sum b_{2\&4}$$

Subject to:
Behavior constraints:

$$g_1(X) : \text{Stress constraints,}\ \ |\sigma_i| - 172.43 \le 0, \text{in}\ MPa$$

$$g_2(X) : \text{Displacement constraints,}\ \ |\delta_j| - 100 \le 0, \text{in}\ mm$$

$$g_3(X) : \text{Euler buckling constraints,}\ \ |\sigma_i^{comp}| - \sigma_i^{cr} \le 0,\ \ \text{where}\ \sigma_i^{cr} = \frac{k_i A_i E_i}{L_i^2}$$

$$g_4(X) : f_1 \ge 4, \text{in}\ Hz$$

Side constraints:

$$\text{Cross} - \text{sectional area constraints,}\ \ 1 \le A_i \le 100, \text{cm}^2$$

$$\text{where}\quad i = 1, 2, .., 10; j = 1, 2, .., 6$$

- Assume population size = 05
- Design variables: the continuous cross − sectional areas : $[A^{min}, A^{max}] = [1, 100]\ \text{cm}^2 = ([1, 100] * 10^{-4}\ \text{m}^2)$
- Termination criterion is presented in Eqs. 4.2 and 4.3. Assume function evaluations (FE) ≤ 10.

Step 3: Initiate a randomly generated set of size variables (i.e., population) within its upper and lower bounds.
This population is expressed as follows:

- $X(k) = \{A_1, A_2, .., A_{10}\}$ % A: area of cross section in m^2
- X(1) = [21.30, 69.21, 77.05, 89.43, 19.01, 47.52, 24.60, 62.38, 30.78, 64.85]/10000
- X(2) = [88.54, 13.34, 89.35, 22.09, 3.84, 73.90, 62.37, 10.31, 68.38, 17.27]/10000

- $X(3) = [53.65, 58.95, 39.92, 86.70, 55.27, 53.78, 56.50, 1.00, 17.20, 90.74]/10000$
- $X(4) = [19.23, 3.36, 99.06, 44.39, 1.00, 100.00, 100.00, 83.52, 32.36, 66.10]/10000$
- $X(5) = [43.34, 100.00, 94.45, 86.71, 26.89, 77.40, 25.71, 70.08, 29.13, 72.15]/10000$

Step 4: Go to truss configurations: Generate trusses according to size variables. Percent generation of coordinates and connectivities:

Node coordinates =	18.288	9.144
	18.288	0
	9.144	9.144
	9.144	0
	0	9.144
	0	0
Node connectivities =	3	5
	1	3
	4	6
	2	4
	3	4
	1	2
	4	5
	3	6
	2	3
	1	4

Number elements = 10
Number nodes = 6
$E = 69e9$; % E; modulus of elasticity in Pa
$ro = 2740$; % E; material density in kg/m^3

Step 5: Proceed to finite element analysis.

In the context of a 2D truss, the truss bar is considered a line element with simply supported ends. It is assumed to have a length of L, cross-sectional area of A, modulus of elasticity of E, and mass density of ρ. The bar element experiences forces and displacements.

Compute the stiffness matrix using Eq. 2.2 for each element sequentially and subsequently integrating it into the global stiffness matrix:

$$\text{GDof} = 2 \times \text{Number Nodes} = 12; \%\text{GDof}: \text{total number of degrees of freedom}$$

Step 6:

Compute the mass matrix, force vector, displacement vector, stresses vector, and natural frequency vector using finite element analysis.

Compute the mass matrix using Eq. 2.6 for each element sequentially and subsequently integrating it into the global mass matrix.

Compute the displacement vectors and stress vectors using Eq. 2.5.
Compute the buckling vectors and stress vectors using Eq. 4.4.
Compute the natural frequency vector using Eq. 2.7.

Step 7: Compute natural frequencies, element stresses, nodal displacements, and Euler buckling of the truss using finite element analysis.

```
% Result of population 1
% element stress:
sigma1(1) = [40103545, 2943458, 12034490, 2702190, 8794857, 4286447,
27699530, 9267574, 11103492, 4442206]
sigma2(1) = [29856117, 3650218, 9087108, 2155280, 115878, 5315676,
14646225, 4319659, 8856202, 5508834]
% Euler buckling stress:
buckling_stress (1) = [7030833, 22844361, 25434662, 29521325, 6273609,
15686981, 4060944, 10296378, 5080156, 10703434]
% Natural frequency:
frequency (1) = [8.6117, 29.0267, 30.2932, 61.4739, 72.9411, 82.0559,
100.3205, 148.9555]
% nodal displacements:
displacements1 (1) = [5.7047, 17.4133, 1.9529, 17.9813, 5.3146,
7.7709, 1.5948, 8.9364]
displacements2 (1) = [4.4403, 12.1908, 1.4899, 12.8952, 3.9566,
5.1015, 1.2042, 5.0861]


% Result of population 2
% element stress:
sigma1(2) = [7317961, 12735025, 12685795, 12471857, 18995980, 2299398,
15602715, 27784622, 5696571, 13913278]
sigma2(2) = [6736966, 14411317, 8277198, 11459170, 26575487, 2602064,
6671175, 20729035, 5234023, 15744661]
% Euler buckling stress:
buckling_stress (2) = [29227217, 4404300, 29495264, 7290395, 1266308,
24392852, 10294646, 1701813, 11286353, 2850573]
% Natural frequency:
frequency (2) = [9.8302, 23.4169, 27.4841, 53.5931, 64.2206, 69.9561,
123.8571, 143.6919]
% nodal displacements:
displacements1 (2) = [2.6575, 13.8428, 3.3339, 14.1475, 0.9698,
8.3339, 1.6811, 5.8165]
displacements2 (2) = [2.8026, 10.9376, 2.6155, 11.2824, 0.8928,
6.3869, 1.0969, 2.8651]
```

```
% Result of population 3
% element stress:
sigma1(3) = [8756541, 5592870, 32853474, 1333931, 1650318, 6130810,
21685235, 34558861, 9508719, 5139004]
sigma2(3) = [8756541, 5592870, 32853474, 1333931, 1650318, 6130810,
21685235, 34558861, 9508719, 5139004]
% Euler buckling stress:
buckling_stress (3) = [17710195, 19460054, 13178956, 28619241,
18244769, 17752556, 9324540, 165047, 2838928, 14975618]
% Natural frequency:
frequency (3) = [8.1699, 27.8106, 28.6449, 63.3397, 76.9945, 95.1115,
107.5209, 147.0423]
% nodal displacements:
displacements1 (3) = [1.9016, 17.7188, 4.5306, 18.5313, 1.1604,
10.3200, 4.3538, 10.1013]
displacements2 (3) = [1.8753, 11.9354, 3.0888, 12.7498, 1.1323,
6.0256, 2.9133, 5.7815]


% Result of population 4
% element stress:
sigma1(4) = [37638974, 43989284, 10676644, 6704031, 19077426, 1477851,
8658275, 4715823, 13004617, 3161778]
sigma2(4) = [37638974, 43989284, 10676644, 6704031, 19077426, 1477851,
8658275, 4715823, 13004617, 3161778]
% Euler buckling stress:
buckling_stress (4) = [6348395, 1108971, 32698238, 14652495, 330093,
33009325, 16504663, 13784821, 5341153, 10909900]
% Natural frequency:
frequency (4) = [8.2698, 20.7104, 33.0822, 56.3713, 68.7002, 76.1015,
95.4451, 140.7601]
% nodal displacements:
displacements1 (4) = [10.8175, 16.7801, 2.3033, 16.9760, 4.9880,
6.2379, 1.4149, 3.7097]
displacements2 (4) = [11.6001, 15.1025, 1.6800, 15.3336, 4.7200,
5.7953, 0.8711, 1.6423]


% Result of population 5
% element stress:
sigma1(5) = [21760016, 2232687, 8875694, 2561513, 10254171, 2884695,
21614511, 10045264, 10781045, 4376025]
sigma2(5) = [21760016, 2232687, 8875694, 2561513, 10254171, 2884695,
```

```
21614511, 10045264, 10781045, 4376025]
% Euler buckling stress:
buckling_stress (5) = [14307842, 33009325, 31176789, 28621420,
8875520, 25548451, 4243130, 11567192, 4808526, 11908846]
% Natural frequency:
frequency (5) = [9.7025, 30.3536, 33.4990, 61.9019, 71.9300, 82.7932,
107.8180, 155.5453]
% nodal displacements:
displacements1 (5) = [3.1796, 12.4206, 1.5157, 12.8029, 2.8837,
5.5461, 1.1762, 6.9050]
displacements2 (5) = [2.4735, 8.5562, 1.1736, 9.0152, 2.1183, 3.4419,
0.9026, 3.7877]
```

Step 8: Check for constraint violations in the penalty function. If there is a violation, assign a penalty as per Eqs. 4.2 and 4.3; otherwise, compute the total mass of the truss.

```
% Result of population 1
fit (1)=1487.6355 Kg
fpenalty (1)= 1
fit_fpenalty (1)= 1487.6355 Kg
% Result of population 2
fit (2)= 1320.2695 Kg
fpenalty (2)= 1286.2458
fit_fpenalty (2)= 1698191.0769 Kg


% Result of population 3
fit (3)= 1488.7769 Kg
fpenalty (3)= 10817.2232
fit_fpenalty (3)= 16104432.0276 Kg


% Result of population 4
fit (4)= 1698.1929 Kg
fpenalty (4)= 3490.3476
fit_fpenalty (4)= 5927283.5760 Kg


% Result of population5
fit (5)= 1802.6145 Kg
fpenalty (5)= 1
fit_fpenalty (5)= 1802.6145 Kg
```

Step 9: Assign functional values and proceed to the optimization algorithm.

```
% Obtain functional values
F(X(1)) = fit_fpenalty (1)= 1487.6355 Kg
F(X(2)) =fit_fpenalty (2)= 1698191.0769 Kg
F(X(3)) =fit_fpenalty (3)= 16104432.0276 Kg
F(X(4)) =fit_fpenalty (4)= 5927283.5760 Kg
F(X(5)) =fit_fpenalty (5)= 1802.6145 Kg
% Generate new set of size variables (i.e., population) within its upper
and lower bounds as per the optimization algorithm rules
% iteration 1
X(k) = {A1, A2, .., A10} % A: area of cross section in m2
X(1)= [93.69, 7.79, 12.22, 87.55, 30.89, 1.00, 63.02, 100.00, 1.00,
26.02]/10000
X(2)=[ 42.57, 31.74, 54.47, 13.71, 49.18, 84.92, 64.50, 16.77, 39.79,
63.58] /10000
X(3)= [4.50, 50.21, 59.64, 89.89, 35.10, 77.94, 10.75, 62.78, 11.18,
77.91] /10000
X(4)= [5.26, 100.00, 100.00, 47.51, 92.81, 100.00, 63.90, 100.00,
25.66, 95.02] /10000
X(5)= [20.84, 99.62, 23.28, 75.59, 3.96, 89.62, 40.55, 100.00, 56.25,
48.47] /10000
% functional values after iteration 1
F(X(1)) = fit_fpenalty (1)= 482828.0328 Kg
F(X(2)) =fit_fpenalty (2)= 1377.1992 Kg
F(X(3)) =fit_fpenalty (3)= 1401.1651 Kg
F(X(4)) =fit_fpenalty (4)= 2154.6971 Kg
F(X(5)) =fit_fpenalty (5)= 1683.0150 Kg
% End of iteration 1
% Select best solotions for the next iteration as per the algorithm.
F(X(1)) = 1377.1992 Kg
F(X(2)) = 1401.1651 Kg
F(X(3)) = 1683.0150 Kg
F(X(4)) = 1487.6355 Kg
F(X(5)) = 1802.6145 Kg
```

Step 10: Check the termination criteria. If not satisfied, generate new trusses (i.e., solutions) according to the algorithm and proceed to Step 4.

```
FE = 10 % function evaluations
```

Step 11: Output: the best solution.

X = [42.57, 31.74, 54.47, 13.71, 49.18, 84.92, 64.50, 16.77, 39.79, 63.58] /10000
$F(X)$ = 1377.1992 Kg

The step-by-step procedure for implementing the 10-bar size optimization in MATLAB is presented below:

Step 1:

```
>>   n1=[2 6 10]; % elements connected with node 1
>>  n2=[4 6 9]; % elements connected with node 2
>>  n3=[1 2 5 8 9]; % elements connected with node 3
>>  n4=[3 4 5 7 10]; % elements connected with node 4
>>  n5=[1 7]; % elements connected with node 5
>>  n6=[3 8]; % elements connected with node 6
```

Step 2:

```
>>  fit= sum(((A).*L))* ro + nodal mass; % Objective function
>>  fpenalty=(1+e1*sum(Constraint violation))^e2; % Penalty function
>>  fit_fpenalty =fit*fpenalty; % Penalized objective function
>>  OPTIONS.popsize = 5; % Population size
>>  OPTIONS.numVar = 10; % Number of design variables
>>  PTIONS.maxFE = 10; % max FE; Termination criterion
>>  ll=1*ones(1,10); % Lower bound
>>  ul=100*ones(1,10); % Upper bound
```

Step 3:

```
>>  for popindex = 1 : OPTIONS.popsize
>>  for k = 1 : OPTIONS.numVar
>>  chrom(k) = (ll(k)+ (ul(k) - ll(k)) * rand);
>>  end
>>  Population(popindex).chrom = chrom;
>>  end
```

Step 4:

```
% generation of coordinates and connectivities
>>  nodeCoordinates=[9.144*2 9.144;9.144*2 0;9.144 9.144;9.144 0 ;0
9.144;0 0];% nodal Coordinates in m
>>  elementNodes=[3 5;1 3;4 6;2 4;3 4;1 2;4 5;3 6;2 3;1 4]; % node
connections of elements
>>  numberElements=size(elementNodes,1); % number of Elements
>>  numberNodes=size(nodeCoordinates,1); % number of nodes
>>  xx=nodeCoordinates(:,1);
>>  yy=nodeCoordinates(:,2);
>>  E = 69e9; % E; modulus of elasticity in Pa
>>  ro = 2740; % E; material density in kg/m³
```

Step 5:

```
% computation of the system stiffness matrix
>>   GDof=2*numberNodes; % GDof: total number of degrees of freedom
% calculation of the system stiffness matrix
>>   stiffness=zeros(GDof);
>>   for e=1:numberElements
% elementDof: element degrees of freedom (Dof)
>>        indice=elementNodes(e,:) ;
>>        elementDof=[indice(1)*2-1 indice(1)*2 indice(2)*2-1 indice
(2)*2] ;
>>        xa=xx(indice(2))-xx(indice(1));
>>        ya=yy(indice(2))-yy(indice(1));
>>        L(e)=sqrt(xa*xa+ya*ya); % L: length of bar
>>        C=xa/L(e);
>>        S=ya/L(e);
>>        k1=E*(A(e))/L(e)*[C*C C*S -C*C -C*S; C*S S*S -C*S -S*S;...
   -C*C -C*S C*C C*S;-C*S -S*S C*S S*S];
>>       stiffness(elementDof,elementDof)= ... stiffness(elementDof,
elementDof)+k1;
>>   end
```

Step 6:

```
% mass : mass matrix
>>   mass=zeros(GDof);
>>   for e=1:numberElements
% elementDof: element degrees of freedom (Dof)
>>        indice=elementNodes(e,:) ;
>>        elementDof=[indice(1)*2-1 indice(1)*2 indice(2)*2-1 indice
(2)*2] ;
>>        xa=xx(indice(2))-xx(indice(1));
>>        ya=yy(indice(2))-yy(indice(1));
>>        L(e)=sqrt(xa*xa+ya*ya); % L: length of bar
>>        m1=(ro*(A(e))*L(e)/6)*[2 0 1 0;0 2 0 1;1 0 2 0;0 1 0 2];
>>      mass(elementDof,elementDof)= ... mass(elementDof,elementDof)
+m1; % global mass matrix
>>   end
% force : force vector
>>   force=zeros(GDof,1); % force vector
% applied load Condition 1
>>   force 1 (4)=-44.537e3; % in N; Condition 1
>>   force 1 (8)=-44.537e3; % in N; Condition 1
% applied load Condition 2
```

```
>>   force 2 (4)=-44.537e3; % in N; Condition 2
% displacements: displacement vector
>>   prescribedDof=(9:12); % boundary conditions/ Restricted DOF
>>   activeDof=setdiff([1:GDof]',[prescribedDof]); % active DOF
>>   U1=stiffness(activeDof,activeDof)\force1(activeDof);
>>   displacements1=zeros(GDof,1);
>>   displacements1(activeDof)=U1; % global displacements vector1
>>   U2=stiffness(activeDof,activeDof)\force2(activeDof);
>>   displacements2=zeros(GDof,1);
>>   displacements2(activeDof)=U2; % global displacements vector2
% stress: stress vector
>>   for e=1:numberElements
>>        indice=elementNodes(e,:);
>>        elementDof=[indice(1)*2-1 indice(1)*2 indice(2)*2-1 indice
(2)*2] ;
>>        xa=xx(indice(2))-xx(indice(1));
>>        ya=yy(indice(2))-yy(indice(1));
>>        L=sqrt(xa*xa+ya*ya); % L: length of bar
>>        C=xa/L;
>>        S=ya/L;
>>        sigma1(e)=E/L*[-C -S C S]*displacements1(elementDof);
>>        sigma2(e)=E/L*[-C -S C S]*displacements2(elementDof);
>>   end
% Euler buckling stress: buckling stress vector
>>   buckling_stress=(4*E).*(A./(L.^2));


% Natural frequency: Natural frequency vector
>>   aa=zeros(12,1);aa(3:4)=500;aa(7:8)=500; % lumped masses
>>   mass1=diag(aa); % lumped mass matrix
>>   mass=mass+mass1; % mass matrix
>>   f=eig(stiffness(activeDof,activeDof),mass(activeDof,
activeDof));
>>   f=realsqrt(f);
>>   frequency =f/(2*pi); % Natural frequency vector
```

Step 7:

```
>>   disp('sigma1')
>>   disp('sigma2')
>>   disp('buckling_stress')
>>   disp(' frequency')
>>   disp('displacements1')
>>   disp('displacements2')
```

Step 8:

```
>>   fit= sum(((A).*L))* ro + nodal mass; % total mass
>>   fpenalty= (1+e1*sum(Constraint violation))^e2; % penalty function
>>   fit_fpenalty =fit*fpenalty;
```

Step 9:

```
% iteration 1
% Generate new set of size variables (i.e., population) within its upper
and lower bounds as per the optimization algorithm rules
End of iteration 1
% Select best solotions for the next iteration as per the algorithm.
```

Step 10:

```
>>   while FE ≤ maxFE
>>   ...Algorithms code
>>   end
```

Step 11:

```
>>   disp('X')
>>   disp('fit_fpenalty')
```

We performed 100 independent runs to ensure that the optimization process was robust and reliable. Table 4.2 presents various parameters such as size variables, best mass, mean mass, SD of mass, stress, displacement, and frequency response for each algorithm.

Among the different evolutionary algorithms considered in this study, we found that the ALO, DA, GWO, MVO, SCA, WOA, HTS, TLBO, PVS, and SOS algorithms achieved the best results of 562.6522, 661.2114, 556.0166, 562.4796, 666.2459, 678.4778, 555.9299, 554.3641, 554.3446, and 556.2881 kg, respectively. It is evident from the results that the PVS, TLBO, and HTS algorithms performed better than the other algorithms in obtaining the minimum mean mass. PVS was the most effective in achieving the minimum mean mass, followed by TLBO in second place, and HTS in third place.

To further evaluate the performance of the different evolutionary methods, we used the mean and standard deviation (SD) values of the structural mass. The ALO, DA, GWO, MVO, SCA, WOA, HTS, TLBO, PVS, and SOS algorithms resulted in mean weights of 692.5350, 821.1315, 624.9402, 635.3154, 771.5319, 1011.2094, 625.0795, 617.3169, 615.8329, and 619.0436 kg, respectively. Meanwhile, the SD of weight for ALO, DA, GWO, MVO, SCA, WOA, HTS, TLBO, PVS, and SOS algorithms are 56.4250, 113.7960, 41.5766, 41.0732, 42.2198, 173.3757, 37.8395, 46.6647, 40.3093, and 44.8839, respectively.

Table 4.2 Size optimization of the 10-bar truss (continuous section)

Element no.	ALO	DA	GWO	MVO	SCA	WOA	HTS	TLBO	PVS	SOS
1	2.5793	7.5210	2.5882	2.5889	2.8420	2.8509	2.5951	2.5840	2.5837	2.5968
2	1.0000	2.2747	1.0213	1.0000	7.2446	4.6334	1.0029	1.0000	1.0003	1.0000
3	63.9642	52.5551	63.6662	63.9386	67.3803	63.8570	63.6712	63.6196	63.6187	63.6585
4	33.2824	28.3003	32.5719	33.5160	31.2453	32.1435	32.8393	32.8518	32.5688	33.4815
5	36.8979	1.0000	33.6645	35.3989	34.8365	29.4101	33.8994	33.8652	33.5925	34.2392
6	1.0000	1.7835	1.0170	1.0058	3.6795	6.9381	1.0070	1.0000	1.0001	1.0000
7	18.0313	4.7401	17.9066	18.4766	22.5716	26.9774	17.9470	17.8395	17.8246	18.0804
8	1.0000	60.3031	1.0259	1.0000	1.1106	1.0000	1.0135	1.0000	1.0003	1.0000
9	7.0251	2.1782	5.5226	6.4457	13.6309	11.9918	5.9100	6.0662	5.5408	6.9074
10	26.1798	44.8551	28.8746	27.1671	38.1466	44.1717	28.0910	27.6806	28.6086	26.3953
Mass (kg)	562.6522	661.2114	556.0166	562.4796	666.2459	678.4778	555.9299	554.3641	554.3446	556.2881
σ^{max} *(MPa)*	172.4292	171.6175	172.0743	171.5084	154.7574	152.7403	171.6401	172.4016	172.4198	171.4496
δ^{max} *(mm)*	69.1981	82.3117	73.4562	70.0865	50.8693	48.7573	72.1845	71.9660	73.6340	69.3780
f_1 *(Hz)*	5.5762	5.1113	5.4184	5.5394	6.2718	6.3069	5.4650	5.4745	5.4135	5.5710
Mean	692.5350	821.1315	624.9402	635.3154	771.5319	1011.2094	625.0795	617.3169	615.8329	619.0436
SD	56.4250	113.7960	41.5766	41.0732	42.2198	173.3757	37.8395	46.6647	40.3093	44.8839

Based on the statistical analysis of the results, we can conclude that the GWO algorithm was the best in obtaining a lower mean weight, while HTS performed better in obtaining a lower SD of the weight of the truss. These findings provide valuable insights into the effectiveness of the different evolutionary methods in optimizing the size of the 10-bar truss.

4.3.2 *Size Optimization of the 14-Bar Truss with Continuous Cross Sections*

In this section, the primary focus is on optimizing the 14-bar truss for size optimization, which involves utilizing 14 continuous size variables. A ground structure of second benchmark truss, the 14-bar truss, with load and boundary conditions is shown in Fig. 4.2. This problem has been optimized in (Tang et al., 2005) for TSS optimization with 14 continuous cross-sectional areas and 8 continuous shape variables subjected to static constraints. However, this problem has never been solved with consideration of static and dynamic constraints.

In this problem, the continuous cross-sectional areas (A_i, where $i = 1, 2, \ldots, 14$) are considered over [1, 100] cm^2. All details of design variables as well as load conditions, design constraints, and other relevant data for the test problem are given in Table 4.3.

The outcomes are based on 100 independent runs, with the best results displayed in Table 4.4. Various algorithms such as ALO, DA, GWO, MVO, SCA, WOA, HTS, TLBO, PVS, and SOS were employed, and their respective best performances are shown in kg. ALO, DA, GWO, MVO, SCA, WOA, HTS, TLBO, PVS, and SOS algorithms achieved 807.7191, 1014.0330, 744.8846, 762.6514, 873.3889, 1055.7500, 745.7420, 738.9470, 739.6401, and 748.9644 kg, respectively. The

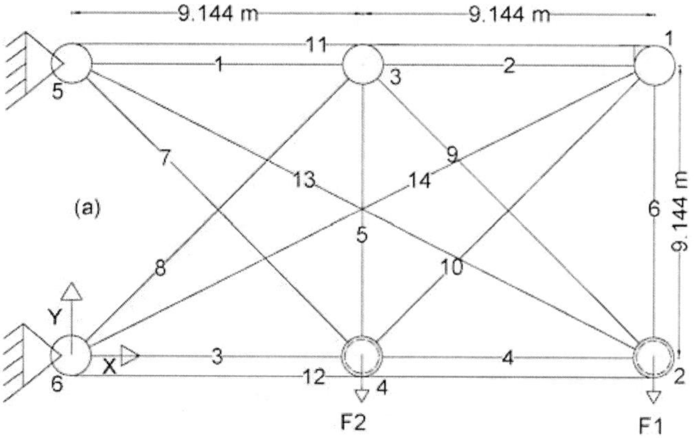

Fig. 4.2 The 14-bar truss

Table 4.3 Design parameters of the 14-bar truss

Size variables: $X_i = A_i,\ i = 1, 2, \ldots, 14$	
Multiple load conditions	
Condition 1 : $F_1 = 44.537$ KN, $F_2 = 44.537$ KN	Condition 2 : $F_1 = 44.537$ KN, $F_2 = 0$ KN
Lumped mass on node 2 and 4 are 500 kg	
Stress and displacement constraints: $\sigma_i^{max} = 172.43$ MPa and $\delta_{y_5 \& y_6}^{max} = 100$ mm (0.1 m)	
Natural frequency constraint: $f_1 \geq 4$ Hz	
The continuous cross − sectional areas : $[A^{min}, A^{max}] = [1, 100]\ cm^2\ ([1, 100] * 10^{-4}\ m^2)$	
Material properties: $E = 69$ GPa and $\rho = 2740 kg/m^3$	

results indicate that TLBO and PVS algorithms are the most effective in producing a lightweight truss. TLBO was found to be the most effective in achieving the minimum mean mass, followed by PVS in second place, and GWO in third place. These outcomes have significant implications in the field of engineering, where lightweight structures are essential for the optimization of various systems.

Furthermore, the structural mean mass and SD values were employed for the statistical evaluation of the performance of various evolutionary methods. The mean weights of the truss achieved by the ALO, DA, GWO, MVO, SCA, WOA, HTS, TLBO, PVS, and SOS algorithms are 1207.6045, 1610.8547, 831.3401, 955.0830, 1071.0554, 2140.0382, 939.9534, 947.5392, 869.1719, and 926.2849 kg, respectively. Similarly, the SD of the weights achieved by these algorithms are 215.4468, 300.6802, 119.1292, 171.0657, 177.3218, 443.0294, 175.0427, 181.5019, 138.1304, and 169.5377 kg, respectively. Based on the results, it is evident that the GWO algorithm outperforms the other algorithms in terms of obtaining a better mean and SD of the weight of the truss.

Overall, the findings presented in this section are crucial in advancing the field of engineering and optimizing various systems by providing insights into the most effective algorithms for achieving lightweight structures.

4.3.3 *Size Optimization of the 15-Bar Truss with Continuous Cross Sections*

The focus of this section is on optimizing the 15-bar truss for size optimization, which involves the use of 15 continuous size variables. The ground structure of this truss is presented in Fig. 4.3. This problem is adapted from Rahami et al. (2008) and Ahrari et al. (2014). Previously, the researchers had investigated this problem with static constraints (i.e., stress, buckling, and displacement) but not examined with simultaneous static and dynamic (i.e., natural frequency) constraints. This problem is modified to introduce the effect of multi-load conditions, lumped masses, natural frequency constraints, and continuous cross-sectional area.

Table 4.4 Size optimization of the 14-bar truss (continuous section)

Element no.	ALO	DA	GWO	MVO	SCA	WOA	HTS	TLBO	PVS	SOS
1	1.1307	1.0091	1.1322	1.0225	1.3525	6.5742	1.1696	1.0009	1.0109	1.7169
2	1.0591	1.4057	4.6166	3.1345	1.1950	1.0000	4.0724	3.5694	3.1644	5.0151
3	76.3215	54.7664	60.4616	61.1082	59.2706	73.1945	60.9710	60.2546	60.1064	60.1509
4	1.1418	1.0282	1.0904	1.0000	2.4675	1.0000	1.0246	1.0078	1.0029	1.0000
5	18.3747	32.1872	13.9814	13.8318	38.9164	2.2088	14.7731	14.0343	14.5852	15.2672
6	2.1247	1.3471	8.4419	9.8208	4.0634	9.6759	6.0104	6.3749	5.3317	4.7853
7	19.9242	19.8970	16.3438	22.5716	17.3598	24.1956	16.2310	16.1180	16.1043	17.5731
8	1.0000	2.1642	1.0415	1.0807	2.0088	38.5930	1.0000	1.0009	1.0006	1.0062
9	1.0000	1.9837	1.0127	1.0000	2.5802	1.0000	1.0122	1.0006	1.0027	1.0000
10	52.9599	59.3766	57.3431	57.9547	51.0855	59.3604	57.1051	56.8322	56.4539	56.2213
11	1.8786	1.0337	2.0415	2.8895	1.5012	4.8893	1.9504	2.2376	2.1535	1.0000
12	48.0632	86.1422	39.9556	37.7046	58.9226	41.2188	41.3185	40.8868	41.9058	42.6768
13	1.0000	2.8311	1.0008	1.0000	1.1093	1.0000	1.0030	1.0000	1.0000	1.0000
14	1.0000	1.0529	1.0111	1.0091	1.2260	21.0825	1.0345	1.0004	1.0000	1.0000
Mass (kg)	807.7191	1014.0330	744.8846	762.6514	873.3889	1055.7500	745.7420	738.9470	739.6401	748.9644
σ^{max}	150.7458	172.4300	171.9938	143.2971	162.5225	89.1940	170.9013	171.7212	172.2498	168.8961
δ^{max}	1878.5583	67.9851	60.3073	50.9782	64.2327	37.6601	62.7998	62.4801	64.3495	64.2670
f_1	5.1815	5.0748	5.5938	6.0644	5.3628	6.9350	5.5196	5.5348	5.4698	5.4662
Mean	1207.6045	1610.8547	831.3401	955.0830	1071.0554	2140.0382	939.9534	947.5392	869.1719	926.2849
SD	215.4468	300.6802	119.1292	171.0657	177.3218	443.0294	175.0427	181.5019	138.1304	169.5377

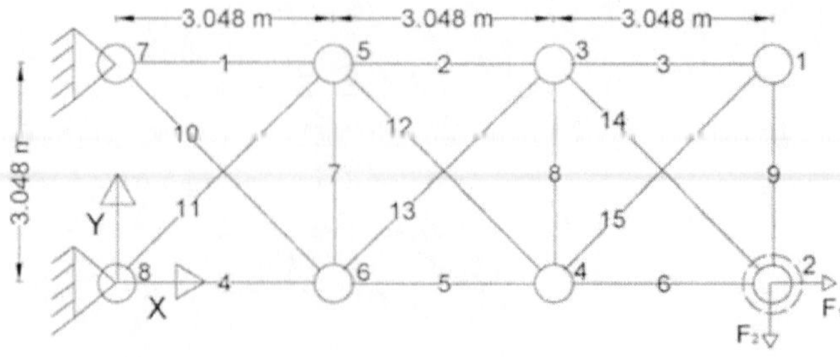

Fig. 4.3 The 15-bar truss

Table 4.5 Design parameters of the 15-bar truss

Size variables: $X_i = A_i$, $i = 1, 2, \ldots, 15$	
Multiple load conditions	
Condition 1 : $F_1 = -44.537$ KN, $F_2 = 0$ KN	Condition 2 : $F_1 = 0$ KN, $F_2 = 4.4537$ KN
Lumped mass on node 2 is 500 kg	
Stress and displacement constraints: $\sigma_i^{max} = 172.43$ MPa and $\delta_{x_2 \& y_2}^{max} = 100$ mm (0.1 m)	
Natural frequency constraint: $f_1 \geq 4$ Hz	
The continuous cross − sectional areas : $[A^{min}, A^{max}] = [1, 50]$ cm^2 ($[1, 50] * 10^{-4}$ m^2)	
Material properties: $E = 69$ GPa and $\rho = 2740$kg/m^3	

In this problem, the continuous cross-sectional areas (A_i, where $i = 1,2,\ldots, 15$) are considered over [1, 50] cm^2. Design considerations and material properties are tabulated in Table 4.5.

Table 4.6 presents the most effective solutions obtained in this study, which demonstrate that the ALO, DA, GWO, MVO, SCA, WOA, HTS, TLBO, PVS, and SOS algorithms produced optimized trusses with a minimum weight of 128.0254, 146.0234, 113.0469, 114.9422, 152.6426, 155.9305, 113.8195, 112.8336, 112.6401, and 115.1230 kg, respectively, while adhering to all constraints. It is worth noting that the PVS algorithm resulted in the lightest structure, followed by TLBO in second place, and GWO in third place.

These results have significant implications for the field of engineering, where lightweight structures are essential for optimizing various systems. The findings suggest that the PVS algorithm may be the most effective choice for achieving a lightweight truss, while TLBO and GWO algorithms can also produce high-performing structures. By optimizing the weight of the truss, engineers can design more efficient systems with improved performance and reduced material costs.

In the field of evolutionary methods, it is crucial to evaluate their performance in order to determine the effectiveness of different algorithms. To accomplish this, researchers often use statistical analysis to compare the results produced by various algorithms. One common way to do this is by calculating the mean and standard deviation of a given metric. In this case, the metric being evaluated was the structural mass.

Table 4.6 Size optimization of the 15-bar truss (continuous section)

Element no.	ALO	DA	GWO	MVO	SCA	WOA	HTS	TLBO	PVS	SOS
1	6.1260	11.1945	6.1241	6.2307	15.1729	6.0204	6.0872	6.0257	6.0309	6.1175
2	5.9903	6.1267	4.6278	4.7350	16.1687	31.5450	4.6189	4.6362	4.5467	4.7943
3	7.6094	2.4225	2.3809	2.3400	3.0525	2.3441	2.3259	2.3564	2.3523	2.8712
4	12.0525	13.3385	12.0741	12.4093	19.9343	18.4560	12.1098	12.0699	12.0705	12.0613
5	14.5134	14.8977	12.2398	12.2191	21.5440	17.0554	12.3461	12.2129	12.2055	13.1892
6	12.5673	12.1931	12.1084	13.7827	12.4759	13.6971	12.2071	12.1164	12.1122	12.4875
7	1.0000	1.6866	1.0026	1.0000	1.6958	1.0000	1.0456	1.0011	1.0011	1.0061
8	1.0000	1.0000	1.0084	1.0000	1.3733	1.0000	1.0307	1.0131	1.0013	1.0000
9	2.7114	2.3624	2.3767	2.7674	3.2635	3.0810	2.3428	2.3799	2.3490	2.6258
10	4.5892	18.8862	4.9466	4.9807	5.6204	7.2207	5.1031	4.9617	4.9469	4.9753
11	3.8451	6.1761	3.5330	3.4836	3.8278	3.9545	3.6116	3.5070	3.5073	3.5871
12	12.1498	5.7650	5.1632	5.1014	7.0746	7.4098	5.1208	5.1160	5.1398	5.4003
13	1.0000	1.0000	1.7259	1.8669	1.4338	1.0000	1.9402	1.7510	1.7010	1.5572
14	4.7130	8.3770	5.2353	5.2144	6.1449	8.4310	5.3580	5.1958	5.1822	5.1404
15	3.2813	3.4448	3.1000	2.8645	4.3213	3.5315	3.1035	3.0847	3.0757	3.2388
Mass (kg)	128.0254	146.0234	113.0469	114.9422	152.6426	155.9305	113.8195	112.8336	112.6401	115.1230
σ^{max} (MPa)	35.8060	35.5735	36.1401	36.1549	34.6044	31.7396	35.8109	36.2145	36.2407	35.8162
δ^{max} (mm)	7.2933	7.0400	7.8277	7.5916	4.9676	5.6302	7.7930	7.8388	7.8425	7.5912
f_1 (Hz)	5.7313	6.3634	5.5766	5.6168	7.0851	6.1600	5.6059	5.5636	5.5515	5.6018
Mean	159.1585	211.0599	113.9629	121.9248	177.7725	246.6050	115.5839	115.9005	112.7702	117.6808
SD	17.8226	32.0740	0.5145	6.6524	9.8684	46.9447	0.8749	3.7406	0.1693	1.6216

To determine the mean and standard deviation values of the structural mass, several evolutionary methods were tested. These methods included the ALO, DA, GWO, MVO, SCA, WOA, HTS, TLBO, PVS, and SOS algorithms. After analyzing the results, it was found that the mean weights for these algorithms ranged from 112.7702 kg to 246.6050 kg.

Interestingly, the PVS algorithm was found to have the best performance in terms of both the mean weight and the best standard deviation of weight. Specifically, the mean weight produced by the PVS algorithm was 112.7702 kg, which was the lowest among all the algorithms tested. Additionally, the standard deviation of weight produced by the PVS algorithm was only 0.1693, which was the best among all the algorithms tested.

It is worth noting, however, that the second-best performance was also observed with the PVS algorithm, in terms of the mean weight and the best SD of weight. This indicates that the PVS algorithm is a strong performer when it comes to evaluating structural mass using evolutionary methods.

In conclusion, this study provides valuable insight into the performance of different evolutionary methods in the field of structural mass evaluation. While there were several algorithms that performed well, the PVS algorithm emerged as the top performer in terms of both mean weight and standard deviation of weight. These findings can help inform future research and development in this important field.

4.3.4 Size Optimization of the 24-Bar Truss with Continuous Cross Sections

This section of the study delves into the size optimization of the 24-bar truss using 24 continuous design variables, which presents a number of challenges. The ground structure of the fourth benchmark problem, the 24-bar truss, is depicted in Fig. 4.4. This truss problem was first investigated by Ringertz (1986) using a branch and bound algorithm for topology optimization of a truss with static constraints, while Xu et al. (2003) used for testing a 1D search method, and Kaveh and Zolghadr (2013) employed it for testing the search performance of charged system search and particle swarm optimization algorithms by considering continuous design variables. In this problem, the continuous cross-sectional areas (A_i, where $i = 1, 2, \ldots, 24$) are considered over $[1, 40]$ cm^2.The truss is subject to two external load cases as given in Table 4.7. Truss element cross-sectional areas of the 24-bars are treated as continuous variables, and nodal masses, material properties, and other required data are also shown in the table. The truss is subjected to a nonstructural lumped mass of 500 kg at node 3.

The study then goes on to present the results of the size optimization process using various metaheuristic algorithms in 100 independent runs. The best algorithm is determined by identifying the one with the lowest mass value that meets all of the structural constraints.

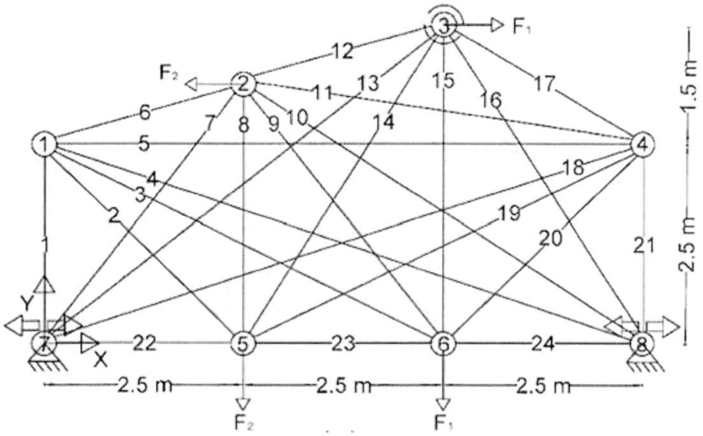

Fig. 4.4 The 24-bar truss

Table 4.7 Design parameters of the 24-bar truss

Size variables: $X_i = A_i$, $i = 1, 2, \ldots, 24$	
Multiple load conditions:	
Condition 1 : $F_1 = 50$ KN, $F_2 = 0$ KN	Condition 2 : $F_1 = 0$ KN, $F_2 = 50$ KN
Lumped mass on node 3 is 500 kg	
Stress and displacement constraints: $\sigma_i^{max} = 172.43$ MPa and $\delta_{y_5 \& y_6}^{max} = 10$ mm $\left(10^{-2}\text{ m}\right)$	
Natural frequency constraint: $f_1 \geq 30$ Hz	
The continuous cross − sectional areas : $[A^{min}, A^{max}] = [1, 40]$ cm^2 ($[1, 40] * 10^{-4}$ m^2)	
Material properties: $E = 69$ GPa and $\rho = 2740$kg/m^3	

According to the results presented in Table 4.8, the reported best mass values for the ALO, DA, GWO, MVO, SCA, WOA, HTS, TLBO, PVS, and SOS algorithms are 233.463 kg, 247.628 kg, 199.751 kg, 208.705 kg, 274.462 kg, 316.937 kg, 223.571 kg, 203.348 kg, 198.598 kg, and 202.008 kg, respectively.

Interestingly, the PVS algorithm was found to produce the lowest mass value among all the algorithms tested. The GWO and SOS algorithms also performed well, producing mass values that were very close to that of the PVS algorithm.

These findings shed light on the efficacy of different metaheuristic algorithms in optimizing the size of the 24-bar truss. Overall, the PVS algorithm was found to be the most effective, followed closely by the GWO and SOS algorithms. These results can inform future research in this area and help guide the development of more efficient optimization techniques.

In order to evaluate the performance of the metaheuristic methods used in this study, the average mass and its standard deviation (SD) were calculated. The mean mass values for each algorithm, including ALO, DA, GWO, MVO, SCA, WOA, HTS, TLBO, PVS, and SOS, were determined to be 300.477 kg, 390.489 kg, 203.661 kg, 236.790 kg, 331.812 kg, 468.575 kg, 244.077 kg, 227.806 kg,

Table 4.8 Size optimization of the 24-bar truss (continuous section)

Element no.	ALO	DA	GWO	MVO	SCA	WOA	HTS	TLBO	PVS	SOS
1	5.066	3.926	4.807	6.010	6.704	11.124	9.431	5.081	4.649	5.715
2	1.118	1.365	1.016	1.151	1.182	10.833	3.702	1.052	1.007	1.000
3	1.000	1.000	1.000	1.008	2.034	1.335	1.122	1.002	1.002	1.002
4	8.332	12.334	8.211	8.216	13.959	13.572	7.417	8.798	8.250	7.624
5	1.000	1.000	1.005	1.000	1.335	8.935	1.364	1.001	1.000	1.004
6	5.040	5.588	5.039	5.273	8.566	10.485	9.766	5.151	4.948	5.672
7	16.684	24.756	17.237	19.229	22.930	14.958	15.876	18.649	17.356	16.567
8	3.061	13.583	3.886	3.296	17.368	10.530	6.643	3.323	4.186	2.247
9	1.000	1.000	1.027	1.021	1.051	1.829	1.022	1.264	1.000	1.000
10	10.136	8.204	10.069	10.799	8.049	15.066	8.704	11.401	10.044	9.429
11	2.613	1.000	1.014	1.043	1.379	12.237	1.037	1.115	1.016	1.000
12	1.685	1.000	1.142	1.013	1.302	3.846	1.139	1.001	1.082	1.002
13	12.389	13.108	12.734	14.858	12.955	19.345	11.856	12.941	12.774	13.244
14	1.000	1.000	1.239	1.211	4.000	2.330	2.566	1.153	1.376	1.001
15	2.911	9.052	4.864	3.917	7.932	10.404	6.229	3.476	4.742	6.506
16	14.770	20.822	19.836	17.928	21.911	17.703	18.274	18.458	19.798	20.106
17	15.758	6.581	6.764	7.192	9.621	9.793	11.128	7.168	6.280	7.114
18	12.738	7.674	8.454	8.029	9.833	11.075	8.254	8.323	8.223	9.467
19	1.000	1.000	1.044	1.021	2.889	2.956	1.069	1.000	1.001	1.000
20	5.385	1.000	1.014	2.759	4.757	7.579	5.745	1.855	1.150	1.134
21	13.650	20.061	5.750	8.731	19.326	6.811	9.726	6.487	5.634	6.986
22	19.797	1.024	1.020	1.524	4.689	6.785	6.864	1.005	1.000	1.000
23	1.029	17.732	1.437	1.590	1.000	3.684	1.456	1.009	1.121	1.124
24	5.599	1.000	1.467	3.243	11.730	3.218	5.380	2.233	1.069	1.464

Mass (kg)	233.463	247.628	199.751	208.705	274.462	316.937	223.571	203.348	198.598	202.008
σ^{max} (MPa)	116.657	57.851	95.672	107.752	41.549	31.954	49.858	108.960	89.340	148.383
δ^{max} (mm)	6.502	3.568	5.494	5.865	2.928	2.701	3.454	6.016	5.448	7.810
f_1 (Hz)	31.014	30.028	30.022	31.700	32.587	39.340	30.497	30.045	30.003	30.119
Mean	300.477	390.489	203.661	236.790	331.812	468.575	244.077	227.806	205.506	216.960
SD	31.325	49.704	1.652	15.128	21.093	68.657	11.593	17.268	4.230	11.174

205.506 kg, and 216.960 kg, respectively. Additionally, the SD values for each algorithm were calculated to be 31.325 kg, 49.704 kg, 1.652 kg, 15.128 kg, 21.093 kg, 68.657 kg, 11.593 kg, 17.268 kg, 4.230 kg, and 11.174 kg, respectively.

The results of the study demonstrate that the GWO algorithm performs the best in terms of achieving the lowest mean mass and the lowest SD. This suggests that the GWO algorithm has a superior convergence rate and consistency compared to the other algorithms tested. Meanwhile, the PVS algorithm was found to perform second best in terms of achieving the lowest mean mass and the lowest SD.

Based on these findings, it can be concluded that the GWO and PVS algorithms are the most suitable for this particular design problem. The results of this study can be used to guide future research in this area and help inform the development of more efficient optimization techniques.

4.3.5 *Size Optimization of the 20-Bar Truss with Continuous Cross Sections*

In this section, the focus is on the optimization of the 20-bar truss for size optimization using 20 continuous sections. Figure 4.5 signifies the ground structure of the 20-bar truss. This benchmark problem was firstly framed for the topology and size optimization by Xu et al. (2003) and then tested by Kaveh and Zolghadr (2013) for continuous optimization.

In this problem, the continuous cross-sectional areas (A_i, where $i = 1,2,\ldots,20$) are considered over [1, 100] cm^2. Table 4.9 gives all details of design variables, loading conditions, structural safety constraints, and other relevant data for the test problem. For this problem, no lumped masses are added to the truss.

According to the results presented in Table 4.10, the results for the ALO, DA, GWO, MVO, SCA, WOA, HTS, TLBO, PVS, and SOS algorithms are reported as 222.4818, 289.7974, 205.0657, 213.9475, 310.5999, 337.2046, 225.4577, 210.1644, 209.3130, and 212.1035 kg, respectively. The GWO algorithm achieves

Fig. 4.5 The 20-bar truss

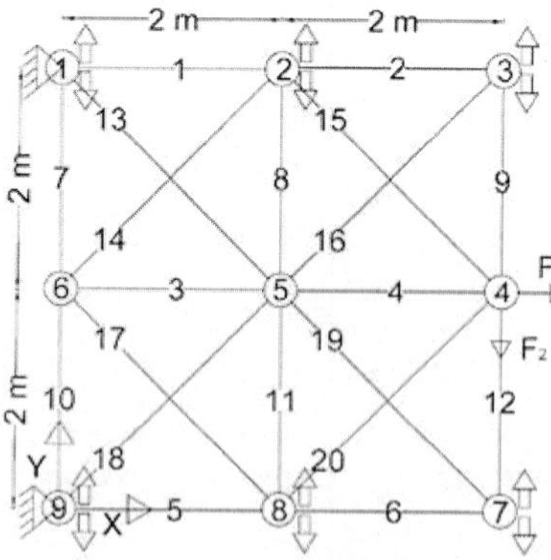

Table 4.9 Design parameters of the 20-bar truss

Size variables: $X_i = A_i$, $i = 1, 2, \ldots, 20$	
Multiple load conditions	
Condition 1 : $F_1 = 500$ KN, $F_2 = 0$ KN	Condition 2 : $F_1 = 0$ KN, $F_2 = 500$ KN
No lumped mass	
Stress and displacement constraints: $\sigma_i^{max} = 172.43$ MPa and $\delta_{y_4}^{max} = 60$ mm $(6 \times 10^{-2}$ m$)$	
Natural frequency constraints: $f_1 \geq 60$ Hz and $f_2 \geq 100$ Hz	
The continuous cross − sectional areas : $[A^{min}, A^{max}] = [1, 100]$ cm^2 ($[1, 100] * 10^{-4}$ m^2)	
Material properties: $E = 69$ GPa and $\rho = 2740$kg/m^3	

the least structural mass, with the PVS and TLBO algorithms coming in a close second and third, respectively. These results indicate that the GWO algorithm is the most effective in terms of reducing the mass of the truss while maintaining structural integrity. The PVS and TLBO algorithms also performed well and can be considered as viable alternatives to the GWO algorithm for this design problem.

In the context of metaheuristic optimization methods, statistical analysis is a crucial step in evaluating the effectiveness of different algorithms in solving a particular optimization problem. The average mass and standard deviation are important metrics used to assess the performance of these algorithms. The average mass represents the mean value of the mass obtained from multiple runs of each algorithm, while the standard deviation measures the variability of the results obtained from different runs. A low standard deviation indicates that the results obtained from different runs are relatively consistent, indicating that the algorithm is reliable and stable. Analysis of the available results shows that the ALO, DA, GWO, MVO, SCA, WOA, HTS, TLBO, PVS, and SOS algorithms have mass mean values of 319.5906, 394.0374, 211.3255, 247.2300, 386.8633, 482.2698, 237.8769, 242.4180, 222.2840, and 238.7597 kg, respectively, with corresponding SDs of 21.8294, 41.1317, 16.4935, 15.9310, 12.7556, 42.5808, 19.5985, 16.1044, 17.5640, and 16.5501.

Based on the available results, it can be observed that the different metaheuristic algorithms performed differently in terms of their average mass and standard deviation. The GWO algorithm achieved the lowest mass mean value and the best SD, indicating that it is the most effective algorithm for this optimization problem. The PVS algorithm also performed well, achieving the second lowest mass mean value and the best SD.

Therefore, based on the statistical outcomes, it can be concluded that the GWO algorithm is significantly superior to the other algorithms in solving this particular optimization problem. However, it is worth noting that the performance of different algorithms may vary depending on the optimization problem being solved. Therefore, it is essential to evaluate the performance of different algorithms in different contexts to select the most effective one for a particular optimization problem.

Table 4.10 Size optimization if the 20-bar truss (continuous section)

Element no.	ALO	DA	GWO	MVO	SCA	WOA	HTS	TLBO	PVS	SOS
1	21.8294	41.1317	16.4935	15.9310	12.7556	42.5808	19.5985	16.1044	17.5640	16.5501
2	1.0130	44.8902	1.5726	1.0672	5.4929	1.0000	6.7307	3.9900	5.1060	3.8090
3	1.4071	1.0000	1.1064	1.1331	4.9070	8.1900	2.9504	1.0005	1.1639	1.0000
4	11.4882	20.8295	13.5045	13.8893	18.7658	35.1619	16.2259	13.9646	15.3193	13.0819
5	19.1237	18.8302	19.7782	20.6787	22.5967	32.4246	21.0764	24.3540	19.6006	19.2881
6	8.1456	10.9274	6.2497	11.5152	23.2066	27.4161	6.3680	4.1112	7.3147	4.4703
7	1.0002	10.2882	1.0972	1.0509	3.4737	37.8107	2.2596	5.9202	1.0151	3.1700
8	17.7034	20.2042	17.0189	18.8147	34.6002	40.9859	14.3731	15.8008	14.4578	18.4264
9	1.1900	9.5805	1.5881	1.0552	2.1666	1.0000	6.0238	3.0996	5.3188	2.8475
10	10.5358	9.3905	7.9868	8.9433	7.1752	8.8835	10.3190	5.3045	9.1034	6.0467
11	18.6245	13.9877	20.2504	31.1263	58.5840	9.4439	19.9675	23.5867	13.9595	21.0005
12	9.1416	34.6040	6.3646	5.9726	13.9256	35.6762	8.5086	3.9250	7.2600	5.8016
13	15.1387	19.4008	18.0580	19.1541	44.2623	22.5618	17.4002	18.3623	16.4398	18.0048
14	13.9388	10.4007	11.4723	11.1942	8.7272	22.2453	15.6104	11.0524	12.7198	10.7854
15	37.9685	9.8930	20.4196	19.8074	15.6371	29.6361	14.3124	16.7190	14.9281	20.8889
16	8.3341	27.1365	10.6327	8.8693	18.1472	7.3429	19.5205	14.7708	18.2231	13.8493
17	2.1403	1.0000	1.3766	1.1941	8.1935	8.5781	4.5369	1.0018	2.5871	1.0000
18	31.8796	41.3373	31.1132	30.5980	33.1512	30.6158	32.5550	32.0013	31.2001	31.9728
19	5.4270	12.2527	3.0939	2.8386	1.1802	35.1813	2.7580	1.0000	4.1799	1.4198
20	28.4819	27.8115	30.4625	31.5878	66.5848	22.4869	31.1222	32.5366	28.8809	36.0342
Mass (kg)	222.4818	289.7974	205.0657	213.9475	310.5999	337.2046	225.4577	210.1644	209.3130	212.1035
σ^{max} (MPa)	172.4299	172.4300	172.3608	172.0801	161.6728	116.0483	170.2424	172.4253	172.3804	172.0619
δ^{max} (mm)	20.8471	18.6897	23.0996	22.5151	16.7085	15.1869	21.8899	22.8504	23.6674	22.2080
f_1 (Hz)	93.9538	86.4006	100.3072	99.5260	94.9406	90.5406	98.7399	101.0045	97.0271	97.9544
f_2 (Hz)	171.8464	161.1215	175.9233	170.8943	150.8725	160.8784	175.4879	176.6941	175.4664	174.0282
Mean	319.5906	394.0374	211.3255	247.2300	386.8633	482.2698	237.8769	242.4180	222.2840	238.7597
SD	36.0423	42.0448	3.4447	19.4460	25.1806	71.1059	8.4228	19.9559	6.5592	13.3848

4.3.6 Size Optimization of the 72-Bar 3D Truss with Continuous Cross Sections

This section presents the optimization of the 72-bar truss using 16 continuous variables for size optimization. The ground structure of the sixth test problem, the 72-bar truss, is signified in Fig. 4.6. This 3D truss was applied by Kaveh and Zolghadr (2013) for continuous optimization.

In this problem, the continuous cross-sectional areas of groups (G_i, where i = 1,2,. . .,16) are considered over [1, 30] cm^2. Table 4.11 shows all details of material properties, design variables, loading conditions, lumped mass, and other relevant data of the test problem. The elements are clustered into 16 groups (i.e., G_1 [A_1–A_4], G_2 [A_5–A_{12}], G_3 [A_{13}–A_{16}], G_4 [A_{17}–A_{18}], G_5 [A_{19}–A_{22}], G_6 [A_{23}–A_{30}], G_7 [A_{31}–A_{34}], G_8 [A_{35}–A_{36}], G_9 [A_{37}–A_{40}], G_{10} [A_{41}–A_{48}], G_{11} [A_{49}–A_{52}], G_{12} [A_{53}–A_{54}], G_{13} [A_{55}–A_{58}], G_{14} [A_{59}–A_{66}], G_{15} [A_{67}–A_{70}], and G_{16} [A_{71}–A_{72}]) by considering structural symmetry as per Kaveh and Zolghadr (2013). The truss has four nonstructural lumped masses of 2270 kg added to all the four top nodes (node numbers 1–4).

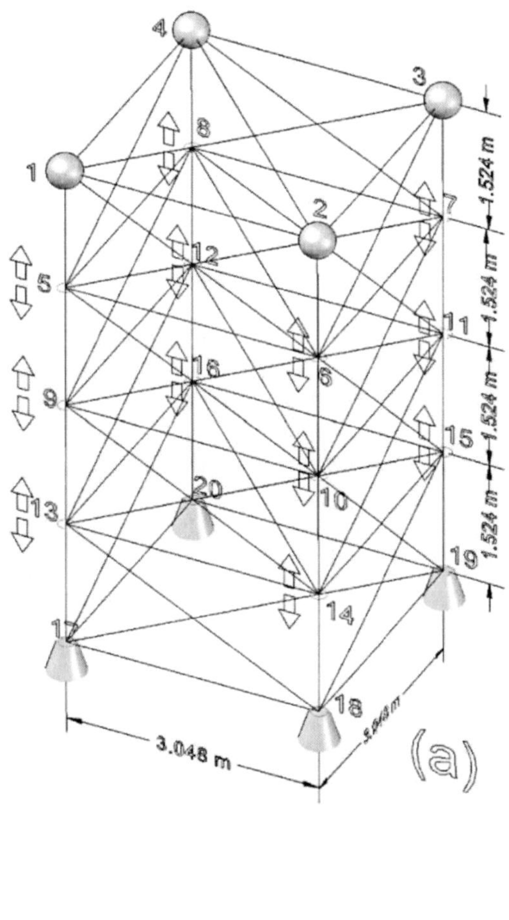

Fig. 4.6 The 72-bar 3D truss

Table 4.11 Design parameters of the 72-bar truss

Size variables: G_i, $i = 1, 2, \ldots, 16$; $G =$ group number				
Multiple load conditions				
Condition 1	$F_{1x} = 22.25$ KN	$F_{1y} = 22.25$ KN		$F_{1z} = -22.25$ KN
Condition 2	$F_{1z} = -22.25$ KN	$F_{2z} = -22.25$ KN	$F_{3z} = -22.25$ KN	$F_{4z} = -22.25$ KN
Lumped masses at nodes 1, 2, 3, and 4 are 2270 kg				
Stress and displacement constraints: $\sigma_i^{max} = 172.375$ MPa and $\delta_{x_j/y_j}^{max} = 6.35$ mm $(6.35 \times 10^{-3}$ m$)$ (for node, j = 1, 2, 3, and 4 along *x*- and *y*-axes)				
Natural frequency constraints: $f_1 \geq 4$ Hz and $f_3 \geq 6$ Hz				
The continuous cross – sectional areas : $[A^{min}, A^{max}] = [1, 30]$ cm^2($[1, 30] * 10^{-4}$ m^2)				
Material properties: $E = 68.95$ GPa and $\rho = 2767.99$kg/m^3				

To assess the performance of various algorithms, we conducted 100 independent runs of size optimization for the 72-bar truss and compared the results in Table 4.12. Our goal was to identify the algorithm that produced the lowest mass value while satisfying all structural constraints, and we found that the ALO, DA, GWO, MVO, SCA, WOA, HTS, TLBO, PVS, and SOS algorithms were the most effective.

The best mass values reported for these algorithms were 484.3573, 608.8999, 470.9994, 476.3906, 575.1474, 633.9965, 473.2674, 471.0249, 470.5528, and 475.3283 kg, respectively. Among these algorithms, the PVS algorithm had the lowest structural mass, followed by the GWO algorithm and TLBO algorithm in third place. These results demonstrate the effectiveness of the metaheuristic algorithms in achieving the minimum mass value for the 72-bar truss while satisfying all structural constraints.

In this research study, various metaheuristic methods were evaluated based on their statistical performance in terms of average mass and standard deviation. The ten algorithms that were examined in this study were ALO, DA, GWO, MVO, SCA, WOA, HTS, TLBO, PVS, and SOS. The mean mass values obtained for these algorithms were 759.7721, 1030.2015, 473.2981, 537.1026, 636.1164, 1100.3489, 480.5615, 478.8744, 472.1863, and 484.7760 kg, respectively, while their corresponding standard deviations were 123.6783, 223.8227, 1.4212, 49.8139, 36.3345, 283.1382, 4.3989, 8.3690, 1.1019, and 6.1296.

Upon analyzing the results, the PVS algorithm was found to have demonstrated the best performance as it had the lowest mass mean value and the best standard deviation. The GWO algorithm also showed promising results as it came in second for both measures. Therefore, for this particular design problem, the PVS and GWO algorithms were found to be the better options in terms of both convergence rate and consistency.

Overall, this study highlights the importance of selecting the appropriate metaheuristic method based on its statistical performance in solving design problems. The findings of this study can aid researchers and practitioners in selecting the most effective algorithm to use in their respective design tasks.

Table 4.12 Size optimization of the 72-bar truss (continuous section)

Element group no	ALO	DA	GWO	MVO	SCA	WOA	HTS	TLBO	PVS	SOS
1	4.8963	7.6989	4.2742	4.1277	9.1679	7.7140	5.1169	4.1076	4.3693	4.1921
2	7.5039	13.7821	7.8604	8.1647	7.3827	7.7329	7.0133	7.5174	8.2680	7.5029
3	5.9404	8.7055	4.4863	4.7953	14.9577	11.9867	4.5429	4.4919	4.4611	4.5752
4	10.2634	11.3339	7.7202	7.9885	8.9725	13.0041	7.9326	7.8279	7.7289	7.9533
5	6.8620	8.8370	6.9223	5.6769	13.3485	4.6201	8.0275	6.7724	8.4141	6.8823
6	7.2571	7.9668	8.1713	7.2065	7.0501	9.5035	8.6550	7.8845	7.8250	8.6390
7	1.0000	1.3412	1.0606	1.0000	6.9612	5.9670	1.0592	1.0042	1.0003	1.0000
8	2.7576	11.6800	1.5118	1.2010	2.8700	1.2153	1.2236	1.2513	1.3857	1.9291
9	13.7001	8.6644	14.0954	16.0711	7.2807	16.2908	13.4573	13.4419	13.2837	11.4857
10	8.7469	6.7061	7.8954	8.7344	8.0234	6.1719	8.3257	8.4787	7.9007	8.0371
11	1.0013	1.5667	1.2428	1.2166	1.1056	15.6380	1.1995	1.2652	1.2869	1.3353
12	1.6305	2.9356	2.3938	2.4506	1.5682	7.1191	2.4245	2.4002	2.4791	2.5465
13	17.0502	21.6000	16.9236	17.8269	27.9147	20.3277	15.8238	17.8550	16.0416	20.7009
14	8.9661	6.8501	8.2923	8.2626	11.4858	10.1975	8.5114	8.3842	8.2413	8.1011
15	1.0001	17.7826	1.0450	1.3266	1.5497	3.4357	1.0015	1.0751	1.0090	1.0000
16	1.0213	1.0000	1.0275	1.0379	2.8087	9.1875	1.0146	1.0660	1.0080	1.0000
Mass (kg)	484.3573	608.8999	470.9994	476.3906	575.1474	633.9965	473.2674	471.0249	470.5528	475.3283
σ^{max} (MPa)	40.7095	25.2157	46.5649	48.5970	22.0339	26.9375	39.9069	48.5431	45.3722	47.1649
δ^{max} (mm)	3.0868	3.0289	3.1698	3.1002	2.9843	2.8643	3.2239	3.1568	3.2077	3.1237
f_1 (Hz)	4.0000	4.0000	4.0000	4.0058	4.0730	4.0595	4.0017	4.0009	4.0006	4.0127
f_3 (Hz)	6.0000	6.0000	6.0003	6.0010	6.0384	6.0000	6.0084	6.0003	6.0006	6.0004
Mean	759.7721	1030.2015	473.2981	537.1026	636.1164	1100.3489	480.5615	478.8744	472.1863	484.7760
SD	123.6783	223.8227	1.4212	49.8139	36.3345	283.1382	4.3989	8.3690	1.1019	6.1296

4.3.7 *Size Optimization of the 39-Bar Truss with Continuous Cross Sections*

This section details the optimization of the 39-bar truss using 21 continuous variables for size optimization. The seventh benchmark problem, the two-tier 39-bar planar truss, displayed in Fig. 4.7, is considered to investigate large size truss subjected to stress, buckling, displacement, and fundamental frequency constraints with continuous and continuous design variables. This problem was initiated by Deb and Gulati (2001) and followed by many researchers (Ahrari et al., 2014) with static constraints, but has never been studied with simultaneous static and dynamic constraints. This problem is modified to investigate the effect of multi-load conditions, lumped masses, and natural frequency constraint.

In this problem, the continuous cross-sectional areas of groups (G_i, where $i = 1, 2, \ldots, 21$) are considered over [1, 50] cm^2. Design considerations such as design variables, multi-load conditions, lumped masses, constraints, and material properties are presented in Table 4.13. The elements are grouped into 21 groups by seeing symmetry about the vertical middle plane similar to Deb and Gulati (2001). The truss is subjected to three nonstructural lumped masses of 500 kg at nodes 2, 3, and 4. Two different loading conditions are considered to fulfill the multi-load condition as shown in Fig. 4.7.

To evaluate the performance of different optimization algorithms, the truss was subjected to 100 runs, and the design variables and corresponding truss weights were recorded for the best designs. Table 4.14 presents the results, indicating that the ALO, DA, GWO, MVO, SCA, WOA, HTS, TLBO, PVS, and SOS algorithms

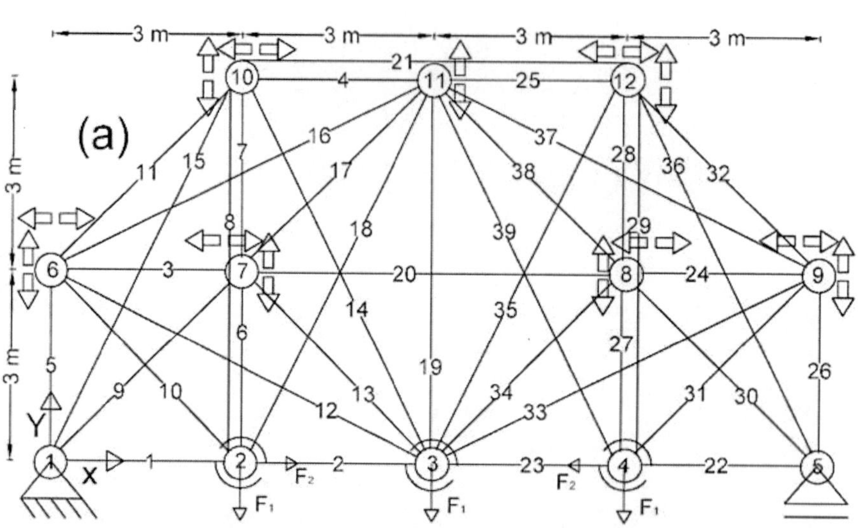

Fig. 4.7 The 39-bar truss

Table 4.13 Design parameters of the 39-bar truss

Size variables: $X_i = G_i$, $i = 1, 2, \ldots, 21$	
Multiple load conditions	
Condition 1 : $F_1 = 100\ KN$, $F_2 = 0\ KN$	Condition 2 : $F_1 = 0\ KN$, $F_2 = 50\ KN$
Lumped masses on nodes 2, 3, and 4 are 500 kg	
Stress and displacement constraints: $\sigma_i^{max} = 172.43$ MPa and $\delta_{x_j/y_j}^{max} = 50$ mm (0.05 m) (for node, $j = 2$, 3 and 4 along x- and y-axes)	
Natural frequency constraints: $f_1 \geq 4$ Hz	
The continuous cross − sectional areas : $[A^{min}, A^{max}] = [1, 50]\ \text{cm}^2$ $([1, 50] * 10^{-4}\ \text{m}^2)$	
Material properties: $E = 69\ GPa$ and $\rho = 2740\text{kg/m}^3$	

produced the best weights of 602.3027, 683.4948, 517.8574, 540.0351, 720.0120, 719.1296, 552.8566, 522.2055, 523.6126, and 547.4524 kg, respectively.

Among the ten algorithms evaluated, the GWO algorithm produced the lowest weight of 602.3027 kg, followed by the TLBO algorithm at 517.8574 kg, and the PVS algorithm at 522.2055 kg. The DA, GWO, MVO, SCA, WOA, HTS, and SOS algorithms also produced competitive weights ranging from 517.8574 kg to 720.0120 kg.

It is noteworthy that the GWO algorithm produced the best truss weight among all the algorithms except for the ALO and TLBO algorithms. These findings indicate that the ALO, TLBO, and GWO algorithms are promising optimization methods for truss design problems. However, further analysis is necessary to investigate their performance in other design scenarios.

Overall, the results of this study demonstrate the effectiveness of different optimization algorithms for truss design problems. The findings can serve as a guide for designers and engineers in selecting the most appropriate optimization algorithm for their specific design tasks.

In order to accurately assess the performance of the various metaheuristic algorithms, this study evaluated their average mass and standard deviation. The algorithms that were evaluated included ALO, DA, GWO, MVO, SCA, WOA, HTS, TLBO, PVS, and SOS. The analysis revealed that the mean mass values for these algorithms were 726.4545, 886.6181, 529.1717, 596.5996, 797.3408, 1174.6214, 604.6696, 584.3053, 549.8662, and 575.6052 kg, respectively. Their corresponding standard deviations were 74.4845, 88.4850, 7.6053, 27.0595, 45.0094, 190.1176, 28.5475, 38.4717, 17.9570, and 19.7378.

The GWO algorithm achieved the best mean weight of 529.1717 kg and SD of 7.6053 kg. The PVS algorithm closely followed with a mean weight of 549.8662 kg and SD of 17.9570 kg. These findings indicate that the GWO and PVS algorithms are the top-performing algorithms for achieving both convergence rate and consistency in this specific truss design problem. Upon analyzing the results, it can be concluded that the GWO algorithm performs better than the other algorithms in terms of achieving a better mean and standard deviation of the weight of the truss. The PVS algorithm, on the other hand, performed the second best in terms of

Table 4.14 Size optimization of the 39-bar truss (continuous section)

Group no.	ALO	DA	GWO	MVO	SCA	WOA	HTS	TLBO	PVS	SOS
1	4.1926	4.3739	1.0195	1.0126	1.9416	4.5940	1.0871	1.0000	1.0033	1.0135
2	15.5634	38.8658	17.4446	22.5637	16.6759	29.9241	18.5995	22.7302	15.6166	18.7010
3	2.3056	3.8068	6.3174	11.0361	6.0707	1.0000	4.3498	3.0064	2.1263	2.2681
4	12.3981	8.2204	5.0587	7.6610	11.5351	8.1287	8.9078	5.9037	6.4582	19.2819
5	15.6790	32.1805	20.3727	23.7733	23.0640	30.6349	20.5376	23.2202	19.7788	28.0062
6	16.2860	1.4316	1.2505	2.2219	2.0435	1.0000	4.4287	1.1397	3.2787	1.2664
7	2.9366	1.0000	2.1211	2.2862	1.7614	14.4046	6.1831	1.1290	5.2947	1.2892
8	8.3424	5.1713	1.4305	1.6458	2.0485	11.3865	2.9581	1.6807	2.5726	2.3667
9	14.5840	16.7444	7.1204	9.3914	27.5968	6.5276	2.8289	9.7170	2.4010	8.5568
10	1.1370	1.1333	4.9798	4.6709	8.5564	6.7337	6.5285	4.3138	5.2597	6.3613
11	14.5794	18.4691	13.7684	20.3894	22.0184	18.5705	19.8172	15.8857	17.4694	18.7241
12	6.0231	6.7714	5.9800	4.5181	9.7583	14.6814	6.0711	5.7999	6.5507	6.2515
13	1.0002	17.9213	1.0282	1.1528	3.1696	6.1595	1.3236	1.0002	1.0134	1.3109
14	1.0000	1.0000	1.0051	1.3153	1.4120	1.0000	2.2908	1.0423	1.5193	1.0677
15	31.0114	30.3447	19.7368	17.4694	18.9437	30.7257	25.0963	16.6275	24.0354	15.0685
16	27.1812	28.2317	37.3112	33.1491	37.2327	31.9008	29.5540	35.0649	31.0521	34.0750
17	1.0000	1.0000	1.2357	1.0257	14.8433	2.7751	1.0171	1.1879	1.0245	1.2604
18	5.0667	2.8732	5.1879	4.0723	12.0902	3.6002	4.8207	4.1038	4.7776	4.9669
19	1.0000	3.7423	2.1633	1.0000	1.5111	1.0000	1.1803	1.0012	1.0151	1.1596
20	18.2248	20.4586	6.0428	6.7443	19.6366	23.1390	7.5307	12.9836	8.1057	13.9299
21	21.7608	21.8884	19.2822	23.2626	27.7629	23.5278	24.8220	20.2048	23.8512	16.2805
Mass (kg)	602.3027	683.4948	517.8574	540.0351	720.0120	719.1296	552.8566	522.2055	523.6126	547.4524
σ^{max} (MPa)	135.9472	159.9612	156.6192	168.8990	110.9946	73.0470	153.9637	168.3275	161.5572	134.8521
δ^{max} (mm)	19.4916	14.0396	18.2186	20.2042	12.8318	9.3217	15.9070	20.2461	18.0387	18.1315
f_1 (Hz)	11.5901	12.0871	8.0559	8.3295	10.9398	12.6398	8.1003	8.1197	7.7449	8.2856
Mean	726.4545	886.6181	529.1717	596.5996	797.3408	1174.6214	604.6696	584.3053	549.8662	575.6052
SD	74.4845	88.4850	7.6053	27.0595	45.0094	190.1176	28.5475	38.4717	17.9570	19.7378

achieving a better mean and standard deviation of the weight of the truss. Therefore, the GWO and PVS algorithms were found to be the best options for both convergence rate and consistency for this particular design problem.

It is important to note that the choice of algorithm ultimately depends on the specific requirements of the design problem at hand. While the GWO and PVS algorithms performed well in this study, it is possible that other algorithms may perform better for different design problems. Therefore, designers should carefully evaluate the available metaheuristic algorithms and choose the one that is best suited for their specific design requirements.

4.3.8 Size Optimization of the 45-Bar Truss with Continuous Cross Sections

In this section, the 45-bar truss underwent size optimization using 31 size variables. The ninth benchmark problem, the 45-bar truss is considered to investigate complex structure subjected to stress, buckling, displacement, and fundamental frequency constraints with continuous design variables. The ground structure of this truss with the multiple loads and the boundary condition is illustrated Fig. 4.8. This problem was initiated by Deb and Gulati (2001) and with static constraints, but has never

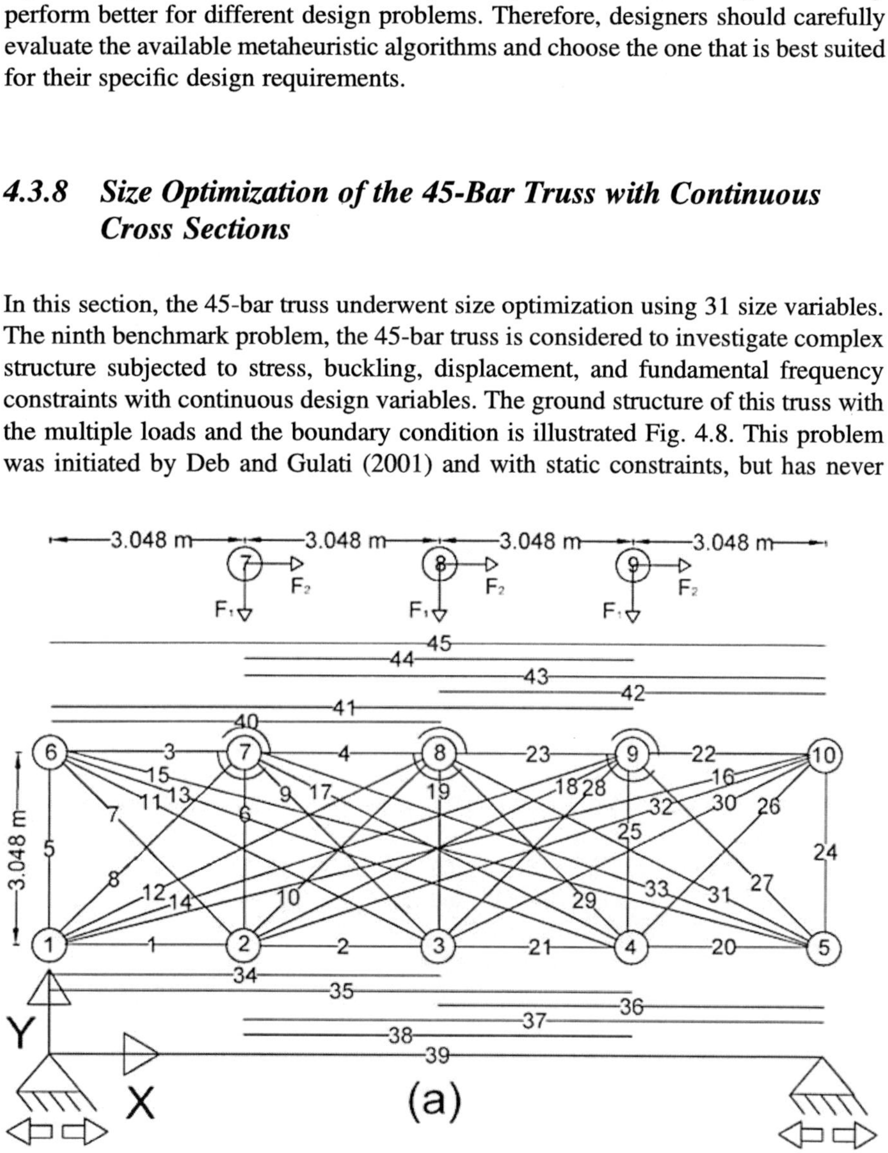

Fig. 4.8 The 45-bar truss

Table 4.15 Design parameters of the 45-bar truss

Size variables: $X_i = G_i$, $i = 1, 2, \ldots, 31$; $G =$ group number Shape variables: $-2.5\text{ m} \leq \xi_{x_j/y_j} \leq 2.5\text{ m}$, for $x_1 = -x_5$ and $y_1 = y_5$ with respect to their original coordinates	
Multiple load conditions	
Condition 1 : $F_1 = 44.537$ KN	Condition 2 : $F_2 = 4.4537$ KN
Stress and displacement constraints: $\sigma_i^{max} = 172.43$ MPa $\delta_{x_j/y_j}^{max} = 5$ mm (0.005 m) (for nodes, $j = 7, 8$, and 9 along x- and y-axes)	
Lumped masses on nodes 7, 8, and 9 are 500 kg	
Natural frequency constraints: $f_1 \geq 15$ Hz	
The continuous cross − sectional areas : $[A^{min}, A^{max}] = [1, 50]$ cm^2 ($[1, 50] * 10^{-4}$ m^2)	
Material properties: $E = 69$ GPa and $\rho = 2740\text{kg/m}^3$	

been examined with static and dynamic constraints simultaneously. This problem is modified to introduce the effect of multi-load conditions, lumped masses, natural frequency and constraint sections.

In this problem, the continuous cross-sectional areas of groups (G_i, where $i =$ 1,2,. . .,31) are considered over [1, 50] cm^2. Design considerations such as design variables, multi-load conditions, lumped masses, constraints, and material properties are presented in Table 4.15. The 28 elements (element numbers 1 to 14 and 20 to 33) are grouped into 14 groups by seeing symmetry about the vertical middle plane while remaining elements are considered distinct. Hence, size variable numbers shrink to 31. The truss is subjected to three nonstructural lumped masses of 500 kg at nodes 7, 8, and 9. Two different loading conditions are considered to fulfill multi-load condition as shown in Fig. 4.8.

Table 4.16 presents the areas of elements and corresponding truss weights for the best designs achieved through 100 runs using various algorithms, including ALO, DA, GWO, MVO, SCA, WOA, HTS, TLBO, PVS, and SOS. The minimum weight attained by each algorithm was as follows: 570.0026, 716.5188, 407.2329, 458.0537, 597.7005, 745.6150, 541.5890, 424.0364, 411.6108, and 425.0109 kg, respectively.

The minimum weight attained by each algorithm was provided in the passage, with GWO, PVS, and TLBO being the top three performers. Notably, their optimal topologies were quite similar. In terms of minimum weight, GWO was the most effective algorithm, followed by PVS in second place and TLBO in third place.

Overall, the results indicate that the GWO, PVS, and TLBO algorithms were successful in generating lightweight structures for the 45-bar truss. Additionally, the similarity in their optimal topologies suggests that these algorithms may share some common underlying principles that contribute to their effectiveness. This information could be valuable for future research and development of optimization algorithms for similar structures.

Table 4.16 Size optimization of the 45-bar truss (continuous section)

Group no.	ALO	DA	GWO	MVO	SCA	WOA	HTS	TLBO	PVS	SOS
1	1.2500	1.0000	2.2652	6.1057	10.7445	1.1246	1.1035	2.7101	2.0208	3.4838
2	33.2914	1.0000	1.0600	5.0196	9.0137	13.9405	5.1846	4.7711	1.0134	1.0000
3	1.0001	1.0000	1.1683	1.0176	2.0152	1.0000	2.2096	1.0287	1.1266	1.0362
4	11.5035	6.6273	2.6917	1.8824	5.8221	25.7374	8.2213	7.6689	15.8370	14.3984
5	14.0222	6.2444	3.7560	2.3972	5.2566	18.8708	5.8263	2.8853	3.6073	4.4199
6	7.4513	12.4972	1.1422	5.3167	1.4118	4.5690	17.5101	1.0971	1.0197	1.2193
7	1.0066	1.0000	1.0018	1.0034	1.0000	1.0000	1.0649	1.0482	1.0201	1.0005
8	26.5255	29.8809	18.8501	22.9808	26.1683	24.0138	25.1159	16.0506	21.8032	19.0751
9	1.0000	1.0000	1.5725	4.5949	2.4439	1.0000	1.6465	1.0669	1.0045	1.0295
10	10.5265	50.0000	4.0360	4.7343	7.2828	10.5494	6.7725	3.9649	4.3223	4.6027
11	1.0025	1.0000	1.0064	1.0082	3.5658	1.0000	1.7659	1.0057	1.0078	1.0000
12	19.2181	16.5138	22.1449	21.1289	21.2332	21.2735	20.2300	22.3468	21.9369	22.1669
13	1.0000	1.1173	1.0252	1.0045	1.0242	10.2451	1.3684	1.0009	1.0098	1.0028
14	17.7524	20.7003	22.9132	19.5652	26.4043	17.8330	19.6173	23.2919	20.2857	21.2365
15	6.8623	4.4142	3.6038	4.6379	7.0675	5.0077	8.2748	6.5172	4.1854	3.3323
16	2.0414	3.7934	3.9947	3.8096	3.1953	3.8229	2.7591	5.5090	4.1700	2.8229
17	1.0000	1.0000	2.1002	2.7891	10.3920	1.5517	7.7364	1.8146	1.4411	1.2910
18	9.1864	3.1055	1.5074	4.6404	4.0549	26.4335	2.7109	2.6132	1.1146	1.8533
19	5.0051	4.6130	2.3210	5.9017	17.2509	16.9807	6.2977	4.0633	2.4994	4.3431
20	1.0000	1.0052	1.0505	1.1104	1.5898	5.8840	1.9111	1.0845	1.1460	1.0000
21	1.0000	1.0085	1.0004	1.0659	2.1470	18.3672	1.0000	1.6562	1.0546	1.0000
22	1.4306	1.0234	1.8871	1.7397	1.2580	16.4398	3.4371	1.0473	1.1902	1.2014
23	2.1842	43.3960	1.2893	1.3292	4.2601	9.6534	2.1404	1.0001	1.1691	1.1222
24	10.7034	44.9089	1.0217	1.0000	5.5155	1.0000	11.0767	1.0197	1.0752	1.4203
25	1.0000	1.0000	1.6895	1.0721	2.8507	6.9090	1.9154	1.3299	1.0221	1.1231

(continued)

Table 4.16 (continued)

Group no.	ALO	DA	GWO	MVO	SCA	WOA	HTS	TLBO	PVS	SOS
26	1.0000	1.0000	1.4039	3.0850	5.7423	5.1167	9.7037	1.4564	1.3862	1.1708
27	24.4804	23.8380	9.9135	9.3436	12.3350	10.9375	11.9843	8.3329	8.4689	12.8987
28	1.0070	7.7085	1.5091	1.8071	1.2180	1.0000	1.7392	2.1032	1.4819	1.9521
29	14.6560	7.6049	8.9435	10.6759	37.4575	27.9361	18.3624	8.2639	8.5001	13.7947
30	30.4472	29.2931	9.5975	29.9673	8.9052	15.8379	20.6239	7.6656	8.0636	7.3741
31	1.0000	1.0000	4.3313	3.3063	2.3454	13.3924	4.2550	5.2659	5.1005	3.7060
Mass (kg)	570.0026	716.5188	407.2329	458.0537	597.7005	745.6150	541.5890	424.0364	411.6108	425.0109
σ^{max} (MPa)	29.0519	28.7889	28.6469	26.2729	27.5025	28.1598	24.5148	33.0689	26.1106	29.2570
δ^{max} (mm)	3.0256	2.5825	3.4864	3.2144	2.9603	2.9341	3.0950	3.5036	3.4414	3.4169
f_1 (Hz)	27.5066	28.0636	26.0508	27.2694	27.7306	27.2392	27.5898	25.6225	26.0984	26.1229
Mean	707.7729	969.4915	414.5002	532.9115	689.1013	1238.6808	614.1146	522.6060	452.1238	469.3024
SD	77.9451	121.6691	3.7702	40.5397	44.8436	210.9988	36.1323	64.4740	26.7185	34.6123

In addition, the results presented in Table 4.16 provide statistical information for the proposed study conducted over 100 independent runs. The table displays the mean mass values and standard deviations (SD) for each of the 10 optimization algorithms used in the study.

With the available results, it can be seen that ALO, DA, GWO, MVO, SCA, WOA, HTS, TLBO, PVS, and SOS algorithms have the mass mean value of 707.7729, 969.4915, 414.5002, 532.9115, 689.1013, 1238.6808, 614.1146, 522.6060, 452.1238, and 469.3024 kg, respectively, and SD of 77.9451, 121.6691, 3.7702, 40.5397, 44.8436, 210.9988, 36.1323, 64.4740, 26.7185, and 34.612,3 respectively.

The mean mass values ranged from 414.5002 kg for GWO to 1238.6808 kg for WOA, while the SD values ranged from 3.7702 for GWO to 210.9988 for WOA. The ALO and DA algorithms had the highest mean mass values, while GWO and PVS had the lowest mean mass values. Moreover, the standard deviations of GWO and PVS were the lowest among all the algorithms, indicating that their results were more consistent across the independent runs.

It can be concluded that the GWO and PVS algorithms performed best in terms of both convergence rate and consistency.

4.3.9 *Size Optimization of the 25-Bar 3D Truss with Continuous Cross Sections*

Using eight size variables, size optimization was performed on the 25-bar truss in this section. The ninth benchmark problem, the 25-bar truss is considered to investigate space truss subjected to stress, buckling, displacement, and fundamental frequency constraints with continuous design variables. The initial layout of this space truss is depicted in Fig. 4.9. This problem was investigated by Deb and Gulati (2001) and Ahrari et al. (2014) with static constraints, but has never been examined with simultaneous static and dynamic constraints. This problem is modified to introduce the effect of multi-load conditions, lumped masses, natural frequency constraint, and continuous area.

In this problem, the continuous cross-sectional areas of groups (G_i, where $i =$ 1, 2,. . ., 8) are considered over [1, 50] cm^2. Design considerations such as design variables, load conditions, lumped masses, constraints, and material properties are presented in Table 4.17. Element grouping is adopted because of structural symmetry in this problem according to the table. Therefore, this truss only has eight size variables, while topology variables are incorporated with it as discussed earlier. The truss is subjected to two nonstructural lumped masses of 500 kg at nodes 1 and 2. Two different loading conditions are considered as depicted in the table.

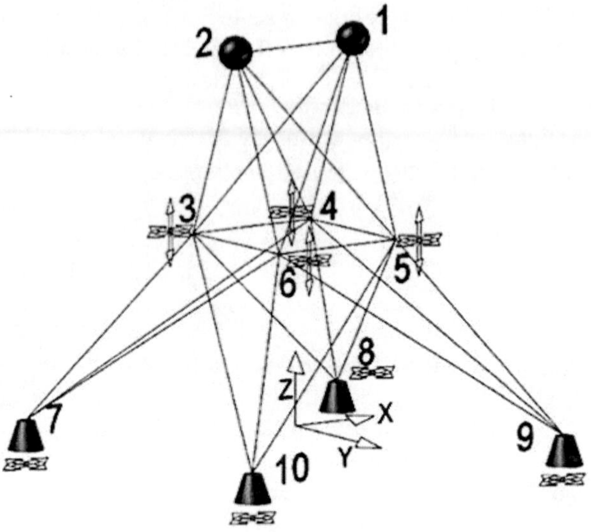

Fig. 4.9 The 25-bar 3D truss: (**a**) ground structure and (**b**) optimal structure

Table 4.18 displays the element areas and corresponding truss weights for the best designs achieved using various algorithms, including ALO, DA, GWO, MVO, SCA, WOA, HTS, TLBO, PVS, and SOS, in 100 runs. The lowest weight achieved by each algorithm is as follows: 384.2495, 395.9368, 384.1224, 384.5962, 413.1713, 401.7247, 383.6164, 383.5887, 383.5887, and 383.6674 kg, respectively.

The algorithm that satisfies all structural constraints and yields the minimum weight is deemed the best run. PVS and TLBO algorithms resulted in similar minimum mass values of 383.5887 kg, while the mass values obtained by ALO, DA, GWO, MVO, SCA, WOA, HTS, and SOS ranged from 383.6164 to 413.1713 kg.

To assess the performance of the algorithms statistically, the mean and standard deviation (SD) of the structural mass values were calculated. The ALO, DA, GWO, MVO, SCA, WOA, HTS, TLBO, PVS, and SOS algorithms yielded mass mean values of 410.9680, 439.0713, 388.1758, 392.5535, 447.9697, 474.5781, 383.7014, 383.5939, 383.6030, and 384.1060 kg, respectively, with corresponding SDs of 15.3732, 27.4426, 2.5528, 7.9276, 17.4846, 36.0876, 0.0692, 0.0170, 0.0386, and 0.5878.

Among the algorithms evaluated, the TLBO algorithm was found to be the most effective in achieving the lowest mass mean and SD values, indicating a superior convergence rate and consistency. The PVS algorithm was found to be the second-best algorithm in achieving the lowest mass mean and SD values. These findings suggest that the TLBO and PVS algorithms are optimal for this design problem.

Overall, the study provided valuable insights into the performance of various optimization algorithms in truss design problems. The results can guide engineers in selecting the most effective algorithm for a given design problem, ultimately leading to more efficient and cost-effective designs.

Table 4.17 Design parameters of the 25-bar 3D truss

Size variables: $X_i = G_i$, $i = 1, 2, \ldots, 8$; $G =$ group number				
Multiple load conditions				
Condition 1	$F_{y_1} = 89.074$ kN	$F_{z_1} = -22.25$ kN	$F_{y_2} = -89.074$ kN	$F_{z_2} = -22.25$ kN
Condition 2	$F_{x_1} = 4.454$ kN	$F_{y_1} = -44.537$ kN	$F_{z_1} = -44.537$ kN	$F_{y_2} = -44.537$ kN
	$F_{z_2} = -44.537$ kN	$F_{x_3} = 2.227$ kN	$F_{x_6} = 2.672$ kN	
Lumped masses at nodes 1 and 2 are 500 kg				
Stress and displacement constraints: $\sigma_i^{max} = 172.375$ MPa and $\delta_{x_j/y_j/z_j}^{max} = 8.9$ mm (0.0089 m) (for node, $j = 1,2,\ldots, 6$ along x-, y-, and z-axes)				
Natural frequency constraints: $f_1 \geq 15$ Hz				
The continuous cross − sectional areas : $[A^{min}, A^{max}] = [1, 50]$ cm^2 ($[1, 50] * 10^{-4}$ m^2)				
Material properties: $E = 69$ GPa and $\rho = 2740 \text{kg/m}^3$				
Element grouping				
Group number	Element number (end nodes)		Group number	Element number (end nodes)
G_1	1(1,2)		G_5	12(3, 4), 13(5,6)
G_2	2(1, 4), 3(2, 3), 4(1, 5), 5(2,6)		G_6	14(3, 10), 15(6,7), 16(4,9), 17(5, 8)
G_3	6(2, 5), 7(2, 4), 8(1, 3), 9(1,6)		G_7	18(3, 8), 19(4,7), 20(6,9), 21(5,10)
G_4	10(3,6), 11(4,5)		G_8	22(3, 7), 23(4,8), 24(5,9), 25(6,10)

Table 4.18 The 25-bar 3D truss size optimization (continuous section)

Element group no	ALO	DA	GWO	MVO	SCA	WOA	HTS	TLBO	PVS	SOS
1	1.0240	1.8661	1.0672	1.4393	1.0453	1.0000	1.0000	1.0000	1.0000	1.0086
2	16.0945	16.3340	16.1543	16.1315	17.3270	16.4320	16.0926	16.0910	16.0910	16.0906
3	15.0975	15.0199	15.0955	15.2645	16.0832	15.1845	15.1004	15.0992	15.0992	15.1156
4	1.0001	2.6110	1.0109	1.0000	1.3340	1.0000	1.0000	1.0000	1.0000	1.0000
5	3.1078	5.8974	2.3557	2.3817	5.3909	5.8808	2.3306	2.3300	2.3300	2.3335
6	12.3086	13.0095	12.3094	12.3113	13.3345	12.5776	12.3077	12.3064	12.3064	12.3084
7	19.7164	19.8577	19.7135	19.7087	21.1603	22.2880	19.7172	19.7156	19.7156	19.7182
8	17.3803	17.8562	17.4943	17.4565	19.0803	17.0532	17.4315	17.4304	17.4304	17.4307
Mass (kg)	384.2495	395.9368	384.1224	384.5962	413.1713	401.7247	383.6164	383.5887	383.5887	383.6674
$\sigma^{\max}$ *(MPa)*	56.6244	56.3334	56.5631	55.8492	53.0712	56.5071	56.6266	56.6310	56.6310	56.5735
$\delta^{\max}$ *(mm)*	7.7248	7.5735	7.7125	7.6974	7.1158	7.5889	7.7277	7.7282	7.7282	7.7257
f_1 *(Hz)*	15.3694	15.5518	15.3998	15.4016	15.9962	15.3797	15.3810	15.3804	15.3804	15.3819
Mean	410.9680	439.0713	388.1758	392.5535	447.9697	474.5781	383.7014	383.5939	383.6030	384.1060
SD	15.3732	27.4426	2.5528	7.9276	17.4846	36.0876	0.0692	0.0170	0.0386	0.5878

4.3.10 *Size Optimization of the 39-Bar 3D Truss with Continuous Cross Sections*

In this particular section, the focus is on comparing the performance of various optimization algorithms for the size optimization of a 39-bar 3D truss with a continuous section. A ground structure of the 39 elements 3D truss is shown in Fig. 4.10. Element cross-sectional areas are grouped into 11 sets in view of symmetry about *x-z* and *y-z* planes. This truss has been studied by Deb and Gulati (2001) and Luh and Lin (2008) with static constants but has never been investigated with simultaneous static and dynamic constraints. Thus, this problem is modified to study the effect of multi-load conditions, lumped masses, natural frequency constraint, and continuous area. Moreover, a constantly lumped mass of 500 kg is attached at top nodes (nodes 1 and 2) as presented in Fig. 4.10.

In this problem, the continuous cross-sectional areas of groups (G_i, where $i = 1,2, \ldots, 11$) are considered over [1, 50] cm^2. The design parameters such as material properties, loading, design variable bounds, constraints, etc. of this truss are tabulated in Table 4.19.

The results presented in this section are based on 100 independent runs of the optimization process, and the performance of the algorithms is summarized in Table 4.20. The table presents various parameters such as size variables, best mass, mean mass, SD of mass, stress, displacement, and frequency response for each algorithm. The best run in each case is the one that achieves the minimum mass while satisfying all the structural constraints. To evaluate the performance of the different evolutionary methods statistically, the mean and SD values of the structural mass are used. The best results obtained by ALO, DA, GWO, MVO, SCA, WOA,

Fig. 4.10 The 39-bar 3D truss

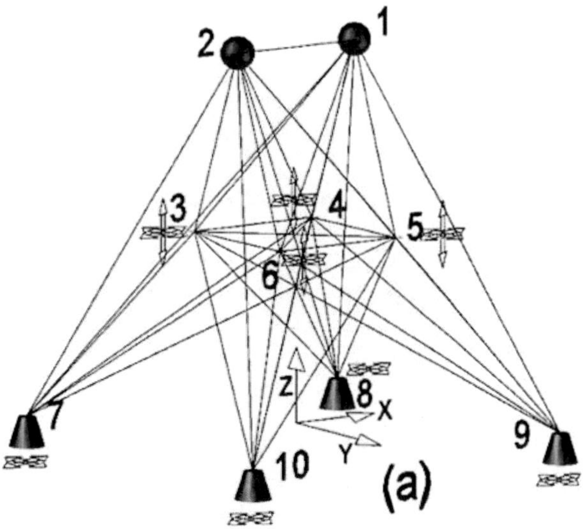

Table 4.19 Design parameters of the 39-bar 3D truss

Size variables: $X_i = G_i$, $i = 1, 2, \ldots, 11$; G = group number			
load condition:	$F_{z_1} = -222.685\ kN$		$F_{z_2} = -222.685\ kN$
Lumped masses at nodes 1 and 2 are 500 kg			
Stress and displacement constraints: $\sigma_i^{max} = 172.375$ MPa and $\delta_{x_j/y_j/z_j}^{max} = 8.9$ mm (0.0089 m) (for node, $j = 1, 2, \ldots, 6$ along x-, y-and z-axes)			
Natural frequency constraints: $f_1 \geq 15\ Hz$			
The continuous cross − sectional areas : $[A^{min}, A^{max}] = [1, 50]$ cm^2 ($[1, 50] * 10^{-4}$ m^2)			
Material properties: $E = 69\ GPa$ and $\rho = 2740 \text{kg/m}^3$			
Element grouping			
Group number	Element number (end nodes)	Group number	Element number (end nodes)
G_1	1(1,2)	G_7	22(3,7), 23(4,8), 24(5,9), 25(6,10)
G_2	2(1,4), 3(2,3), 4(1,5), 5(2,6)	G_8	26(5,7), 27(6,8), 28(3,9), 29(4,10)
G_3	6(2,5), 7(2,4), 8(1,3), 9(1,6)	G_9	30(3,5), 31(4,6)
G_4	10(3,6), 11(4,5),12(3,4), 13(5,6)	G_{10}	32(1,7), 33(1,10), 34(2,9), 35(2,8)
G_5	14(3,10), 15(6,7), 16(4,9), 17(5,8)	G_{11}	36(2,7), 37(2,10), 38(1,8), 39(1,9)
G_6	18(3,8), 19(4,7), 20(6,9), 21(5,10)		

HTS, TLBO, PVS, and SOS algorithms are 615.8052, 673.5938, 597.2903, 597.3578, 690.6195, 741.9855, 594.2252, 593.0043, 593.1909, and 594.6904 kg, respectively.

From the results presented in Table 4.20, it is observed that the TLBO algorithm performs the best with the lowest structural mass. The PVS and HTS algorithms come a close second and third, respectively, with slightly higher structural mass values.

Overall, this study provides valuable insights into the performance of various optimization algorithms for size optimization of a 39-bar 3D truss with a continuous section. The results can guide engineers in selecting the most effective algorithm for similar design problems, leading to more efficient and cost-effective designs.

Metaheuristic methods are used in solving complex optimization problems that may be difficult to solve using traditional methods. To evaluate the true performance of these methods, various measures are taken into account, including the average mass and standard deviation. These measures help to determine how well the algorithms are performing in terms of convergence rate and consistency.

In this context, the ALO, DA, GWO, MVO, SCA, WOA, HTS, TLBO, PVS, and SOS algorithms were evaluated based on their average mass and standard deviation. The mean mass values for these algorithms were found to be 695.9976, 788.9464,

Table 4.20 The 39-bar 3D truss size optimization (continuous section)

Element group no	ALO	DA	GWO	MVO	SCA	WOA	HTS	TLBO	PVS	SOS
1	19.7562	1.0000	1.4342	1.6581	13.1215	24.9612	1.0057	1.0004	1.0004	1.0000
2	5.3424	11.2481	4.3091	4.4174	7.3596	6.5045	4.2573	4.1786	4.1893	4.2956
3	6.5211	9.3512	6.8826	6.8500	8.3785	7.1002	6.7415	6.7201	6.7367	6.8006
4	1.0000	10.7596	1.0675	1.0000	1.0000	7.0029	1.0116	1.0001	1.0000	1.0000
5	6.3335	6.9041	5.7851	5.8087	6.4924	9.9371	5.7036	5.6590	5.6599	5.7494
6	7.3768	8.4436	7.6611	7.7264	10.9808	16.2284	7.5844	7.5552	7.5611	7.6784
7	8.3705	18.1029	8.1445	8.1620	15.1861	11.8443	8.1613	8.0723	8.0970	8.1723
8	4.8208	6.5174	5.2878	5.1292	5.7079	6.3042	5.1768	5.1410	5.1490	5.1297
9	1.0000	1.0162	1.3536	1.0568	1.2734	11.7476	1.0000	1.0001	1.0022	1.0000
10	28.8049	29.5084	30.4378	30.4220	34.0970	31.8139	30.6177	30.6140	30.6085	30.6193
11	27.0090	21.2640	24.5239	24.6317	24.4255	24.9464	24.3397	24.3532	24.3473	24.2997
Mass (kg)	615.8052	673.5938	597.2903	597.3578	690.6195	741.9855	594.2252	593.0043	593.1909	594.6904
σ^{max} *(MPa)*	2.2239	46.6185	33.6784	32.6895	23.2409	26.0246	35.8148	35.7086	35.7271	35.7244
δ^{max} *(mm)*	23.9497	2.1168	2.2505	2.2477	1.9349	2.0277	2.2538	2.2565	2.2560	2.2523
f_1 *(Hz)*	31.3792	32.3631	31.1236	31.1726	32.9503	32.1855	31.1102	31.0851	31.0868	31.1120
Mean	695.9976	788.9464	603.8533	630.1323	749.1221	880.1007	600.9179	596.3765	597.1080	603.2365
SD	45.8796	75.8523	3.8673	23.7435	31.0764	82.3005	7.7990	7.3568	3.2797	7.3270

603.8533, 630.1323, 749.1221, 880.1007, 600.9179, 596.3765, 597.1080, and 603.2365 kg, respectively. The corresponding standard deviations were 45.8796, 75.8523, 3.8673, 23.7435, 31.0764, 82.3005, 7.7990, 7.3568, 3.2797, and 7.3270, respectively.

The TLBO algorithm was found to have the best mean mass value, while the PVS algorithm was the best in achieving the lowest standard deviation. Additionally, the PVS algorithm achieved the second-best mean mass value, whereas the TLBO algorithm demonstrated the second-best standard deviation.

Based on these results, it can be concluded that the SOS and PVS algorithms are the best in terms of convergence rate and consistency for this particular optimization problem. Therefore, they can be used effectively to solve similar problems and obtain optimal solutions.

4.3.11 A Comprehensive Analysis

Table 4.21 presents a comprehensive analysis of the best results obtained from ten benchmark problems using proposed algorithms in 100 independent runs. The table showcases minimum mass and offers a detailed evaluation of overall, average, and individual ranks, as well as the best and worst results for each algorithm.

After analyzing the results in Table 4.21, it was discovered that the PVS algorithm achieved the minimum mass and outperformed all other algorithms in terms of overall rank. The TLBO, GWO, and HTS algorithms followed with the second-, third-, and fourth-best performances, respectively.

Moreover, the PVS algorithm provided the highest number of best solutions, including five best solutions, four second-best solutions, and one third-best solution without any worse solutions. Similarly, the TLBO algorithm had three best solutions, four second-best solutions, and two third-best solutions, with no worse solutions.

However, the WOA algorithm presented the worst solutions for eight problems, emphasizing the significance of choosing the most appropriate algorithm for specific problems to achieve optimal results.

Overall, the analysis of Table 4.21 provides valuable insights into the performance of proposed algorithms for ten benchmark problems, aiding researchers and practitioners in selecting the most suitable algorithm for similar problems and improving the optimization process' efficiency.

Table 4.22 presents the mean mass achieved by the proposed algorithms for ten benchmark problems in 100 independent runs. Like Table 4.22, it displays the overall rank, average rank, individual ranks, and best and worst results for each algorithm.

Upon reviewing the results in Table 4.22, it was observed that the GWO algorithm outperformed all other algorithms in terms of overall rank to obtain the mean mass. It was followed by the PVS, SOS, and HTS algorithms, respectively.

Table 4.21 Minimum mass obtained using size optimization with continuous sections

	ALO	DA	GWO	MVO	SCA	WOA	HTS	TLBO	PVS	SOS
10-bar truss	562.7	661.2	556	562.5	666.2	678.5	555.9	554.4	**554.3**	556.3
14-bar truss	807.7	1014	744.9	762.7	873.4	1055.8	745.7	**738.9**	739.6	749
15-bar truss	128	146	113	114.9	152.6	155.9	113.8	112.8	**112.6**	115.1
24-bar truss	233.5	247.6	199.8	208.7	274.5	316.9	223.6	203.3	**198.6**	202
20-bar truss	222.5	289.8	**205.1**	213.9	310.6	337.2	225.5	210.2	209.3	212.1
72-bar 3D truss	484.4	608.9	471	476.4	575.1	634	473.3	471	**470.6**	475.3
39-bar truss	602.3	683.5	**517.9**	540	720	719.1	552.9	522.2	523.6	547.5
45-bar truss	570	716.5	407.2	458.1	597.7	745.6	541.6	424	411.6	425
25-bar 3D truss	384.2	395.9	384.1	384.6	413.2	401.7	**383.6**	**383.6**	**383.6**	383.7
39-bar 3D truss	615.8	673.6	597.3	597.4	690.6	742	594.2	**593.0**	593.2	594.7
Overall rank	7	8	3	6	9	10	4	2	1	5
Count best	0	0	3	0	0	0	1	3	5	0
Count 2nd best	0	0	2	0	0	0	0	4	4	0
Count 3rd best	0	0	2	0	0	0	2	2	1	1
Count worst	0	0	0	0	2	8	0	0	0	0

Note: Bold indicates the best solution

Moreover, the GWO algorithm had eight best solutions without any worse solution count. Similarly, the PVS algorithm provided one best solutions, six second-best solutions, and three third-best solution, with no worse solutions. Conversely, the SOS algorithm had three best solutions, one second-best solution, four third-best solutions, and no worse solutions.

In contrast, the WOA algorithm had the highest number of worst solutions for all ten benchmark problems. These results suggest that the performance of the WOA, DA, and SCA algorithms may not be optimal for certain types of problems.

Table 4.22 Mean mass obtained using size optimization with continuous sections

	ALO	DA	GWO	MVO	SCA	WOA	HTS	TLBO	PVS	SOS
10-bar truss	692.5	821.1	624.9	635.3	771.5	1011.2	625.1	617.3	**615.8**	619
14-bar truss	1207.6	1610.9	**831.3**	955.1	1071.1	2140	940	947.5	869.2	926.3
15-bar truss	159.2	211.1	114	121.9	177.8	246.6	115.6	115.9	**112.8**	117.7
24-bar truss	300.5	390.5	**203.7**	236.8	331.8	468.6	244.1	227.8	205.5	217
20-bar truss	319.6	394	**211.3**	247.2	386.9	482.3	237.9	242.4	222.3	238.8
72-bar 3D truss	759.8	1030.2	473.3	537.1	636.1	1100.3	480.6	478.9	**472.2**	484.8
39-bar truss	726.5	886.6	**529.2**	596.6	797.3	1174.6	604.7	584.3	549.9	575.6
45-bar truss	707.8	969.5	**414.5**	532.9	689.1	1238.7	614.1	522.6	452.1	469.3
25-bar 3D truss	411	439.1	388.2	392.6	448	474.6	383.7	**383.6**	**383.6**	384.1
39-bar 3D truss	696	788.9	603.9	630.1	749.1	880.1	600.9	**596.4**	597.1	603.2
Overall rank	7	9	2	6	8	10	5	3	1	4
Count best	0	0	5	0	0	0	0	2	4	0
Count 2nd best	0	0	2	0	0	0	0	1	6	0
Count 3rd best	0	0	0	0	0	0	4	1	0	5
Count worst	0	0	0	0	0	10	0	0	0	0

Note: Bold indicates the best solution

4.3.12 The Friedman Rank Test

The results obtained from the experimentation reveal that the PVS, GWO, and TLBO algorithms exhibit better performance compared to the other algorithms under consideration. These algorithms demonstrate promising results in terms of their ability to optimize the given test problems. However, in order to establish a clear ranking among all the algorithms, it is important to conduct statistical tests that can provide a robust and objective evaluation.

Draa proposes the use of the Wilcoxon signed-rank test as an alternative method for comparing two related sample algorithms. This test is useful for making pairwise comparisons between algorithms. However, it may not be suitable for ranking all the algorithms when multiple comparisons are involved. In contrast, the Friedman rank test is a nonparametric statistical test that allows for simultaneous comparison of multiple algorithms and provides individual ranks for each algorithm based on their performance.

Therefore, in this study, the Friedman rank test is utilized to assess the performance of the ALO, DA, GWO, MVO, SCA, WOA, HTS, TLBO, PVS, and SOS algorithms for the continuous section test problems and presented in Table 4.23. The mean mass and success rate of each algorithm are considered as the evaluation criteria. The statistical analysis is performed by conducting Friedman's rank test and allocating respective ranks to the optimizers based on their performance.

Table 4.23 The Friedman test

	ALO	DA	GWO	MVO	SCA	WOA	HTS	TLBO	PVS	SOS
10-bar truss	649	859	393	431	839	969	372	334	289	365
14-bar truss	709	886.5	261	460	606	955.5	441	436	320	425
15-bar truss	728	907	229	562	802	963	377	337	101	494
24-bar truss	719	898	147	481	796	980	545	418	181	335
20-bar truss	705	863	113	474	845	977	431	435	240	417
72-bar 3D truss	400	464	96	292	357	472	200	161	69	239
39-bar truss	365	438	65	237	400	493	248	202	117	185
45-bar truss	370	457	50	230	362	490	301	216	122	152
25-bar 3D truss	353	420	269	287	445	474	149	81	73	199
39-bar 3D truss	361	436	213	285	413	487	160	92	116	187
Overall Friedman value	5359	6628.5	1836	3739	5865	7260.5	3224	2712	1628	2998
Overall Friedman rank	7	9	2	6	8	10	5	3	1	4

Friedman's rank-sum test is applied as a comprehensive statistical review of all the considered optimizers. This test allows for a thorough comparison of the algorithms and helps in identifying the top-performing algorithms based on their overall ranks. The significance level is set at 95% to ensure a reliable and statistically significant evaluation.

Based on the results obtained from Friedman's rank test at the 95% significance level, it is observed that the PVS algorithm outperforms the other algorithms, securing the top rank. Following closely, the GWO and TLBO algorithms are ranked second and third, respectively, based on their performance in optimizing the test problems. This information provides valuable insights into the relative performance of the algorithms and helps in identifying the most effective algorithms for the given optimization tasks.

4.4 Results and Discussion on Size Optimization with Discrete Cross Sections

This section utilizes ten benchmark problems to evaluate the efficacy of algorithms for optimizing discrete cross sections. These benchmarks comprise of diverse truss structures including the 10-bar, 14-bar, 15-bar, 24-bar, 20-bar, 72-bar 3D, 39-bar, 45-bar, 25-bar 3D, and 39-bar 3D trusses. The suggested algorithms are employed to address each truss problem through 100 independent runs, taking into account the stochastic nature of metaheuristics. The population size for all the problems is fixed at 50, and the total number of function evaluations (FEs) for each run is 20,000. The key objective is to gauge the capabilities of the selected algorithms in resolving discrete size optimization problems.

4.4.1 Size Optimization of the 10-Bar Truss with Discrete Cross Sections

This section is dedicated to the optimization of a 10-bar truss, aiming to minimize its size using 10 discrete size variables. Figure 4.11 illustrates the ground structure of the benchmark truss, considering the applied load and boundary conditions. Numerous researchers have examined this problem, focusing on static constraints such as elemental stress and nodal displacement. Some notable studies include those conducted by Hajela and Lee (1995), Deb and Gulati (2001), Richardson et al. (2012), and Miguel et al. (2013). Additionally, several researchers, such as Wei et al. (2005), Gomes (2011), Wei at al. (2011), Kaveh and Zolghadr (2011), Miguel and Miguel (2012), Kaveh and Zolghadr (2012), and Farshchin et al. (2016), have also investigated the optimization of both size and shape of this truss with respect to dynamic constraints, specifically the natural frequencies.

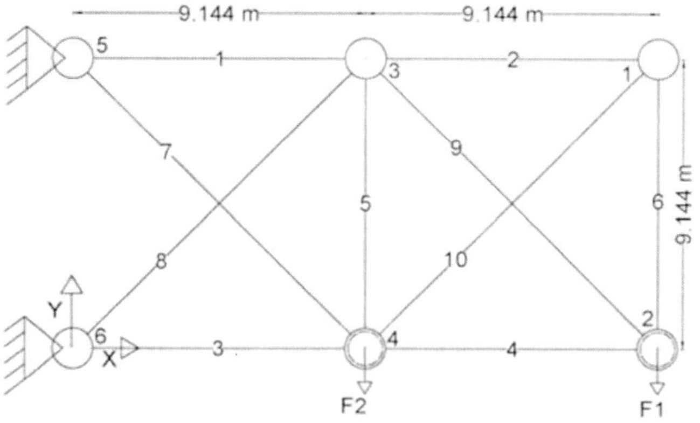

Fig. 4.11 The 10-bar truss

Table 4.24 Design parameters of the 10-bar truss

Size variables: $X_i = A_i$, $i = 1, 2, \ldots, 10$	
Multiple load conditions	
Condition 1 : $F_1 = 44.537\ KN$, $F_2 = 44.537\ KN$	Condition 2 : $F_1 = 44.537\ KN$, $F_2 = 0\ KN$
Lumped masses on node 2 and 4 are 500 kg	
Stress and displacement constraints: $\sigma_i^{\max} = 172.43$ MPa and $\delta_{y_5\&y_6}^{\max} = 100$ mm (0.1 m)	
Natural frequency constraint: $f_1 \geq 4$ Hz	
The discrete cross − sectional areas : [1,100] cm^2in increments of 1 cm^2	
Material properties: $E = 69\ GPa$ and $\rho = 2740\text{kg/m}^3$	

In previous studies, researchers have explored this problem separately with either static constraints, such as stress, buckling, and displacement, or dynamic constraints, specifically natural frequencies. However, they have not examined the problem considering both static and dynamic constraints simultaneously. In this particular study, the problem has been modified to incorporate various additional factors. These include the consideration of multiple load conditions, the inclusion of lumped masses, the imposition of natural frequency constraints, and the introduction of discrete cross-sectional area variables. By incorporating these modifications, the study aims to capture a more comprehensive understanding of the optimization problem and its practical implications. In this problem, the discrete cross-sectional areas (A_i, where $i = 1, 2, \ldots, 10$) are considered over [1, 100] cm^2 in increments of 1 cm^2. Thus, discrete design variables may take any integer values within [1, 100] cm^2. Table 4.24 shows all details of material properties, design variables, loading conditions, lumped mass, and other important data of the test problem.

Table 4.25 displays the outcomes of the size optimization of the truss, with discrete cross sections, using various algorithms in 100 independent runs. The best run achieved the minimum mass value while satisfying all the structural constraints. The table also presents the mean and standard deviation (SD) values of the structural mass, which will be utilized to statistically evaluate the performance of different evolutionary methods.

Table 4.25 Size optimization of the 10-bar truss (discrete section)

Element no.	ALO	DA	GWO	MVO	SCA	WOA	HTS	TLBO	PVS	SOS
1	3	3	3	3	3	3	3	3	3	3
2	1	16	4	1	3	1	1	1	1	1
3	73	68	64	64	69	65	64	64	64	64
4	35	15	15	33	39	34	33	33	33	33
5	38	16	17	35	43	65	34	34	34	34
6	1	3	4	1	2	2	1	1	1	1
7	21	23	21	22	21	23	22	22	22	22
8	1	1	1	1	2	1	1	1	1	1
9	12	1	1	6	12	10	6	6	6	6
10	22	58	57	28	31	27	28	28	28	28
Mass (kg)	606.7458	627.2499	581.5438	575.2127	662.2220	672.0658	572.7073	572.7073	572.7073	572.7073
σ^{max} (MPa)	148.4696	146.9063	148.1106	148.1666	148.2479	147.1826	148.1962	148.1962	148.1962	148.1962
δ^{max} (mm)	55.5151	62.1029	68.4344	65.3560	53.5360	55.6126	65.3895	65.3895	65.3895	65.3895
f_1 (Hz)	6.1852	5.8575	5.5356	5.7362	6.1955	6.0974	5.7356	5.7356	5.7356	5.7356
Mean	725.9935	845.0171	651.5703	639.6097	770.1767	999.2790	631.7011	649.0318	639.0650	630.3029
SD	62.3234	122.5552	23.7013	37.5427	47.7770	167.5741	37.6028	58.4325	34.8554	40.5673

Based on the table, the ALO, DA, GWO, MVO, SCA, WOA, HTS, TLBO, PVS, and SOS algorithms all produce excellent outcomes with minimum mass values of 606.7458, 627.2499, 581.5438, 575.2127, 662.2220, 672.0658, 572.7073, 572.7073, 572.7073, and 572.7073 kg, respectively. Additionally, the results show that the HTS, TLBO, PVS, and SOS algorithms all achieve a similar optimum mass of 572.7073 kg.

To evaluate the actual performance of the metaheuristic methods, the mean mass and its standard deviation (SD) are considered. Based on the available results, it is evident that the ALO, DA, GWO, MVO, SCA, WOA, HTS, TLBO, PVS, and SOS algorithms have mean mass values of 725.9935, 845.0171, 651.5703, 639.6097, 770.1767, 999.2790, 631.7011, 649.0318, 639.0650, and 630.3029 kg, respectively, with corresponding SD values of 62.3234, 122.5552, 23.7013, 37.5427, 47.7770, 167.5741, 37.6028, 58.4325, 34.8554, and 40.5673.

The GWO and PVS algorithms perform the best and second-best, respectively, achieving the lowest mean mass and SD values. Therefore, for this design problem, the GWO and PVS algorithms are the superior algorithms in terms of convergence rate and consistency.

4.4.2 Size Optimization of the 14-Bar Truss with Discrete Cross Sections

The subsequent passage discusses the optimization of a 14-bar truss using various evolutionary algorithms. Figure 4.12 presents the ground structure of the second benchmark truss, referred to as the 14-bar truss, along with its corresponding load and boundary conditions. Previous studies by Tang et al. (2005), Rahami et al. (2008), and Miguel et al. (2013) have focused on optimizing this problem using TSS optimization, considering 15 discrete cross-sectional areas and 8 continuous shape

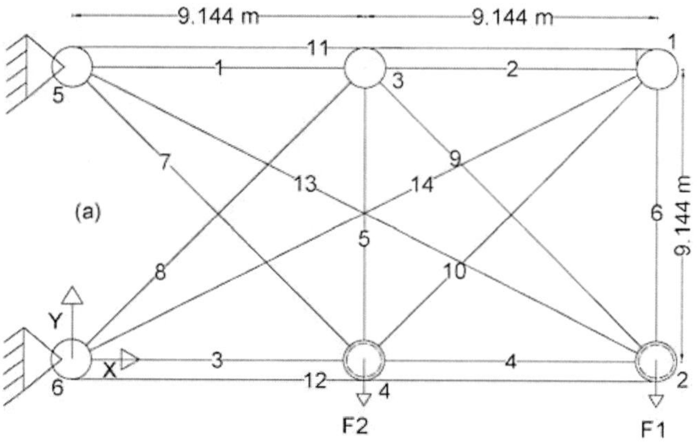

Fig. 4.12 The 14-bar truss

Table 4.26 Design parameters of the 14-bar truss

Size variables: $X_i = A_i$, $i = 1, 2, \ldots, 14$	
Multiple load conditions	
Condition 1 : $F_1 = 44.537$ KN, $F_2 = 44.537$ KN	Condition 2 : $F_1 = 44.537$ KN, $F_2 = 0$ KN
Lumped mass on node 2 and 4 are 500 kg	
Stress and displacement constraints: $\sigma_i^{max} = 172.43$ MPa and $\delta_{y_5\&y_6}^{max} = 100$ mm (0.1 m)	
Natural frequency constraint: $f_1 \geq 4$ Hz	
The discrete cross − sectional areas : [1,100] cm^{2}in increments of 1 cm^2	
Material properties: $E = 69$ GPa and $\rho = 2740$kg/m^3	

variables while adhering to static constraints. However, this problem has not been addressed with the simultaneous consideration of static and dynamic constraints. A comprehensive set of details pertaining to the design variables, load conditions, design constraints, and other pertinent data for the test problem are provided in Table 4.26. In this problem, the discrete cross-sectional areas (A_i, where $i = 1, 2, \ldots, 14$) are considered over [1, 100] cm^2 in increments of 1 cm^2. Thus, discrete design variables may take any integer values within [1, 100] cm^2.

The truss has 14 discrete size variables, and the results of 100 independent runs of each algorithm are presented in Table 4.27. The best run is the one that gives the minimum mass value while satisfying all structural constraints. The mean and standard deviation values of the structural mass are given to evaluate the performance of the algorithms statistically.

Based on the table, the ALO, DA, GWO, MVO, SCA, WOA, HTS, TLBO, PVS, and SOS algorithms all produce excellent outcomes with minimum mass values of 801.7392, 893.7672, 748.9465, 761.0439, 875.5230, 1278.1950, 748.9465, 755.4251, 748.9465, and 751.4520 kg, respectively. Additionally, the results show that the PVS and GWO algorithms all achieve a similar optimum mass of 748.9465 kg.

To evaluate the actual performance of the metaheuristic methods, the mean mass and its standard deviation (SD) are considered. Based on the available results, it is evident that the ALO, DA, GWO, MVO, SCA, WOA, HTS, TLBO, PVS, and SOS algorithms have mean mass values of 1226.3418, 1609.5341, 847.6027, 930.7329, 1034.1764, 2203.6815, 928.1314, 1038.4392, 881.0885, and 924.5203 kg, respectively, with corresponding SD values of 230.1394, 281.6952, 143.1742, 146.2338, 106.1154, 449.4998, 168.5480, 206.4794, 125.5877, and 153.6273.

The average mass and its standard deviation are used to assess the true performance of the metaheuristic methods. The results indicate that the PVS, GWO, and SCA algorithms perform better, achieving the lowest mass mean and SD values. Therefore, these are the most suitable algorithms for this design problem in terms of both convergence rate and consistency.

In summary, this section presents the optimization of the 14-bar truss using different evolutionary algorithms. The results show that the PVS and GWO algorithms are the most effective algorithm for this problem.

Table 4.27 Size optimization of the 14-bar truss (discrete section)

Element no.	ALO	DA	GWO	MVO	SCA	WOA	HTS	TLBO	PVS	SOS
1	1	2	1	2	1	28	1	1	1	1
2	2	7	2	4	4	1	2	2	2	2
3	66	57	61	62	70	47	61	64	61	60
4	3	1	1	1	1	26	1	1	1	1
5	15	36	15	16	21	1	15	16	15	16
6	8	1	7	8	6	11	7	5	7	4
7	17	22	17	19	22	9	17	17	17	18
8	1	1	1	1	4	47	1	1	1	1
9	1	5	1	1	1	1	1	1	1	1
10	62	38	57	57	69	57	57	56	57	56
11	2	1	2	1	3	1	2	2	2	2
12	45	66	42	41	43	68	42	43	42	44
13	1	1	1	1	1	3	1	1	1	1
14	1	5	1	1	2	35	1	1	1	1
Mass (kg)	801.7392	893.7672	748.9465	761.0439	875.5230	1278.1950	748.9465	755.4251	748.9465	751.4520
σ^{max}	168.6856	156.8080	169.2601	149.3530	141.4962	167.3240	169.2601	169.5458	169.2601	169.0080
δ^{max}	62.3148	60.4980	63.5843	58.4427	52.2368	52.9694	63.5843	65.5318	63.5843	66.5901
f_1	5.4273	5.6035	5.4496	5.6695	5.9811	5.5102	5.4496	5.4069	5.4496	5.3800
Mean	1226.3418	1609.5341	847.6027	930.7329	1034.1764	2203.6815	928.1314	1038.4392	881.0885	924.5203
SD	230.1394	281.6952	143.1742	146.2338	106.1154	449.4998	168.5480	206.4794	125.5877	153.6273

4.4.3 Size Optimization of the 15-Bar Truss with Discrete Cross Sections

Figure 4.13 displays the ground structure of the 15-bar truss with discrete cross sections. This specific problem has been adapted from Rahami et al. (2008) and Ahrari et al. (2014). Previous investigations conducted by the researchers focused solely on static constraints such as stress, buckling, and displacement. However, they did not explore the effects of simultaneous static and dynamic constraints, specifically the natural frequency. To address this gap, the problem has been modified to incorporate multi-load conditions, lumped masses, natural frequency constraints, and discrete cross-sectional area. All design considerations and material properties can be found in Table 4.28. In this problem, the discrete cross-sectional areas (A_i, where $i = 1, 2, \ldots, 15$) are considered over [1, 100] cm^2 in increments of 1 cm^2. Thus, discrete design variables may take any integer values within [1, 100] cm^2.

Table 4.29 displays the results obtained from the size optimization of the 15-bar truss with discrete cross sections using various algorithms in 100 independent runs. The best run is determined based on the algorithm that yields the lowest mass value while satisfying all the structural constraints. The mean and standard deviation (SD) values of the structural mass are also provided to statistically evaluate the performance of different evolutionary methods.

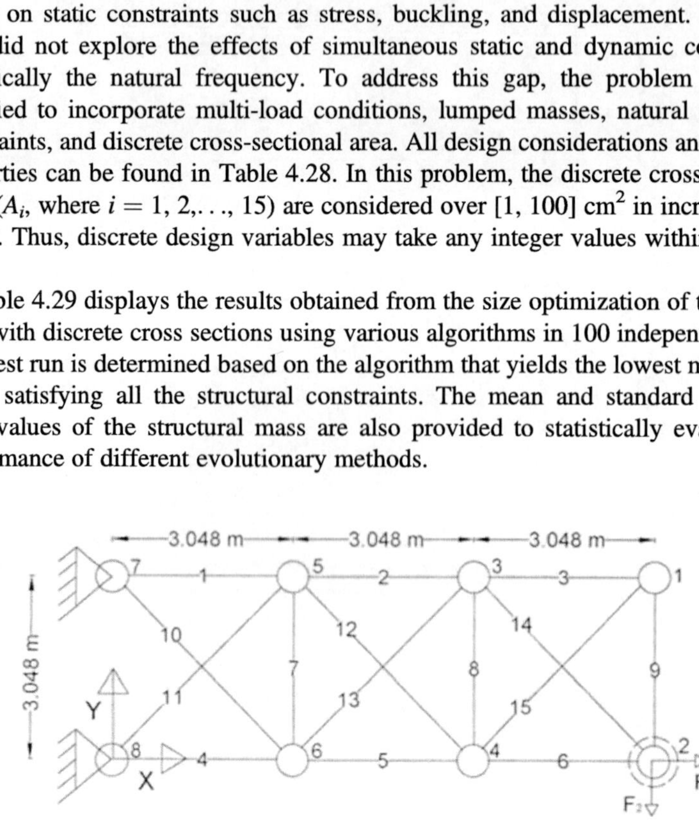

Fig. 4.13 The 15-bar truss

Table 4.28 Design parameters of the 15-bar truss

Size variables: $X_i = A_i$, $i = 1, 2, \ldots, 15$	
Multiple load conditions	
Condition 1 : $F_1 = -44.537$ KN, $F_2 = 0$ KN	Condition 2 : $F_1 = 0$ KN, $F_2 = 4.4537$ KN
Lumped mass on node 2 is 500 kg	
Stress and displacement constraints: $\sigma_i^{max} = 172.43$ MPa and $\delta_{x_2 \& y_2}^{max} = 100$ mm (0.1 m)	
Natural frequency constraint: $f_1 \geq 4$ Hz	
The discrete cross $-$ sectional areas : $[1, 50]$ cm^{2}in increments of 1 cm^2	
Material properties: $E = 69$ GPa and $\rho = 2740$kg/m^3	

Table 4.29 Size optimization of the 15-bar truss (discrete section)

Element no.	ALO	DA	GWO	MVO	SCA	WOA	HTS	TLBO	PVS	SOS
1	8	7	7	6	9	12	7	7	6	6
2	7	5	5	5	5	5	5	5	5	5
3	2	4	3	3	4	3	3	3	3	3
4	13	17	13	13	12	12	13	13	13	13
5	13	13	13	13	25	13	13	13	13	13
6	19	22	13	14	13	14	13	13	14	14
7	3	2	1	1	3	1	1	1	1	1
8	1	2	1	1	1	1	1	1	1	1
9	2	4	3	3	9	28	3	3	3	3
10	5	6	5	6	7	10	5	5	6	6
11	6	4	4	4	8	16	4	4	4	4
12	4	7	6	5	11	6	6	6	5	5
13	14	9	2	2	2	1	2	2	2	2
14	6	5	5	6	9	7	5	5	6	6
15	2	15	4	3	6	6	4	4	3	3
Mass (kg)	140.4904	157.8014	119.9821	119.9821	158.4339	168.6584	119.9821	119.9821	119.9821	119.9821
σ^{max} *(MPa)*	33.5020	33.2921	34.0539	34.0539	34.6504	35.1864	34.0539	34.0539	34.0539	34.0539
δ^{max} *(mm)*	6.7121	5.8546	7.1885	7.1885	6.2313	7.0991	7.1885	7.1885	7.1885	7.1885
f_1 *(Hz)*	6.3498	6.3391	5.8522	5.8522	6.5552	6.3752	5.8522	5.8522	5.8522	5.8522
Mean	171.5412	216.9791	120.4372	127.9900	176.9352	243.8100	122.1254	142.2685	122.1096	122.3245
SD	16.5082	34.9411	0.6483	6.2701	9.2079	48.1363	1.9165	16.2378	1.7897	2.4141

The ALO, DA, GWO, MVO, SCA, WOA, HTS, TLBO, PVS, and SOS algorithms report the best results with minimum mass values of 140.4904, 157.8014, 119.9821, 119.9821, 158.4339, 168.6584, 119.9821, 119.9821, 119.9821, and 119.9821 kg, respectively. The GWO, MVO, HTS, PVS, TLBO, and SOS algorithms produce the same solution of 119.9821 kg, which is the lowest mass value. Therefore, these algorithms are the top performers for this design problem.

After analyzing the results presented in Table 4.29, it can be concluded that to assess the performance of the metaheuristic methods, the mean mass and its SD are crucial. The algorithms, ALO, DA, GWO, MVO, SCA, WOA, HTS, TLBO, PVS, and SOS, achieved mean mass values of 171.5412, 216.9791, 120.4372, 127.9900, 176.9352, 243.8100, 122.1254, 142.2685, 122.1096, and 122.3245 kg, respectively, with corresponding SD values of 16.5082, 34.9411, 0.6483, 6.2701, 9.2079, 48.1363, 1.9165, 16.2378, 1.7897, and 2.4141. It is evident that the GWO and PVS algorithms perform the best as they achieve the lowest mean mass and SD values. Therefore, for this design problem, the GWO and PVS algorithms are considered to be the better options, as they provide both a good convergence rate and consistency.

4.4.4 Size Optimization of the 24-Bar Truss with Discrete Cross Sections

This section focuses on the optimization of the 24-bar truss using proposed algorithms, employing 24 discrete size variables. The ground structure of the benchmark problem, the 24-bar truss, is illustrated in Fig. 4.14. Initially investigated by Ringertz (1986) using a branch and bound algorithm for topology optimization with static constraints, Xu et al. (2003) later utilized it for testing a 1D search method. Kaveh

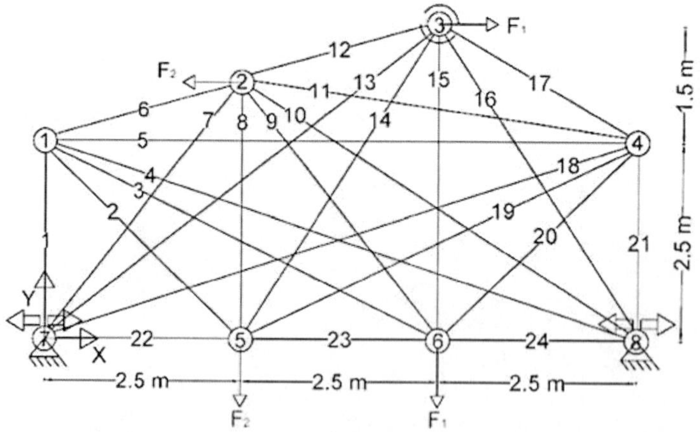

Fig. 4.14 The 24-bar truss

Table 4.30 Design parameters of the 24-bar truss

Size variables: $X_i = A_i$, $i = 1, 2, \ldots, 24$	
Multiple load conditions:	
Condition 1 : $F_1 = 50$ KN, $F_2 = 0$ KN	Condition 2 : $F_1 = 0$ KN, $F_2 = 50$ KN
Lumped mass on node 3 is 500 kg	
Stress and displacement constraints: $\sigma_i^{max} = 172.43$ MPa and $\delta_{y_5 \& y_6}^{max} = 10$ mm $\left(10^{-2}\ \text{m}\right)$	
Natural frequency constraint: $f_1 \geq 30$ Hz	
The discrete cross − sectional areas : $[1, 40]$ cm^2 in increments of 1 cm^2	
Material properties: $E = 69$ GPa and $\rho = 2740\text{kg/m}^3$	

and Zolghadr (2013) employed the truss problem to assess the search performance of charged system search and particle swarm optimization algorithms, considering continuous design variables. The truss is subjected to two external load cases as specified in Table 4.30. The cross-sectional areas of the truss elements (24-bars) are considered as discrete variables. In this problem, the discrete cross-sectional areas (A_i, where $i = 1, 2, \ldots, 24$) are considered over [1, 40] cm^2 in increments of 1 cm^2. Thus, discrete design variables may take any integer values within [1, 40] cm^2.

Additionally, a nonstructural lumped mass of 500 kg is applied at node 3. The table includes information about nodal masses, material properties, and other relevant data.

Table 4.31 displays the outcomes of the size optimization of the 24-bar truss with discrete cross sections, using various algorithms in 100 independent runs. The results indicate that the best run delivers the lowest mass value while satisfying all structural constraints. The mean and standard deviation (SD) values of the structural mass are provided to statistically evaluate the performance of different evolutionary techniques. Among the ALO, DA, GWO, MVO, SCA, WOA, HTS, TLBO, PVS, and SOS algorithms, the best reported results are 259.7219, 265.8764, 203.0364, 212.5595, 297.7128, 342.7003, 227.3010, 210.8793, 202.9632, and 206.3402 kg, respectively. PVS produces the lowest structural mass outcome, with GWO and SOS algorithms providing similar solutions as the second-best alternatives.

To assess the statistical performance of the metaheuristic methods, we analyze the average mass and its SD. The mean mass values and corresponding SDs achieved by the ALO, DA, GWO, MVO, SCA, WOA, HTS, TLBO, PVS, and SOS algorithms were 321.9689/36.5382, 386.2188/46.7519, 206.3391/2.2034, 242.4514/19.4787, 329.5287/17.8495, 502.5487/64.9527, 250.7054/12.7640, 275.0739/38.6776, 217.8674/7.6862, and 223.5708/14.5712 kg, respectively. Therefore, based on the analysis of the average mass and SD, the GWO and PVS algorithms are the best and second-best options, respectively, for this optimization problem as they achieve the lowest mean mass and the lowest SD.

Table 4.31 Size optimization of the 24-bar truss (discrete section)

Element no.	ALO	DA	GWO	MVO	SCA	WOA	HTS	TLBO	PVS	SOS
1	13	9	5	10	6	28	7	6	5	7
2	3	1	1	3	1	5	2	1	1	2
3	1	1	1	1	1	2	1	1	1	1
4	6	6	8	8	7	6	9	8	9	8
5	1	1	1	1	1	1	1	1	1	1
6	6	5	5	8	18	10	6	5	6	8
7	19	21	18	15	33	28	18	18	17	15
8	11	4	4	3	12	21	6	4	3	2
9	1	1	1	1	7	1	1	1	1	1
10	8	14	10	10	27	8	13	9	11	10
11	1	2	1	1	1	11	1	1	1	1
12	1	1	1	1	1	1	2	1	1	1
13	16	13	13	13	13	12	12	13	13	13
14	1	2	1	1	1	13	1	1	1	1
15	10	9	5	5	12	15	4	7	3	5
16	17	14	19	19	23	17	19	20	18	20
17	21	24	7	7	33	9	14	7	8	7
18	5	8	8	9	5	16	8	9	9	10
19	1	1	1	1	1	8	1	1	1	1
20	8	9	2	2	4	13	2	1	2	1
21	12	28	8	7	9	16	11	16	7	7
22	30	4	1	4	1	1	7	3	1	4
23	1	23	2	7	3	5	8	1	1	1
24	16	9	4	3	1	22	2	1	3	2

Mass (kg)	259.7219	265.8764	203.0364	212.5595	297.7128	342.7003	227.3010	210.8793	202.9632	206.3402
$\sigma^{\max}$ (*MPa*)	38.6093	88.2618	95.8324	98.0302	37.7787	26.6904	82.8259	95.7590	119.8736	140.8183
$\delta^{\max}$ (*mm*)	2.5563	4.9471	5.4940	5.3726	2.4765	1.8511	5.4254	5.4517	6.6224	7.3313
f_1 (*Hz*)	32.3845	30.9565	30.0028	30.0729	30.0818	34.9167	30.8352	30.0446	30.0253	30.0069
Mean	321.9689	386.2188	206.3391	242.4514	329.5287	502.5487	250.7054	275.0739	217.8674	223.5708
SD	36.5382	46.7519	2.2034	19.4787	17.8495	64.9527	12.7640	38.6776	7.6862	14.5712

4.4.5 Size Optimization of the 20-Bar Truss with Discrete Cross Sections

Figure 4.15 represents the ground structure of the 20-bar truss with discrete cross sections. This particular benchmark problem was initially introduced for topology and size optimization by Xu et al. (2003) and subsequently examined by Kaveh and Zolghadr (2013) for continuous optimization. In this problem, the discrete cross-sectional areas (A_i, where $i = 1, 2, \ldots, 20$) are considered over [1, 100] cm^2 in increments of 1 cm^2. Thus, discrete design variables may take any integer values within [1, 100] cm^2. All pertinent information regarding design variables, loading conditions, structural safety constraints, and other relevant data for the test problem can be found in Table 4.32. It is worth noting that no lumped masses have been incorporated into the truss for this specific problem.

Table 4.33 provides a comparison of the results obtained from various metaheuristic optimization algorithms applied to solve the 20-bar truss size problem with discrete cross-sections. The minimum mass value while satisfying all the structural constraints was obtained from the best run. The mean and SD values of the structural mass were calculated and will be used to statistically evaluate the performance of different evolutionary methods. The best results reported for the ALO, DA, GWO, MVO, SCA, WOA, HTS, TLBO, PVS, and SOS algorithms were 246.1333, 296.2283, 207.2186, 218.5386, 348.9370, 343.9499, 224.5115,

Fig. 4.15 The 20-bar truss

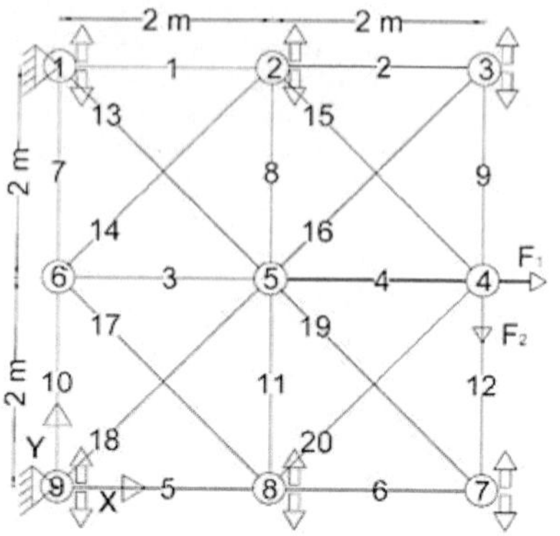

Table 4.32 Design parameters of the 20-bar truss

Size variables: $X_i = A_i$, $i = 1, 2, \ldots, 20$	
Multiple load conditions	
Condition 1 : $F_1 = 500$ KN, $F_2 = 0$ KN	Condition 2 : $F_1 = 0$ KN, $F_2 = 500$ KN
No lumped mass	
Stress and displacement constraints: $\sigma_i^{max} = 172.43$ MPa and $\delta_{y_4}^{max} = 60$ mm $(6 \times 10^{-2}$ m$)$	
Natural frequency constraints: $f_1 \geq 60$ Hz and $f_2 \geq 100$ Hz	
The discrete cross − sectional areas : [1,100] cm^{2}in increments of 1 cm^2	
Material properties: $E = 69$ GPa and $\rho = 2740$kg/m^3	

Table 4.33 Size optimization of the 20-bar truss (discrete section)

Element no.	ALO	DA	GWO	MVO	SCA	WOA	HTS	TLBO	PVS	SOS
1	19	20	16	16	55	16	18	20	20	19
2	5	13	1	4	42	36	4	5	4	5
3	1	7	1	1	18	1	1	1	3	1
4	14	28	13	15	25	28	17	14	16	15
5	21	40	21	21	54	40	22	27	21	23
6	20	42	4	4	21	32	12	7	8	7
7	3	14	1	2	5	1	3	2	1	1
8	14	11	18	16	9	15	14	15	14	17
9	5	10	1	4	14	12	5	5	4	6
10	11	28	8	9	12	12	10	25	12	8
11	25	10	26	30	32	33	14	22	13	19
12	7	12	5	8	6	48	11	6	8	12
13	17	22	19	19	25	19	20	17	15	17
14	16	18	12	11	23	15	15	13	17	11
15	15	11	21	16	8	16	14	18	16	16
16	17	21	9	18	32	34	17	17	18	18
17	10	18	2	1	10	1	5	2	6	1
18	29	27	31	32	48	45	29	32	30	37
19	1	14	1	1	2	40	9	3	5	3
20	52	27	33	34	37	22	30	30	29	31
Mass (kg)	246.1333	296.2283	207.2186	218.5386	348.9370	343.9499	224.5115	228.9506	218.3505	221.7325
σ^{max} (MPa)	169.0725	155.3644	172.4257	171.4468	166.7920	170.0624	170.9400	170.8769	170.7607	171.9585
δ^{max} (mm)	21.1997	18.5994	22.5515	22.5133	14.8237	16.7932	22.1713	21.4585	23.0397	21.9935

(continued)

Table 4.33 (continued)

Element no.	ALO	DA	GWO	MVO	SCA	WOA	HTS	TLBO	PVS	SOS
f_1 (Hz)	90.6378	92.2009	102.5280	97.5212	99.1389	74.8206	96.7500	98.8951	97.2957	95.5629
f_2 (Hz)	162.0882	168.1419	171.0070	170.4646	192.1740	148.1407	172.1960	172.3449	174.3996	178.5556
Mean	330.5669	409.5704	214.0419	245.5682	386.2671	496.8543	241.2443	269.0910	229.9030	238.9011
SD	33.8663	52.3567	3.7098	13.4824	26.5772	77.1958	7.7102	28.1174	7.0838	13.5702

228.9506, 218.3505, and 221.7325 kg, respectively. It is worth noting that the GWO and PVS algorithms resulted in the least structural masses.

To more accurately evaluate the performance of the metaheuristic methods, we will consider their average mass and standard deviation (SD). Based on the available results, it can be observed that the ALO, DA, GWO, MVO, SCA, WOA, HTS, TLBO, PVS, and SOS algorithms have mean mass values of 330.5669, 409.5704, 214.0419, 245.5682, 386.2671, 496.8543, 241.2443, 269.0910, 229.9030, and 238.9011 kg, respectively, with corresponding SD values of 33.8663, 52.3567, 3.7098, 13.4824, 26.5772, 77.1958, 7.7102, 28.1174, 7.0838, and 13.5702. The GWO and PVS algorithms have the best mean mass values and the lowest SD, indicating that they are the superior algorithms in terms of both convergence rate and consistency for this design problem.

4.4.6 Size Optimization of the 72-Bar 3D Truss with Discrete Cross Sections

In this section, we focus on the optimization of the 72-bar truss using 16 discrete variables for size optimization. The ground structure of the test problem, known as the 72-bar truss, is depicted in Fig. 4.16. This three-dimensional truss was utilized by Kaveh and Zolghadr (2013) for continuous optimization purposes. In this problem, the discrete cross-sectional areas (G_i, where $i = 1, 2,\ldots, 16$) are considered over [1, 30] cm^2 in increments of 1 cm^2. Thus, discrete design variables may take any integer values within [1, 30] cm^2. Table 4.34 provides comprehensive information regarding material properties, design variables, loading conditions, lumped mass, and other relevant data associated with the test problem. The elements are clustered into 16 groups (i.e., G_1 [A_1–A_4], G_2 [A_5–A_{12}], G_3 [A_{13}–A_{16}], G_4 [A_{17}–A_{18}], G_5 [A_{19}–A_{22}], G_6 [A_{23}–A_{30}], G_7 [A_{31}–A_{34}], G_8 [A_{35}–A_{36}], G_9 [A_{37}–A_{40}], G_{10} [A_{41}–A_{48}], G_{11} [A_{49}–A_{52}], G_{12} [A_{53}–A_{54}], G_{13} [A_{55}–A_{58}], G_{14} [A_{59}–A_{66}], G_{15} [A_{67}–A_{70}], and G_{16} [A_{71}–A_{72}]) by considering structural symmetry as per Kaveh and Zolghadr (2013). The truss has four nonstructural lumped masses of 2270 kg added to all the four top nodes (node numbers 1–4).

The outcomes of the size optimization of the 72-bar truss using discrete cross sections for various algorithms in 100 independent runs are presented in Table 4.35. The optimal run is determined by finding the minimum mass value that meets all structural requirements. The average and standard deviation (SD) of the structural mass are given to assess the evolutionary approaches' performance statistically. The ALO, DA, GWO, MVO, SCA, WOA, HTS, TLBO, PVS, and SOS algorithms have reported the best results of 529.7608, 604.0033, 479.5989, 485.3599, 565.1052, 651.6939, 481.2863, 485.3599, 481.2863, and 481.9852 kg, respectively. According to the findings, the GWO algorithm has provided the most optimal solution, with a mass value of 447.5989 kg, while the PVS and HTS algorithms have generated identical solutions, resulting in the second-best structural mass.

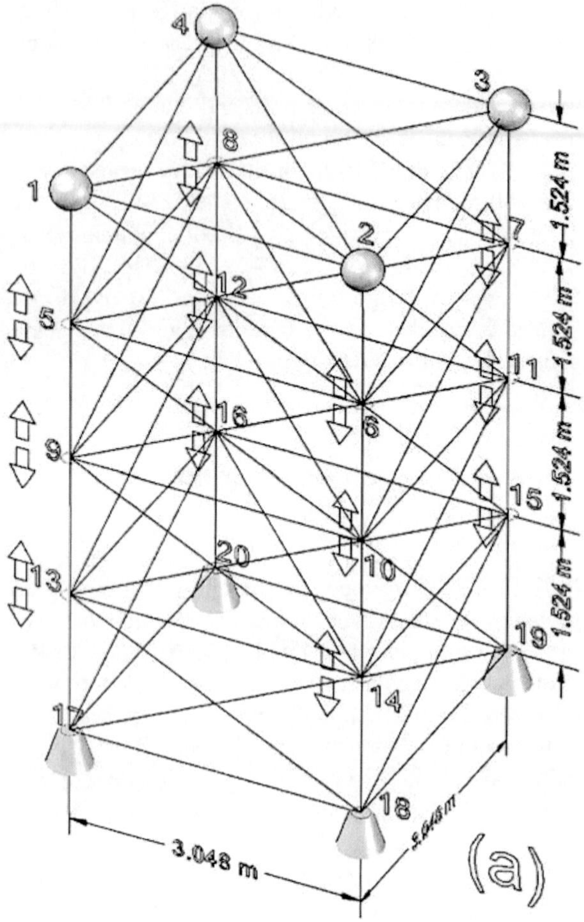

Fig. 4.16 The 72-bar 3D truss

Table 4.34 Design parameters of the 72-bar truss

Size variables: G_i, $i = 1, 2, \ldots, 16$; $G =$ group number				
Multiple load conditions				
Condition 1	$F_{1x} = 22.25$ KN	$F_{1y} = 22.25$ KN		$F_{1z} = -22.25$ KN
Condition 2	$F_{1z} = -22.25$ KN	$F_{2z} = -22.25$ KN	$F_{3z} = -22.25$ KN	$F_{4z} = -22.25$ KN
Lumped masses at nodes 1, 2, 3, and 4 are 2270 kg				
Stress and displacement constraints: $\sigma_i^{\max} = 172.375$ MPa and $\delta_{x_j/y_j}^{\max} = 6.35$ mm $(6.35 \times 10^{-3}$ m$)$ (for node, $j = 1, 2, 3$, and 4 along x- and y-axes)				
Natural frequency constraints: $f_1 \geq 4$ Hz and $f_3 \geq 6$ Hz				
The discrete cross − sectional areas : $[1, 30]$ cm^2 in increments of 1 cm^2				
Material properties: $E = 68.95$ GPa and $\rho = 2767.99$ kg/m^3				

Table 4.35 Size optimization of the 72-bar truss (discrete section)

Element group no.	ALO	DA	GWO	MVO	SCA	WOA	HTS	TLBO	PVS	SOS
1	9	7	6	6	8	20	5	5	7	6
2	7	8	7	8	8	8	9	8	7	8
	5	10	5	5	5	21	5	5	5	5
4	18	7	8	9	8	13	8	8	8	8
5	26	10	7	8	8	9	7	8	9	7
6	8	6	8	8	11	7	7	10	8	7
7	1	1	1	1	2	1	2	2	1	1
8	1	7	2	2	2	1	2	3	2	3
9	12	10	14	12	15	11	13	15	13	13
10	8	10	9	10	7	19	8	7	9	9
11	1	1	1	2	2	1	1	1	1	1
12	2	24	2	2	4	13	2	2	2	2
13	13	20	16	17	17	16	17	15	15	17
14	10	10	9	7	12	7	9	8	9	9
15	1	6	1	1	5	1	1	1	1	1
16	1	7	1	1	7	1	1	1	1	1
Mass (kg)	529.7608	604.0033	479.5989	485.3599	565.1052	651.6939	481.2863	485.3599	481.2863	481.9852
σ^{max} (MPa)	25.1229	28.6341	34.4773	34.1158	26.4301	23.5804	39.4011	39.3146	29.7395	33.9208
δ^{max} (mm)	3.2254	3.0493	3.1834	3.1697	2.9310	2.9544	3.1443	3.1693	3.2387	3.1614
f_1 (Hz)	4.0257	4.0692	4.0071	4.0076	4.2111	4.0387	4.0003	4.0095	4.0114	4.0108
f_3 (Hz)	6.0193	6.0272	6.0433	6.0218	6.3567	6.1913	6.0394	6.0206	6.0426	6.0416
Mean	774.4527	1116.2141	481.9568	547.9789	645.3111	1039.7903	489.8840	530.8089	485.1778	492.0734
SD	106.5866	233.4360	1.9968	52.6350	38.8403	279.1663	7.4682	33.4891	2.9359	6.6347

To accurately assess the statistical performance of the proposed metaheuristic methods, the mean mass and its standard deviation (SD) were considered. Based on the results obtained, it can be observed that the ALO, DA, GWO, MVO, SCA, WOA, HTS, TLBO, PVS, and SOS algorithms have mean mass values of 774.4527, 1116.2141, 481.9568, 547.9789, 645.3111, 1039.7903, 489.8840, 530.8089, 485.1778, and 492.0734 kg, respectively, with corresponding SD values of 106.5866, 233.4360, 1.9968, 52.6350, 38.8403, 279.1663, 7.4682, 33.4891, 2.9359, and 6.6347. The GWO and PVS algorithms performed the best and second-best, respectively, in terms of achieving the lowest mass mean and SD values. Hence, the GWO and PVS algorithms are the preferred choices for this design problem as they offer both a fast convergence rate and consistent results.

4.4.7 Size Optimization of the 39-Bar Truss with Discrete Cross Sections

In this section, we provide a detailed explanation of the optimization process for a 39-bar truss. The truss is optimized for size using 21 discrete variables. We focus on the seventh benchmark problem, which involves a two-tier 39-bar planar truss illustrated in Fig. 4.17. The purpose of this problem is to examine the behavior of a large truss under stress, buckling, displacement, and fundamental frequency constraints. Both continuous and discrete design variables are considered in this investigation. The problem was initially introduced by Deb and Gulati in 2001 and has been further explored by various researchers, including Luh and Lin in 2008, Wu and Tseng in 2010, and Ahrari et al. in 2014. Previous studies mainly focused on static constraints, whereas our research aims to incorporate both static and dynamic constraints simultaneously. We have made modifications to the problem setup to study the impact of multiple load conditions, lumped masses, and a natural frequency constraint.

In this problem, the discrete cross-sectional areas (G_i, where $i = 1, 2, \ldots, 21$) are considered over [1, 30] cm^2 in increments of 1 cm^2. Thus, discrete design variables may take any integer values within [1, 50] cm^2. In Table 4.36, we present the design considerations, including design variables, multi-load conditions, lumped masses, constraints, and material properties. To simplify the analysis, the elements of the truss are divided into 21 groups, following the symmetry observed about the vertical middle plane as demonstrated by Deb and Gulati (2001). The truss is subjected to three nonstructural lumped masses of 500 kg located at nodes 2, 3, and 4. Additionally, two different loading conditions are applied to satisfy the multi-load condition, as depicted in Fig. 4.17.

Table 4.37 shows the performance of different algorithms in the size optimization of the 39-bar truss with discrete cross sections in 100 independent runs. The minimum mass value while satisfying all the given structural constraints was achieved by the best run. The mean and standard deviation (SD) values of structural

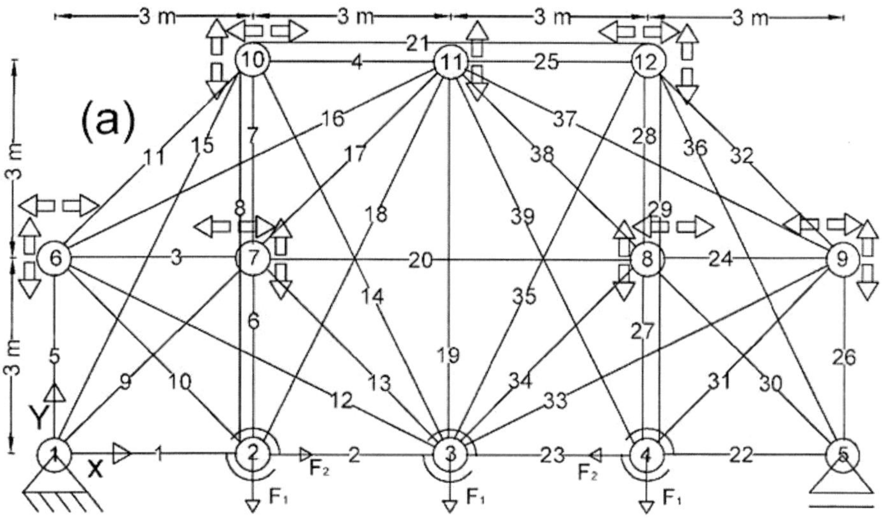

Fig. 4.17 The 39-bar truss

Table 4.36 Design parameters of the 39-bar truss

Size variables: $X_i = G_i$, $i = 1, 2, \ldots, 21$	
Multiple load conditions	
Condition 1 : $F_1 = 100$ *KN*, $F_2 = 0$ *KN*	Condition 2 : $F_1 = 0$ *KN*, $F_2 = 50$ *KN*
Lumped masses on nodes 2, 3, and 4 are 500 kg	
Stress and displacement constraints: $\sigma_i^{\max} = 172.43$ MPa and $\delta_{x_j/y_j}^{\max} = 50$ mm (0.05 m) (for node, $j = 2$, 3 and 4 along x- and y-axes)	
Natural frequency constraints: $f_1 \geq 4$ Hz	
The discrete cross − sectional areas : $[A^{\min}, A^{\max}] = [1, 50]$ cm^{2}in increments of 1 cm^2	
Material properties: $E = 69$ *GPa* and $\rho = 2740$kg/m^3	

mass were provided to statistically evaluate the performance of various evolutionary methods. The best mass results obtained by ALO, DA, GWO, MVO, SCA, WOA, HTS, TLBO, PVS, and SOS algorithms are 635.4329, 701.1410, 522.6491, 544.6392, 720.0876, 813.1218, 566.5718, 566.9635, 534.5416, and 548.9495 kg, respectively. It is evident from the results that the GWO algorithm yielded the least structural mass, while PVS is the second best.

Table 4.37 evaluates the performance of various metaheuristic methods in the size optimization of the 39-bar truss with discrete cross sections. The algorithm that provides the minimum mass value while satisfying all the stated structural constraints is considered the best run. The mean and SD values of structural mass are provided to statistically evaluate the performance of the different evolutionary methods.

Table 4.37 Size optimization of the 39-bar truss (discrete section)

Element no.	ALO	DA	GWO	MVO	SCA	WOA	HTS	TLBO	PVS	SOS
1	1	1	1	1	8	1	1	1	1	1
2	31	16	13	38	29	23	42	23	18	16
3	2	10	3	3	2	22	1	2	4	4
4	32	5	5	5	16	18	9	10	7	7
5	28	20	20	21	17	20	22	22	21	25
6	2	5	1	3	2	19	1	3	1	8
7	2	20	2	1	7	1	1	4	5	9
8	2	7	2	2	10	12	2	5	4	1
9	7	7	6	12	12	27	8	2	2	3
10	13	9	4	10	10	10	6	6	6	4
11	14	23	15	18	28	22	18	22	18	19
12	5	7	7	4	6	5	4	6	6	6
13	1	3	1	3	4	1	1	1	1	1
14	1	14	1	1	2	13	4	2	2	1
15	17	28	22	10	34	25	21	26	24	25
16	36	23	37	31	25	29	30	30	31	30
17	1	1	1	1	13	4	2	3	1	1
18	5	5	6	3	4	11	3	5	5	5
19	4	1	1	1	5	1	1	3	1	1
20	32	15	9	23	17	29	12	5	7	7
21	18	50	21	23	25	24	26	23	24	28
Mass (kg)	635.4329	701.1410	522.6491	544.6392	720.0876	813.1218	566.5718	566.9635	534.5416	548.9495
$\sigma^{\max}$ (MPa)	112.1145	154.9390	161.5237	165.6774	58.1327	171.5701	169.3983	163.0839	151.0149	170.4049
$\delta^{\max}$ (mm)	15.3399	16.2664	19.0775	18.6799	10.2702	16.9789	16.0349	16.1268	17.0446	19.5420
f_1 (Hz)	8.2829	8.6268	7.8648	8.6004	15.2506	9.8315	8.1535	7.8656	7.6862	7.8382
Mean	773.8653	921.501	537.6654	626.6921	806.2057	1204.933	618.7048	663.3968	575.3016	597.8774
SD	76.2612	107.8494	12.0917	48.1990	44.7018	177.7042	30.4231	59.0415	23.6903	28.5770

Based on the results, the ALO, DA, GWO, MVO, SCA, WOA, HTS, TLBO, PVS, and SOS algorithms have mass mean values of 773.8653, 921.5010, 537.6654, 626.6921, 806.2057, 1204.9332, 618.7048, 663.3968, 575.3016, and 597.8774 kg, respectively, with corresponding SD values of 76.2612, 107.8494, 12.0917, 48.1990, 44.7018, 177.7042, 30.4231, 59.0415, 23.6903, and 28.5770 kg. The GWO algorithm performs the best in achieving the lowest mass mean and SD values, followed by the PVS and SOS algorithms.

Therefore, the GWO, PVS, and SOS algorithms are considered better choices for both convergence rate and consistency for this design problem.

4.4.8 Size Optimization of the 45-Bar Truss with Discrete Cross Sections

This section focuses on the size optimization of a 45-bar truss using 31 size variables. The analysis revolves around the ninth benchmark problem, which involves a complex structure subjected to stress, buckling, displacement, and fundamental frequency constraints. Both continuous and discrete design variables are considered in this investigation. The ground structure of the truss, along with multiple loads and boundary conditions, is depicted in Fig. 4.18. Initially proposed by Deb and Gulati in 2001, this problem has previously been explored with static constraints alone,

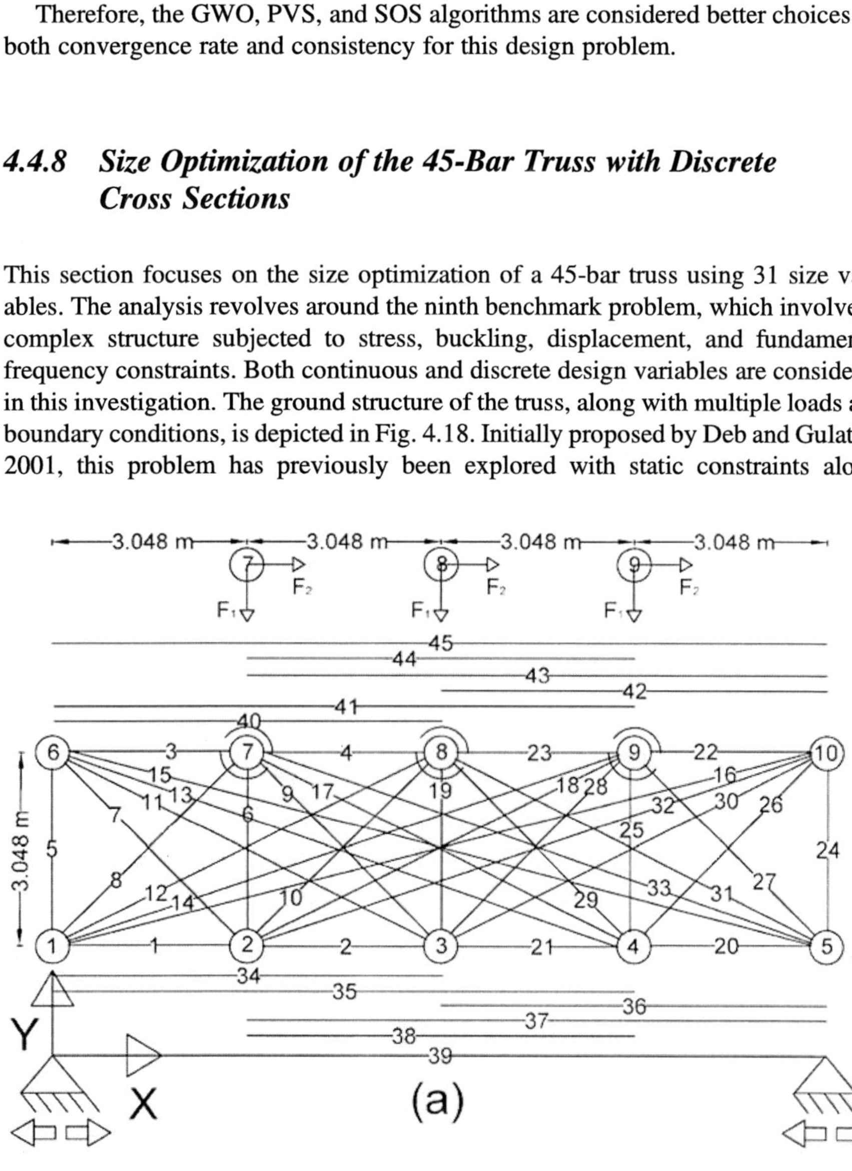

Fig. 4.18 The 45-bar truss

Table 4.38 Design parameters of the 45-bar truss

Size variables: $X_i = G_i$, $i = 1, 2, \ldots, 31$; $G =$ group number Shape variables: $-2.5\ \text{m} \leq \xi_{x_j/y_j} \leq 2.5\ \text{m}$, for $x_1 = -x_5$ and $y_1 = y_5$ with respect to their original coordinates	
Multiple load conditions	
Condition 1 : $F_1 = 44.537$ KN	Condition 2 : $F_2 = 4.4537$ KN
Stress and displacement constraints: $\sigma_i^{\max} = 172.43$ MPa $\delta_{x_j/y_j}^{\max} = 5$ mm (0.005 m) (for nodes, $j = 7$, 8, and 9 along x- and y-axes)	
Lumped masses on nodes 7, 8, and 9 are 500 kg	
Natural frequency constraints: $f_1 \geq 15$ Hz	
The discrete cross − sectional areas : $[A^{\min}, A^{\max}] = [1, 50]\ \text{cm}^2$in increments of 1 cm^2	
Material properties: $E = 69$ GPa and $\rho = 2740\text{kg/m}^3$	

without considering dynamic constraints simultaneously. To address this limitation, we have modified the problem to incorporate the effects of multi-load conditions, lumped masses, natural frequency constraints, and discrete sections.

In this problem, the discrete cross-sectional areas (G_i, where $i = 1, 2, \ldots, 31$) are considered over [1, 50] cm^2 in increments of 1 cm^2. Thus, discrete design variables may take any integer values within [1, 50] cm^2. In Table 4.38, we present the various design considerations, including design variables, multi-load conditions, lumped masses, constraints, and material properties. Symmetry about the vertical middle plane is observed for 28 elements (element numbers 1 to 14 and 20 to 33), which allows us to group them into 14 groups. The remaining elements are treated as distinct, resulting in a reduction in the number of size variables to 31. Additionally, three nonstructural lumped masses of 500 kg are applied at nodes 7, 8, and 9. To fulfill the multi-load condition, two different loading conditions are considered, as illustrated in Fig. 4.18.

In this section, the size optimization of the 45-bar truss with 31 size variables is presented, and the results obtained with various algorithms are shown in Table 4.39. The best run achieved the minimum mass value while satisfying all the structural constraints. The mean and SD values of the structural mass are also provided, and they will be used to evaluate the statistical performance of the different evolutionary methods. The ALO, DA, GWO, MVO, SCA, WOA, HTS, TLBO, PVS, and SOS algorithms were tested, and the best mass mean values obtained were 587.3273, 692.5331, 409.7544, 437.2813, 573.0231, 704.3031, 546.8008, 471.4876, 424.3267, and 414.6528 kg, respectively. It is evident from the results that the GWO, SOS, and PVS algorithms achieved the least structural mass, with GWO being the best performer, followed by the SOS and PVS algorithms.

To evaluate the true statistical performance of the metaheuristic methods, the average mass and its standard deviation (SD) were considered. Based on the available results, it can be concluded that ALO, DA, GWO, MVO, SCA, WOA, HTS, TLBO, PVS, and SOS algorithms have the mean mass values of 757.5820, 972.4318, 419.7305, 553.4498, 707.4119, 1184.3318, 631.3488, 626.0456,

Table 4.39 Size optimization of the 45-bar truss (discrete section)

Group no.	ALO	DA	GWO	MVO	SCA	WOA	HTS	TLBO	PVS	SOS
1	2	7	3	3	5	13	8	4	3	4
2	15	12	1	1	2	1	18	2	1	1
3	1	1	1	1	1	2	1	1	1	1
4	28	5	4	4	3	16	3	18	7	4
5	13	10	3	3	18	2	4	3	4	6
6	6	1	1	1	5	17	3	1	1	1
7	1	5	1	1	1	1	2	1	1	1
8	27	32	22	21	31	36	28	27	19	21
9	1	1	1	1	1	3	1	2	2	1
10	12	6	5	6	13	15	7	6	5	5
11	1	1	1	1	1	1	2	1	1	1
12	19	26	22	22	27	27	21	21	22	22
13	1	1	1	1	1	1	2	1	1	1
14	17	24	22	21	19	17	20	19	22	22
15	4	3	3	3	6	3	5	3	4	3
16	5	2	4	3	1	16	5	3	4	1
17	2	6	2	2	9	4	4	3	2	3
18	1	8	1	1	3	1	5	2	3	1
19	12	34	3	3	50	5	36	8	3	3
20	8	1	1	1	2	10	8	1	3	1
21	2	3	1	1	2	6	2	8	1	1
22	1	1	2	8	1	1	1	1	2	1
23	7	1	1	1	6	1	1	1	1	2
24	16	1	1	2	1	2	3	2	2	1
25	1	1	1	1	6	18	1	1	2	1

(continued)

Table 4.39 (continued)

Group no.	ALO	DA	GWO	MVO	SCA	WOA	HTS	TLBO	PVS	SOS
26	7	1	1	2	9	31	6	17	2	1
27	6	22	12	16	12	6	13	10	9	12
28	17	50	1	5	1	1	4	1	2	1
29	6	50	10	14	14	8	13	16	9	17
30	39	9	11	16	19	28	15	10	11	10
31	6	1	3	1	1	17	7	1	5	1
Mass (kg)	587.3273	692.5331	409.7544	437.2813	573.0231	704.3031	546.8008	471.4876	424.3267	414.6528
$\sigma^{\max}$ (MPa)	27.5408	17.4290	25.8859	27.3961	23.5973	18.2661	21.7139	26.9747	29.0877	28.9485
$\delta^{\max}$ (mm)	2.9915	2.5713	3.4121	3.3820	2.5790	2.2955	3.0237	3.2562	3.4284	3.4041
f_1 (Hz)	27.4265	28.7373	26.3074	26.2646	29.2963	31.2410	27.2507	26.7153	26.1777	26.3093
Mean	757.5820	972.4318	419.7305	553.4498	707.4119	1184.3318	631.3488	626.0456	488.3093	483.7550
SD	93.9935	118.5188	6.1229	51.8030	54.5057	226.2410	37.0469	93.2132	36.4157	37.0395

488.3093, and 483.7550 kg, respectively, and SD values of 93.9935, 118.5188, 6.1229, 51.8030, 54.5057, 226.2410, 37.0469, 93.2132, 36.4157, and 37.0395, respectively.

Based on the results presented in Table 4.39, it can be concluded that the GWO algorithm outperforms other metaheuristic algorithms in terms of both convergence rate and consistency in the size optimization of the 45-bar truss with discrete cross sections. PVS and SOS algorithms also show good performance in achieving low mean mass values and SDs. However, it is important to note that the choice of the optimization algorithm should be made based on the specific problem and the designer's objectives, as the best algorithm for one problem may not be the best for another.

4.4.9 Size Optimization of the 25-Bar 3D Truss with Discrete Cross Sections

In this section, size optimization was performed on the 25-bar truss using eight size variables. The purpose of this investigation is to study the behavior of a space truss under various constraints, including stress, buckling, displacement, and fundamental frequency. The initial configuration of the truss can be seen in Fig. 4.19. Previous studies by Deb and Gulati (2001) and Ahrari et al. (2014) have explored this problem with static constraints only, but this study introduces the consideration of both static and dynamic constraints. To incorporate multiple load conditions, lumped masses, natural frequency constraint, and discrete area, modifications were made to the original problem.

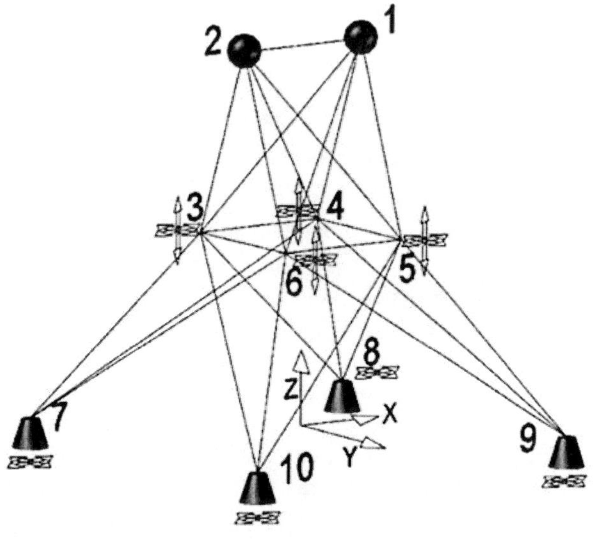

Fig. 4.19 The 25-bar 3D truss: (**a**) ground structure and (**b**) optimal structure

In this problem, the discrete cross-sectional areas (G_i, where $i = 1, 2, \ldots, 8$) are considered over [1, 50] cm^2 in increments of 1 cm^2. Thus, discrete design variables may take any integer values within [1, 50] cm^2. Detailed information regarding design variables, load conditions, lumped masses, constraints, and material properties can be found in Table 4.40. Due to structural symmetry, element grouping is applied to the truss, resulting in only eight size variables. It should be noted that topology variables are also included, as discussed earlier. The truss is subjected to two nonstructural lumped masses of 500 kg located at nodes 1 and 2. Two different loading conditions are considered, as described in the table.

In Table 4.41, the results of optimizing the 25-bar truss with discrete cross sections using different algorithms in 100 independent runs are displayed. The minimum mass value that satisfies all structural constraints was achieved in the most favorable run. Additionally, the table includes the average and standard deviation values of the structural mass, which can be used to statistically assess the effectiveness of the various evolutionary techniques. The ALO, DA, GWO, MVO, SCA, WOA, HTS, TLBO, PVS, and SOS algorithms were tested, and their best mass mean values were 394.7270, 402.3180, 393.6831, 393.6831, 423.4515, 416.9514, 393.6831, 393.6831, 393.6831, and 393.6831 kg, respectively. Among these algorithms, the GWO, MVO, HTS, TLBO, PVS, and SOS algorithms all produced the same optimal result of 393.6831 kg for the truss mass.

To evaluate the performance of various metaheuristic methods, the mean mass and standard deviation were considered. The mean mass values for the ALO, DA, GWO, MVO, SCA, WOA, HTS, TLBO, PVS, and SOS algorithms were 424.0738, 440.3866, 397.1982, 402.0912, 449.0666, 471.5895, 393.9904, 396.9697, 394.2240, and 393.8722 kg, respectively. The corresponding standard deviations were 19.0568, 27.4416, 1.9951, 10.3832, 14.7487, 37.7155, 0.5237, 4.5810, 0.5923, and 0.4377, respectively. Based on these results, the SOS, HTS, and PVS algorithms are the best options for solving this particular design problem due to their superior convergence rate and consistency, ranking first, second, and third, respectively.

4.4.10 Size Optimization of the 39-Bar 3D Truss with Discrete Cross Sections

This section focuses on the comparison of different optimization algorithms for the size optimization of a 39-bar 3D truss with a continuous section. Figure 4.20 illustrates the ground structure of the 39-element 3D truss. To account for symmetry about the *x*-*z* and *y*-*z* planes, the element cross-sectional areas are grouped into 11 sets. Previous studies by Deb and Gulati (2001) and Luh and Lin (2008) have examined this truss with static constraints only, but this investigation extends the analysis to include both static and dynamic constraints. Consequently, the problem is modified to incorporate multi-load conditions, lumped masses, a natural frequency constraint, and discrete area considerations. Additionally, a lumped mass of 500 kg is consistently attached to the top nodes (nodes 1 and 2) as depicted in Fig. 4.20.

Table 4.40 Design parameters of the 25-bar 3D truss

Size variables: $X_i = G_i$, $i = 1, 2, \ldots, 8$; G = group number				
Multiple load conditions				
Condition 1	$F_{y_1} = 89.074$ kN	$F_{z_1} = -22.25$ kN	$F_{y_2} = -89.074$ kN	$F_{z_2} = -22.25$ kN
Condition 2	$F_{x_1} = 4.454$ kN	$F_{y_1} = -44.537$ kN	$F_{z_1} = -44.537$ kN	$F_{y_2} = -44.537$ kN
	$F_{z_2} = -44.537$ kN	$F_{x_3} = 2.227$ kN	$F_{x_6} = 2.672$ kN	
Lumped masses at nodes 1 and 2 are 500 kg				
Stress and displacement constraints: $\sigma_i^{max} = 172.375$ MPa and $\delta_{x_j/y_j/z_j}^{max} = 8.9$ mm (0.0089 m) (for node, $j = 1, 2, \ldots, 6$ along x-, y-, and z-axes)				
Natural frequency constraints: $f_1 \geq 15$ Hz				
The discrete cross − sectional areas : $[A^{min}, A^{max}] = [1, 50]$ cm^{2}in increments of 1 cm^2				
Material properties: $E = 69$ GPa and $\rho = 2740$kg/m^3				
Element grouping				
Group number	Element number (end nodes)		Group number	Element number (end nodes)
G_1	1(1,2)		G_5	12(3, 4), 13(5,6)
G_2	2(1, 4), 3(2, 3), 4(1, 5), 5(2,6)		G_6	14(3, 10), 15(6,7), 16(4,9), 17(5, 8)
G_3	6(2, 5), 7(2, 4), 8(1, 3), 9(1,6)		G_7	18(3, 8), 19(4,7), 20(6,9), 21(5,10)
G_4	10(3,6), 11(4,5)		G_8	22(3, 7), 23(4,8), 24(5,9), 25(6,10)

Table 4.41 Size optimization of the 25-bar 3D truss (discrete section)

Element group no	ALO	DA	GWO	MVO	SCA	WOA	HTS	TLBO	PVS	SOS
1	1	1	1	1	1	1	1	1	1	1
2	16	16	16	16	17	16	16	16	16	16
3	16	16	16	16	19	17	16	16	16	16
4	1	2	1	1	3	1	1	1	1	1
5	4	9	3	3	3	26	3	3	3	3
6	13	13	13	13	13	13	13	13	13	13
7	20	21	20	20	23	20	20	20	20	20
8	18	17	18	18	18	17	18	18	18	18
Mass (kg)	394.7270	402.3180	393.6831	393.6831	423.4515	416.9514	393.6831	393.6831	393.6831	393.6831
$\sigma^{\max}$ (MPa)	53.9188	54.0972	53.9186	53.9186	46.2567	51.2048	53.9186	53.9186	53.9186	53.9186
$\delta^{\max}$ (mm)	7.4497	7.5412	7.4606	7.4606	6.9530	7.4456	7.4606	7.4606	7.4606	7.4606
f_1 (Hz)	15.6035	15.4234	15.6036	15.6036	15.9706	15.4432	15.6036	15.6036	15.6036	15.6036
Mean	424.0738	440.3866	397.1982	402.0912	449.0666	471.5895	393.9904	396.9697	394.2240	393.8722
SD	19.0568	27.4416	1.9951	10.3832	14.7487	37.7155	0.5237	4.5810	0.5923	0.4377

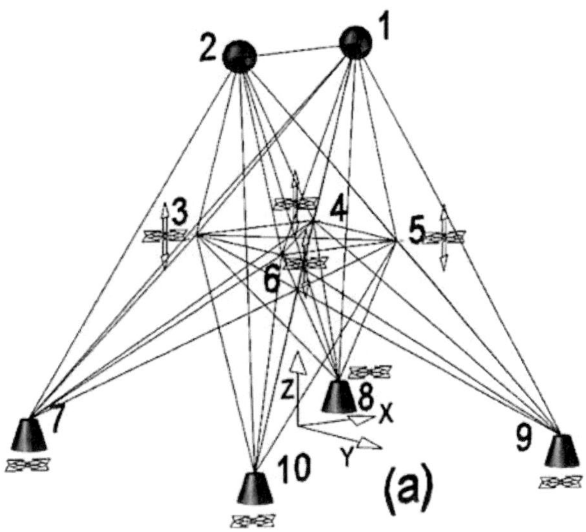

Fig. 4.20 The 39-bar 3D truss

In this problem, the discrete cross-sectional areas (G_i, where $i = 1, 2, \ldots, 11$) are considered over [1, 50] cm^2 in increments of 1 cm^2. Thus, discrete design variables may take any integer values within [1, 50] cm^2. The design parameters, such as material properties, loading conditions, design variable bounds, constraints, and more, for this truss can be found in Table 4.42.

Various algorithms were utilized to optimize the 39-bar 3D truss with 11 discrete cross sections, and the outcomes of 100 independent runs are displayed in Table 4.43. The most favorable run resulted in the minimum mass value while satisfying all the structural constraints. Additionally, the average and standard deviation values of the structural mass are provided and will be utilized to statistically evaluate the effectiveness of different evolutionary techniques. Optimal results with masses of 622.1855, 664.2218, 608.3270, 608.3270, 671.7652, 684.1119, 608.3270, 608.3270, 608.3270, and 608.3270 kg were obtained by the ALO, DA, GWO, MVO, SCA, WOA, HTS, TLBO, PVS, and SOS algorithms, respectively. It is evident that the GWO, MVO, HTS, TLBO, PVS, and SOS algorithms produced an identical solution with the lowest structural mass.

To evaluate the effectiveness of different metaheuristic algorithms, their mean mass and standard deviation were computed. Results showed that the average masses produced by the ALO, DA, GWO, MVO, SCA, WOA, HTS, TLBO, PVS, and SOS algorithms were 717.9839, 785.1641, 615.4971, 638.3832, 748.2769, 900.6919, 618.3751, 652.3191, 626.8995, and 614.4562 kg, respectively. Their corresponding standard deviations were 48.3326, 71.7645, 6.1560, 22.2488, 33.1772, 107.2711, 14.8039, 31.9472, 16.9438, and 7.7294. Based on these results, the SOS algorithm was found to be the most effective in achieving the lowest average mass, followed by the GWO algorithm. Additionally, the GWO algorithm was found to have the lowest standard deviation, while the PVS algorithm produced the second lowest standard deviation.

Table 4.42 Design parameters of the 39-bar 3D truss

Size variables: $X_i = G_i$, $i = 1, 2, \ldots, 11$; $G =$ group number			
Load condition:	$F_{z_1} = -222.685\ kN$		$F_{z_2} = -222.685\ kN$
Lumped masses at nodes 1 and 2 are 500 kg			
Stress and displacement constraints: $\sigma_i^{\max} = 172.375$ MPa and $\delta_{x_j/y_j/z_j}^{\max} = 8.9$ mm (0.0089 m) (for node, $j = 1, 2, \ldots, 6$ along x-, y-and z-axes)			
Natural frequency constraints: $f_1 \geq 15\ Hz$			
The discrete cross − sectional areas : $[A^{\min}, A^{\max}] = [1, 50]$ cm^{2}in increments of 1 cm^2			
Material properties: $E = 69\ GPa$ and $\rho = 2740 kg/m^3$			
Element grouping			
Group number	Element number (end nodes)	Group number	Element number (end nodes)
G_1	1(1,2)	G_7	22(3,7), 23(4,8), 24(5,9), 25(6,10)
G_2	2(1,4), 3(2,3), 4(1,5), 5(2,6)	G_8	26(5,7), 27(6,8), 28(3,9), 29(4,10)
G_3	6(2,5), 7(2,4), 8(1,3), 9(1,6)	G_9	30(3,5), 31(4,6)
G_4	10(3,6), 11(4,5),12(3,4), 13(5,6)	G_{10}	32(1,7), 33(1,10), 34(2,9), 35(2,8)
G_5	14(3,10), 15(6,7), 16(4,9), 17(5,8)	G_{11}	36(2,7), 37(2,10), 38(1,8), 39(1,9)
G_6	18(3,8), 19(4,7), 20(6,9), 21(5,10)		

Based on the statistical analysis of the results in Table 4.43, it can be concluded that the GWO, PVS, and SOS algorithms are the most effective methods for the size optimization of the 45-bar truss with discrete cross sections. These algorithms were found to consistently produce the lowest mean mass values while maintaining a low standard deviation, indicating good convergence and consistency in the optimization process. However, it is important to note that other factors, such as computational efficiency and ease of implementation, should also be considered when selecting an optimization algorithm for a particular problem.

4.4.11 A Comprehensive Analysis

Table 4.44 provides a detailed analysis of the best results achieved by proposed algorithms in 100 independent runs on ten benchmark problems. The table lists the minimum mass and offers a comprehensive evaluation of overall, average, and individual ranks, as well as the best and worst results for each algorithm.

Upon reviewing the results in Table 4.44, it was found that the GWO and PVS algorithms outperformed all other algorithms in terms of overall rank to achieve the minimum mass. The SOS, HTS, and MVO algorithms followed with the second-, third-, and fourth-best performances, respectively.

Table 4.43 Size optimization of the 39-bar 3D truss (discrete section)

Element group no	ALO	DA	GWO	MVO	SCA	WOA	HTS	TLBO	PVS	SOS
1	4	11	1	1	3	1	1	1	1	1
2	5	8	5	5	5	5	4	4	4	4
3	8	6	7	7	10	8	7	7	7	7
4	1	1	1	1	4	12	1	1	1	1
5	6	6	7	7	9	13	6	6	6	6
6	8	9	8	8	11	9	8	8	8	8
7	11	8	9	9	14	8	8	8	8	8
8	6	4	5	5	6	8	5	5	5	5
9	1	1	1	1	3	1	1	1	1	1
10	30	38	31	31	29	31	31	31	31	31
11	25	25	24	24	25	24	24	24	24	24
Mass (kg)	622.1855	664.2218	608.3270	608.3270	671.7652	684.1119	608.3270	608.3270	608.3270	608.3270
σ^{max} (MPa)	24.7751	20.5495	35.4013	35.4013	30.4050	34.9765	35.4013	35.4013	35.4013	35.4013
δ^{max} (mm)	2.1706	1.9442	2.2147	2.2147	2.1038	2.1622	2.2147	2.2147	2.2147	2.2147
f_1 (Hz)	31.4602	33.3445	31.3788	31.3788	31.4217	31.4308	31.3788	31.3788	31.3788	31.3788
Mean	717.9839	785.1641	615.4971	638.3832	748.2769	900.6919	618.3751	652.3191	626.8995	614.4562
SD	48.3326	71.7645	6.1560	22.2488	33.1772	107.2711	14.8039	31.9472	16.9438	7.7294

Table 4.44 Minimum mass obtained using size optimization with discrete sections

	ALO	DA	GWO	MVO	SCA	WOA	HTS	TLBO	PVS	SOS
10-bar truss	606.7	627.2	581.5	575.2	662.2	672.1	**572.7**	**572.7**	**572.7**	**572.7**
14-bar truss	801.7	893.8	**748.9**	761.0	875.5	1278.2	**748.9**	755.4	**748.9**	751.5
15-bar truss	140.5	157.8	**120.0**	**120.0**	158.4	168.7	**120.0**	**120.0**	**120.0**	**120.0**
24-bar truss	259.7	265.9	**203.0**	212.6	297.7	342.7	227.3	210.9	**203.0**	206.3
20-bar truss	246.1	296.2	**207.2**	218.5	348.9	343.9	224.5	229.0	218.4	221.7
72-bar 3D truss	529.8	604.0	**479.6**	485.4	565.1	651.7	481.3	485.4	481.3	482.0
39-bar truss	635.4	701.1	**522.6**	544.6	720.1	813.1	566.6	567.0	534.5	548.9
45-bar truss	587.3	692.5	**409.8**	437.3	573.0	704.3	546.8	471.5	424.3	414.7
25-bar 3D truss	394.7	402.3	**393.7**	**393.7**	423.5	417.0	**393.7**	**393.7**	**393.7**	**393.7**
39-bar 3D truss	622.2	664.2	**608.3**	**608.3**	671.8	684.1	**608.3**	**608.3**	**608.3**	**608.3**
Overall rank	7	8	1	5	9	10	4	6	1	3
Count best	0	0	9	3	0	0	5	4	6	4
Count 2nd best	0	0	0	0	0	0	1	0	3	1
Count 3rd best	0	0	0	2	0	0	0	0	1	1
Count worst	0	0	0	0	2	8	0	0	0	0

Note: Bold indicates the best solution

Furthermore, the GWO algorithm produced the highest number of best solutions, including nine best solutions, without any worse solutions. Similarly, the PVS algorithm provided six best solutions, three second-best solutions, and one third-best solution, with no worse solutions.

However, the WOA algorithm produced the worst solutions for eight problems, highlighting the importance of selecting the appropriate algorithm for specific problems to achieve optimal results.

Overall, the analysis of Table 4.44 provides valuable insights into the performance of proposed algorithms for ten benchmark problems. These findings can assist researchers and practitioners in selecting the most suitable algorithm for similar problems and improving the optimization process' efficiency.

Table 4.25 presents the mean mass achieved by the proposed algorithms for ten benchmark problems in 100 independent runs. Like Table 4.44, it displays the overall rank, average rank, individual ranks, best and worst results for each algorithm.

Upon reviewing the results in Table 4.45, it was observed that the GWO algorithm outperformed all other algorithms in terms of overall rank to obtain the mean mass. It was followed by the PVS, SOS, and HTS algorithms, respectively.

Moreover, the GWO algorithm had eight best solutions without any worse solution count. Similarly, the PVS algorithm provided one best solutions, six second-best solutions, and three third-best solution, with no worse solutions. Conversely, the SOS algorithm had three best solutions, one second-best solution, four third-best solutions, and no worse solutions.

In contrast, the WOA algorithm had the highest number of worst solutions for all ten benchmark problems. These results suggest that the performance of the WOA, DA, and SCA algorithms may not be optimal for certain types of problems.

4.4.12 The Friedman Rank Test

After conducting a comprehensive analysis of the results, it can be concluded that the GWO, SOS, and PVS algorithms demonstrate significantly better performance compared to other algorithms. However, to establish a more comprehensive and reliable ranking of all the algorithms based on the obtained results from the proposed method in comparison to other algorithms as suggested by Draa (2015), statistical tests such as the Friedman rank test are necessary.

The Friedman rank test is performed on the mean mass and success rate of the ALO, DA, GWO, MVO, SCA, WOA, HTS, TLBO, PVS, and SOS algorithms, respectively. The results of the Friedman test are then normalized based on the best value obtained, which ensures a fair and unbiased comparison among the algorithms. The algorithms are ranked based on the normalized values, providing a clear understanding of their relative performance.

Table 4.46 presents the detailed results of the Friedman rank test for the discrete sections of the test problems. The table provides valuable insights into the performance of each algorithm and their rankings based on the obtained results.

The findings of the Friedman rank test reveal that the GWO algorithm consistently secures the top rank in terms of obtaining the best mean mass, followed by the PVS and SOS algorithms. This information highlights the superior performance of these algorithms compared to others, indicating their effectiveness in optimizing the test problems in the discrete sections.

Table 4.45 Mean mass obtained using size optimization with discrete sections

	ALO	DA	GWO	MVO	SCA	WOA	HTS	TLBO	PVS	SOS
10-bar truss	726.0	845.0	651.6	639.6	770.2	999.3	631.7	649.0	639.1	**630.3**
14-bar truss	1226.3	1609.5	**847.6**	930.7	1034.2	2203.7	928.1	1038.4	881.1	924.5
15-bar truss	171.5	217.0	**120.4**	128.0	176.9	243.8	122.1	142.3	122.1	122.3
24-bar truss	322.0	386.2	**206.3**	242.5	329.5	502.5	250.7	275.1	217.9	223.6
20-bar truss	330.6	409.6	**214.0**	245.6	386.3	496.9	241.2	269.1	229.9	238.9
72-bar 3D truss	529.8	604.0	**479.6**	485.4	565.1	651.7	481.3	485.4	481.3	482.0
39-bar truss	773.9	921.5	**537.7**	626.7	806.2	1204.9	618.7	663.4	575.3	597.9
45-bar truss	757.6	972.4	**419.7**	553.4	707.4	1184.3	631.3	626.0	488.3	483.8
25-bar 3D truss	424.1	440.4	397.2	402.1	449.1	471.6	394.0	397.0	394.2	**393.9**
39-bar 3D truss	718.0	785.2	615.5	638.4	748.3	900.7	618.4	652.3	626.9	**614.5**
Overall rank	7	9	1	5	8	10	4	6	2	3
Count best	0	0	8	1	0	0	1	1	1	3
Count 2nd best	0	0	0	0	0	0	4	0	6	1
Count 3rd best	0	0	0	0	0	0	0	0	3	4
Count worst	0	0	0	0	0	10	0	0	0	0

Note: Bold indicates the best solution

Table 4.46 The Friedman rank test

	ALO	DA	GWO	MVO	SCA	WOA	HTS	TLBO	PVS	SOS
10-bar truss	350	427	199.5	183.5	397	481	156.5	212	172.5	171
14-bar truss	350	438	130	207	300	493	216	237	183	196
15-bar truss	366	451	76	234	395	476	134.5	299	160	158.5
24-bar truss	368	440	55	209	380	496	247	287	122	146
20-bar truss	356	442	51	215	414	484	202	280	130	176
72-bar 3D truss	408	476	66	280	354	456	153	262	112.5	182.5
39-bar truss	370	439	53	214	389	490	217	268	131	179
45-bar truss	376	453	52	201	353	489	280	275	138	133
25-bar 3D truss	361.5	408	247	269.5	438	471	113	201.5	136	104.5
39-bar 3D truss	365	433	127.5	222.5	405	485	138	261.5	192.5	120
Overall Friedman value	3670.5	4407	1057	2235.5	3825	4821	1857	2583	1477.5	1566.5
Overall Friedman rank	7	9	1	5	8	10	4	6	2	3

Note: Bold indicates the best solution

4.5 Multi-objective Optimization for Structure Design

Optimizing structures stand as a critical challenge in today's engineering designs due to the array of objectives constrained by various factors. A multi-objective optimization design issue involves two primary aims: first, attaining a set of nondominant solutions that closely align with the true Pareto front (PF) and, second, acquiring a well-dispersed set of solutions along the PF (Angelo et al., 2015; Kumar et al. 2021a, 2022b). Hence, the multi-objective optimization methods yield a range of trade-off solutions, enabling the designer to select one based on the prioritization of objectives.

Engineering design challenges involving multi-objective structure optimization often encompass diverse objectives, like weight minimization and maximum nodal deflection. To tackle these complex multi-objective problems, metaheuristics have shown remarkable performance in recent years. Over the last two decades, a significant number of multi-objective metaheuristics (MOMHs) have emerged, including NSGA-II, multi-objective ant colony system (MOACS) (Angelo et al., 2015), multi-objective ant system (MOAS) (Angelo et al., 2012), multi-objective heat transfer search algorithm (MOHTS) (Tejani et al., 2019a, b, c; Kumar et al., 2020, 2022a), multi-objective teaching-learning-based optimization algorithm (MOTLBO) (Kumar et al., 2021a), multi-objective symbiotic organisms search (MOSOS) multi-objective passing vehicle search (Kumar et al., 2021b), multi-objective multiverse optimizer (MOMVO) (Kumar et al., 2023), and several others. However, these methods can encounter issues such as slow convergence rates, the need for fine-tuning problem-specific parameters, inability to guarantee global solutions, getting trapped in local optima, high computation time, and complexity (Tejani et al., 2018d, 2019b). Additionally, an effective MH should strike a balance between search exploration and exploitation rates, a balance that often appears lacking in many algorithms (Kumar et al., 2019).

4.5.1 Mathematical Formulation of Multi-objective Structure Optimization Problem

Multi-objective design problems are typically framed to identify a decision set that complies with multiple constraints while simultaneously enhancing various objectives. In this particular section, the primary aim is to minimize the weight of the structure as the first objective function while also considering the maximum nodal deflection as the second objective function.

The mathematical formulation for the multi-objective truss optimization problem is as follows:

$$\text{Find,} \quad X = \{A_1, A_2, .., A_m\} \tag{3}$$

to minimize the structure mass and maximum nodal deflection

$$f_1(X) = \sum_{i=1}^{m} A_i \rho_i L_i \text{ and } f_2(X) = \max(|\delta_j|)$$

Subjected to:
Behavior constraints:

$$g(X) : \text{stress constraints}, \quad |\sigma_i| - \sigma_i^{\max} \leq 0$$

Side constraints:

$$\text{Cross} - \text{sectional area constraints}, \quad A_i^{\min} \leq A_i \leq A_i^{\max}$$

where

$$i = 1, 2, .., m; j = 1, 2, .., n$$

Here, A_i, ρ_i, L_i, and σ_i stand for cross sectional-area, mass density, element length, and stress, on the element "*i*" respectively. δ_j is value of nodal displacement value of node "*j*" Superscripts "max" and "min" denote the maximum and minimum allowable limit, respectively.

The penalty function technique proves effective for managing constraints in optimization problems, yet it may yield a sizable search space and numerical instability if the penalty parameters are not meticulously selected. Employed to address constraints, this technique operates by assigning a penalty. When constraints remain unviolated, the penalty reduces to zero; however, it increases based on specific criteria if constraints are violated (Deb & Gulati, 2001; Kaveh & Zolghadr, 2013).

$$\text{Penalized } F(X) = F(X) * F_{\text{penalty}} \tag{2.12}$$

$$\text{where } F_{\text{penalty}} = (1 + \varepsilon_1 * C)^{\varepsilon_2}, \quad C = \sum_{i=1}^{q} C_i, \quad C_i = \left|1 - \frac{p_i}{p_i^*}\right| \tag{3.14}$$

where p_i is the level of constraint violation having the bound as p_i^*. The parameter q is a number of active constraints. The variables ε_1 and ε_2 are predetermined by a user. In this study, the values of both ε_1 and ε_2 are set as 3, which were obtained from experimenting their effect on the balance of the exploitation-exploration balance (Tejani, et al. 2023).

The 942-bar tower truss, referred to as the tower structure, represents the benchmark analyzed in this study and illustrated in Fig. 4.21. The material properties include a density of 0.1 lb/in^3 (2 767.99 g/m^3) and a modulus of elasticity of 104 ksi (7171 MN/m^2). The loading assumptions involve vertical loads along the z-axis: -3 kips (-13.34 KN), -6 kips (-26.69 KN), and -9 kips (4.00 KN) at nodes in the first, second, and third sections, respectively. Additionally, lateral loads along the y-axis are 1 kip at each truss node, while lateral loads along the x-axis are 1.5 kips (6.67 KN) and 1.0 kip (4.45 KN) at nodes on the left and right sides, respectively. Tensile and compressive stresses are constrained to 25 ksi (17.58 MN/m^2). Elemental cross sections are chosen from a range of 200 discrete values, [1, 2, 3, ..., 200] in^2 ([1, 2, 3, ..., 200]* 6.4516 cm^2). To account for structural similarity based on prior research (Angelo et al., 2012, 2015), the bar elementals are categorized into 59 groups.

This study conducts 100 individual runs of algorithm for considered test example. Each run involves a population size of 100 and encompasses 50,000 functional evaluations, as specified by Kumar et al. (2021b).

Three algorithms were employed to analyze truss structures through multi-objective optimization, tackling the challenge of optimizing a 942-bar truss with discrete sections as design variables. The optimal Pareto fronts achieved via the multi-objective symbiotic organism search (MOSOS), multi-objective passing vehicle search (MOPVS), and multi-objective heat transfer search (MOHTS) were compared with benchmark algorithms like multi-objective ant system (MOAS) and multi-objective ant colony system (MOACS). These algorithms strike a balance between search intensification and diversification, yielding nondominated solutions while maintaining diversity in the Pareto fronts. Additionally, performance indicators such as hypervolume were utilized for precise performance evaluation. The study also includes Friedman's rank test for statistical analysis of the experimental work.

The Pareto fronts results of the 942-bar example are displayed in Table 4.47. Mean values from the MOAS, MOACS, MOSOS, MOHTS, and MOPVS algorithms are 60,795,099.20, 61,174,938.07, 71,950,729.01, 76,064,884.48, and 76,279,259.43, respectively. Similarly, the STD values from the PFHV test are 60795099, 61174938, 71950729, 76064884, and 76279259 for the MOAS, MOACS, MOSOS, MOHTS, and MOPVS algorithms, respectively. MOPVS demonstrates superior convergence rate and consistency. Furthermore, to ensure a fair comparison among all algorithms, a statistical test (Friedman rank test) is conducted at a 95% significance level. The test values obtained by MOAS, MOACS, MOSOS, MOHTS, and MOPVS are 152, 148, 300, 413, and 487, respectively. Based on ranking, MOPVS outperforms others, followed by the MOHTS and MOSOS algorithms.

Fig. 4.21 The 942-bar truss

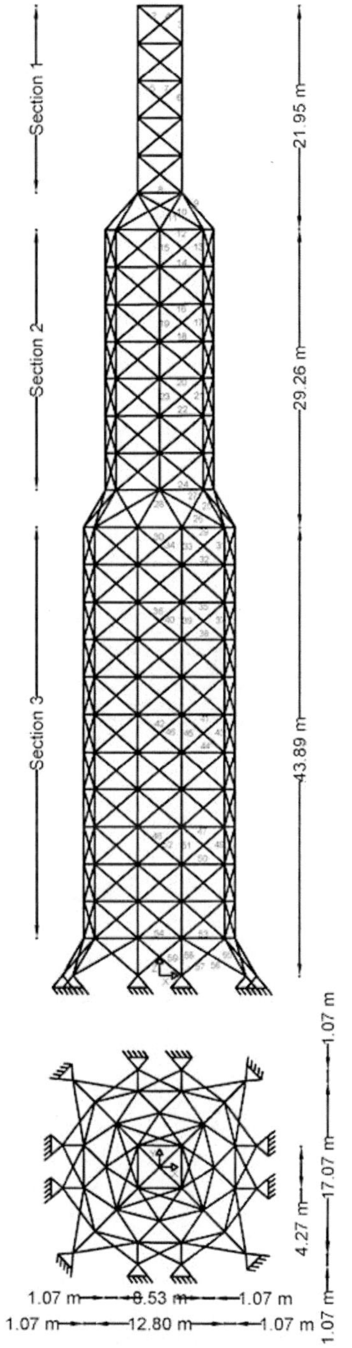

Table 4.47 The hypervolume of the 942-bar truss

Algorithms	MOAS	MOACS	MOSOS	MOHTS	MOPVS
Min	52814995	57425914	70108177	75780010	76013417
Max	67170666	63616875	73682237	76306770	76536083
Mean	60795099	61174938	71950729	76064884	76279259
STD	4084186	1116526	696708	113801	122569
Friedman test	152	148	300	413	487
Friedman rank	4	5	3	2	1

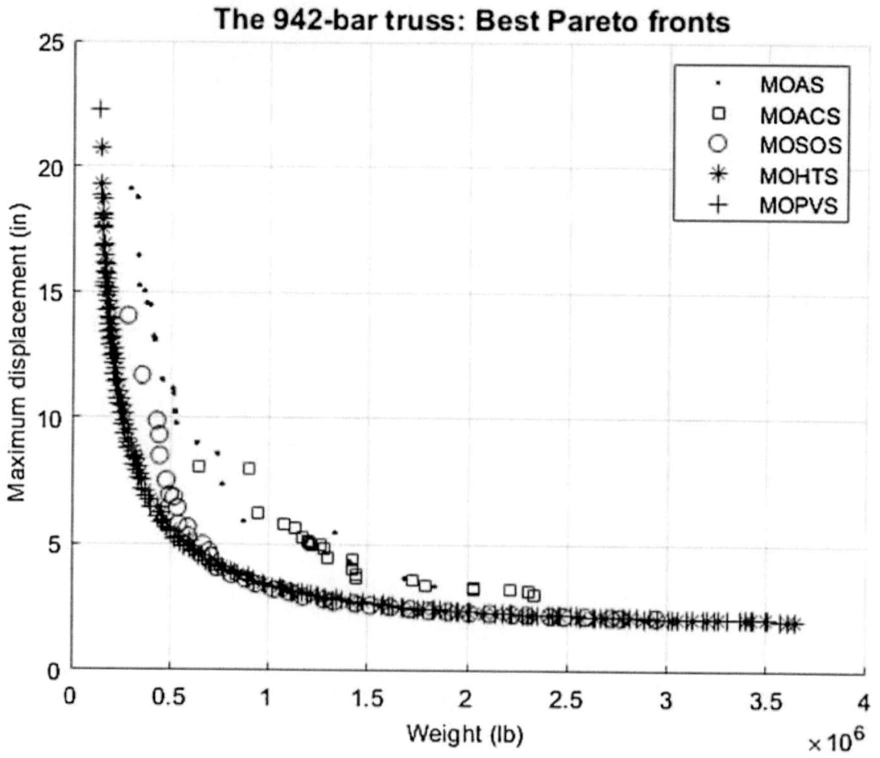

Fig. 4.22 Best Pareto fronts of the 942-bar truss

Figure 4.22 depicts the best Pareto fronts for the 942-bar example derived from all the algorithms under consideration across 100 runs. Observing the figure, it becomes evident that the solutions generated by MOAS, MOACS, and MOSOS appear scattered and discontinuous. In contrast, the Pareto fronts from MOPVS and MOHTS display a smooth, consistently distributed nature, showcasing a stable array of diverse solutions. In a comprehensive assessment, MOPVS emerges as the superior algorithm among the considered ones for effectively addressing the challenges posed by the multi-objective large structure design problem.

4.6 The CEC2014 Benchmark Functions

In this section, we are using ten benchmark functions from the CEC2014 special session on single-objective real-parameter numerical optimization (Liang et al., 2014) to show the algorithm results. These functions are listed in Table 4.47 and fall into four groups: unimodal (F1–F3), multimodal (F4–F6), hybrid (F7–F8), and composition functions (F9–F10). To check the results, we are comparing the basic, modified, and improved algorithms. We are working with 30-dimensional functions within a search range of [-100, 100]. Our algorithms use a population size of 50, and we have set FE_{max} at 150,000. All the results come from running each test function independently 30 times (Table 4.48).

Table 4.49 displays the comparative minimum, maximum, mean, and standard deviation (SD) of fitness values of CEC2014 for the respective considered algorithms. Employing statistical tests is crucial to ascertain the significance and effectiveness of these algorithms. Hence, the Friedman rank test is executed on these algorithms. The outcomes of the Friedman test are ranked based on the normalized Friedman rank values. Overall, the SOS algorithm demonstrates the best performance, followed by the PVS and MVO algorithms, respectively, among the considered algorithms. Figure 4.23 presents the graphical representation of the normalized fitness values obtained using these algorithms.

Table 4.48 The CEC2014 benchmark functions (Liang et al., 2014)

Test function	Optimum value
F1: rotated high conditioned elliptic function	100
F2: rotated bent cigar function	200
F3: rotated discus function	300
F4: shifted and rotated Rosenbrock function	400
F5: shifted and rotated Ackley's function	500
F6: shifted and rotated Weierstrass function	600
F7: hybrid function 1	1700
F8: hybrid function 2	1800
F9: composition function 1	2300
F10: composition function 2	2400

Table 4.49 Results of the CEC2014 benchmark functions

F1	DA	MVO	SCA	WOA	ALO	HTS	PVS	SOS	GWO	TLBO
Minimum	58565780	2553172	117320600	20764280	3909515	20241650	356407	742321	17397500	544284
Maximum	471056300	8086422	299010600	116252200	11940810	65857310	1314715	6884409	139180500	6318889
Mean	236879378	4442254	209659910	46701934	6975804	42486932	653219	3215624	69132940	2855202
SD	139907334	1831582	55258537	27696340	2486152	16913839	356992	1556268	47665518	2020151
F2	DA	MVO	SCA	WOA	ALO	HTS	PVS	SOS	GWO	TLBO
Minimum	427919600	26910	14746420000	10002460	1586	1172297000	282	65133	319445200	3415
Maximum	6891164000	84130	22494760000	118318000	23340	7501994000	2284	671328	4775068000	38920
Mean	2847946520	43587	17490803000	35357405	11399	4519134400	1455	302165	2444525000	22422
SD	2451775239	16108	2574878090	38888312	7718	2111519691	650	175613	1562343306	12516
F3	DA	MVO	SCA	WOA	ALO	HTS	PVS	SOS	GWO	TLBO
Minimum	51007	448	35862	17481	22751	3598	363	1767	21391	350
Maximum	129346	993	48294	87506	70241	13604	10938	11418	36891	4695
Mean	88600	672	41152	50593	43734	8329	2114	4645	30582	1621
SD	23992	145	3559	26685	17239	3240	3167	2735	4571	1629
F4	DA	MVO	SCA	WOA	ALO	HTS	PVS	SOS	GWO	TLBO
Minimum	606.98	463.48	1140.39	565.41	419.74	702.48	466.37	474.11	553.60	468.36
Maximum	1076.44	580.23	1586.79	714.85	549.12	1235.45	544.28	592.34	881.73	612.81
Mean	818.47	514.49	1386.20	619.78	518.74	971.12	504.79	520.65	627.15	532.04
SD	140.27	32.40	113.93	57.85	41.30	177.51	32.17	39.50	104.33	50.01
F5	DA	MVO	SCA	WOA	ALO	HTS	PVS	SOS	GWO	TLBO
Minimum	520.75	520.05	520.83	520.15	520.00	520.50	520.88	520.16	520.93	520.74
Maximum	521.05	520.24	521.04	520.74	520.37	520.58	521.01	520.33	521.05	520.95
Mean	520.97	520.09	520.94	520.44	520.09	520.54	520.94	520.26	520.99	520.87
SD	0.08	0.06	0.06	0.16	0.14	0.03	0.04	0.05	0.03	0.07
F6	DA	MVO	SCA	WOA	ALO	HTS	PVS	SOS	GWO	TLBO
Minimum	627.46	604.09	630.77	632.92	612.56	610.50	613.95	609.61	611.17	622.35
Maximum	637.96	614.86	640.72	640.44	629.48	622.25	620.68	617.28	622.25	630.02
Mean	634.52	610.36	637.82	635.31	622.08	618.15	617.22	613.06	616.06	625.99

SD	3.38	3.61	3.06	2.71	4.92	2.88	2.06	2.17	3.36	2.41
F7	DA	MVO	SCA	WOA	ALO	HTS	PVS	SOS	GWO	TLBO
Minimum	909536.60	38303.40	1881665.00	2166853.00	16587.23	97713.45	33201.82	178085.90	55680.36	6619.57
Maximum	41244400.00	478752.30	17175590.00	17048760.00	607682.20	1679102.00	135111.90	896813.20	10289240.00	237370.20
Mean	9764114.16	249524.23	7723997.60	7402637.20	240189.58	689303.20	84215.14	388513.13	3061148.95	76115.04
SD	11858120.12	166078.37	4144957.60	5270508.69	157631.68	524401.14	28790.39	219012.43	3490400.94	73179.79
F8	DA	MVO	SCA	WOA	ALO	HTS	PVS	SOS	GWO	TLBO
Minimum	3453.40	2750.16	93398860.00	3070.49	2139.65	2405.67	1844.34	1878.42	2627.03	2073.25
Maximum	3098282.00	28755.69	423036900.00	18749.62	18686.11	5999088.00	5723.41	10791.01	2641.35	21066.18
Mean	783890.49	12032.07	203479872.00	7370.96	4941.97	618831.73	2888.27	4291.75	2633.80	5609.60
SD	1106462.25	9974.74	103209826.26	4794.52	5183.64	1890578.05	1256.50	3618.23	5.02	5740.27
F9	DA	MVO	SCA	WOA	ALO	HTS	PVS	SOS	GWO	TLBO
Minimum	2650.36	2615.64	2654.62	2500.00	2615.60	2622.35	2615.24	2500.00	2627.03	2615.24
Maximum	2714.01	2618.02	2689.63	2643.56	2618.53	2672.58	2615.24	2615.27	2641.35	2615.25
Mean	2674.64	2616.56	2672.19	2623.57	2616.72	2636.13	2615.24	2557.92	2633.80	2615.24
SD	20.35	0.90	11.53	43.63	0.92	15.62	0.00	60.44	5.02	0.00
F10	DA	MVO	SCA	WOA	ALO	HTS	PVS	SOS	GWO	TLBO
Minimum	2602.82	2602.30	2600.04	2600.71	2639.68	2613.82	2623.83	2600.02	2600.00	2600.00
Maximum	2656.76	2641.83	2600.54	2620.55	2656.03	2659.96	2649.52	2600.16	2600.01	2600.00
Mean	2648.19	2631.69	2600.24	2608.96	2648.23	2641.11	2628.39	2600.09	2600.01	2600.00
SD	15.76	12.80	0.18	5.64	6.33	13.20	7.60	0.04	0.00	0.00
	DA	MVO	SCA	WOA	ALO	HTS	PVS	SOS	GWO	TLBO
Overall Friedman test	183	63.5	169.5	137	92.5	141	56	55	82	66
Normalized Friedman value	1.000	0.3470	0.9262	0.7486	0.5055	0.7705	0.3060	0.3005	0.6967	0.3607
Friedman rank	10	3	9	7	5	8	2	1	6	4

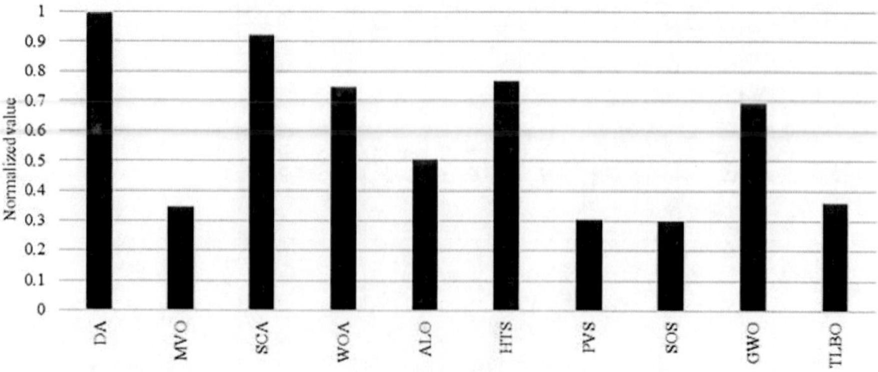

Fig. 4.23 The Friedman rank test of mean fitness values obtained the algorithms

References

Achtziger, W. (1999a). Local stability of trusses in the context of topology optimization. Part I: Exact modelling. *Structural Optimization, 17*(4), 235–246. https://doi.org/10.1007/s001580050056

Achtziger, W. (1999b). Local stability of trusses in the context of topology optimization part II: A numerical approach. *Structural Optimization, 17*(4), 247–258. https://doi.org/10.1007/s001580050056

Ahrari, A., Atai, A. A., & Deb, K. (2014). Simultaneous topology, shape and size optimization of truss structures by fully stressed design based on evolution strategy. *Engineering Optimization, 47*(8), 37–41. https://doi.org/10.1080/0305215X.2014.947972

Angelo, J. S., Barbosa, H. J. C., & Bernardino, H. S. (2012). *Multi-objective ant colony approaches for structural optimization problems*. Proc. of the Eleventh International Conference on Computational Structures Technology.

Angelo, J. S., Bernardino, H. S., & Barbosa, H. J. C. (2015). Ant colony approaches for multiobjective structural optimization problems with a cardinality constraint. *Advances in Engineering Software, 80*, 101–115. https://doi.org/10.1016/j.advengsoft.2014.09.015

Bennage, W. A., & Dhingra, A. K. (1995). Optimization of truss topology using tabu search. *International Journal for Numerical Methods in Engineering, 38*(23), 4035–4052. https://doi.org/10.1002/nme.1620382308

Cheng, M. Y., & Prayogo, D. (2014). Symbiotic organisms search: A new metaheuristic optimization algorithm. *Computers and Structures, 139*, 98–112. https://doi.org/10.1016/j.compstruc.2014.03.007

Deb, K., & Gulati, S. (2001). Design of truss-structures for minimum weight using genetic algorithms. *Finite Elements in Analysis and Design, 37*(5), 447–465. https://doi.org/10.1016/S0168-874X(00)00057-3

Fenton, M., Mcnally, C., Byrne, J., Hemberg, E., Mcdermott, J., & Neill, M. O. (2014). Automation in construction automatic innovative truss design using grammatical evolution. In *Automation in construction* (Vol. 39, pp. 59–69). Elsevier B.V. https://doi.org/10.1016/j.autcon.2013.11.009

Gonçalves, M. S., Lopez, R. H., & Miguel, L. F. F. (2015). Search group algorithm: A new metaheuristic method for the optimization of truss structures. *Computers & Structures, 153*, 165–184. https://doi.org/10.1016/j.compstruc.2015.03.003

Hajirasouliha, I., Pilakoutas, K., & Moghaddam, H. (2011). Topology optimization for the seismic design of truss-like structures. In *Computers and structures* (Vol. 89, pp. 702–711). Elsevier Ltd. https://doi.org/10.1016/j.compstruc.2011.02.003

Jin, P., & De-yu, W. (2006). Topology optimization of truss structure with fundamental frequency and frequency domain dynamic response constraints. *Acta Mechanica Solida Sinica, 19*(3) The Chinese Society of Theoretical and Applied Mechanics, 231–240. https://doi.org/10.1007/s10338-006-0628-2

Kaveh, A., & Zolghadr, A. (2013). Topology optimization of trusses considering static and dynamic constraints using the CSS. *Applied Soft Computing Journal, 13*(5) Elsevier B.V, 2727–2734. https://doi.org/10.1016/j.asoc.2012.11.014

Kumar, S., Tejani, G. G., & Mirjalili, S. (2019). Modified symbiotic organisms search for structural optimization. *Engineering Computations, 35*, 1269–1296. https://doi.org/10.1007/s00366-018-0662-y

Kumar, S., Tejani, G. G., Pholdee, N., et al. (2020). Multi-objective modified heat transfer search for truss optimization. *Engineering Computations, 37*, 3439. https://doi.org/10.1007/s00366-020-01010-1

Kumar, S., Jangir, P., Tejani, G. G., et al. (2021a). MOPGO: A new physics-based multi-objective plasma generation optimizer for solving structural optimization problems. *IEEE Access, 1–1*, 84982. https://doi.org/10.1109/access.2021.3087739

Kumar, S., Tejani, G. G., Pholdee, N., et al. (2021b). Multi-objective teaching-learning-based optimization for structure optimization. *Smart Science, 1*, 1–12. https://doi.org/10.1080/23080477.2021.1975074

Kumar, S., Tejani, G. G., Pholdee, N., & Bureerat, S. (2022). Performance enhancement of meta-heuristics through random mutation and simulated annealing-based selection for concurrent topology and sizing optimization of truss structures. *Soft Computing, 26*, 5661–5683. https://doi.org/10.1007/s00500-022-06930-2

Kumar, S., Jangir, P., Tejani, G. G., et al. (2022a). A decomposition based multi-objective heat transfer search algorithm for structure optimization. *Knowledge-Based Systems, 253*, 109591. https://doi.org/10.1016/j.knosys.2022.109591

Kumar, S., Jangir, P., Tejani, G. G., et al. (2022b). MOTEO: A novel physics-based multiobjective thermal exchange optimization algorithm to design truss structures. *Knowledge-Based Systems, 242*, 108422. https://doi.org/10.1016/j.knosys.2022.108422

Kumar, S., Panagant, N., Tejani, G. G., et al. (2023). A two-archive multi-objective multi-verse optimizer for truss design. *Knowledge-Based Systems, 270*, 110529. https://doi.org/10.1016/j.knosys.2023.110529

Liang, J. J., Qu, B. Y., & Suganthan, P. N. (2014). *Problem definitions and evaluation criteria for the CEC 2014 special session and competition on single objective real-parameter numerical optimization*. Computational Intelligence Laboratory, Zhengzhou University, Zhengzhou China and Technical Report, Nanyang Technological University.

Mela, K. (2014). Resolving issues with member buckling in truss topology optimization using a mixed variable approach. *Structural and Multidisciplinary Optimization, 50*(6), 1037–1049. https://doi.org/10.1007/s00158-014-1095-x

Mirjalili, S. (2015). The ant lion optimizer. *Advances in Engineering Software, 83*. Elsevier Ltd, 80–98. https://doi.org/10.1016/j.advengsoft.2015.01.010

Mirjalili, S. (2016a). Dragonfly algorithm: A new meta-heuristic optimization technique for solving single-objective, discrete, and multi-objective problems. *Neural Computing and Applications, 27*(4) Springer, 1053–1073. https://doi.org/10.1007/s00521-015-1920-1

Mirjalili, S. (2016b). SCA: A sine cosine algorithm for solving optimization problems. *Knowledge-Based Systems, 96*, 120. https://doi.org/10.1016/j.knosys.2015.12.022

Mirjalili, S., & Lewis, A. (2016). The whale optimization algorithm. *Advances in Engineering Software, 95*. Elsevier Ltd, 51–67. https://doi.org/10.1016/j.advengsoft.2016.01.008

Mirjalili, S., Mirjalili, S. M., & Lewis, A. (2014). Grey Wolf Optimizer. *Advances in Engineering Software, 69*. Elsevier Ltd, 46–61. https://doi.org/10.1016/j.advengsoft.2013.12.007

Mirjalili, S., Mirjalili, S. M., & Hatamlou, A. (2016). Multi-verse optimizer: A nature-inspired algorithm for global optimization. *Neural Computing and Applications, 27*(2), 495–513. https://doi.org/10.1007/s00521-015-1870-7

Mortazavi, A., & Toğan, V. (2016). Simultaneous size, shape, and topology optimization of truss structures using integrated particle swarm optimizer. *Structural and Multidisciplinary Optimization, 54*, 715. https://doi.org/10.1007/s00158-016-1449-7

Nakamura, T., & Ohsaki, M. (1992). A natural generator of optimum topology of plane trusses for specified fundamental-frequency. *Computer Methods in Applied Mechanics and Engineering, 94*(1), 113–129. https://doi.org/10.1016/0045-7825(92)90159-h

Noilublao, N., & Bureerat, S. (2011). Simultaneous topology, shape and sizing optimisation of a three-dimensional slender truss tower using multiobjective evolutionary algorithms. *Computers and Structures, 89*(23–24) Elsevier Ltd, 2531–2538. https://doi.org/10.1016/j.compstruc.2011.08.010

Patel, V. K., & Savsani, V. J. (2015). Heat Transfer Search (HTS): A novel optimization algorithm. *Information Sciences, 324*. Elsevier Ltd, 217–246. https://doi.org/10.1016/j.ins.2015.06.044

Rajeev, S., & Krishnamoorthy, C. S. (1997). Genetic algorithms-based methodologies for design optimization of trusses. *Journal of Structural Engineering, 123*(3), 350. https://doi.org/10.1061/(ASCE)0733-9445(1997)123:3(350)

Rao, R. V., Savsani, V. J., & Vakharia, D. P. (2011). Teaching–learning-based optimization: a novel method for constrained mechanical design optimization problems. *Computer-Aided Design, 43*(3) Elsevier Ltd, 303–315. https://doi.org/10.1016/j.cad.2010.12.015

Rao, R. V., Savsani, V. J., & Vakharia, D. P. (2012). Teaching–learning-based optimization: An optimization method for continuous non-linear large scale problems. *Information Sciences, 183*(1) Elsevier Inc, 1–15. https://doi.org/10.1016/j.ins.2011.08.006

Ringertz, U. L. F., & Torbjörn. (1986). A branch and bound algorithm for topology optimization of truss structures. *Engineering Optimization, 10*, 111–124. https://doi.org/10.1080/03052158608902532

Rozvany, G. I. N. (1996). Difficulties in truss topology optimization with stress and local buckling constraints. *Structural Optimization, 11*(2), 134–136. https://doi.org/10.1007/BF01376857

Savsani, P. V., & Savsani, V. J. (2015). Passing vehicle search (PVS): A novel metaheuristic algorithm. *Applied Mathematical Modelling, 40*(5–6), 3951–3978. https://doi.org/10.1016/j.apm.2015.10.040

Savsani, V. J., Tejani, G. G., & Patel, V. K. (2016). Truss topology optimization with static and dynamic constraints using modified subpopulation teaching–learning-based optimization. *Engineering Optimization, 48*, 1–17. https://doi.org/10.1080/0305215X.2016.1150468

Savsani, V. J., Tejani, G. G., Patel, V. K., & Savsani, P. (2017). Modified meta-heuristics using random mutation for truss topology optimization with static and dynamic constraints. *Journal of Computational Design and Engineering, 4*, 106–130. https://doi.org/10.1080/0305215X.2016.1150468

Smith, O. (1996). An interactive system for truss topology design. *Advances in Engineering Software, 27*(96), 167–178. https://doi.org/10.1016/0965-9978(96)00010-5

Souza, R. R., Miguel, L. F. F., Lopez, R. H., Miguel, L. F. F., & Torii, A. J. (2016). A procedure for the size, shape and topology optimization of transmission line tower structures. *Engineering Structures, 111*. Elsevier Ltd, 162–184. https://doi.org/10.1016/j.engstruct.2015.12.005

Stolpe, M., & Svanberg, K. (2003). A note on stress-constrained truss topology optimization. *Structural and Multidisciplinary Optimization, 25*, 62–64.

Tang, W., Tong, L., & Gu, Y. (2005). Improved genetic algorithm for design optimization of truss structures with sizing, shape and topology variables. *International Journal for Numerical Methods in Engineering, 62*, 1737–1762. https://doi.org/10.1002/nme.1244

Tejani, G. G., Savsani, V. J., & Patel, V. K. (2016a). Modified sub-population teaching-learning-based optimization for design of truss structures with natural frequency constraints. *Mechanics Based Design of Structures and Machines, 44*, 495–513. https://doi.org/10.1080/15397734.2015.1124023

Tejani, G. G., Savsani, V. J., & Patel, V. K. (2016b). Adaptive symbiotic organisms search (SOS) algorithm for structural design optimization. *Journal of Computational Design and Engineering, 3*, 226–249. https://doi.org/10.1016/j.jcde.2016.02.003

Tejani, G. G., Savsani, V. J., & Patel, V. K. (2017a). Modified sub-population based heat transfer search algorithm for structural optimization. *International Journal of Applied Metaheuristic Computing, 8*, 1–23. https://doi.org/10.4018/IJAMC.2017070101

Tejani, G. G., Savsani, V. J., Patel, V. K., & Bureerat, S. (2017b). Topology, shape, and size optimization of truss structures using modified teaching-learning based optimization. *Adv Comput Des, 2*, 313–331. https://doi.org/10.12989/acd.2017.2.4.313

Tejani, G. G., Savsani, V. J., Bureerat, S., & Patel, V. K. (2018a). Topology and size optimization of trusses with static and dynamic bounds by modified symbiotic organisms search. *Journal of Computing in Civil Engineering, 32*, 1–11. https://doi.org/10.1061/(ASCE)CP.1943-5487.0000741

Tejani, G. G., Savsani, V. J., Patel, V. K., & Savsani, P. V. (2018b). Size, shape, and topology optimization of planar and space trusses using mutation-based improved metaheuristics. *Journal of Computational Design and Engineering, 5*, 198–214. https://doi.org/10.1016/j.jcde.2017.10.001

Tejani, G. G., Pholdee, N., Bureerat, S., et al. (2018c). Multiobjective adaptive symbiotic organisms search for truss optimization problems. *Knowledge-Based Systems, 161*, 398–414. https://doi.org/10.1016/j.knosys.2018.08.005

Tejani, G. G., Savsani, V. J., Patel, V. K., & Mirjalili, S. (2018d). An improved heat transfer search algorithm for unconstrained optimization problems. *Journal of Computational Design and Engineering, 6*, 13. https://doi.org/10.1016/j.jcde.2018.04.003

Tejani, G. G., Savsani, V. J., Patel, V. K., & Mirjalili, S. (2018e). Truss optimization with natural frequency bounds using improved symbiotic organisms search. *Knowledge-Based Systems, 143*, 162–178. https://doi.org/10.1016/j.knosys.2017.12.012

Tejani, G. G., Savsani, V. J., Bureerat, S., et al. (2019a). Topology optimization of truss subjected to static and dynamic constraints by integrating simulated annealing into passing vehicle search algorithms. *Engineering Computations, 35*, 1–19.

Tejani, G. G., Kumar, S., & Gandomi, A. H. (2019b). Multi-objective heat transfer search algorithm for truss optimization. *Engineering Computations, 37*, 641. https://doi.org/10.1007/s00366-019-00846-6

Tejani, G. G., Pholdee, N., Bureerat, S., et al. (2019c). Structural optimization using multi-objective modified adaptive symbiotic organisms search. *Expert Systems with Applications, 125*, 425–441. https://doi.org/10.1016/j.eswa.2019.01.068

Xu, B., Jiang, J., Tong, W., & Wu, K. (2003). Topology group concept for truss topology optimization with frequency constraints. *Journal of Sound and Vibration, 261*(5), 911–925. https://doi.org/10.1016/S0022-460X(02)01021-0

Zhou, M. (1996). Difficulties in truss topology optimization with stress and local buckling constraints. *Structural Optimization., 11*, 134–136.

Chapter 5
Topology and Size Optimization

5.1 Introduction

The investigation of structural optimization, particularly in truss sizing, has placed a significant emphasis on design criteria such as structural strength and rigidity, as evidenced by studies conducted by Ringertz (1986), Deb and Gulati (2001), and Ahrari et al. (2014). However, limited research has been conducted to incorporate dynamic structural parameters such as natural frequency in the design problem, as shown by studies conducted by Bennage and Dhingra (1995), Xu et al. (2003), Jin and De-yu (2006), Noilublao and Bureerat (2011), Kaveh and Zolghadr (2013), Tejani et al. (2016a, b, 2017a, b, 2018c). Natural frequency is an essential design criterion for structures under dynamic loadings since it measures dynamic structural stiffness. Most engineering structures are exposed to various sources of dynamic excitations, leading to unwanted vibration and noise (Hajirasouliha et al., 2011). To avoid undesirable vibration phenomena, structural resonance and natural frequency constraints should be included in structural design problems (Kaveh & Zolghadr, 2013; Nakamura & Ohsaki, 1992; Xu et al., 2003). However, this constraint often leads to non-convex feasible regions in structural optimization problems. Furthermore, buckling can complicate truss topology optimization (TTO), producing non-convex feasible regions for the optimization problem (Achtziger, 1999a, b; Kumar et al., 2022; Mela, 2014; Rozvany, 1996; Savsani et al., 2016; Savsani et al., 2017; Smith, 1996; Stolpe & Svanberg, 2003; Tejani et al., 2018a, b, d; Zhou, 1996). The simultaneous inclusion of natural frequencies and buckling constraints in structural optimization problems may result in difficulty in solving the problem and high computational effort (Xu et al., 2003). Moreover, the simultaneous consideration of topology, shape, and size (TSS) or topology and size (TS) optimization adds a higher level of complexity to the problem (Tejani, 2017). However, these constraints cannot be avoided to ensure the usability of structures. To prevent ill-conditioning in structural finite element analysis, kinematic instability and illegitimate trusses should be detected before conducting structural analyses.

V. Savsani et al., *Truss Optimization*, https://doi.org/10.1007/978-3-031-49295-2_5

Metaheuristic methods are the most popular optimizer in various engineering optimization problems due to their simplicity, robustness, and gradient-free nature. With thousands of these methods introduced in recent decades, the ALO, DA, GWO, MVO, SCA, WOA, HTS, TLBO, PVS, and SOS algorithms have not been examined for TSS/TS problems subjected to static and dynamic constraints with multi-load conditions. This lack of examination motivated us to incorporate these basic algorithms and assess their effectiveness in solving TTO problems. In this book, these algorithms are tested on ten distinct trusses, with the objective function being structural mass and design constraints including stress, displacement, buckling, and frequency as constraints, along with multi-load conditions.

5.2 Truss Topology Optimization

Truss structural optimization can be classified into three types: size optimization, shape optimization, and topology optimization (Tejani et al., 2017a, b). The choice of optimization type depends on the specific design variables. Topological design variables are used to determine the initial configuration or layout of the structure. This type of design is challenging and computationally demanding since the goal is to find the best possible initial configuration, resulting in a large-scale design problem. TTO aims to minimize mass while searching for the optimal topology and element cross-sectional areas in the final design, leading researchers and engineers to investigate various aspects of TTO problems.

To address TTO, the ground structure method is widely used due to its simplicity and flexibility. This approach involves defining a ground structure consisting of user-defined truss elements or ground elements, which are typically connections between nodes chosen based on personal preferences or construction feasibility. The TTO process involves removing or maintaining the ground elements until the best element connection is found. However, removing or keeping ground elements may lead to a singular global stiffness matrix, causing an optimization run to fail before reaching an optimal solution. One way to avoid this situation is to assign significantly small values for cross-sectional areas of deleted elements, but this adds unnecessary computing time when assembling a global stiffness matrix. Furthermore, this strategy may not be suitable for truss optimization with natural frequency constraints, as microelements can affect the calculated natural frequency values. Therefore, this study uses a restructuring approach instead of microelements, where the finite element model is restructured based on the existence of elements controlled by topological design variables. A binary string with 0/1 bits represents the removed/existing elements, and an additional type of design variable is used to determine element cross-sectional areas, reducing the analysis of removed elements. Although most studies ignore concentrated masses at the nodes, these should be considered since they affect the overall structural mass.

Researchers and engineers have explored the optimization of truss sizing using metaheuristic methods and have considered various constraints, including natural frequency, displacement, stress, and buckling. However, many researchers have focused on sizing optimization with static loading constraints, such as stress and displacement. Some researchers have proposed combined shape and sizing designs for trusses with only natural frequency constraints (Gholizadeh & Barzegar, 2013; Kaveh & Zolghadr, 2014; Li & Liu, 2011; Pholdee & Bureerat, 2014; Singh et al., 2022). Although there have been several studies on shape and sizing optimization with natural frequency constraints, research on truss topology optimization with natural frequency constraints has been somewhat limited (Bai et al., 2009; Gonçalves et al., 2015; Jin & De-yu, 2006; Kaveh & Zolghadr, 2013; Kumar et al., 2019, 2020; Noilublao & Bureerat, 2011; Ohsaki et al., 1999; Xu et al., 2003). Therefore, topological truss design with natural frequency constraints can be considered a new area for exploration.

Researchers have employed two design approaches to address truss topology and sizing optimization (TS), namely the two-stage and single-stage strategies. In the two-stage approach, ground elements are created with a constant cross-sectional area, and the topological design is performed by deleting or maintaining some ground elements without changing the truss elements' cross-sectional areas or nodal coordinates. Once the optimal topology is obtained, sizing and/or shape optimization is activated to search for the optimal element cross-sectional areas. However, this approach may not achieve a global optimum solution. On the other hand, the single-stage approach simultaneously searches for topology and size (TS) or topology, shape, and size (TSS) variables in one optimization run, but it requires more computational efforts. In this chapter, the proposed methods for TSS and TS optimization will combine the ground structure method and finite element model restructuring. Ten truss design test problems will be presented to validate the proposed methods.

5.3 Problem Formulation

The ground structure approach is a commonly employed method in topology optimization problems that enables results to be compared fairly to benchmark problems. This method involves representing all possible connections between nodes as the ground structure. An optimization algorithm is then applied to determine the existence of each element required to form an optimal topology. The goal of a typical TTO problem is to minimize the mass of the truss, subject to various design constraints such as displacement, stress, buckling instability, natural frequencies, and kinematic stability. In cases where nodal and element masses exist, they are also included in the objective function. The topology optimization problem can be expressed as follows:

$$\text{Find}, X = \{X_1, X_2, .., X_m\} \tag{5.1}$$

to minimize, mass of truss,

$$F(X) = \sum_{i=1}^{m} B_i X_i \rho_i L_i + \sum_{j=1}^{n} b_j, B_i = \begin{cases} 0, \textit{if } X_i < \text{Critical area} \\ 1, \textit{if } X_i \geq \text{Critical area} \end{cases}$$

Subjected to:

$$\begin{aligned}
&g_1(X) : \text{Stress constraints}, \ |B_i\sigma_i| - \sigma_i^{\max} \leq 0 \\
&g_2(X) : \text{Displacement constraints}, \ |\delta_j| - \delta_j^{\max} \leq 0 \\
&g_3(X) : \text{Euler buckling constraints}, \ |B_i\sigma_i^{\text{comp}}| - \sigma_i^{cr} \leq 0, \ \text{where } \sigma_i^{cr} = \tfrac{k_i A_i E_i}{L_i^2} \\
&g_4(X) : f_r - f_r^{\min} \geq 0 \ \text{for some natural frequencies} \\
&g_5(X) : \text{Cross} - \text{sectional area constraints}, \ X_i^{\min} \leq X_i \leq X_i^{\max} \\
&g_6 : \text{Check on kinematic stability} \\
&g_7 : \text{Check on validity of structure} \\
&\text{where}, \ i = 1, 2, \ldots, m; j = 1, 2, \ldots, n
\end{aligned}$$

where B_i is a binary element (0 and 1 for deleting and remaining the *i*th element, respectively). $X_i, \rho_i, L_i, E_i, \sigma_i$, and σ_i^{cr} stand for design variable, mass density, element length, Young's modulus, stress, and critical buckling stress on the element "*i*" in that order. δ_j and b_j are values of nodal displacement and mass values of node "*j*," respectively. f_r is structural natural frequent of the *r*th mode. The superscripts "max," "min," and "comp" denote maximum allowable stress, minimum allowable stress, and compressive stress, respectively. k_i is the Euler buckling coefficient calculated from elements' cross-sections.

It is worth mentioning that kinematic stability (g_7) is considered a design constraint as it can potentially avoid an unstable condition during structural analysis. The criteria for kinematic instability, as proposed by Deb and Gulati (2001), are based on two steps, presented in Chap. 2.

A truss topology is considered unusable (g_8) if none of its elements are connected to nodes with applied loads, nodes with boundary conditions, and unchangeable nodes predefined by the user (Li & Liu, 2011).

The penalty function technique is employed to manage the constraints. This technique penalizes violated solutions by adding positive penalty terms to their objective function values, with the aim of minimizing them. The computation of the penalty value can be done by following the methodology described by Deb and Gulati (2001) and Kaveh and Zolghadr (2013).

$$\text{Penalized } F(X) = \begin{cases} 10^9, \text{if } g_8 \text{ is violated} \\ 10^8, \text{if } g_7 \text{ is violated with DOF} \\ 10^7, \text{if } g_6 \text{ is violated with positive definiteness} \\ F(X) * F_{\text{penalty}}, \text{otherwise} \end{cases} \tag{5.2}$$

where

$$F_{\text{penalty}} = (1 + \varepsilon_1 * C)^{\varepsilon_2}, C = \sum_{i=1}^{q} C_i, C_i = \left| 1 - \frac{p_i}{p_i^*} \right| \tag{5.3}$$

where p_i is the level of constraint violation having the bound as p_i^*. The parameter q is a number of active constraints. The variables ε_1 and ε_2 are predetermined by a user. In this study, the values of both ε_1 and ε_2 are set as 1.5, which were obtained from experimenting their effect on the balance of the exploitation–exploration balance.

The dragonfly algorithm (DA) (Mirjalili, 2016a), multiverse optimizer (MVO) (Mirjalili et al., 2016), sine cosine algorithm (SCA) (Mirjalili, 2016b), whale optimization algorithm (WOA) (Mirjalili & Lewis, 2016), ant lion optimizer (ALO) (Mirjalili, 2015), heat transfer search (HTS) (Patel & Savsani, 2015), passing vehicle search (PVS) (Savsani & Savsani, 2015), symbiotic organisms search (SOS) (Cheng & Prayogo, 2014), gray wolf optimizer (GWO) (Mirjalili et al., 2014), and teaching–learning-based optimization (TLBO) (Rao et al., 2011, 2012) algorithms are used for TTO.

5.4 Results and Discussion on Truss Topology Optimization with Continuous Cross-Sections

Ten different benchmark problems are employed in this section to evaluate the effectiveness of the algorithms being considered for continuous cross-sections. These benchmark problems consist of various truss structures, such as the 10-bar, 14-bar, 15-bar, 24-bar, 20-bar, 72-bar 3D, 39-bar, 45-bar, 25-bar 3D, and 39-bar 3D trusses. The proposed algorithms are utilized to tackle each truss problem by conducting 100 independent runs, acknowledging the stochastic nature of metaheuristics. The population size remains constant at 50 for all the problems, while the total number of function evaluations (FE) per run is set to 20,000. The primary goal is to assess the performance of the chosen algorithms in solving the truss TTO optimization problems that are continuous in nature.

5.4.1 Topology Optimization of the 10-Bar Truss with Continuous Cross-Sections

In this section, the optimization of a 10-bar truss for the TTO approach is performed using ten continuous size and topology variables. The ground structure of the first benchmark truss, along with the load and boundary conditions, is illustrated in Fig. 5.1. Previous studies by Deb and Gulati (2001), and, Richardson et al.

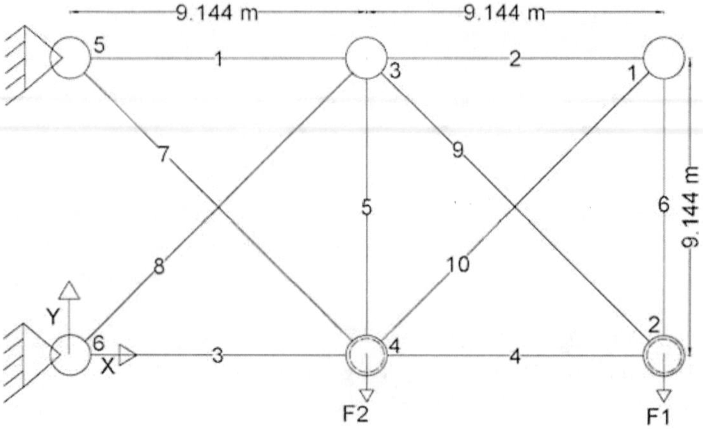

Fig. 5.1 The 10-bar truss

Table 5.1 Design parameters of the 10-bar truss

Size variables: $X_i = A_i$, $i = 1, 2, \ldots, 10$	
Multiple load conditions:	
Condition 1 : $F_1 = 44.537$ kN, $F_2 = 44.537$ kN	Condition 2 : $F_1 = 44.537$ kN, $F_2 = 0$ kN
Lumped masses on nodes 2 and 4 are 500 kg	
Stress and displacement constraints: $\sigma_i^{\max} = 172.43$ MPa and $\delta_{y_5 \& y_6}^{\max} = 100$ mm (0.1 m)	
Natural frequency constraint: $f_1 \geq 4$ Hz	
The continuous cross-sectional areas: $[A^{\min}, A^{\max}] = [-100, 100]\ \text{cm}^2 = ([-100, 100] \times 10^{-4}\ \text{m}^2)$ and critical area = 1 cm^2	
Material properties: $E = 69$ GPa and $\rho = 2740\text{kg/m}^3$	

(2012), have investigated this problem focusing on static constraints such as elemental stress and nodal displacement.

While previous investigations have focused on either static constraints (stress, buckling, displacement) or dynamic constraints (natural frequencies), this study introduces simultaneous consideration of both static and dynamic constraints. To achieve this, the problem is modified to incorporate multi-load conditions, lumped masses, natural frequency constraints, and continuous cross-sectional areas. In this problem, the continuous cross-sectional areas (A_i, where $i = 1, 2, \ldots, 10$) are considered within the range of $[-100, 100]$ cm^2, where a zero or negative value indicates removal of the element. For further details on material properties, design variables, loading conditions, lumped masses, and other important data related to the test problem, refer to Table 5.1.

- The 10-bar truss is considered as a demonstration of the truss topology optimization procedure. The step-by-step procedure for implementing the topology optimization with a sample calculation is outlined below:

Step 1: Begin by defining the ground structure of the truss. This involves generating nodes over a predefined design domain and assigning all possible element connections, as well as material properties, loading, and boundary conditions.

- Figure 5.1 illustrates the ground structure of the 10-bar truss, showcasing the loading and boundary conditions, while Table 5.1 provides the material properties.

Step 2: Define the objective function, population size, design variables, constraints, algorithm controlling parameters, and termination criterion are defined.

- The objective function and constraints of the 10-bar truss is:

$$\text{Find}, X = \{A_1, A_2, .., A_{10}\} \tag{5.4}$$

to minimize, mass of truss,

$$F(X) = \sum_{i=1}^{10} B_i X_i \rho_i\, L_i + \sum b_{2\&4}\ B_i = \begin{cases} 0, \text{if } X_i < \text{Critical area} \\ 1, \text{if } X_i \geq \text{Critical area} \end{cases}$$

Subject to:
Behavior constraints:

$g_1(X)$: Stress constraints, $|\sigma_i| - 172.43 \leq 0$, in MPa
$g_2(X)$: Displacement constraints, $|\delta_j| - 100 \leq 0$, in mm
$g_3(X)$: Euler buckling constraints, $\left|\sigma_i^{\text{comp}}\right| - \sigma_i^{cr} \leq 0$, where $\sigma_i^{cr} = \frac{k_i A_i E_i}{L_i^2}$
$g_4(X)$: $f_1 \geq 4$, in Hz

Side constraints:

Cross-sectional area constraints, $-100 \leq A_i \leq 100$ cm^2 and critical area $= 1$ cm^2 where $i = 1, 2, \ldots, 10; j = 1, 2, \ldots, 6$

- Assume population size = 05
- Design variables: The continuous cross-sectional areas: $[A^{min}, A^{max}] = [-100, 100]$ cm^2 = ($[1, 100] \times 10^{-4}$ m^2) and critical area = 1 cm$^{2.}$
- Termination criterion is presented in Eqs. (5.2) and (5.3); assume function evaluations (FE) ≤ 10

Step 3: Initiate a randomly generated set of topology variables (i.e., population) within its upper and lower bounds.

- $X(k) = \{A_1, A_2, .., A_{10}\}$ `%A: area of cross section in m`^{2}
- $X(1) = [26.69, 8.61, 33.36, 100, -22.84, 20.47, 0.77, -100, 95.27, 16.22]/10000$

- $X(2)$ = [11.71, −3.13, 99.87, 47.97, 43.93, 70.74, 4.85, −100, 54.47, −69.20]/10000
- $X(3)$ = [−33.72, 6.87, 90.54, 100, 40.12, 13.27, 9.02, −66.13, 26.78, 17.43]/10000
- $X(4)$ = [27.49, −2.47, 74.91, 100, 22.06, 35.78, 10.23, −100, 100, −17.82]/10000
- $X(5)$ = [44.16, 24.83, 48.32, 70.56, 40.70, 18.21, 12.81, −100, 21.72, −42.41]/10000

Step 4: Go to truss configurations: Generate trusses according to topology variables.

% generation of coordinates and connectivities

	18.288	9.144
	18.288	0
Node coordinates =	9.144	9.144
	9.144	0
	0	9.144
	0	0

	3	5
	1	3
	4	6
	2	4
Node connectivities =	3	4
	1	2
	4	5
	3	6
	2	3
	1	4

Number Elements = 10
Number Nodes = 6
E = 69e9; % E; modulus of elasticity in Pa
ro = 2740; % E; material density in kg/m^3
% computation of B_i & Active Design variables

B(1)=[1, 1, 1, 1, 0, 1, 0, 0, 1, 1]

- $X(1)$= [26.69, 8.61, 33.36, 100.00, 0, 20.47, 0, 0, 95.27, 16.22]/10000

B(2) = [1, 1, 1, 1, 1, 1, 1, 0, 1, 1]

- $X(2)$= [42.42, 7.73, 48.94, 100.00, 21.18, 22.11, 11.43, 0.00, 7.21, 7.94]/10000

B(3) = [0, 1, 1, 1, 1, 1, 1, 0, 1, 1]

- $X(3)$= [0.00, 6.87, 90.54, 100.00, 40.12, 13.27, 9.02, 0.00, 26.78, 17.43]/10000

B(4) = [1, 0, 1, 1, 1, 1, 1, 0, 1, 0]

- *X*(4)= [27.49, 0, 74.91, 100.00, 22.06, 35.78, 10.23, 0, 100.00, 0]/10000

B(5) = [1, 1, 1, 1, 1, 1, 1, 0, 1, 0]

- *X*(5)= [44.16, 24.83, 48.32, 70.56, 40.70, 18.21, 12.81, −100.00, 21.72, −42.41]/10000

Step 5: Identify invalid trusses, which are those that lack connections to unchangeable nodes (i.e., nodes subjected to load or support), as discussed in Sect. 2.2. Evaluate the degree of freedom of valid trusses as per Eq. 2.11. If a truss is invalid or its degree of freedom is a positive number, assign a large penalty to the objective function and proceed to Step 9.

```
% population 1
DOF(1) = 1 & eigan_value = not applicable % Truss is invalid
fit_fpenalty = 100000000 kg => go to Step 9.
% population 2
DOF(2) = 0 & & eigan_value (2) > 0% Truss is valid
% population 3
DOF(3) = 0 & eigan_value (3) < 0 % Truss is invalid
fit_fpenalty = 10000000 kg => go to Step 9.
% population 4
DOF(4) = 1 & eigan_value = not applicable % Truss is invalid
fit_fpenalty = 100000000 kg => go to Step 9.
% population 5
DOF(5) = 0 & eigan_value (2) > 0% Truss is valid
```

Step 6: If the truss is a valid structure, proceed to finite element analysis. Compute the mass matrix, force vector, displacement vector, stresses vector, and natural frequency vector using finite element analysis.

In the context of a 2D truss, the truss bar is considered a line element with simply supported ends. It is assumed to have a length of L, cross-sectional area of A, modulus of elasticity of E, and mass density of ρ. The bar element experiences forces and displacements.

Compute the stiffness matrix using Eq. 2.2 for each element sequentially and subsequently integrating it into the global stiffness matrix.

```
GDof = 2 x Number Nodes = 12; % GDof: total number of degrees of freedom
% computation of the system stiffness matrix
```

Compute the mass matrix using Eq. 2.6 for each element sequentially and subsequently integrating it into the global mass matrix.

Compute the displacement vectors and stresses vectors using Eq. 2.5.

Compute the buckling vectors and stresses vectors using Eq. 5.4.

Compute the natural frequency vector using Eq. 2.7.

Step 7: Compute natural frequencies, element stresses, nodal displacements, and Euler buckling of the truss using finite element analysis.

```
% Result of population 1
% element stress:
sigma1(1) = Not applicable as tuss is invalid
sigma2(1) = Not applicable as tuss is invalid
% Euler buckling stress:
buckling_stress (1) = Not applicable as tuss is invalid
% Natural frequency:
frequency (1) = Not applicable as tuss is invalid
% nodal displacements:
displacements1 (1) = Not applicable as tuss is invalid
displacements2 (1) = Not applicable as tuss is invalid

% Result of population 2
% element stress:
sigma1(2) = [10500132, 26414345, 27298883, 2411106, 11384356, 9239166,
110195455, 0, 47289294, 36360269]
sigma2(2) = [10500132, 26414345, 18199255, 2411106, 11384356, 9239166,
55097728, 0, 47289294, 36360269]
% Euler buckling stress:
buckling_stress (2) = [14001122, 2552577, 16156005, 33009325, 6991083,
7297700, 1886726, 0, 1190077, 1311223]
% Natural frequency:
frequency (2) = [5.8168, 16.8497, 26.6538, 68.4210, 84.9916, 86.6687,
122.6441, 129.2618]
% nodal displacements:
displacements1 (2) = [4.8920, 50.9710, 3.9372, 52.1954, 1.3915,
34.3330, 3.6177, 32.8243]
displacements2 (2) = [4.8920, 33.9559, 2.7313, 35.1803, 1.3915,
18.5238, 2.4118, 17.0151]

% Result of population 3
% element stress:
sigma1(3) = Not applicable as tuss is invalid
sigma2(3) = Not applicable as tuss is invalid
% Euler buckling stress:
buckling_stress (3) = Not applicable as tuss is invalid
% Natural frequency:
frequency (3) = Not applicable as tuss is invalid
% nodal displacements:
```

```
displacements1 (3) = Not applicable as tuss is invalid
displacements2 (3) = Not applicable as tuss is invalid

% Result of population 4
% element stress:
sigma1(4) = Not applicable as tuss is invalid
sigma2(4) = Not applicable as tuss is invalid
% Euler buckling stress:
buckling_stress (4) = Not applicable as tuss is invalid
% Natural frequency:
frequency (4) = Not applicable as tuss is invalid
% nodal displacements:
displacements1 (4) = Not applicable as tuss is invalid
displacements2 (4) = Not applicable as tuss is invalid

% Result of population 5
% element stress:
sigma1(5) = [10085062, 0, 27653331, 6311897, 10941490, 0, 98302397,
0, 28996048, 0]
sigma2(5) = [10085062, 0, 27653331, 6311897, 10941490, 0, 98302397,
0, 28996048, 0]
% Euler buckling stress:
buckling_stress (5) = [14577365, 8196064, 15948925, 23291513,
13436345, 6012411, 2114991, 0, 3585121, 0]
% Natural frequency:
frequency (5) = [6.1920, 18.6593, 27.9942, 68.4063, 74.8524, 94.6274,
107.6639, 155.3504]
% nodal displacements:
displacements1 (5) = [1.3365, 44.6919, 4.5011, 44.6919, 1.3365,
31.1691, 3.6647, 29.7191]
displacements2 (5) = [1.3365, 29.2216, 3.2796, 29.2216, 1.3365,
16.9203, 2.4431, 15.4703]
```

Step 8: Check for constraint violations in the penalty function. If there is a violation, assign a penalty as per Eqs. 5.2 and 5.3, otherwise compute the total mass of the truss.

```
% Result of population 1
fit (1) = Not applicable as tuss is invalid
fpenalty (1) = Not applicable as tuss is invalid
fit_fpenalty (1) = 100000000 kg

% Result of population 2
fit (2) = 1320.2695 Kg
```

```
fpenalty (2) = 731.4739
fit_fpenalty (2) = 659202.0254 Kg

% Result of population 3
fit (3) = Not applicable as tuss is invalid
fpenalty (3) = Not applicable as tuss is invalid
fit_fpenalty (3) = 10000000 kg

% Result of population 4
fit (4) = Not applicable as tuss is invalid
fpenalty (4) = Not applicable as tuss is invalid
fit_fpenalty (4) = 100000000 kg

% Result of population 5
fit (5) = 770.6842
fpenalty (5) = 12.3214
fit_fpenalty (5) = 9495.9369 kg
```

Step 9: Assign functional values and proceed to the optimization algorithm

```
% Obtain functional values
F(X(1)) = fit_fpenalty (1) = 100000000 kg
F(X(2)) =fit_fpenalty (2) = 659202.0254 Kg
F(X(3)) =fit_fpenalty (3) = 10000000 kg
F(X(4)) =fit_fpenalty (4) = 100000000 kg
F(X(5)) =fit_fpenalty (5) = 100000000 kg
% Generate new set of variables (i.e., population) within its upper and
lower bounds as per the optimization algorithm rules
% iteration 1
```

- $X(k) = \{A_1, A_2, .., A_{10}\}$ `% A: area of cross section in` m^2.
- $X(1) = [45.25, 22.79, 56.55, 100, 12.11, 26.15, 1.30, 0.00, 22.12, 33.12]/10000$
- $X(2) = [70.99, 10.25, 62.24, 5.78, 29.40, -7.69, 14.12, -70.90, 37.97, -11.20]/10000$
- $X(3) = [33.98, 15.67, 69.13, 76.91, 25.70, 8.84, 4.89, -100, 28.20, -49.52]/10000$
- $X(4) = [69.39, 17.30, 98.70, 100, 19.37, 37.70, 6.49, -100.00, 34.09, -39.07]/10000$
- $X(5) = [41.71, 5.87, 100.00, 23.65, 34.56, 17.92, 19.96, -92.93, 13.82, -24.55]/10000$

```
% functional values after iteration 1
F(X(1)) = fit_fpenalty (1) = 335537.8218 Kg
F(X(2)) =fit_fpenalty (2) = 9495.9369 kg
```

```
F(X(3)) =fit_fpenalty (3)= 20766.9676 Kg
F(X(4)) =fit_fpenalty (4)= 51042.1810 Kg
F(X(5)) =fit_fpenalty (5)= 28154.5751 Kg
% End of iteration 1
% Select best solotions for the next iteration as per the algorithm.
F(X(1)) = fit_fpenalty (1)= 335537.8218 Kg
F(X(2)) =fit_fpenalty (2)= 555584.992 Kg
F(X(3)) =fit_fpenalty (3)= 20766.9676 Kg
F(X(4)) =fit_fpenalty (4)= 51042.1810 Kg
F(X(5)) =fit_fpenalty (5)= 28154.5751 Kg
```

Step 10: Check the termination criteria. If not satisfied, generate new trusses (i.e., solutions) according to the algorithm and proceed to Step 4.

```
FE = 10 % function evaluations
```

Step 11: Output: The best solution.

```
X= [44.16, 24.83, 48.32, 70.56, 40.70, 18.21, 12.81, -100.00, 21.72,
-42.41] /10000
F(X) = 9495.9369 Kg
```

The step-by-step procedure for implementing the 10-bar topology optimization in Matlab is presented below:

Step 1:

```
>> n1=[2 6 10]; % elements connected with node 1
>> n2=[4 6 9]; % elements connected with node 2
>> n3=[1 2 5 8 9]; % elements connected with node 3
>> n4=[3 4 5 7 10]; % elements connected with node 4
>> n5=[1 7]; % elements connected with node 5
>> n6=[3 8]; % elements connected with node 6
```

Step 2:

```
>> fit= sum(((A).*L))* ro + nodal mass; % Objective function
>> fpenalty=(1+e1*sum(Constraint violation))^e2; % Penalty function
>> fit_fpenalty =fit*fpenalty; % Penalized objective function
>> OPTIONS.popsize = 5; % Total population size
>> OPTIONS.numVar = 10; % Number of variables
>> PTIONS.maxFE = 10; % Termination criterion
```

```
>> ll=-100*ones(1,10); % Lower bound
>> ul=100*ones(1,10); % Upper bound
```

Step 3:

```
>> for popindex = 1 : OPTIONS.popsize
>> for k = 1 : OPTIONS.numVar
>> chrom(k) = (ll(k)+ (ul(k) - ll(k)) * rand);
>> end
>> Population(popindex).chrom = chrom;
>> end
```

Step 4:

```
% generation of coordinates and connectivities
>> nodeCoordinates=[9.144*2 9.144;9.144*2 0;9.144 9.144;9.144 0 ;0
9.144;0 0];% nodal Coordinates in m
>> elementNodes=[3 5;1 3;4 6;2 4;3 4;1 2;4 5;3 6;2 3;1 4]; % connections
at elements
>> numberElements=size(elementNodes,1); % number of Elements
>> numberNodes=size(nodeCoordinates,1); % number of nodes
>> xx=nodeCoordinates(:,1);
>> yy=nodeCoordinates(:,2);
>> E = 69e9; % E; modulus of elasticity in Pa
>> ro = 2740; % E; material density in kg/m^3
% computation of B_i
>> B=x;
>> B(B>=1e-4)=1;
>> B(B<1e-4)=0; % 0 = absent node, 1 = present
```

Step 5:

```
% basic nodes for supports and load
>> node_topo=[max((B(n1))),max((B(n2))),max((B(n3))),max((B(n4))),
max((B(n5))),max((B(n6)))]; % 0 = absent node, 1 = present
>> DOF=2*sum(node_topo)-sum(B)-4; % DOF=2n-m-nl, Grubler's criterion
n=node, m=link, nl=4=DOF lost at suppot
>> if
>> sum(B(n1))==0||sum(B(n2))==0||sum(B(n3))==0||sum(B(n4))==0||
sum(B(n5)
)==0||sum(B(n6))==0
>> fit=1e9;
>> elseif DOF>0.1
>> fit=1e8; % Go to step 8
```

```
>> else
>> ...% Go to step 6
>> end
```

Step 6:

```
% computation of the system stiffness matrix
>> GDof=2*numberNodes; % GDof: total number of degrees of freedom
% calculation of the system stiffness matrix
>> stiffness=zeros(GDof);
>> for e=1:numberElements
% elementDof: element degrees of freedom (Dof)
>> indice=elementNodes(e,:) ;
>> elementDof=[indice(1)*2-1 indice(1)*2 indice(2)*2-1 indice(2)*2]
;
>> xa=xx(indice(2))-xx(indice(1));
>> ya=yy(indice(2))-yy(indice(1));
>> L(e)=sqrt(xa*xa+ya*ya); % L: length of bar
>> C=xa/L(e);
>> S=ya/L(e);
>> k1=E*(B(e)*A(e))/L(e)*[C*C C*S -C*C -C*S; ...
C*S S*S -C*S -S*S;-C*C -C*S C*C C*S;-C*S -S*S C*S S*S];
>> stiffness(elementDof,elementDof)= ... stiffness(elementDof,
elementDof)+k1;
>> end

% mass : mass matrix
>> mass=zeros(GDof);
>> for e=1:numberElements
% elementDof: element degrees of freedom (Dof)
>> indice=elementNodes(e,:) ;
>> elementDof=[indice(1)*2-1 indice(1)*2 indice(2)*2-1 indice(2)*2];
>> xa=xx(indice(2))-xx(indice(1));
>> ya=yy(indice(2))-yy(indice(1));
>> L(e)=sqrt(xa*xa+ya*ya); % L: length of bar
>> m1=(ro*(A(e))*L(e)/6)*[2 0 1 0;0 2 0 1;1 0 2 0;0 1 0 2];
>> mass(elementDof,elementDof)= ... mass(elementDof,elementDof)+m1;
>> end
% force : force vector
>> force=zeros(GDof,1); % force vector
% applied load Condition 1
>> force 1 (4)=-44.537e3; % in N; Condition 1
>> force 1 (8)=-44.537e3; % in N; Condition 1
```

```
% applied load Condition 2
>> force 2 (4)=-44.537e3; % in N; Condition 2

% displacements: displacement vector
>> prescribedDof=(9:12); % boundary conditions; Restricted DOF
>> activeDof=setdiff([1:GDof]', [prescribedDof]); % active DOF
>> U1=stiffness(activeDof,activeDof)\force1(activeDof);
>> displacements1=zeros(GDof,1);
>> displacements1(activeDof)=U1; % global displacements vector1
>> U2=stiffness(activeDof,activeDof)\force2(activeDof);
>> displacements2=zeros(GDof,1);
>> displacements2(activeDof)=U2; % global displacements vector2

% stress: stress vector
>> for e=1:numberElements
>> indice=elementNodes(e,:);
>> elementDof=[indice(1)*2-1 indice(1)*2 indice(2)*2-1 indice(2)*2]
;
>> xa=xx(indice(2))-xx(indice(1));
>> ya=yy(indice(2))-yy(indice(1));
>> L=sqrt(xa*xa+ya*ya); % L: length of bar
>> C=xa/L;
>> S=ya/L;
>> sigma1(e)=E/L*[-C -S C S]*displacements1(elementDof);
>> sigma2(e)=E/L*[-C -S C S]*displacements2(elementDof);
>> end

% Euler buckling stress: buckling stress vector
>> buckling_stress=(4*E*B).*(A./(L.^2));

% Natural frequency: Natural frequency vector
>> aa=zeros(12,1);aa(3:4)=500;aa(7:8)=500; % lumped masses
>> mass1=diag(aa); % lumped mass matrix
>> mass=mass+mass1; % mass matrix
>> f=eig(stiffness(activeDof,activeDof),mass(activeDof,
activeDof));
>> f=realsqrt(f);
>> frequency =f/(2*pi); % Natural frequency vector
```

Step 7:

```
>> disp('sigma1')
>> disp('sigma2')
```

```
>> disp('buckling_stress')
>> disp('frequency')
>> disp('displacements1')
>> disp('displacements2')
```

Step 8:

```
>> fit= sum(((A).*L.*B))* ro + nodal mass; % total mass
>> fpenalty=(1+e1*sum(Constraint violation))^e2; % penalty function
>> fit_fpenalty =fit*fpenalty;
```

Step 9:

```
% iteration 1
% Generate new set of size variables (i.e., population) within its upper
and lower bounds as per the optimization algorithm rules
End of iteration 1
% Select best solotions for the next iteration as per the algorithm.
```

Step 10:

```
>> while FE ≤ maxFE
>> ... Algorithms code
>> end
```

Step 11: Output: The best solution.

```
>> disp('X')
>> disp('fit_fpenalty')
```

The obtained results are based on 100 independent runs, and Table 5.2 shows the best results achieved. The table indicates that elements 2, 6, 8, and 10 are removed in all approaches, so they are not mentioned. It should be noted that all approaches have identified the same topologies. The algorithms ALO, DA, GWO, MVO, SCA, WOA, HTS, TLBO, PVS, and SOS have achieved the best results of 423.5682, 459.3181, 424.1185, 423.7610, 436.9897, 514.3427, 423.5624, 423.5295, 423.5367, and 423.5815 kg, respectively. Consequently, the TLBO and PVS algorithms have performed better among the considered algorithms to obtain the minimum mean mass.

The mean and standard deviation (SD) values of the structural mass are used for the statistical evaluation of the performance of the different evolutionary methods. The ALO, DA, GWO, MVO, SCA, WOA, HTS, TLBO, PVS, and SOS algorithms have resulted in the mean weight of 574.9293, 820.0753, 525.6897, 514.7999, 530.0982, 951.8673, 561.3710, 478.0661, 469.8529, and 461.1166 kg, respectively.

Table 5.2 Truss topology optimization of the 10-bar truss (continuous section)

Element no.	ALO	DA	GWO	MVO	SCA	WOA	HTS	TLBO	PVS	SOS
1	3.5696	2.8598	3.3042	3.9155	5.6538	2.9671	3.6851	3.7714	3.7251	3.9433
3	63.6214	72.4579	63.6578	63.6399	64.2890	90.3735	63.6230	63.6214	63.6221	63.6223
4	36.7321	36.7318	36.7322	36.7325	37.7794	43.2050	36.7334	36.7318	36.7327	36.7335
5	36.7318	40.3902	36.8198	36.7505	37.7533	36.7318	36.7349	36.7318	36.7324	36.7323
7	7.6842	10.6030	7.8897	7.8002	8.8250	11.8100	7.6071	7.5371	7.5693	7.3412
9	5.3444	4.1821	5.3938	5.0116	4.5829	3.7700	5.3340	5.3381	5.3392	5.4249
Mass (kg)	423.5682	459.3181	424.1185	423.7610	436.9897	514.3427	423.5624	423.5295	423.5367	423.5815
σ^{max} (MPa)	163.9333	155.7347	159.6634	161.4954	142.7418	167.0685	165.5949	167.1328	166.4218	171.5927
δ^{max} (mm)	99.9996	99.9994	99.9028	99.9645	93.3321	99.3342	99.9823	99.9997	99.9992	99.9990
f_1 (Hz)	4.8028	4.8204	4.8083	4.8090	4.9806	4.8042	4.8016	4.7995	4.8003	4.7934
Mean	574.9293	820.0753	525.6897	514.7999	530.0982	951.8673	561.3710	478.0661	469.8529	461.1166
SD	308.7656	580.9469	308.6173	304.5681	56.9181	244.7879	474.9811	55.6729	51.1988	43.3092

Meanwhile, the SD of weight for ALO, DA, GWO, MVO, SCA, WOA, HTS, TLBO, PVS, and SOS algorithms are 308.7656, 580.9469, 308.6173, 304.5681, 56.9181, 244.7879, 474.9811, 55.6729, 51.1988, and 43.3092 kg, respectively. From the results, it can be observed that the SOS algorithm performs better in obtaining a better mean and SD of the weight of the truss.

The findings of the study demonstrate that topology optimization is a highly effective approach for achieving lighter truss designs compared to size optimization. Specifically, the optimal truss design obtained through topology optimization was found to be 23.6% lighter compared to the optimal truss design achieved through size optimization.

This highlights the importance of considering the topology of the truss in the design process, as it can have a significant impact on the overall weight and structural performance of the truss. By allowing the optimization algorithm to modify the topology of the truss, engineers can explore a wider range of possible designs and identify solutions that may not have been possible through traditional design methods.

Overall, the results of this study underscore the value of topology optimization as a powerful tool for achieving lightweight and efficient truss designs. By incorporating this approach into the design process, engineers can create trusses that are not only lighter but also stronger, more durable, and more cost-effective.

5.4.2 *Topology Optimization of the 14-Bar Truss with Continuous Cross-Sections*

In this section, the main objective is to optimize the 14-bar truss for both size and topology optimization. The optimization process involves the utilization of 14 continuous size variables. The ground structure of the second benchmark truss, referred to as the 14-bar truss, along with the load and boundary conditions, is depicted in Fig. 5.2. Previous studies have focused on TSS (topology, size, and shape) optimization of this problem. These studies utilized 14 continuous cross-sectional areas and eight continuous shape variables, considering static constraints. However, no previous investigation has addressed the simultaneous consideration of static and dynamic constraints for this problem.

For detailed information regarding design variables, load conditions, design constraints, and other relevant data related to the test problem, refer to Table 5.3. In this problem, the continuous cross-sectional areas (A_i, where $i = 1, 2, \ldots, 14$) are considered within the range of $[-100, 100]$ cm^2, where a zero or negative value indicates removal of the element.

This section involves optimizing the 14-bar truss for TTO, using 14 continuous size and topology variables. The results are based on 100 independent runs, with the best results displayed in Table 5.4. The table indicates that elements 1, 2, 5, 6, 8, 9, 10, 11, 12, and 14 were removed in all approaches and are therefore not included in the table. It is worth noting that all approaches identified the same topologies. The best performance achieved by various algorithms, such as ALO, DA, GWO, MVO, SCA, WOA, HTS, TLBO, PVS, and SOS, were 363.6204, 367.6268, 363.8773,

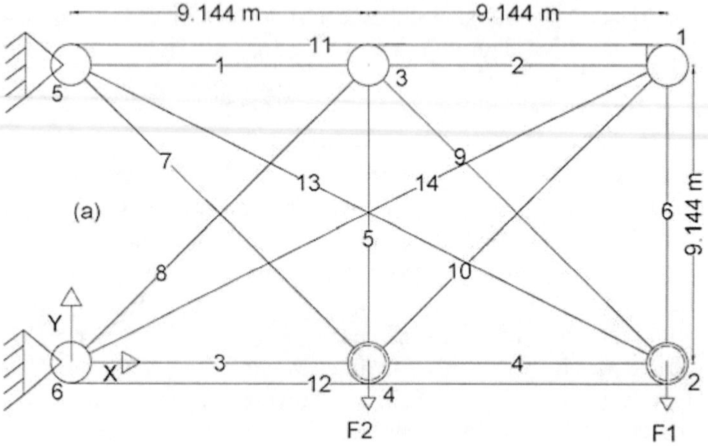

Fig. 5.2 The 14-bar truss

Table 5.3 Design parameters of the 14-bar truss

Size variables: $X_i = A_i$, $i = 1, 2, \ldots, 14$	
Multiple load conditions:	
Condition 1 : $F_1 = 44.537$ kN, $F_2 = 44.537$ kN	Condition 2 : $F_1 = 44.537$ kN, $F_2 = 0$ kN
Lumped mass on nodes 2 and 4 are 500 kg	
Stress and displacement constraints: $\sigma_i^{max} = 172.43$ MPa and $\delta_{y_5 \& y_6}^{max} = 100$ mm (0.1 m)	
Natural frequency constraint: $f_1 \geq 4$ Hz	
The continuous cross-sectional areas: $[A^{min}, A^{max}] = [-100, 100]$ cm^2 = ($[-100, 100] \times 10^{-4}$ m^2) and critical area = 1 cm^2	
Material properties: $E = 69$ GPa and $\rho = 2740$kg/m^3	

363.7171, 378.5574, 378.8441, 363.6564, 363.6202, 363.6202, and 363.6218 kg, respectively. From the results, it is evident that TLBO and PVS algorithms are the most effective metaheuristics in producing a lightweight truss.

The structural mass mean and SD values are utilized for statistical evaluation of the performance of various evolutionary methods. The mean weights of the truss achieved by the ALO, DA, GWO, MVO, SCA, WOA, HTS, TLBO, PVS, and SOS algorithms are 682.6968, 768.2996, 598.0133, 520.9313, 453.4004, 1270.6459, 550.5072, 448.6450, 419.3410, and 422.4828 kg, respectively. Similarly, the SD of the weights achieved by these algorithms are 216.9401, 222.8007, 1086.4086, 164.4870, 95.1238, 580.9986, 181.8424, 116.9639, 80.6689, and 96.1710 kg, respectively. Based on the results, it can be observed that the PVS algorithm outperforms the other algorithms in terms of obtaining a better mean and SD of the weight of the truss.

The findings of the study clearly demonstrate that topology optimization is a highly effective approach for achieving significantly lighter truss designs when compared to size optimization methods. Specifically, the optimal truss design

Table 5.4 Truss topology optimization of the 14-bar truss (continuous section)

Element no.	ALO	DA	GWO	MVO	SCA	WOA	HTS	TLBO	PVS	SOS
3	63.6213	63.6213	63.6295	63.6339	63.7631	63.6213	63.6214	63.6213	63.6213	63.6215
4	51.9467	52.0213	51.9621	51.9563	52.8068	57.0352	51.9493	51.9466	51.9466	51.9467
7	3.6528	4.7234	3.6928	3.6561	6.5543	3.8973	3.6603	3.6528	3.6528	3.6529
13	7.3410	7.3456	7.3510	7.3462	7.7241	7.6281	7.3415	7.3410	7.3410	7.3411
Mass (kg)	363.6204	367.6268	363.8773	363.7171	378.5574	378.8441	363.6564	363.6202	363.6202	363.6218
σ^{max}	172.4300	135.5740	170.5635	172.2715	128.9319	161.6112	172.0768	172.4300	172.4300	172.4247
δ^{max}	100.0000	99.9369	99.8753	99.9346	95.4559	96.2119	99.9938	100.0000	100.0000	99.9988
f_1	4.5251	4.5291	4.5280	4.5265	4.6313	4.5941	4.5253	4.5251	4.5251	4.5251
Mean	682.6968	768.2996	598.0133	520.9313	453.4004	1270.6459	550.5072	448.6450	419.3410	422.4828
SD	216.9401	222.8007	1086.4086	164.4870	95.1238	580.9986	181.8424	116.9639	80.6689	96.1710

obtained through topology optimization was found to be 50.8% lighter than the optimal truss design achieved through size optimization.

These results highlight the importance of considering the topology of the truss in the design process, as it can have a profound impact on the truss's weight and structural performance. By allowing the optimization algorithm to modify the topology of the truss, engineers can explore a wider range of possible designs and identify solutions that may not have been possible through traditional design methods.

Overall, the findings of the study underscore the value of topology optimization as a powerful tool for achieving lightweight and efficient truss designs. By incorporating this approach into the design process, engineers can create trusses that are not only lighter but also stronger, more durable, and more cost-effective, making them an attractive solution for a wide range of structural applications.

5.4.3 *Topology Optimization of the 15-Bar Truss with Continuous Cross-Sections*

In this section, the main focus is on optimizing the 15-bar truss for both size and topology optimization, utilizing 15 continuous size variables. The ground structure of this truss can be seen in Fig. 5.3. This problem is adapted from the works of Rahami et al. (2008) and Ahrari et al. (2014). Previous studies on this problem have mainly considered static constraints, such as stress, buckling, and displacement. However, simultaneous consideration of static and dynamic constraints, specifically natural frequency constraints, has not been explored.

To address this gap, the problem is modified to incorporate multi-load conditions, lumped masses, natural frequency constraints, and continuous cross-sectional areas. In this problem, the continuous cross-sectional areas (A_i, where $i = 1, 2, \ldots, 15$) are considered within the range of $[-50, 50]$ cm^2, where a zero or negative value indicates removal of the element. Design considerations and material properties are detailed in Table 5.5, providing important information for the optimization process.

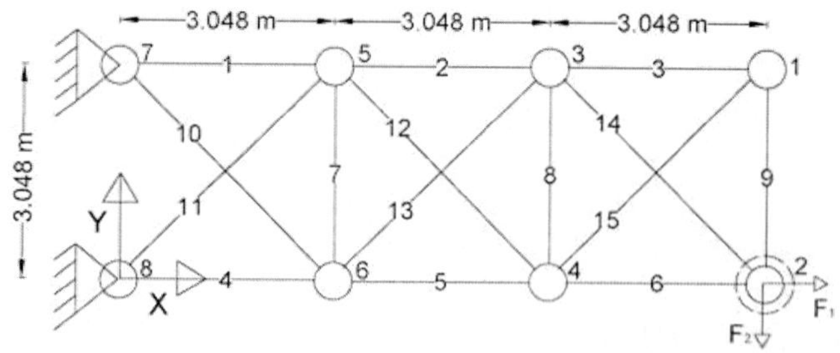

Fig. 5.3 The 15-bar truss

Table 5.5 Design parameters of the 15-bar truss

Size variables: $X_i = A_i$, $i = 1, 2, \ldots, 15$	
Multiple load conditions:	
Condition 1 : $F_1 = -44.537$ kN, $F_2 = 0$ kN	Condition 2 : $F_1 = 0$ kN, $F_2 = 4.4537$ kN
Lumped mass on node 2 is 500 kg	
Stress and displacement constraints: $\sigma_i^{max} = 172.43$ MPa and $\delta_{x_2\&y_2}^{max} = 100$ mm (0.1 m)	
Natural frequency constraint: $f_1 \geq 4$ Hz	
The continuous cross-sectional areas: $[A^{min}, A^{max}] = [-50, 50]$ cm^2 = ($[-50, 50] \times 10^{-4}$ m^2) and critical area = 1 cm^2	
Material properties: $E = 69$ GPa and $\rho = 2740$kg/m^3	

Table 5.6 presents the most effective solutions obtained in this study, which demonstrate that the ALO, DA, GWO, MVO, SCA, WOA, HTS, TLBO, PVS, and SOS algorithms produced optimized trusses with a minimum weight of 97.7022, 117.4506, 94.1896, 94.4608, 117.5747, 146.6506, 94.8382, 93.7101, 93.6813, and 95.6295 kg, respectively, while adhering to all constraints. Notably, the PVS algorithm resulted in the lightest structure. Furthermore, the initial truss, consisting of 15 bars and 8 nodes, was optimized to only 10 bars and 7 nodes for the best solutions, highlighting the importance of a well-designed topology in achieving minimal weight.

To evaluate the evolutionary methods' performance statistically, the mean and SD values of the structural mass were calculated. The ALO, DA, GWO, MVO, SCA, WOA, HTS, TLBO, PVS, and SOS algorithms produced mean weights of 135.3269, 1.20E+07, 9.00E+06, 2.40E+07, 1.00E+06, 249.3707, 158.8889, 1.00E+06, 100.0660, and 101.6690 kg, respectively. Meanwhile, the SD of weight was 33.5626, 3.27E+07, 2.88E+07, 4.29E+07, 1.00E+07, 66.5149, 230.9441, 1.00E+07, 10.4440, and 6.5126 kg for the ALO, DA, GWO, MVO, SCA, WOA, HTS, TLBO, PVS, and SOS algorithms, respectively. These results demonstrate that the PVS algorithm performs better in terms of the mean weight, while the SOS algorithm gives the best SD of weight.

The findings of the study suggest that utilizing the topology optimization approach leads to trusses that are significantly lighter than those resulting from size optimization approaches. In particular, the top-performing solutions attained via topology optimization produced trusses that were 16.8% lighter compared to those resulting from size optimization.

5.4.4 Topology Optimization of the 24-Bar Truss with Continuous Cross-Sections

This section focuses on the size and topology optimization of the 24-bar truss, which poses several challenges due to the large number of design variables involved. The

Table 5.6 Truss topology optimization of the 15-bar truss (continuous section)

Element no.	ALO	DA	GWO	MVO	SCA	WOA	HTS	TLBO	PVS	SOS
1	7.5524	14.5338	5.4891	5.5448	7.2332	11.1556	5.5439	5.4773	5.4766	5.4942
2	5.4757	9.5829	5.5873	5.4927	7.3873	6.7343	5.5774	5.4758	5.4766	5.5535
3	—	—	—	—	—	11.7660	—	—	—	—
4	12.2451	19.3948	12.2452	12.2593	19.4926	13.9730	12.4906	12.2610	12.2446	12.8420
5	12.2442	13.1151	12.2467	12.8480	12.9463	13.5996	12.3519	12.2441	12.2446	12.4860
6	12.2467	12.2439	12.2711	12.3320	13.1018	20.7696	12.3505	12.2445	12.2443	12.8485
7	3.8722	—	1.1079	1.0270	12.1835	—	1.2795	1.0016	1.0008	1.5413
8	1.5165	5.1956	1.2063	1.0574	—	2.7006	1.1494	1.0006	1.0000	1.0258
10	—	6.5877	6.5190	6.5207	—	7.0473	6.5345	6.5130	6.5126	6.5340
11	5.6276	—	—	—	5.1808	1.0000	—	—	—	—
12	—	3.2757	—	—	2.8896	—	—	—	—	—
13	1.9138	1.0079	1.5791	1.5097	2.7306	6.9386	1.5524	1.5321	1.5247	1.4816
14	6.5485	6.5654	6.5527	6.5618	7.9579	7.0710	6.6961	6.5168	6.5126	6.6962
15	—	—	—	—	—	11.1795	—	—	—	—
Mass (kg)	97.7022	117.4506	94.1896	94.4608	117.5747	146.6506	94.8382	93.7101	93.6813	95.6295
σ^{max} (MPa)	36.3739	62.4923	39.8863	41.7208	34.4012	32.7489	40.5725	41.1101	41.3085	42.5110
δ^{max} (mm)	9.9999	1.1490	9.8358	9.9963	8.1917	6.6178	9.8337	9.9796	9.9997	9.9772
f_1 (Hz)	4.6557	4.6569	4.6940	4.6562	5.1495	6.0133	4.6939	4.6610	4.6564	4.6595
Mean	135.3269	1.20E+07	9.00E+06	2.40E+07	1.00E+06	249.3707	158.8889	1.00E+06	100.0660	101.6690
SD	33.5626	3.27E+07	2.88E+07	4.29E+07	1.00E+07	66.5149	230.9441	1.00E+07	10.4440	6.5126

ground structure of the fourth benchmark problem, known as the 24-bar truss, is illustrated in Fig. 5.4. This truss problem was initially investigated by Ringertz (1986) using a Branch and Bound algorithm for topology optimization under static constraints. Subsequently, Xu et al. (2003) utilized it to test a 1-D search method, while Kaveh and Zolghadr (2013) employed it to evaluate the search performance of charged system search and particle swarm optimization algorithms, considering continuous design variables.

The truss is subjected to two external load cases, as specified in Table 5.7. In this optimization problem, the cross-sectional areas of the truss elements are treated as continuous variables. Additionally, a nonstructural lumped mass of 500 kg is applied at node 3. Nodal masses, material properties, and other relevant data can also be found in the table. In this problem, the continuous cross-sectional areas (A_i, where $i = 1, 2, \ldots, 24$) are considered within the range of [−40, 40] cm^2, where a zero or negative value indicates removal of the element.

This section explores the simultaneous TTO optimization of a 24-bar truss using 24 continuous design variables. Table 5.8 presents the results of TTO optimization

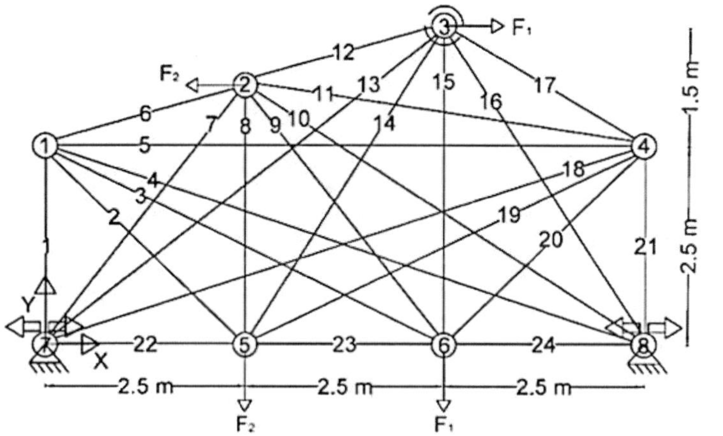

Fig. 5.4 The 24-bar truss

Table 5.7 Design parameters of the 24-bar truss

Size variables: $X_i = A_i$, $i = 1, 2, \ldots, 24$	
Multiple load conditions:	
Condition 1 : $F_1 = 50$ kN, $F_2 = 0$ kN	Condition 2 : $F_1 = 0$ kN, $F_2 = 50$ kN
Lumped mass on node 3 is 500 kg	
Stress and displacement constraints: $\sigma_i^{max} = 172.43$ MPa and $\delta_{y_5\&y_6}^{max} = 10$ mm $\left(10^{-2}\text{ m}\right)$	
Natural frequency constraint: $f_1 \geq 30$ Hz	
The continuous cross-sectional areas: $[A^{min}, A^{max}] = [-40, 40]$ cm^2 = ($[-40, 40] \times 10^{-4}$ m^2) and critical area = 1 cm^2	
Material properties: $E = 69$ GPa and $\rho = 2740$kg/m^3	

using various metaheuristic algorithms in 100 independent runs. The algorithm with the lowest mass value that meets all the structural constraints is considered the best. The reported best results for ALO, DA, GWO, MVO, SCA, WOA, HTS, TLBO, PVS, and SOS algorithms are 129.4230, 160.2325, 120.5914, 120.7168, 149.5898, 208.5086, 124.8393, 120.9519, 119.1468, and 122.0408 kg, respectively. The PVS algorithm produces the lowest mass value, followed closely by the GWO and MVO algorithms. The mass values for these three algorithms are nearly similar, and their topologies are identical. The best topology found by all approaches consists of only 9 out of 24 elements.

To evaluate the performance of the metaheuristic methods, the average mass and its standard deviation (SD) are calculated. The mean mass values for ALO, DA, GWO, MVO, SCA, WOA, HTS, TLBO, PVS, and SOS algorithms are 206.4122, 267.7236, 149.8978, 162.6256, 189.3643, 387.7904, 196.3363, 171.1830, 135.5368, and 145.3252 kg, respectively. The SD values for these algorithms are 46.7999, 62.3702, 31.8295, 41.6670, 22.9531, 95.5260, 67.4874, 38.6586, 13.0909, and 17.5308, respectively. The PVS algorithm performs the best in terms of achieving the lowest mean mass and the lowest SD, indicating its superior convergence rate and consistency. Therefore, the PVS algorithm is the most suitable for this design problem.

The study suggests that topology optimization is a more effective approach compared to size optimization in designing lighter trusses. Specifically, the best solutions obtained through topology optimization were 40.0% lighter compared to the truss obtained through size optimization. This finding highlights the importance of considering the topology of the structure in addition to its size, as optimizing the shape and distribution of the material can result in significant weight savings while still maintaining structural integrity.

5.4.5 Topology Optimization of the 20-Bar Truss with Continuous Cross-Sections

In this section, the main objective is to optimize the 20-bar truss for size and topology optimization, utilizing 20 continuous sections. The ground structure of the 20-bar truss is represented in Fig. 5.5. This benchmark problem was initially formulated for topology and size optimization by Xu et al. (2003) and subsequently tested for continuous optimization by Kaveh and Zolghadr (2013).

For this test problem, Table 5.9 provides detailed information regarding design variables, loading conditions, structural safety constraints, and other relevant data. In this problem, the continuous cross-sectional areas (A_i, where $i = 1, 2, \ldots, 20$) are considered within the range of $[-100, 100]$ cm^2, where a zero or negative value

Table 5.8 Truss topology optimization of the 24-bar truss (continuous section)

Element no.	ALO	DA	GWO	MVO	SCA	WOA	HTS	TLBO	PVS	SOS
1	—	—	—	—	—	9.8353	—	—	—	—
2	—	—	—	—	—	5.4392	—	—	—	—
6	—	—	—	—	—	21.3984	—	—	—	—
7	19.7666	21.2790	18.6246	19.1584	23.1366	—	19.7155	19.1324	19.0544	19.9798
8	2.5782	7.1796	4.2752	2.9758	17.6726	7.1030	2.9238	2.9007	2.9452	2.9600
9	—	—	—	1.8302	2.1009	4.4516	3.5006	2.4472	1.7875	2.1241
10	1.1907	1.0467	—	—	—	—	—	—	—	—
11	—	—	—	—	—	20.5349	—	—	—	—
12	—	—	1.0367	3.8139	4.5707	—	3.7243	3.8780	4.2694	4.3774
13	16.9841	17.9761	15.4180	13.2223	14.2225	18.3001	12.8764	13.3283	13.6452	13.9458
14	2.2142	—	1.2612	—	—	—	—	—	—	1.0058
15	6.6139	9.6123	3.3326	3.7283	3.9178	—	5.3618	3.6588	3.1750	3.3389
16	23.8913	25.2513	23.9192	23.9352	26.1457	17.8723	24.1583	23.8881	23.9150	24.1150
17	—	5.3172	—	—	—	12.1527	—	—	—	—
19	—	—	—	—	—	11.6584	—	—	—	—
20	—	6.3164	—	—	—	7.4776	—	—	—	—
21	—	—	—	—	—	10.4344	—	—	—	—
22	1.1715	—	1.5539	1.0115	—	—	—	2.2506	1.0371	—
23	—	1.2455	—	—	9.6971	—	2.2630	2.2498	—	—
24	2.0212	10.6469	1.1779	4.1560	3.7202	6.2226	3.2817	—	1.4835	1.2364
Mass (kg)	129.4230	160.2325	120.5914	120.7168	149.5898	208.5086	124.8393	120.9519	119.1468	122.0408
σ^{max} (MPa)	76.0724	114.3617	150.0333	126.5753	119.8407	129.4350	90.1684	128.7158	146.4827	217.0027
δ^{max} (mm)	5.5125	4.0804	9.8521	8.7496	7.5725	7.0006	8.9592	8.9637	9.1763	8.7777
f_1 (Hz)	30.0000	30.0000	30.2286	30.0011	30.9915	30.6130	30.8209	30.0006	30.0039	30.0745
Mean	206.4122	267.7236	149.8978	162.6256	189.3643	387.7904	196.3363	171.1830	135.5368	145.3252
SD	46.7999	62.3702	31.8295	41.6670	22.9531	95.5260	67.4874	38.6586	13.0909	17.5308

Fig. 5.5 The 20-bar truss

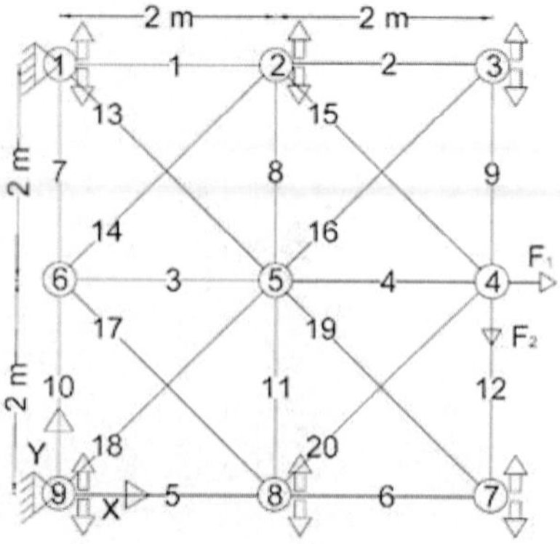

indicates removal of the element. It is important to note that no lumped masses are added to the truss for this particular problem.

This section details the optimization of a 20-bar truss for TTO using 20 continuous sections, and the outcomes obtained from different algorithms are presented in Table 5.10. The algorithm producing the lowest mass value while fulfilling all structural constraints is regarded as the best run. The best results for the ALO, DA, GWO, MVO, SCA, WOA, HTS, TLBO, PVS, and SOS algorithms are 155.4343, 176.8710, 151.1047, 150.7904, 180.3246, 252.8557, 153.4043, 150.8207, 150.6504, and 151.7222 kg, respectively. The PVS algorithm yields the least structural mass, with the MVO and TLBO algorithms coming in a close second and third, respectively. The best topologies generated by the GWO, MVO, HTS, TLBO, PVS, and SOS algorithms are similar but differ from those obtained from other approaches. It is worth noting that all approaches' topologies consist of only eight elements.

To evaluate the metaheuristic methods' statistical performance, the average mass and standard deviation (SD) are analyzed. Based on the available results, it can be seen that the ALO, DA, GWO, MVO, SCA, WOA, HTS, TLBO, PVS, and SOS algorithms have mass mean values of 260.6816, 1.01E+06, 1.66E+06, 1.30E+07, 243.6225, 431.5270, 673.5769, 427.5994, 169.6687, and 176.9714 kg, respectively, with SDs of 44.1233, 1.00E+07, 1.19E+07, 3.38E+07, 35.7250, 105.3987, 2433.7939, 1639.6005, 17.3286, and 22.4792, respectively. The PVS algorithm performs the best in terms of achieving the lowest mass mean value and the best SD. According to the statistical outcomes, the PVS algorithm is significantly superior to other algorithms.

The findings of the study indicate that topology optimization is a more effective approach for designing lighter trusses compared to size optimization methods. The best solutions generated through topology optimization produced trusses that were 26.5% lighter than those created through size optimization. This result highlights the importance of considering the structural design's topology as a variable when

Table 5.9 Design parameters of the 20-bar truss

Size variables: $X_i = A_i$, $i = 1, 2, \ldots, 20$	
Multiple load conditions:	
Condition 1 : $F_1 = 500$ kN, $F_2 = 0$ kN	Condition 2 : $F_1 = 0$ kN, $F_2 = 500$ kN
No lumped mass	
Stress and displacement constraints: $\sigma_i^{max} = 172.43$ MPa and $\delta_{y_4}^{max} = 60$ mm $\left(6 \times 10^{-2}\ \text{m}\right)$	
Natural frequency constraints: $f_1 \geq 60$ Hz and $f_2 \geq 100$ Hz	
The continuous cross-sectional areas: $[A^{min}, A^{max}] = [-100, 100]$ cm^2 = ($[-100, 100] \times 10^{-4}$ m^2) and critical area = 1 cm^2	
Material properties: $E = 69$ GPa and $\rho = 2740$kg/m^3	

optimizing for weight reduction. By examining the truss's shape and the distribution of material, topology optimization can identify the most efficient design configuration that meets the structural requirements while minimizing weight.

5.4.6 Topology Optimization of the 72-Bar 3D Truss with Continuous Cross-Sections

In this section, the optimization process focuses on the size and topology optimization of the 72-bar truss using 16 continuous variables. The ground structure of the sixth test problem, known as the 72-bar truss, is illustrated in Fig. 5.6. This three-dimensional truss was previously utilized by Kaveh and Zolghadr (2013) for continuous optimization.

Table 5.11 provides comprehensive details regarding material properties, design variables, loading conditions, lumped mass, and other relevant data for the test problem. The elements of the truss are clustered into 16 groups, denoted as G_1 (A_1–A_4), G_2 (A_5–A_{12}), G_3 (A_{13}–A_{16}), G_4 (A_{17}–A_{18}), G_5 (A_{19}–A_{22}), G_6 (A_{23}–A_{30}), G_7 (A_{31}–A_{34}), G_8 (A_{35}–A_{36}), G_9 (A_{37}–A_{40}), G_{10} (A_{41}–A_{48}), G_{11} (A_{49}–A_{52}), G_{12} (A_{53}–A_{54}), G_{13} (A_{55}–A_{58}), G_{14} (A_{59}–A_{66}), G_{15} (A_{67}–A_{70}), and G_{16} (A_{71}–A_{72}). This grouping is based on structural symmetry, as described by Kaveh and Zolghadr (2013). In this problem, the continuous cross-sectional areas of 16 groups (G_i, where $i = 1, 2, \ldots, 16$) are considered within the range of $[-30, 30]$ cm^2, where a zero or negative value indicates removal of the element.

Additionally, the truss includes four nonstructural lumped masses of 2270 kg each, which are added to the top nodes (node numbers 1–4). These masses are not considered part of the main structural design.

In this section, we optimized a 72-bar truss for TTO using 16 continuous size and topology variables. Table 5.12 summarizes the results of 100 independent runs of TS optimization for the 72-bar truss with continuous cross-sections, comparing the performance of various algorithms. We found that the best run yielded the minimum mass value while satisfying all structural constraints. The best results were achieved by the ALO, DA, GWO, MVO, SCA, WOA, HTS, TLBO, PVS, and SOS

Table 5.10 Truss topology optimization of the 20-bar truss (continuous section)

Element no.	ALO	DA	GWO	MVO	SCA	WOA	HTS	TLBO	PVS	SOS
1	28.9978	20.8126	14.6126	14.5680	—	21.0702	15.3756	14.5019	14.4992	14.8866
3	—	—	—	—	—	18.6501	—	—	—	—
4	28.9994	—	—	—	35.4260	—	—	—	—	—
5	—	19.0347	19.1640	19.0932	39.3192	27.0275	20.5633	19.0467	19.0411	19.3873
6	—	—	—	—	—	32.5951	—	—	—	—
8	29.0050	21.0832	19.1078	19.0956	—	41.1548	19.2708	19.2462	19.0358	19.1534
10	—	13.9234	—	—	—	—	—	—	—	—
11	—	20.7444	19.2159	19.0429	38.4950	32.2462	19.9492	19.0375	19.0352	19.2827
12	—	—	—	—	—	19.2402	—	—	—	—
13	20.5042	20.5042	20.6179	20.5142	48.1841	27.7126	20.7430	20.5289	20.5073	20.5324
14	—	—	—	—	—	1.0000	—	—	—	—
15	41.0083	20.8462	20.5370	20.5065	—	21.4071	20.8040	20.5123	20.5086	20.6036
17	—	2.8513	—	—	—	—	—	—	—	—
18	45.2724	37.9322	32.1133	32.0329	22.6022	52.9920	32.0929	32.0473	32.0132	32.4429
19	—	—	—	—	—	4.3451	—	—	—	—
20	—	33.3300	32.0155	32.0371	49.5625	31.4463	32.4484	32.0183	32.0142	32.0705
Mass (kg)	155.4343	176.8710	151.1047	150.7904	180.3246	252.8557	153.4043	150.8207	150.6504	151.7222
σ^{max} (MPa)	172.4300	172.4300	172.1547	172.4102	156.4245	159.4189	170.4445	172.3907	172.4227	172.1926
δ^{max} (mm)	34.0425	22.7997	24.5043	24.5756	28.3198	18.1416	24.0326	24.5728	24.6048	24.3925
f_1 (Hz)	117.3226	113.0636	117.0414	116.8890	114.6260	84.6592	116.6395	116.9185	116.8946	116.9538
f_2 (Hz)	157.9742	152.4402	186.1054	185.8798	152.4202	107.7825	188.3601	185.7788	185.7134	186.7119
Mean	260.6816	1.01E+06	1.66E+06	1.30E+07	243.6225	431.5270	673.5769	427.5994	169.6687	176.9714
SD	44.1233	1.00E+07	1.19E+07	3.38E+07	35.7250	105.3987	2433.7939	1639.6005	17.3286	22.4792

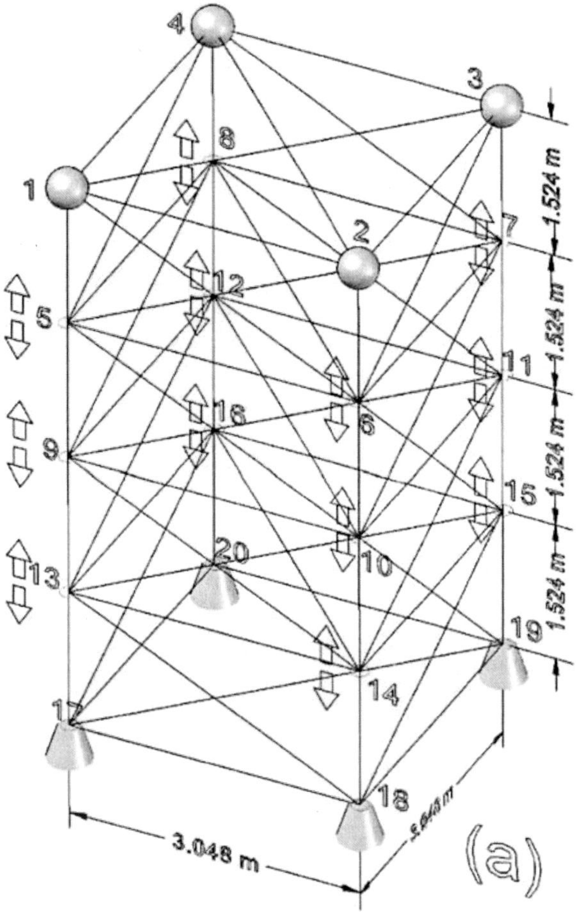

Fig. 5.6 The 72-bar 3D truss

Table 5.11 Design parameters of the 72-bar truss

Size variables: G_i, $i = 1, 2, \ldots, 16$; $G =$ Group number				
Multiple load conditions:				
Condition 1:	$F_{1x} = 22.25$ kN	$F_{1y} = 22.25$ kN		$F_{1z} = -22.25$ kN
Condition 2:	$F_{1z} = -22.25$ kN	$F_{2z} = -22.25$ kN	$F_{3z} = -22.25$ kN	$F_{4z} = -22.25$ kN
Lumped masses at nodes 1, 2, 3, and 4 are 2270 kg				
Stress and displacement constraints: $\sigma_i^{max} = 172.375$ MPa and $\delta_{x_j/y_j}^{max} = 6.35$ mm $\left(6.35 \times 10^{-3}\ \text{m}\right)$ (for node, $j = 1, 2, 3$, and 4 along x- and y-axes)				
Natural frequency constraints: $f_1 \geq 4$ Hz and $f_3 \geq 6$ Hz				
The continuous cross-sectional areas: $[A^{min}, A^{max}] = [-30, 30]$ cm^2 = $([-30, 30] \times 10^{-4}$ m$^2)$ and critical area = 1 cm^2				
Material properties: $E = 68.95$ GPa and $\rho = 2767.99$kg/m^3				

algorithms, with reported mass values of 452.2808, 505.5763, 436.3522, 446.6385, 510.6434, 559.7465, 444.6792, 438.9908, 436.1486, and 441.1139 kg, respectively. The PVS algorithm had the lowest structural mass, with the GWO algorithm coming in a close second.

To evaluate the true statistical performance of the metaheuristic methods, we considered the average mass and standard deviation (SD). Based on the results available, we concluded that the ALO, DA, GWO, MVO, SCA, WOA, HTS, TLBO, PVS, and SOS algorithms had mass mean values of 792.8080, 1050.6924, 454.3264, 7.00E+06, 642.4582, 1377.7487, 766.3641, 498.2148, 464.9480, and 459.7845 kg, respectively, with SDs of 290.8837, 825.4387, 21.7597, 2.56E+07, 64.9807, 550.0149, 584.8062, 208.1630, 48.5852, and 21.5205, respectively. The GWO algorithm performed best in achieving the lowest mass mean value, while the SOS algorithm had the best SD. Therefore, for this design problem, the PVS, GWO, and SOS algorithms are the actual better algorithms for both convergence rate and consistency.

The findings of the study indicate that topology optimization is more effective than size optimization approaches in designing lightweight trusses. In particular, the top-performing solutions generated through topology optimization resulted in a truss that was 7.3% lighter compared to the truss obtained through size optimization.

5.4.7 *Topology Optimization of the 39-Bar Truss with Continuous Cross-Sections*

In this section, the optimization focuses on both size and topology optimization of the 39-bar truss using 21 continuous variables. The seventh benchmark problem, known as the two-tier 39-bar planar truss, is depicted in Fig. 5.7. The purpose of this problem is to investigate the behavior of a large-sized truss under stress, buckling, displacement, and fundamental frequency constraints, considering discrete design variables.

Initial work on this problem was carried out by Deb and Gulati (2001), and subsequent studies by researchers such as Ahrari et al. (2014) have mainly focused on static constraints. However, the problem has not been explored with simultaneous consideration of static and dynamic constraints.

To address this, the problem is modified to incorporate the effects of multi-load conditions, lumped masses, and natural frequency constraints. Detailed information about design variables, multi-load conditions, lumped masses, constraints, and material properties can be found in Table 5.13. The truss elements are grouped into 21 groups, considering symmetry about the vertical middle plane, following the approach of Deb and Gulati (2001). In this problem, the continuous cross-sectional areas of 16 groups (G_i, where $i = 1, 2, \ldots, 21$) are considered within the range of [−50, 50] cm^2, where a zero or negative value indicates removal of the element.

Table 5.12 Truss topology optimization of the 72-bar 3D truss (continuous section)

Element group no.	ALO	DA	GWO	MVO	SCA	WOA	HTS	TLBO	PVS	SOS
1	12.9362	11.4642	5.0020	6.3905	12.6592	7.2713	4.6756	4.3581	4.9132	4.3453
2	10.2368	12.6935	10.2477	10.2919	10.8496	10.5279	10.2503	10.2347	10.2433	10.2611
4	—	—	—	—	10.7202	—	—	—	—	—
5	7.7564	6.7550	9.4262	11.1297	9.9112	4.8885	7.5901	7.3034	9.1374	7.9759
6	8.0973	10.0986	7.6565	8.0984	7.4965	7.1352	7.7230	7.3128	7.9584	7.7433
7	—	—	—	—	—	—	—	2.5179	—	2.5037
8	3.2812	8.7539	3.9016	3.4815	—	6.9912	4.5592	5.0254	3.9120	5.0989
9	13.0114	18.0588	12.3594	11.9886	9.9127	30.9154	12.9293	13.1124	13.3378	15.4644
10	8.2663	7.0075	7.9054	8.2735	12.2540	18.0660	7.2258	7.6076	7.7213	7.8137
11	—	6.8934	—	—	—	—	4.3095	—	—	—
13	15.4625	18.9079	15.7612	13.7271	17.9400	27.6807	17.1605	17.3255	15.1423	15.1693
14	8.0517	5.7556	7.0415	8.5033	9.1527	7.1633	7.6372	7.6384	6.9519	7.0486
15	—	—	2.1774	—	—	—	—	—	2.0654	—
Mass (kg)	452.2808	505.5763	436.3522	446.6385	510.6434	559.7465	444.6792	438.9908	436.1486	441.1139
σ^{max} (MPa)	106.7526	54.4177	99.9349	104.2959	73.8746	91.8129	74.9600	69.2091	99.6246	68.0465
δ^{max} (mm)	6.1997	4.4770	6.1044	6.1658	4.2977	4.9206	5.3409	4.9844	6.0828	4.9679
f_1 (Hz)	4.0466	4.1614	4.0023	4.0084	4.1549	4.4509	4.0008	4.0000	4.0005	4.0068
f_3 (Hz)	6.0000	6.0000	6.0004	6.0010	6.3129	6.2212	6.0010	6.0001	6.0005	6.0009
Mean	792.8080	1050.6924	454.3264	7.00E+06	642.4582	1377.7487	766.3641	498.2148	464.9480	459.7845
SD	290.8837	825.4387	21.7597	2.56E+07	64.9807	550.0149	584.8062	208.1630	48.5852	21.5205

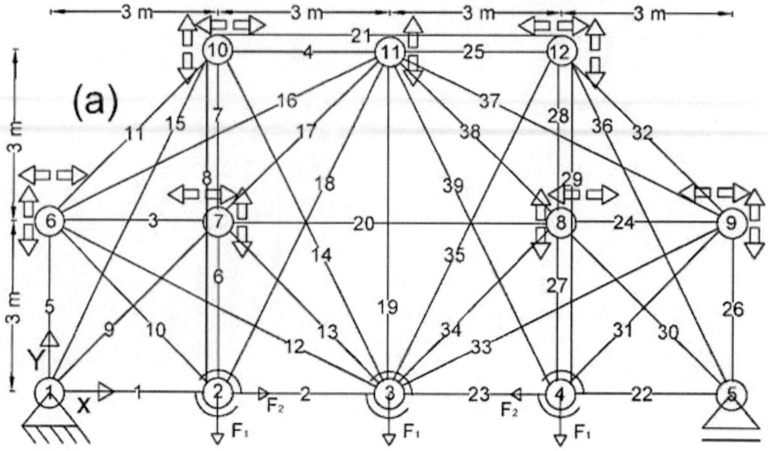

Fig. 5.7 The 39-bar truss

Table 5.13 Design parameters of the 39-bar truss

Size variables: $X_i = G_i$, $i = 1, 2, \ldots, 21$	
Multiple load conditions:	
Condition 1 : $F_1 = 100$ kN, $F_2 = 0$ kN	Condition 2 : $F_1 = 0$ kN, $F_2 = 50$ kN
Lumped masses on nodes 2, 3, and 4 are 500 kg	
Stress and displacement constraints: $\sigma_i^{max} = 172.43$ MPa and $\delta_{x_j/y_j}^{max} = 50$ mm (0.05 m) (for node, $j = 2$, 3, and 4 along x- and y-axes)	
Natural frequency constraints: $f_1 \geq 4$ Hz	
The continuous cross-sectional areas: $[A^{min}, A^{max}] = [-50, 50]$ cm^2 = ($[-50, 50] \times 10^{-4}$ m^2) and critical area = 1 cm^2	
Material properties: $E = 69$ GPa and $\rho = 2740$kg/m^3	

Furthermore, the truss is subjected to three nonstructural lumped masses of 500 kg each, placed at nodes 2, 3, and 4. To satisfy the multi-load condition, two different loading conditions are considered, as shown in Fig. 5.7.

In this section, a 39-bar truss was optimized using 21 continuous size and topology variables for TTO optimization. Table 5.14 displays the design variables and corresponding truss weights for the best designs over 100 runs. The results show that the ALO, DA, GWO, MVO, SCA, WOA, HTS, TLBO, PVS, and SOS algorithms produced the best weights of 302.9106, 360.7307, 235.5432, 236.3183, 323.5875, 427.0381, 262.5679, 235.1135, 233.2562, and 237.6321 kg, respectively. It is worth noting that the PVS algorithm's best topology only consists of 17 elements out of 39.

To accurately evaluate the metaheuristic methods' performance, the average mass and its standard deviation (SD) were taken into consideration. Based on the available results, it can be concluded that the ALO, DA, GWO, MVO, SCA, WOA, HTS, TLBO, PVS, and SOS algorithms have a mass mean value of 452.3114, 575.9249,

Table 5.14 Truss topology optimization of the 39-bar truss (continuous section)

Group no.	ALO	DA	GWO	MVO	SCA	WOA	HTS	TLBO	PVS	SOS
1	9.6076	7.6230	1.6569	1.0286	1.0024	1.6752	1.7360	1.1904	1.0408	1.0010
2	13.3367	12.7688	12.7994	14.4938	18.0294	13.7706	13.1359	12.7811	12.8203	12.7961
3	—	9.2433	—	—	—	—	4.3207	—	—	—
4	16.6743	29.2118	18.1088	18.1362	—	18.3845	19.5668	18.0589	18.1175	18.1250
5	22.1217	25.1725	22.1251	22.2220	24.4961	24.9513	22.2199	22.1555	22.2215	22.3038
6	—	6.5176	—	—	—	12.8245	—	—	—	—
7	2.9094	—	—	—	—	18.7786	2.3470	—	—	—
8	1.4499	2.8567	1.8004	1.4499	—	—	2.2777	1.4701	1.4629	1.5361
9	—	—	—	—	—	14.0315	—	—	—	—
10	6.1853	8.3341	6.1807	6.1693	9.3109	16.5910	6.2066	6.1750	6.1899	6.9162
11	26.3009	26.6558	26.3023	26.4344	—	32.7812	26.7967	26.3320	26.3478	26.4182
13	2.0504	—	—	—	—	—	—	—	—	—
14	—	3.3926	3.3048	3.2854	—	18.6908	3.2713	3.2474	3.2674	3.3076
16	—	—	—	—	46.1078	—	—	1.0018	—	1.0264
17	15.1849	—	—	—	—	1.0000	—	—	—	—
18	—	—	—	—	2.3245		—	—	—	—
19	9.8896	29.6066	1.2421	1.0970	7.5050	—	1.3463	—	1.0224	—
Mass (kg)	302.9106	360.7307	235.5432	236.3183	323.5875	427.0381	262.5679	235.1135	233.2562	237.6321
σ^{max} (MPa)	172.4299	164.7737	171.6074	172.4298	151.8876	63.9299	170.8913	172.1427	171.3520	169.0081
δ^{max} (mm)	25.0402	26.7248	28.1862	26.6743	27.2225	12.8868	29.3518	27.1635	26.9735	26.3047
f_1 (Hz)	10.0736	11.8263	7.2926	5.8510	5.6413	7.3323	7.4182	6.2530	5.8735	5.7581
Mean	452.3114	575.9249	315.6692	340.2469	432.4706	852.8247	371.0790	372.4602	300.3748	313.5924
SD	83.19219	96.55983	48.2254	70.17548	48.04747	202.1064	77.43036	88.51629	43.31807	35.52744

315.6692, 340.2469, 432.4706, 852.8247, 371.0790, 372.4602, 300.3748, and 313.5924 kg, respectively, with SDs of 83.19219, 96.55983, 48.2254, 70.17548, 48.04747, 202.1064, 77.43036, 88.51629, 43.31807, and 35.52744, respectively. The results suggest that the PVS algorithm performs better in terms of achieving a better mean and SOS for the SD of weight of the truss. Therefore, the best algorithm for both convergence rate and consistency for this design problem is the PVS algorithm.

The findings of the study indicate that topology optimization can yield substantially lighter truss designs compared to size optimization techniques. Specifically, the best-performing solutions generated through topology optimization resulted in trusses that were 55.0% lighter than those obtained through size optimization. This result highlights the effectiveness of topology optimization in reducing material usage while maintaining the structural integrity of the truss. These findings can have significant implications in the design of lightweight and efficient truss structures for various engineering applications, including aerospace, civil, and mechanical engineering.

5.4.8 Topology Optimization of the 45-Bar Truss with Continuous Cross-Sections

In this section, the size optimization of the 45-bar truss was performed using 31 size and topology variables. The ninth benchmark problem, known as the 45-bar truss, was chosen to investigate a complex structure subjected to stress, buckling, displacement, and fundamental frequency constraints, taking into account continuous design variables.

The ground structure of the truss, along with multiple loads and the boundary condition, is depicted in Fig. 5.8. Initially proposed by Deb and Gulati (2001) with static constraints, this problem has not been previously examined with simultaneous consideration of static and dynamic constraints. To address this, the problem is modified to incorporate the effects of multi-load conditions, lumped masses, natural frequency constraints, and continuous sections.

Detailed design considerations, including design variables, multi-load conditions, lumped masses, constraints, and material properties, are presented in Table 5.15. The 28 elements, specifically element numbers 1–14 and 20–33, are grouped into 14 groups based on symmetry about the vertical middle plane, while the remaining elements are treated as distinct. As a result, the number of size variables is reduced to 31. In this problem, the continuous cross-sectional areas of 16 groups (G_i, where $i = 1, 2, \ldots, 31$) are considered within the range of $[-50, 50]$ cm^2, where a zero or negative value indicates removal of the element.

In addition, the truss is subjected to three nonstructural lumped masses of 500 kg each, placed at nodes 7, 8, and 9. To fulfill the multi-load condition, two different loading conditions are considered, as shown in Fig. 5.8.

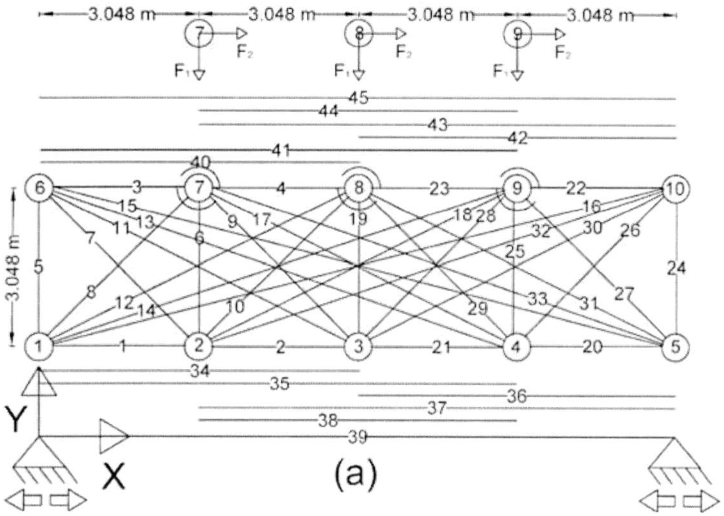

Fig. 5.8 The 45-bar truss

Table 5.15 Design parameters of the 45-bar truss

Size variables: $X_i = G_i$, $i = 1, 2, \ldots, 31$; $G =$ Group number Shape variables: $-2.5\ \text{m} \leq \xi_{x_j/y_j} \leq 2.5$ m, for $x_1 = -x_5$ and $y_1 = y_5$ with respect to their original coordinates	
Multiple load conditions:	
Condition 1 : $F_1 = 44.537$ kN	Condition 2 : $F_2 = 4.4537$ kN
Stress and displacement constraints: $\sigma_i^{\max} = 172.43$ MPa and $\delta_{x_j/y_j}^{\max} = 5$ mm (0.005 m) (for node, $j = 7$, 8, and 9 along x- and y-axes)	
Lumped masses on nodes 7, 8, and 9 are 500 kg	
Natural frequency constraints: $f_1 \geq 15$ Hz	
The continuous cross-sectional areas: $[A^{\min}, A^{\max}] = [-50, 50]\ \text{cm}^2 = ([-50, 50] \times 10^{-4}\ \text{m}^2)$ and critical area $= 1\ \text{cm}^2$	
Material properties: $E = 69$ GPa and $\rho = 2740\text{kg/m}^3$	

The 45-bar truss was optimized for TTO with 31 size and topology variables in this section. Table 5.16 shows the areas of elements and corresponding truss weights for the best designs obtained over 100 runs by using various algorithms, including ALO, DA, GWO, MVO, SCA, WOA, HTS, TLBO, PVS, and SOS. The minimum weight achieved by each algorithm was 148.0654, 235.7855, 148.0984, 148.0769, 165.5466, 148.4876, 163.7102, 148.0654, 148.0655, and 148.0701 kg, respectively. The ALO and TLBO algorithms produced the lightest structures and their best topologies were similar.

Additionally, Table 5.16 presents the statistical results of the proposed study for 100 independent runs. With the available results, it can be seen that ALO, DA, GWO, MVO, SCA, WOA, HTS, TLBO, PVS, and SOS algorithms having the mass

Table 5.16 Truss topology optimization of the 45-bar truss (continuous section)

Gropu no.	ALO	DA	GWO	MVO	SCA	WOA	HTS	TLBO	PVS	SOS
1	—	2.1906	—	—	—	—	—	—	—	—
3	—	—	—	—	—	—	1.4439	—	—	—
4	10.2033	18.1335	10.2226	10.2037	12.0401	10.3321	11.0047	10.2033	10.2033	10.2034
5	—	—	—	—	—	—	1.0476	—	—	—
6	—	10.5592	—	—	—	—	—	—	—	—
8	17.1598	28.2165	17.1620	17.1613	20.6463	17.2832	17.4524	17.1598	17.1598	17.1599
9	—	6.1910	—	—	—	—	—	—	—	—
12	24.1242	24.5928	24.1247	24.1267	26.7142	24.1242	24.2801	24.1242	24.1242	24.1255
18	—	5.4780	—	—	—	—	—	—	—	—
Mass (kg)	148.0654	235.7855	148.0984	148.0769	165.5466	148.4876	163.7102	148.0654	148.0655	148.0701
σ^{max} (MPa)	43.6497	25.0424	43.6467	43.6477	36.9905	43.1055	40.4708	43.6497	43.6497	43.6491
δ^{max} (mm)	4.3091	3.7267	4.3288	4.3088	3.6077	4.2698	4.1468	4.3091	4.3091	4.3091
f_1 (Hz)	15.6441	20.3397	15.6448	15.6448	16.7289	15.6854	15.8693	15.6441	15.6441	15.6444
Mean	300.3706	425.1876	205.0801	218.9916	275.6165	548.5013	274.6866	248.7299	189.1637	187.5750
SD	89.7122	96.7337	51.2368	72.1131	52.1453	209.7979	73.5918	70.0884	37.7913	44.3008

mean value of 300.3706, 425.1876, 205.0801, 218.9916, 275.6165, 548.5013, 274.6866, 248.7299, 189.1637, and 187.5750 kg, respectively, with SDs of 89.7122, 96.7337, 51.2368, 72.1131, 52.1453, 209.7979, 73.5918, 70.0884, 37.7913, and 44.3008, respectively. It can be concluded that the SOS and PVS algorithms performed best in terms of both convergence rate and consistency, with mean mass values of 187.5750 and 189.1637 kg and SD values of 44.3008 and 37.7913, respectively. Therefore, these algorithms are considered the most effective for this design problem.

The results of the study reveal that the topology optimization approach leads to significantly lighter trusses compared to size optimization approaches. Specifically, the best solutions obtained through topology optimization resulted in a truss that was 63.6% lighter compared to the truss obtained through size optimization.

This finding highlights the substantial benefits of using topology optimization in designing truss structures, as it can lead to a drastic reduction in material usage and weight. By allowing the algorithm to adjust the topology of the truss members based on the applied loads and design constraints, topology optimization can find more efficient load paths and remove unnecessary material, resulting in a truss that is significantly lighter compared to the one obtained through traditional size optimization. The reduction in weight can also translate to cost savings in terms of material and manufacturing expenses. Therefore, it can be concluded that topology optimization is a highly effective tool for designing lightweight and cost-effective truss structures.

5.4.9 *Topology Optimization of the 25-Bar 3D Truss with Continuous Cross-Sections*

In this section, the size and topology optimization of the 25-bar truss were performed using eight variables. The ninth benchmark problem, known as the 25-bar truss, was chosen to investigate a space truss subjected to stress, buckling, displacement, and fundamental frequency constraints, with continuous design variables.

The initial layout of the space truss is depicted in Fig. 5.9. Previous studies by Deb and Gulati (2001) and Ahrari et al. (2014) have investigated this problem with static constraints, but it has not been examined with simultaneous consideration of static and dynamic constraints. To address this, the problem is modified to include the effects of multi-load conditions, lumped masses, natural frequency constraints, and continuous area.

Detailed design considerations, including design variables, load conditions, lumped masses, constraints, and material properties, are presented in Table 5.17. Due to structural symmetry, element grouping is adopted for this problem, as indicated in the table. Consequently, the truss only has eight size variables, while topology variables are incorporated with them, as discussed earlier. In this problem, the continuous cross-sectional areas of 16 groups (G_i, where $i = 1, 2, \ldots, 8$) are

Fig. 5.9 The 25-bar 3D truss: (**a**) ground structure and (**b**) optimal structure

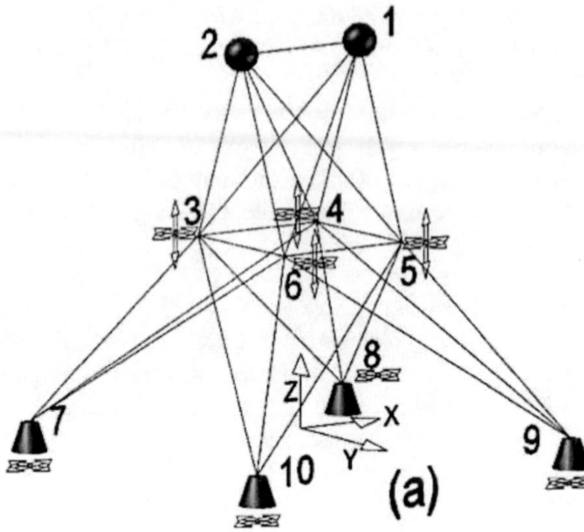

considered within the range of [−50, 50] cm^2, where a zero or negative value indicates removal of the element.

The truss is subjected to two nonstructural lumped masses of 500 kg each, placed at nodes 1 and 2. Two different loading conditions are considered, as depicted in the table.

Table 5.18 presents the outcomes of optimizing the 25-bar truss for TTO with eight continuous sizes using ten different algorithms in 100 independent runs. The algorithm that produced the minimum mass value while satisfying all structural constraints is considered the best run. Among the algorithms, PVS yielded the lowest mass value of 379.6616 kg, while ALO, DA, GWO, MVO, SCA, WOA, HTS, TLBO, and SOS yielded mass values between 379.6618 and 399.3569 kg.

To evaluate the performance of the algorithms statistically, the mean and standard deviation (SD) of the structural mass values were computed. ALO, DA, GWO, MVO, SCA, WOA, HTS, TLBO, PVS, and SOS algorithms resulted in mass mean values of 406.0008, 440.9842, 1.00E+06, 1.00E+06, 443.3417, 502.6210, 451.4840, 391.8926, 379.6675, and 379.7638 kg, respectively, with corresponding SDs of 56.6964, 87.0562, 1.00E+07, 1.00E+07, 16.4971, 94.0916, 153.0280, 67.9597, 0.0590, and 0.1126.

Based on the results, the PVS algorithm is the most effective in achieving the lowest mass mean and SD values, indicating superior convergence rate and consistency. Therefore, the PVS algorithm is the optimal algorithm for this design problem.

Additionally, the inclusion of topology optimization resulted in a truss that was 10.2% lighter than the one obtained through size optimization for the best solutions

Table 5.17 Design parameters of the 25-bar 3D truss

Size variables: $X_i = G_i$, $i = 1, 2, \ldots, 8$; G = Group number				
Multiple load conditions:				
Condition 1:	$F_{y_1} = 89.074$ kN	$F_{z_1} = -22.25$ kN	$F_{y_2} = -89.074$ kN	$F_{z_2} = -22.25$ kN
Condition 2:	$F_{x_1} = 4.454$ kN	$F_{y_1} = -44.537$ kN	$F_{z_1} = -44.537$ kN	$F_{y_2} = -44.537$ kN
	$F_{z_2} = -44.537$ kN	$F_{x_3} = 2.227$ kN	$F_{x_6} = 2.672$ kN	
Lumped masses at nodes 1 and 2 are 500 kg				
Stress and displacement constraints: $\sigma_i^{\max} = 172.375$ MPa and $\delta_{x_j/y_j/z_j}^{\max} = 8.9$ mm (0.0089 m) (for node, $j = 1, 2, \ldots, 6$ along x-, y-, and z-axes)				
Natural frequency constraints: $f_1 \geq 15$ Hz				
The continuous cross-sectional areas: $[A^{min}, A^{max}] = [-50, 50]$ cm^2 = ($[-50, 50] \times 10^{-4}$ m^2) and critical area = 1 cm^2				
Material properties: $E = 69$ GPa and $\rho = 2740$kg/m^3				
Element grouping				
Group number	Element number (end nodes)	Group number		Element number (end nodes)
G_1	1(1,2)	G_5		12(3,4), 13(5,6)
G_2	2(1,4), 3(2,3), 4(1,5), 5(2,6)	G_6		14(3,10), 15(6,7), 16(4,9), 17(5,8)
G_3	6(2,5), 7(2,4), 8(1,3), 9(1,6)	G_7		18(3,8), 19(4,7), 20(6,9), 21(5,10)
G_4	10(3,6), 11(4,5)	G_8		22(3,7), 23(4,8), 24(5,9), 25(6,10)

Table 5.18 Truss topology optimization of the 25-bar 3D truss (continuous section)

Element group no	ALO	DA	GWO	MVO	SCA	WOA	HTS	TLBO	PVS	SOS
1	—	—	—	—	3.8081	—	—	—	—	—
2	15.8118	15.7059	15.8365	15.8116	16.9448	16.7799	15.8118	15.8118	15.8118	15.8118
3	15.2601	16.1326	15.2644	15.2735	14.9967	17.3732	15.2601	15.2601	15.2601	15.2603
6	12.1210	12.1210	12.1514	12.1325	12.4096	12.4389	12.1214	12.1210	12.1210	12.1215
7	19.9211	20.1161	19.9230	19.9260	21.4908	19.9743	19.9211	19.9211	19.9211	19.9217
8	17.5668	17.6070	17.5766	17.5864	18.9134	17.5076	17.5675	17.5668	17.5668	17.5668
Mass (kg)	379.6618	383.0034	379.9634	379.8566	399.3569	391.1128	379.6666	379.6616	379.6616	379.6676
σ^{max} (MPa)	57.2341	54.5642	57.2074	57.1892	55.7839	50.6184	57.2340	57.2341	57.2341	57.2334
δ^{max} (mm)	7.7556	7.6224	7.7505	7.7490	7.4207	7.4337	7.7554	7.7556	7.7556	7.7555
f_1 (Hz)	15.3616	15.4302	15.3674	15.3676	15.7547	15.5902	15.3618	15.3616	15.3616	15.3616
Mean	406.0008	440.9842	1.00E+06	1.00E+06	443.3417	502.6210	451.4840	391.8926	379.6675	379.7638
SD	56.6964	87.0562	1.00E+07	1.00E+07	16.4971	94.0916	153.0280	67.9597	0.0590	0.1126

in each case. These findings suggest that the TTO approach outperforms the size optimization approach in designing lighter trusses.

5.4.10 Topology Optimization of the 39-Bar 3D Truss with Continuous Cross-Sections

In this section, the main focus is to compare the performance of various optimization algorithms for the size optimization of a 39-bar 3D truss with continuous sections. The ground structure of this truss, consisting of 39 elements, is depicted in Fig. 5.10. To account for the symmetry about the *x–z* and *y–z* planes, the element cross-sectional areas are grouped into 11 sets. In this problem, the continuous cross-sectional areas of 16 groups (G_i, where $i = 1, 2, \ldots, 11$) are considered within the range of [−50, 50] cm^2, where a zero or negative value indicates removal of the element.

Previous studies by Deb and Gulati (2001) and Luh and Lin (2008) have investigated this truss with static constraints. However, it has not been examined with simultaneous consideration of static and dynamic constraints. To address this gap, the problem is modified in this study to incorporate multi-load conditions, lumped masses, natural frequency constraints, and continuous cross-sectional areas.

In Fig. 5.10, a constant lumped mass of 500 kg is attached to the top nodes (nodes 1 and 2) of the truss. The design parameters, including material properties, loading conditions, design variable bounds, constraints, and other relevant details, are provided in Table 5.19. This comprehensive table outlines all the necessary information for the truss optimization problem.

Fig. 5.10 The 39-bar 3D truss

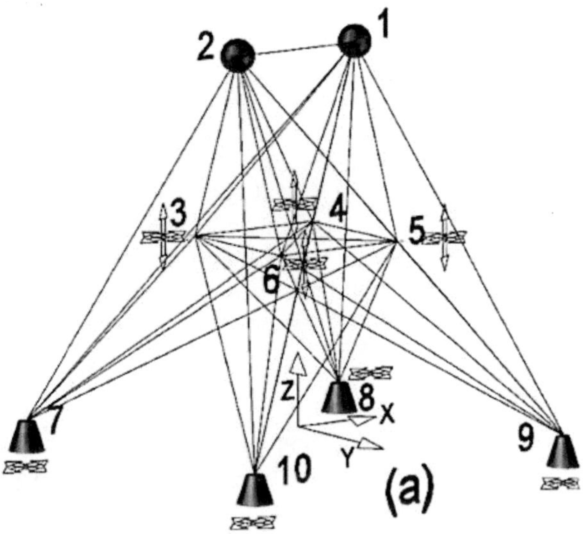

Table 5.19 Design parameters of the 39-bar 3D truss

Size variables: $X_i = G_i$, $i = 1, 2, \ldots, 11$; $G =$ Group number			
Load condition:	$F_{z_1} = -222.685$ kN		$F_{z_2} = -222.685$ kN
Lumped masses at nodes 1 and 2 are 500 kg			
Stress and displacement constraints: $\sigma_i^{max} = 172.375$ MPa and $\delta_{x_j/y_j/z_j}^{max} = 8.9$ mm (0.0089 m) (for node, $j = 1, 2, \ldots, 6$ along x-, y-, and z-axes)			
Natural frequency constraints: $f_1 \geq 15$ Hz			
The continuous cross-sectional areas: $[A^{\min}, A^{\max}] = [-50, 50]$ cm^2 = ($[-50, 50] \times 10^{-4}$ m^2) and critical area = 1 cm^2			
Material properties: $E = 69$ GPa and $\rho = 2740$kg/m^3			
Element grouping			
Group number	Element number (end nodes)	Group number	Element number (end nodes)
G_1	1(1,2)	G_7	22(3,7), 23(4,8), 24(5,9), 25(6,10)
G_2	2(1,4), 3(2,3), 4(1,5), 5(2,6)	G_8	26(5,7), 27(6,8), 28(3,9), 29(4,10)
G_3	6(2,5), 7(2,4), 8(1,3), 9(1,6)	G_9	30(3,5), 31(4,6)
G_4	10(3,6), 11(4,5), 12(3,4), 13(5,6)	G_{10}	32(1,7), 33(1,10), 34(2,9), 35(2,8)
G_5	14(3,10), 15(6,7), 16(4,9), 17(5,8)	G_{11}	36(2,7), 37(2,10), 38(1,8), 39(1,9)
G_6	18(3,8), 19(4,7), 20(6,9), 21(5,10)		

In this section, a comparison of algorithm performance on TTO optimization of a 39-bar 3D truss with continuous section is presented. The results are based on 100 independent runs, and the performance is summarized in Table 5.20, which shows size variables, best mass, mean mass, SD of mass, stress, displacement, and frequency response. It is observed that the best run achieves the minimum mass while satisfying all the stated structural constraints. The mean and SD values of structural mass are used to evaluate the performance of various evolutionary methods statistically. The best results reported by ALO, DA, GWO, MVO, SCA, WOA, HTS, TLBO, PVS, and SOS algorithms are 374.9457, 358.6359, 353.3472, 355.8720, 368.4770, 395.9554, 353.0142, 353.2964, 353.0343, and 353.1012 kg, respectively. The result table shows that the HTS algorithm has the lowest structural mass, with PVS and SOS algorithms being a close second and third, respectively. The topologies found using different approaches consist of only 6 groups out of 11.

To evaluate the true performance of metaheuristic methods, the average mass and its SD are considered. Based on the available results, it can be noted that ALO, DA, GWO, MVO, SCA, WOA, HTS, TLBO, PVS, and SOS algorithms have mass mean values of 450.0405, 482.3661, 422.4684, 429.3367, 439.2900, 539.7457, 412.7281, 408.7038, 393.3833, and 390.4038 kg, respectively, with SDs of 66.2165, 71.3463, 48.7691, 47.6355, 37.8802, 96.2739, 41.1565, 29.4519, 21.9083, and 29.0029,

Table 5.20 Truss topology optimization of the 39-bar 3D truss (continuous section)

Element group no.	ALO	DA	GWO	MVO	SCA	WOA	HTS	TLBO	PVS	SOS
1	15.1833	11.9483	9.4839	14.7071	14.8937	—	9.4453	9.4588	9.4521	9.5662
2	16.9756	3.2354	3.6232	3.7098	2.9574	13.8279	3.6008	3.4501	3.4803	3.5433
3	13.0909	2.2080	2.7077	2.7688	2.6773	5.9996	2.9212	3.1273	2.9532	3.0957
4	—	—	—	—	—	5.3012	—	—	—	—
5	2.6132	—	—	—	—	—	—	—	—	—
6	—	3.4006	1.9937	2.0678	4.6734	—	1.8964	1.8366	1.9732	1.8465
7	24.0283	1.6375	1.6871	1.4014	1.7791	15.2558	1.5895	1.7120	1.5814	1.5793
8	18.7237	—	—	—	—	—	—	—	—	—
10	—	40.3575	40.3595	40.3579	40.5081	32.5106	40.3576	40.3663	40.3578	40.3581
Mass (kg)	374.9457	358.6359	353.3472	355.8720	368.4770	395.9554	353.0142	353.2964	353.0343	353.1012
σ^{max} (MPa)	54.1139	57.0491	71.4839	46.5231	45.8974	34.1957	71.7735	71.6725	71.7240	70.8895
δ^{max} (mm)	6.0394	3.4239	3.4860	3.3784	3.3639	3.0073	3.4874	3.4862	3.4871	3.4835
f_1 (Hz)	15.0000	15.0106	15.0143	15.0042	15.4871	22.0694	15.0001	15.0029	15.0042	15.0010
Mean	450.0405	482.3661	422.4684	429.3367	439.2900	539.7457	412.7281	408.7038	393.3833	390.4038
SD	66.2165	71.3463	48.7691	47.6355	37.8802	96.2739	41.1565	29.4519	21.9083	29.0029

respectively. The SOS algorithm performs best in achieving the best mass mean value, while the PVS algorithm gives the best SD. Therefore, the SOS and PVS algorithms are considered the best algorithms for both convergence rate and consistency for this design problem.

The study demonstrated that topology optimization is a more effective method for designing lighter trusses compared to size optimization approaches. The best solutions obtained through topology optimization resulted in a truss that was 40.5% lighter compared to the truss obtained through size optimization, which is a significant improvement in weight reduction. This indicates that topology optimization enables a more efficient use of material, resulting in a lighter and more optimized design.

5.4.11 A Comprehensive Analysis

A comprehensive analysis of the best results obtained from ten benchmark problems using the proposed algorithms in 100 independent runs is presented in Table 5.21. The table provides details on the minimum mass and offers an in-depth evaluation of overall, average, and individual ranks, as well as the best and worst results for each algorithm.

Upon analyzing the results in Table 5.21, it was revealed that the PVS algorithm achieved the minimum mass and outperformed all other algorithms in terms of overall rank. Following closely were the TLBO, SOS, and GWO algorithms with the second, third, and fourth-best performances, respectively.

Furthermore, the PVS algorithm exhibited outstanding performance by obtaining the best solutions for all ten problems. Similarly, the TLBO algorithm showed five best solutions, two second-best solutions, and one third-best solution, without any worse solutions.

However, the WOA algorithm presented the worst solutions for eight problems, highlighting the importance of selecting the most appropriate algorithm for specific problems to achieve optimal results.

The comprehensive analysis of Table 5.21 offers valuable insights into the performance of the proposed algorithms for ten benchmark problems, which can be useful for researchers and practitioners in selecting the most suitable algorithm for similar problems and enhancing the efficiency of the optimization process.

Table 5.22 showcases the mean mass achieved by the proposed algorithms for ten benchmark problems in 100 independent runs. Similar to Table 5.21, it presents overall rank, average rank, individual ranks, best and worst results for each algorithm.

Table 5.21 Minimum mass obtained using TTO with continuous sections

	ALO	DA	GWO	MVO	SCA	WOA	HTS	TLBO	PVS	SOS
10-bar truss	423.6	459.3	424.1	423.8	437.0	514.3	423.6	**423.5**	**423.5**	423.6
14-bar truss	**363.6**	367.6	363.9	363.7	378.6	378.8	363.7	**363.6**	**363.6**	**363.6**
15-bar truss	97.7	117.5	94.2	94.5	117.6	146.7	94.8	**93.7**	**93.7**	95.6
24-bar truss	129.4	160.2	120.6	120.7	149.6	208.5	124.8	121.0	**119.1**	122.0
20-bar truss	155.4	176.9	151.1	150.8	180.3	252.9	153.4	150.8	**150.7**	151.7
72-bar 3D truss	452.3	505.6	436.4	446.6	510.6	559.7	444.7	439.0	**436.1**	441.1
39-bar truss	302.9	360.7	235.5	236.3	323.6	427.0	262.6	235.1	**233.3**	237.6
45-bar truss	**148.1**	235.8	**148.1**	**148.1**	165.5	148.5	163.7	**148.1**	**148.1**	**148.1**
25-bar 3D truss	**379.7**	383.0	380.0	379.9	399.4	391.1	**379.7**	**379.7**	**379.7**	**379.7**
39-bar 3D truss	374.9	358.6	353.3	355.9	368.5	396.0	**353.0**	353.3	**353.0**	353.1
Overall rank	7	8	4	5	9	10	6	2	1	3
Count best	3	0	1	1	0	0	2	5	10	3
Count second best	0	0	2	1	0	0	0	2	0	0
Count 3nd best	1	0	2	1	0	0	1	1	0	2
Count worst	0	1	0	0	1	8	0	0	0	0

Note: Bold indicates the best solution

Upon analyzing the results in Table 5.22, it was observed that the PVS algorithm outperformed all other algorithms in terms of overall rank, obtaining the highest mean mass. It was followed by the SOS, TLBO, and GWO algorithms, respectively.

Furthermore, the PVS algorithm demonstrated six best solutions, three second-best solutions, and one third-best solution, without any worse solutions. Similarly, the SOS algorithm provided three best solutions and seven second-best solutions, with no worse solutions. On the other hand, the TLBO algorithm had three best solutions, one second-best solution, four third-best solutions, and no worse solutions.

Table 5.22 Mean mass obtained using TTO with continuous sections

	ALO	DA	GWO	MVO	SCA	WOA	HTS	TLBO	PVS	SOS
10-bar truss	574.9	820.1	525.7	514.8	530.1	951.9	561.4	478.1	469.9	**461.1**
14-bar truss	682.7	768.3	598.0	520.9	453.4	1270.6	550.5	448.6	**419.3**	422.5
15-bar truss	135.3	1.20E +07	9.00E +06	2.40E +07	1.00E +06	249.4	158.9	1.00E +06	**100.1**	101.7
24-bar truss	206.4	267.7	149.9	162.6	189.4	387.8	196.3	171.2	**135.5**	145.3
20-bar truss	260.7	1.01E +06	1.66E +06	1.30E +07	243.6	431.5	673.6	427.6	**169.7**	177.0
72-bar 3D truss	792.8	1050.7	**454.3**	7.00E +06	642.5	1377.7	766.4	498.2	464.9	459.8
39-bar truss	452.3	575.9	315.7	340.2	432.5	852.8	371.1	372.5	**300.4**	313.6
45-bar truss	300.4	425.2	205.1	219.0	275.6	548.5	274.7	248.7	189.2	**187.6**
25-bar 3D truss	406.0	441.0	1.00E +06	1.00E +06	443.3	502.6	451.5	391.9	**379.7**	379.8
39-bar 3D truss	450.0	482.4	422.5	429.3	439.3	539.7	412.7	408.7	393.4	**390.4**
Overall rank	7	9	4	7	5	10	6	3	1	2
Count best	0	0	1	0	0	0	0	0	6	3
Count second best	0	0	0	0	0	0	0	0	3	7
Count 3nd best	1	0	3	0	1	0	0	4	1	0
Count worst	0	0	0	3	0	6	0	0	0	0

Note: Bold indicates the best solution

In contrast, the WOA algorithm exhibited the highest number of worst solutions for six benchmark problems. These results suggest that the performance of the WOA, DA, MVO, and ALO algorithms may not be optimal for certain types of problems.

5.5 Results and Discussion on Truss Topology Optimization with Discrete Cross-Sections

This section uses ten different benchmark problems to assess the effectiveness of the algorithms under consideration for discrete cross-sections. The benchmark problems include various truss structures such as the 10-bar, 14-bar, 15-bar, 24-bar, 20-bar, 72-bar 3D, 39-bar, 45-bar, 25-bar 3D, and 39-bar 3D trusses. The proposed algorithms are utilized to tackle each truss problem by conducting 100 independent runs, acknowledging the stochastic nature of metaheuristics. The population size remains constant at 50 for all the problems, while the total number of function evaluations (FE) per run is set to 20,000. The primary objective is to evaluate the performance of the selected algorithms in solving the truss TTO optimization problems that are discrete in nature.

5.5.1 Topology Optimization of the 10-Bar Truss with Discrete Cross-Sections

In this section, the optimization of the 10-bar truss using the TTO approach is carried out. The truss is optimized using ten discrete size and topology variables. The ground structure of the first benchmark truss, along with the load and boundary conditions, is depicted in Fig. 5.11.

Previous studies by Hajela and Lee (1995), Deb and Gulati (2001), Richardson et al. (2012), and Miguel et al. (2013) have investigated this truss problem with a focus on static constraints such as elemental stress and nodal displacement. Additionally, several researchers, including Wei et al. (2005), Gomes (2011), Wei et al.

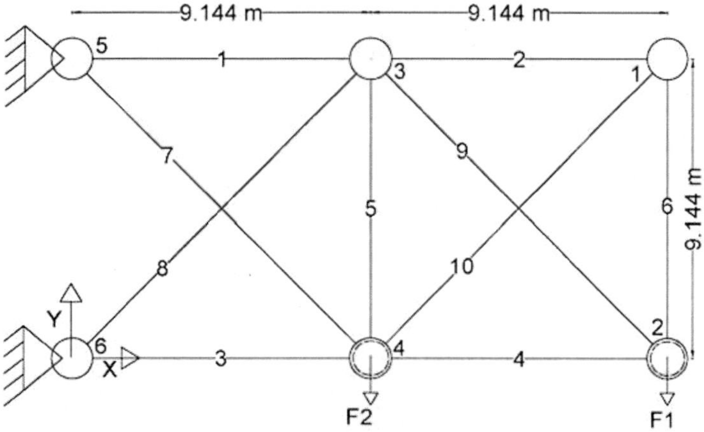

Fig. 5.11 The 10-bar truss

Table 5.23 Design parameters of the 10-bar truss

Size variables: $X_i = A_i$, $i = 1, 2, \ldots, 10$	
Multiple load conditions:	
Condition 1 : $F_1 = 44.537$ kN, $F_2 = 44.537$ kN	Condition 2 : $F_1 = 44.537$ kN, $F_2 = 0$ kN
Lumped masses on nodes 2 and 4 are 500 kg	
Stress and displacement constraints: $\sigma_i^{max} = 172.43$ MPa and $\delta_{y_5 \& y_6}^{max} = 100$ mm (0.1 m)	
Natural frequency constraint: $f_1 \geq 4$ Hz	
The discrete cross-sectional areas: $[A^{min}, A^{max}] = [-100, 100]$ cm^2 = ($[-100, 100] \times 10^{-4}$ m^2) in increments of 1 cm^2 and critical area = 1 cm^2	
Material properties: $E = 69$ GPa and $\rho = 2740$kg/m^3	

(2011), Kaveh and Zolghadr (2011), Miguel and Miguel (2012), Kaveh and Zolghadr (2012), and Farshchin et al. (2016), have explored size and shape optimization of this truss considering dynamic constraints, particularly natural frequencies.

However, in this study, the problem is modified to include both static and dynamic constraints simultaneously. This is achieved by incorporating multi-load conditions, lumped masses, natural frequency constraints, and discrete cross-sectional areas in the optimization problem. The discrete cross-sectional areas (A_i, where $i = 1, 2, \ldots, 10$) are allowed to vary within the range of $[-100, 100]$ cm^2 in increments of 1 cm^2, where a zero or negative value indicates the removal of the corresponding element.

For more detailed information about the material properties, design variables, loading conditions, lumped masses, and other important data related to the test problem, refer to Table 5.23. This table provides comprehensive information necessary for understanding and conducting the truss optimization.

In this section, the optimization of a 10-bar truss is conducted for topology and size optimization using ten discrete size and topology variables. Table 5.24 displays the results obtained from TTO of the truss, with discrete cross-sections, using various algorithms in 100 independent runs. The outcome of this experiment shows that the best run gives the minimum mass value while satisfying all the stated structural constraints. The mean and standard deviation (SD) values of the structural mass are also presented in the table, which will be used to statistically evaluate the performance of different evolutionary methods.

According to the table, the ALO, DA, GWO, MVO, SCA, WOA, HTS, TLBO, PVS, and SOS algorithms produce the best results with minimum mass values of 426.8370, 510.6591, 426.8370, 426.8370, 463.9890, 602.4472, 426.8370, 426.8370, 426.8370, and 426.8370 kg, respectively. Interestingly, all of these methods except WOA exclude elements 2, 5, 6, and 10 in their approaches. Moreover, the results show that these algorithms share a similar optimum weight of 426.8370 kg.

To evaluate the real performance of these metaheuristic methods, the mean mass and its SD are considered. Based on the available results, it can be seen that the ALO, DA, GWO, MVO, SCA, WOA, HTS, TLBO, PVS, and SOS algorithms have the mean mass values of 547.0215, 725.2954, 496.5325, 560.3016, 541.8805,

Table 5.24 Truss topology optimization of the 10-bar truss (discrete section)

Element no.	ALO	DA	GWO	MVO	SCA	WOA	HTS	TLBO	PVS	SOS
1	4	4	4	4	9	15	4	4	4	4
2	—	—	—	—	—	6	—	—	—	—
3	64	64	64	64	68	69	64	64	64	64
4	37	37	37	37	40	13	37	37	37	37
5	37	45	37	37	37	—	37	37	37	37
6	—	—	—	—	—	18	—	—	—	—
7	8	9	8	8	9	11	8	8	8	8
9	5	22	5	5	6	1	5	5	5	5
10	—	—	—	—	—	64	—	—	—	—
Mass (kg)	426.8370	510.6591	426.8370	426.8370	463.9890	602.4472	426.8370	426.8370	426.8370	426.8370
σ^{max} (MPa)	157.4621	157.4621	157.4621	157.4621	157.4621	157.4621	157.4621	157.4621	157.4621	157.4621
δ^{max} (mm)	98.6006	98.6006	98.6006	98.6006	98.6006	98.6006	98.6006	98.6006	98.6006	98.6006
f_1 (Hz)	4.8421	5.5513	4.8421	4.8421	5.3277	5.8722	4.8421	4.8421	4.8421	4.8421
Mean	547.0215	725.2954	496.5325	560.3016	541.8805	944.6175	521.0097	484.5048	466.6023	464.8649
SD	75.9464	156.7443	52.8545	427.9122	53.7790	243.1805	88.0551	61.8279	45.3756	43.4554

944.6175, 521.0097, 484.5048, 466.6023, and 464.8649 kg, respectively, with corresponding SD values of 75.9464, 156.7443, 52.8545, 427.9122, 53.7790, 243.1805, 88.0551, 61.8279, 45.3756, and 43.4554. The SOS and PVS algorithms perform the best, achieving the lowest mean mass and SD values. Therefore, for this design problem, the SOS and PVS algorithms are the actual better algorithms, both in terms of convergence rate and consistency.

This suggests that topology optimization can be a more effective approach to designing lightweight trusses compared to size optimization. The findings of the study show that topology optimization can lead to a substantial reduction in the weight of trusses, with the best solutions obtained through this method resulting in a 25.5% decrease in weight compared to trusses designed through size optimization. This difference in weight reduction can have significant implications for the design and construction of structures, as lighter trusses can lead to cost savings, improved performance, and greater structural efficiency. Overall, the study highlights the importance of considering topology optimization as a viable option for engineers and designers who aim to create more lightweight and efficient truss structures.

5.5.2 *Topology Optimization of the 14-Bar Truss with Discrete Cross-Sections*

In this section, the optimization of the 14-bar truss is performed, considering both size and topology optimization. The optimization process involves the use of 14 discrete size variables. The ground structure of the second benchmark truss, known as the 14-bar truss, is shown in Fig. 5.12, along with the load and boundary conditions.

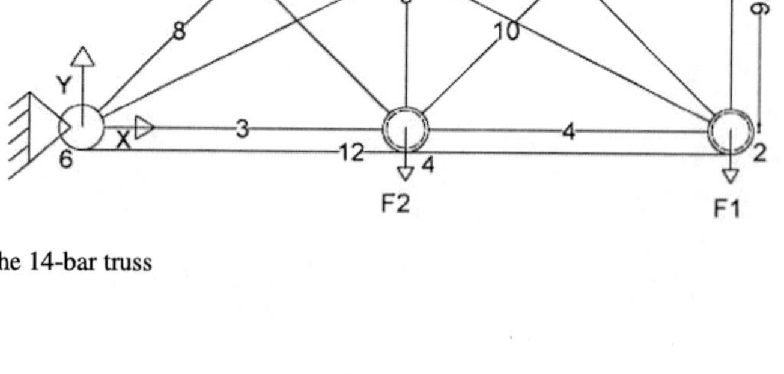

Fig. 5.12 The 14-bar truss

Table 5.25 Design parameters of the 14-bar truss

Size variables: $X_i = A_i$, $i = 1, 2, \ldots, 14$	
Multiple load conditions:	
Condition 1 : $F_1 = 44.537$ kN, $F_2 = 44.537$ kN	Condition 2 : $F_1 = 44.537$ kN, $F_2 = 0$ kN
Lumped mass on nodes 2 and 4 are 500 kg	
Stress and displacement constraints: $\sigma_i^{max} = 172.43$ MPa and $\delta_{y_5\&y_6}^{max} = 100$ mm (0.1 m)	
Natural frequency constraint: $f_1 \geq 4$ Hz	
The discrete cross-sectional areas: $[A^{min}, A^{max}] = [-100, 100]\ \text{cm}^2 = ([-100, 100] \times 10^{-4}\ \text{m}^2)$ in increments of 1 cm^2 and critical area = 1 cm^2	
Material properties: $E = 69$ GPa and $\rho = 2740\text{kg/m}^3$	

Previous studies conducted by Tang et al. (2005), Rahami et al. (2008), and Miguel et al. (2013) have focused on TSS (topology, size, and shape) optimization of this truss problem. These studies utilized 14 discrete cross-sectional areas and eight continuous shape variables, taking into account static constraints. However, none of these previous investigations have addressed the simultaneous consideration of static and dynamic constraints for this particular problem.

The discrete cross-sectional areas (A_i, where $i = 1, 2, \ldots, 14$) are allowed to vary within the range of $[-100, 100]\ \text{cm}^2$ in increments of 1 cm^2, where a zero or negative value indicates the removal of the corresponding element. For detailed information regarding the design variables, load conditions, design constraints, and other relevant data related to the test problem, refer to Table 5.25. This table provides comprehensive details that are essential for understanding and conducting the optimization of the 14-bar truss.

The following passage describes the optimization of a 14-bar truss using different evolutionary algorithms. The truss has 14 discrete size variables, and the results of 100 independent runs of each algorithm are presented in Table 5.26. The best run is the one that gives the minimum mass value while satisfying all structural constraints. The mean and standard deviation values of the structural mass are given to evaluate the performance of the algorithms statistically.

The optimal weight of the truss for the ALO, GWO, MVO, HTS, TLBO, PVS, and SOS algorithms is 369.6249 kg. In these approaches, the same ten elements are removed: 1, 2, 5, 6, 8, 9, 10, 11, 12, and 14. The average mass and its standard deviation are used to assess the true performance of the metaheuristic methods. The results indicate that the PVS algorithm performs best, achieving the lowest mass mean and SCA for SD values. Therefore, it is the most suitable algorithm for this design problem in terms of both convergence rate and consistency.

In summary, this section presents the optimization of a 14-bar truss using different evolutionary algorithms. The results show that the PVS algorithm is the most effective algorithm for this problem.

The results of the study indicate that topology optimization is a promising approach for designing lighter trusses compared to size optimization approaches. The best solutions obtained through topology optimization were found to be 50.6% lighter compared to the truss obtained through size optimization. This means that

Table 5.26 Truss topology optimization of the 14-bar truss (discrete section)

Element no.	ALO	DA	GWO	MVO	SCA	WOA	HTS	TLBO	PVS	SOS
3	64	100	64	64	65	72	64	64	64	64
4	52	52	52	52	58	52	52	52	52	52
7	4	4	4	4	5	11	4	4	4	4
13	8	8	8	8	8	8	8	8	8	8
Mass (kg)	369.6249	459.8213	369.6249	369.6249	390.7063	414.4713	369.6249	369.6249	369.6249	369.6249
σ^{max} (MPa)	157.4621	157.4621	157.4621	157.4621	157.4621	157.4621	157.4621	157.4621	157.4621	157.4621
δ^{max} (mm)	92.5580	92.5580	92.5580	92.5580	92.5580	92.5580	92.5580	92.5580	92.5580	92.5580
f_1 (Hz)	4.6999	4.7401	4.6999	4.6999	4.6947	4.7183	4.6999	4.6999	4.6999	4.6999
Mean	681.8861	772.0058	505.7850	517.0211	449.7459	1332.3791	556.7942	497.3289	419.8813	425.7891
SD	294.4714	205.8923	150.0223	151.8818	60.5212	624.8360	196.1673	141.3824	70.0477	103.2091

topology optimization can significantly reduce the weight of trusses, which can have important implications for various applications such as aerospace, automotive, and civil engineering. It is important to note that this study only looked at a specific set of parameters and constraints, and the results may vary for different design criteria. Nevertheless, the findings suggest that topology optimization can be an effective tool for designing lightweight structures.

5.5.3 *Topology Optimization of the 15-Bar Truss with Discrete Cross-Sections*

In this section, the optimization of the 15-bar truss is conducted, focusing on both size and topology optimization. The optimization process involves the utilization of 15 discrete size variables. The ground structure of this truss can be observed in Fig. 5.13.

This problem is adapted from the previous works of Rahami et al. (2008) and Ahrari et al. (2014). Previous studies on this problem have primarily considered static constraints such as stress, buckling, and displacement. However, the simultaneous consideration of static and dynamic constraints, specifically natural frequency constraints, has not been explored in depth.

The discrete cross-sectional areas (A_i, where $i = 1, 2, \ldots, 15$) are allowed to vary within the range of [−50, 50] cm^2 in increments of 1 cm^2, where a zero or negative value indicates the removal of the corresponding element. To address this research gap, the problem is modified to incorporate several factors, including multi-load conditions, lumped masses, natural frequency constraints, and discrete cross-sectional areas. Design considerations and material properties are provided in Table 5.27, which offers crucial information necessary for the optimization process. This table serves as a valuable resource for understanding the design variables, loading conditions, constraints, and other relevant data associated with the optimization of the 15-bar truss.

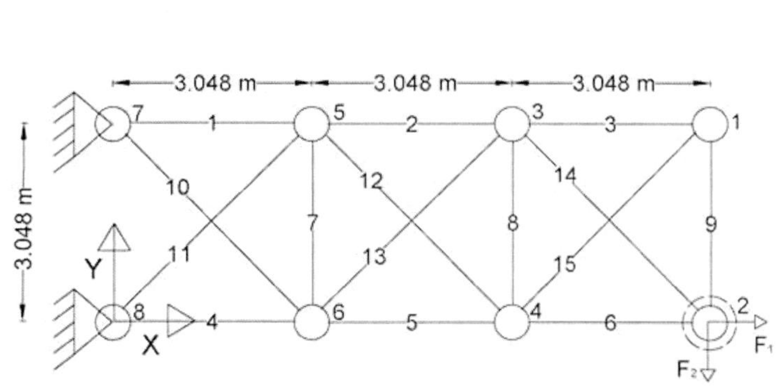

Fig. 5.13 The 15-bar truss

Table 5.27 Design parameters of the 15-bar truss

Size variables: $X_i = A_i$, $i = 1, 2, \ldots, 15$	
Multiple load conditions:	
Condition 1 : $F_1 = -44.537$ kN, $F_2 = 0$ kN	Condition 2 : $F_1 = 0$ kN, $F_2 = 4.4537$ kN
Lumped mass on node 2 is 500 kg	
Stress and displacement constraints: $\sigma_i^{max} = 172.43$ MPa and $\delta_{x_2\&y_2}^{max} = 100$ mm (0.1 m)	
Natural frequency constraint: $f_1 \geq 4$ Hz	
The discrete cross-sectional areas: $[A^{min}, A^{max}] = [-50, 50]$ cm^2 = ($[-50, 50] \times 10^{-4}$ m^2) in increments of 1 cm^2 and critical area = 1 cm^2	
Material properties: $E = 69$ GPa and $\rho = 2740$kg/m^3	

Table 5.28 presents the outcomes of TTO optimization of a 15-bar truss, with discrete cross-sections, using various algorithms in 100 independent runs. The algorithm with the lowest mass value while meeting all the structural constraints is considered the best run. The mean and SD values of the structural mass are also provided to evaluate the performance of different evolutionary methods statistically. The best results reported by ALO, DA, GWO, MVO, SCA, WOA, HTS, TLBO, PVS, and SOS algorithms are 105.5335, 124.1095, 97.9578, 97.9578, 124.3960, 130.1472, 98.1604, 98.1604, 97.9578, and 97.9578 kg, respectively. The lowest mass values are reported by the GWO, MVO, PVS, and SOS algorithms, which all produce the same solution.

To determine the overall performance of the metaheuristic methods, the average mass and its SD are considered. Based on these results, it can be concluded that the PVS algorithm has the best performance, achieving the lowest mass mean value and the best SD. The mass mean values and SDs for all the algorithms are as follows: ALO (151.7644, 32.3378), DA (1.20E+07, 3.28E+07), GWO (6.00E+06, 2.40E+07), MVO (2.40E+07, 4.31E+07), SCA (147.5534, 10.3001), WOA (252.1795, 73.8437), HTS (145.6313, 87.6247), TLBO (122.3045, 34.8650), PVS (104.1138, 5.8307), and SOS (106.1137, 8.0492) kg.

Therefore, it can be concluded that the PVS algorithm is the best option for achieving both a good convergence rate and consistency in this design problem.

This study has found that using the topology optimization approach results in trusses that are significantly lighter than those obtained through size optimization approaches. More precisely, the best solutions obtained through topology optimization resulted in a truss that was 18.4% lighter compared to the truss obtained through size optimization. This indicates that the topology optimization approach can result in more efficient designs that use material more effectively and reduce the weight of the truss structure. This can have important implications for various engineering applications, including the design of aircraft and spacecraft, bridges, and other structures where weight reduction is a critical factor for performance and efficiency.

Table 5.28 Truss topology optimization of the 15-bar truss (discrete section)

Element no.	ALO	DA	GWO	MVO	SCA	WOA	HTS	TLBO	PVS	SOS
1	6	8	7	7	14	7	6	6	7	7
2	13	8	6	6	6	8	6	6	6	6
4	13	16	13	13	13	15	13	13	13	13
5	13	15	13	13	20	15	13	13	13	13
6	13	29	13	13	14	15	13	13	13	13
7	1	—	4	4	12	—	1	1	4	4
8	—	1	1	1	4	13	1	1	1	1
9	—	—	—	—	—	1	—	—	—	—
10	8	—	—	—	—	3	7	7	—	—
11	—	2	3	3	6	2	—	—	3	3
12	1	7	—	—	—	7	—	—	—	—
13	2	5	3	3	4	—	2	2	3	3
14	7	7	7	7	7	7	7	7	7	7
15	—	—	—	—	—	5	—	—	—	—
Mass (kg)	105.5335	124.1095	97.9578	97.9578	124.3960	130.1472	98.1604	98.1604	97.9578	97.9578
σ^{max} (MPa)	34.2592	34.2592	34.2592	34.2592	34.2592	34.2592	34.2592	34.2592	34.2592	34.2592
δ^{max} (mm)	7.8544	7.8544	7.8544	7.8544	7.8544	7.8544	7.8544	7.8544	7.8544	7.8544
f_1 (Hz)	4.6074	4.6074	4.7051	4.7051	5.6642	4.9813	4.9967	4.9967	4.7051	4.7051
Mean	151.7644	1.20E+07	6.00E+06	2.40E+07	147.5534	252.1795	145.6313	122.3045	104.1138	106.1137
SD	32.3378	3.28E+07	2.40E+07	4.31E+07	10.3001	73.8437	87.6247	34.8650	5.8307	8.0492

5.5.4 Topology Optimization of the 24-Bar Truss with Discrete Cross-Sections

In this section, the size and topology optimization of the 24-bar truss is the main focus. This problem presents several challenges due to the large number of design variables involved. The ground structure of the truss, known as the fourth benchmark problem, is depicted in Fig. 5.14.

The 24-bar truss problem was initially investigated by Ringertz (1986) using a Branch and Bound algorithm for topology optimization under static constraints. Subsequently, Xu et al. (2003) used it to test a 1-D search method, and Kaveh and Zolghadr (2013) employed it to evaluate the search performance of charged system search and particle swarm optimization algorithms, considering continuous design variables.

The discrete cross-sectional areas (A_i, where $i = 1, 2, \ldots, 24$) are allowed to vary within the range of $[-40, 40]$ cm^2 in increments of 1 cm^2, where a zero or negative value indicates the removal of the corresponding element. In this optimization problem, the truss is subjected to two external load cases, which are specified in Table 5.29. The cross-sectional areas of the truss elements are treated as discrete variables. Additionally, a nonstructural lumped mass of 500 kg is applied at node 3. Detailed information regarding nodal masses, material properties, and other relevant data can be found in the table, providing important context for the optimization process.

This section demonstrates the TTO of a 24-bar truss through the utilization of proposed algorithms with 24 discrete size and topology variables. Table 5.30 illustrates the outcomes obtained in TS optimization of the 24-bar truss utilizing discrete

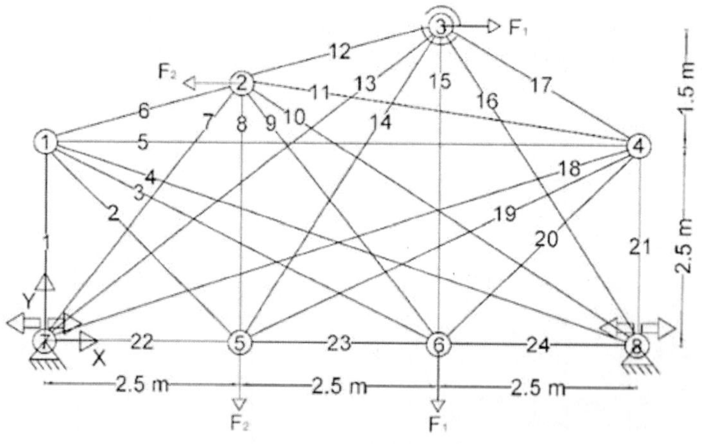

Fig. 5.14 The 24-bar truss

Table 5.29 Design parameters of the 24-bar truss

Size variables: $X_i = A_i$, $i = 1, 2, \ldots, 24$	
Multiple load conditions:	
Condition 1 : $F_1 = 50$ kN, $F_2 = 0$ kN	Condition 2 : $F_1 = 0$ kN, $F_2 = 50$ kN
Lumped mass on node 3 is 500 kg	
Stress and displacement constraints: $\sigma_i^{\max} = 172.43$ MPa and $\delta_{y_5\&y_6}^{\max} = 10$ mm $\left(10^{-2}\ \text{m}\right)$	
Natural frequency constraint: $f_1 \geq 30$ Hz	
The discrete cross-sectional areas: $[A^{\min}, A^{\max}] = [-40, 40]\ \text{cm}^2 = ([-40, 40] \times 10^{-4}\ \text{m}^2)$ in increments of 1 cm^2 and critical area = 1 cm^2	
Material properties: $E = 69$ GPa and $\rho = 2740 \text{kg/m}^3$	

cross-sections for the algorithms in 100 independent runs. The findings reveal that the best run delivers the minimum mass value while satisfying all structural constraints. The mean and standard deviation (SD) values of the structural mass are provided and will be used to statistically evaluate the performance of different evolutionary techniques. Among the ALO, DA, GWO, MVO, SCA, WOA, HTS, TLBO, PVS, and SOS algorithms, the best results reported are 147.9755, 188.6131, 122.8198, 120.7648, 148.5357, 206.9689, 137.8316, 121.7832, 121.2033, and 120.0798 kg, respectively. The outcome from SOS exhibits the least structural mass, with MVO and PVS algorithms providing near solutions as the second-best alternatives.

To evaluate the actual statistical performance of the metaheuristic methods, we consider the average mass and its SD. Based on the available results, we can conclude that the PVS algorithm performed best with a mass mean value of 138.8606 kg and an SD of 141.1555 kg. The other algorithms had mass mean values and SDs as follows: ALO (207.8352, 207.8352), DA (261.3511, 261.3511), GWO (149.5577, 149.5577), MVO (164.1946, 164.1946), SCA (186.2638, 186.2638), WOA (371.5226, 371.5226), HTS (177.0808, 177.0808), TLBO (181.1404, 181.1404), and SOS (141.1555, 141.1555). Therefore, the PVS algorithm is the best optimizer for this test problem, as it achieves the best mass mean value and the lowest SD.

This findings of the study indicate that topology optimization is a more effective approach for designing lighter trusses than size optimization. The best solutions obtained through topology optimization resulted in a truss that was 40.8% lighter than the truss obtained through size optimization. This significant reduction in weight demonstrates the potential benefits of topology optimization in the design of trusses, and may have important implications for industries that rely on lightweight and efficient structures, such as aerospace and automotive engineering.

Table 5.30 Truss topology optimization of the 24-bar truss (discrete section)

Element no.	ALO	DA	GWO	MVO	SCA	WOA	HTS	TLBO	PVS	SOS
1	—	17	—	—	—	—	—	—	—	—
2	—	19	—	—	—	25	—	—	—	—
3	—	11	—	—	—	—	1	—	—	—
6	—	13	—	—	—	4	1	—	—	—
7	21	—	19	19	24	23	21	19	20	19
8	9	—	3	3	5	3	4	3	3	3
9	—	5	2	2	15	—	1	2	2	2
10	1	—	—	—	—	9	2	—	—	—
12	—	4	5	5	19	6	—	5	5	5
13	18	16	14	14	—	14	18	14	14	14
14	—	2	—	—	—	—	—	1	—	—
15	11	13	3	3	13	16	4	4	3	3
16	24	24	24	24	27	25	24	24	24	24
19	—	—	—	—	—	1	—	—	—	—
20	1	—	—	—	—	—	—	—	—	—
21	4	—	—	—	—	2	—	—	—	—
22	1	19	3	2	—	1	3	—	—	—
23	1	—	3	——	3	—	—	—	1	1
24	—	—	—	1	8	21	—	1	1	1
Mass (kg)	147.9755	188.6131	122.8198	120.7648	148.5357	206.9689	137.8316	121.7832	121.2033	120.0798
σ^{max} (MPa)	91.7451	91.7451	91.7451	91.7451	91.7451	91.7451	91.7451	91.7451	91.7451	91.7451
δ^{max} (mm)	3.7791	3.7791	3.7791	3.7791	3.7791	3.7791	3.7791	3.7791	3.7791	3.7791
f_1 (Hz)	30.0938	31.3912	30.5059	30.2826	30.5207	33.0578	30.1274	30.0297	30.2554	30.2559
Mean	207.8352	261.3511	149.5577	164.1946	186.2638	371.5226	177.0808	181.1404	138.8606	141.1555
SD	36.6120	54.7229	24.2195	38.3068	27.8803	87.3270	29.0488	37.4737	11.6859	16.5160

5.5.5 Topology Optimization of the 20-Bar Truss with Discrete Cross-Sections

In this section, the main focus is on the size and topology optimization of the 20-bar truss using 20 discrete sections. The ground structure of the truss can be seen in Fig. 5.15. This benchmark problem was initially formulated for topology and size optimization by Xu et al. (2003) and subsequently tested for continuous optimization by Kaveh and Zolghadr (2013).

The discrete cross-sectional areas (A_i, where $i = 1, 2, \ldots, 20$) are allowed to vary within the range of $[-100, 100]$ cm^2 in increments of 1 cm^2, where a zero or negative value indicates the removal of the corresponding element. Table 5.31 provides detailed information about the design variables, loading conditions, structural safety constraints, and other relevant data for this test problem. It is worth mentioning that no lumped masses are added to the truss in this specific scenario. The table serves as a valuable resource, offering important insights for the optimization process.

This section demonstrates the TTO of the 20-bar truss through the utilization of proposed algorithms with 20 discrete size and topology variables. Table 5.32 presents a comparison of results obtained from various metaheuristic optimizers applied to solve the 20-bar truss problem with discrete cross-sections. The best run obtained the minimum mass while satisfying all structural constraints. The mean and SD values of structural mass were calculated and will be used to statistically evaluate the performance of different evolutionary methods. The best reported results for ALO, DA, GWO, MVO, SCA, WOA, HTS, TLBO, PVS, and SOS algorithms are 157.1498, 202.6373, 162.5388, 155.3468, 194.9747, 282.4375, 157.1498, 157.1498, 154.7988, and 155.3468 kg, respectively. It is worth noting that the PVS algorithm yielded the least structural mass.

To better assess the performance of metaheuristic methods, we consider their average mass and SD. Based on the available results, it can be seen that ALO, DA, GWO, MVO, SCA, WOA, HTS, TLBO, PVS, and SOS algorithms have mean mass values of 263.7888, 2.01E+06, 4.00E+06, 1.20E+07, 258.3247, 486.8263, 4.00E +06, 496.9801, 171.2970, and 172.7819 kg, respectively, with corresponding SD

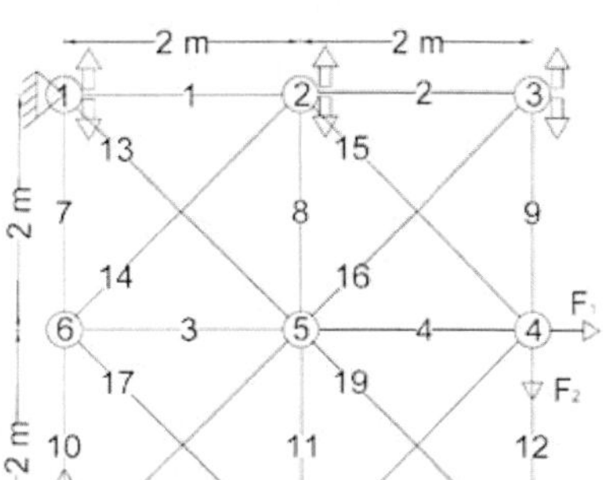

Fig. 5.15 The 20-bar truss

Table 5.31 Design parameters of the 20-bar truss

Size variables: $X_i = A_i$, $i = 1, 2, \ldots, 20$	
Multiple load conditions:	
Condition 1 : $F_1 = 500$ kN, $F_2 = 0$ kN	Condition 2 : $F_1 = 0$ kN, $F_2 = 500$ kN
No lumped mass	
Stress and displacement constraints: $\sigma_i^{max} = 172.43$ MPa and $\delta_{y_4}^{max} = 60$ mm $(6 \times 10^{-2}$ m$)$	
Natural frequency constraints: $f_1 \geq 60$ Hz and $f_2 \geq 100$ Hz	
The discrete cross-sectional areas: $[A^{min}, A^{max}] = [-100, 100]$ cm^2 = ($[-100, 100] \times 10^{-4}$ m^2) in increments of 1 cm^2 and critical area = 1 cm^2	
Material properties: $E = 69$ GPa and $\rho = 2740$kg/m^3	

values of 48.6449, 1.41E+07, 1.98E+07, 3.28E+07, 40.9675, 108.1432, 1.98E+07, 1449.7634, 16.9905, and 16.5635. The PVS algorithm shows the best mean mass value, while the SOS algorithm has the best SD. Therefore, the PVS and SOS algorithms are the actual better algorithms for both convergence rate and consistency for this design problem.

The findings of the study suggest that using the topology optimization approach leads to notably lighter trusses than size optimization approaches. More specifically, the best solutions achieved through topology optimization resulted in trusses that were 25.3% lighter compared to the trusses obtained through size optimization.

5.5.6 *Topology Optimization of the 72-Bar 3D Truss with Discrete Cross-Sections*

In this section, the main objective is to optimize the size and topology of the 72-bar truss using 16 discrete variables. The ground structure of the truss is depicted in Fig. 5.16. This three-dimensional truss problem has been previously studied by Kaveh and Zolghadr (2013) for continuous optimization.

Table 5.33 provides comprehensive details about the material properties, design variables, loading conditions, lumped mass, and other relevant data for this test problem. The elements of the truss are grouped into 16 clusters, labeled as G_1 (A_1–A_4), G_2 (A_5–A_{12}), G_3 (A_{13}–A_{16}), G_4 (A_{17}–A_{18}), G_5 (A_{19}–A_{22}), G_6 (A_{23}–A_{30}), G_7 (A_{31}–A_{34}), G_8 (A_{35}–A_{36}), G_9 (A_{37}–A_{40}), G_{10} (A_{41}–A_{48}), G_{11} (A_{49}–A_{52}), G_{12} (A_{53}–A_{54}), G_{13} (A_{55}–A_{58}), G_{14} (A_{59}–A_{66}), G_{15} (A_{67}–A_{70}), and G_{16} (A_{71}–A_{72}). The grouping is based on structural symmetry considerations as described by Kaveh and Zolghadr (2013). The discrete cross-sectional areas (G_i, where $i = 1, 2, \ldots, 16$) are allowed to vary within the range of $[-30, 30]$ cm^2 in increments of 1 cm^2, where a zero or negative value indicates the removal of the corresponding element.

Furthermore, the truss includes four nonstructural lumped masses, each weighing 2270 kg, which are added to the top nodes (node numbers 1–4). These lumped masses are not considered part of the main structural design but are included for additional analysis.

Table 5.32 Truss topology optimization of the 20-bar truss (discrete section)

Element no.	ALO	DA	GWO	MVO	SCA	WOA	HTS	TLBO	PVS	SOS
1	29	—	15	16	—	31	29	29	15	16
3	—	—	—	—	14	—	—	—	—	—
4	29	29	—	—	38	—	29	29	—	—
5	—	30	20	20	45	29	—	—	20	20
6	—	—	2	—	—	54	—	—	—	—
8	29	—	20	20	—	44	29	29	20	20
9	—	2	—	—	—	—	—	—	—	—
10	—	—	—	—	2	45	—	—	—	—
11	—	32	20	20	38	31	—	—	20	20
12	—	—	3	—	—	9	—	—	—	—
13	21	43	21	21	47	5	21	21	21	21
14	—	—	—	—	—	50	—	—	—	—
15	42	—	21	21	—	31	42	42	21	21
16	—	42	—	—	—	—	—	—	—	—
17	—	—	—	—	—	22	—	—	—	—
18	46	26	33	33	22	—	46	46	33	33
20	—	46	33	33	47	33	—	—	33	33
Mass (kg)	157.1498	202.6373	162.5388	155.3468	194.9747	282.4375	157.1498	157.1498	154.7988	155.3468
σ^{max} (MPa)	172.4138	172.4138	172.4138	172.4138	172.4138	172.4138	172.4138	172.4138	172.4138	172.4138
δ^{max} (mm)	33.6637	33.6637	33.6637	33.6637	33.6637	33.6637	33.6637	33.6637	33.6637	33.6637
f_1 (Hz)	117.1543	64.8126	113.8065	116.7758	113.8353	67.2291	117.1543	117.1543	116.7863	116.7758
f_2 (Hz)	158.6407	120.1737	182.6481	188.1005	144.1768	105.4639	158.6407	158.6407	186.5813	188.1005
Mean	263.7888	2.01E+06	4.00E+06	1.20E+07	258.3247	486.8263	4.00E+06	496.9801	171.2970	172.7819
SD	48.6449	1.41E+07	1.98E+07	3.28E+07	40.9675	108.1432	1.98E+07	1449.7634	16.9905	16.5635

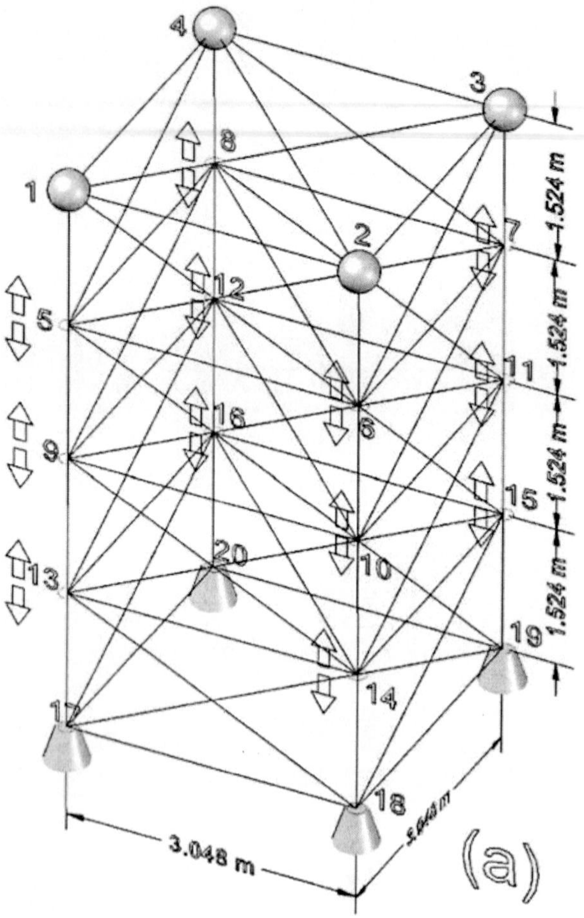

Fig. 5.16 The 72-bar 3D truss

Table 5.33 Design parameters of the 72-bar truss

Size variables: G_i, $i = 1, 2, \ldots, 16$; $G =$ Group number				
Multiple load conditions:				
Condition 1:	$F_{1x} = 22.25$ kN	$F_{1y} = 22.25$ KN		$F_{1z} = -22.25$ kN
Condition 2:	$F_{1z} = -22.25$ kN	$F_{2z} = -22.25$ kN	$F_{3z} = -22.25$ kN	$F_{4z} = -22.25$ kN
Lumped masses at nodes 1, 2, 3, and 4 are 2270 kg				
Stress and displacement constraints: $\sigma_i^{\max} = 172.375$ MPa and $\delta_{x_j/y_j}^{\max} = 6.35$ mm $\left(6.35 \times 10^{-3}\ \text{m}\right)$ (for node, $j = 1, 2, 3$, and 4 along x- and y-axes)				
Natural frequency constraints: $f_1 \geq 4$ Hz and $f_3 \geq 6$ Hz				
The discrete cross-sectional areas: $[A^{\min}, A^{\max}] = [-30, 30]$ cm^2 = ($[-30, 30] \times 10^{-4}$ m^2) in increments of 1 cm^2 and critical area = 1 cm^2				
Material properties: $E = 68.95$ GPa and $\rho = 2767.99$kg/m^3				

Overall, the table and figure provide important information and insights for the optimization process of the 72-bar truss.

This section demonstrates the TTO of the 72-bar 3D truss through the utilization of proposed algorithms with 16 discrete size and topology variables. Table 5.34 presents the results obtained from the TS optimization of the 72-bar truss with discrete cross-sections for different algorithms in 100 independent runs. The best run is identified as the one that yields the minimum mass value while satisfying all the structural constraints. The mean and standard deviation (SD) of the structural mass are provided to evaluate the performance of the evolutionary methods statistically. Specifically, the best results reported by ALO, DA, GWO, MVO, SCA, WOA, HTS, TLBO, PVS, and SOS algorithms are 459.3283, 618.5515, 447.1073, 458.8470, 519.0888, 562.8129, 450.3880, 452.0754, 450.3880, and 450.3880 kg, respectively. The results indicate that the GWO algorithm provides the best solution with a mass value of 447.1073 kg, while the HTS, PVS, and SOS algorithms result in the second-best structural mass with an identical solution.

In order to accurately evaluate the statistical performance of the proposed metaheuristic methods, the average mass and its standard deviation (SD) were taken into consideration. Based on the available results, it can be seen that the ALO, DA, GWO, MVO, SCA, WOA, HTS, TLBO, PVS, and SOS algorithms have mean mass values of 722.3693, 8.00E+06, 4.00E+06, 1.00E+07, 638.7581, 1247.8692, 670.8082, 481.1943, 470.3506, and 459.6306 kg, respectively, with corresponding SD values of 241.0774, 2.74E+07, 1.98E+07, 3.03E+07, 56.3167, 582.4059, 368.6161, 30.2597, 48.1245, and 6.2249. The SOS algorithm performed best in achieving the lowest mass mean and SD values. Therefore, the SOS algorithm is the better choice for both convergence rate and consistency for this design problem.

The findings of the study suggest that the topology optimization approach produces trusses that are noticeably lighter than those produced by size optimization approaches. More specifically, the top-performing solutions generated through topology optimization resulted in trusses that were 6.8% lighter than those created through size optimization.

5.5.7 *Topology Optimization of the 39-Bar Truss with Discrete Cross-Sections*

This section focuses on optimizing the size and topology of a 39-bar truss using 21 discrete variables. The truss, known as the two-tier 39-bar planar truss, is illustrated in Fig. 5.17. The objective of this problem is to analyze the behavior of a large truss under stress, buckling, displacement, and fundamental frequency constraints, taking into account discrete design variables.

Table 5.34 Truss topology optimization of the 72-bar 3D truss (discrete section)

Element group no.	ALO	DA	GWO	MVO	SCA	WOA	HTS	TLBO	PVS	SOS
1	5	11	5	7	5	12	6	5	5	5
2	11	19	11	8	12	9	11	11	11	11
4	—	15	—	10	—	10	—	—	—	—
5	10	39	9	8	6	15	8	10	10	10
6	9	8	7	9	8	12	9	8	8	8
7	3	—	—	—	—	—	—	—	—	—
8	7	14	4	—	7	—	4	4	4	4
9	17	12	13	12	13	8	11	14	11	11
10	7	8	9	9	9	8	8	8	8	9
11	—	—	—	—	—	9	—	—	—	—
12	—	—	—	—	—	10	—	—	—	—
13	13	12	15	15	43	18	16	13	15	15
14	7	8	7	9	8	7	8	9	9	8
15	—	—	3	—	3	7	—	—	—	—
Mass (kg)	459.3283	618.5515	447.1073	458.8470	519.0888	562.8129	450.3880	452.0754	450.3880	450.3880
σ^{max} (MPa)	131.6950	219.3218	117.7873	116.9796	192.7776	271.3296	135.8179	126.6655	116.6868	128.7385
δ^{max} (mm)	58.0034	58.0034	58.0034	58.0034	58.0034	58.0034	58.0034	58.0034	58.0034	58.0034
f_1 (Hz)	4.7136	4.7136	4.7136	4.7136	4.7136	4.7136	4.7136	4.7136	4.7136	4.7136
f_3 (Hz)	6.0496	6.2266	6.0458	6.0029	6.3085	6.2084	6.0177	6.0169	6.0071	6.0064
Mean	722.3693	8.00E+06	4.00E+06	1.00E+07	638.7581	1247.8692	670.8082	481.1943	470.3506	459.6306
SD	241.0774	2.74E+07	1.98E+07	3.03E+07	56.3167	582.4059	368.6161	30.2597	48.1245	6.2249

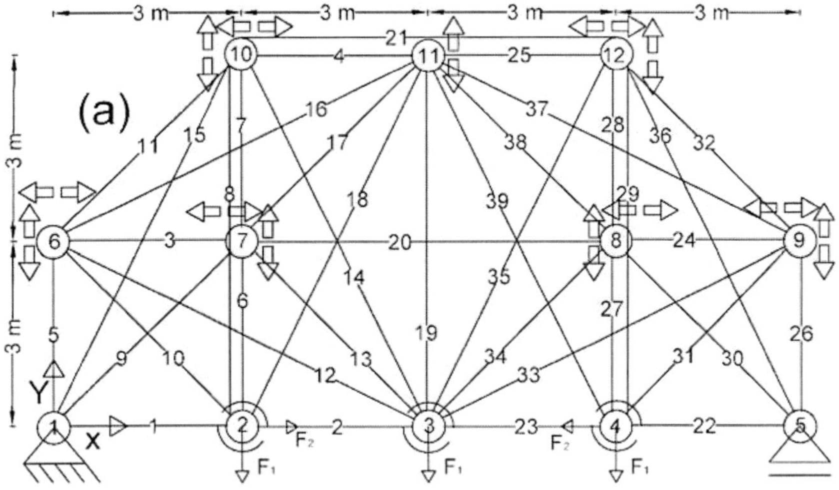

Fig. 5.17 The 39-bar truss

Previous research by Deb and Gulati (2001) initiated work on this problem, while subsequent studies by Luh and Lin (2008), Wu and Tseng (2010), and Ahrari et al. (2014) primarily focused on static constraints. However, the exploration of this problem with simultaneous consideration of both static and dynamic constraints has not been undertaken.

To address this gap, the problem has been modified to incorporate the effects of multi-load conditions, lumped masses, and natural frequency constraints. Detailed information about the design variables, multi-load conditions, lumped masses, constraints, and material properties can be found in Table 5.35. The discrete cross-sectional areas (G_i, where $i = 1, 2, \ldots, 21$) are allowed to vary within the range of [−50, 50] cm^2 in increments of 1 cm^2, where a zero or negative value indicates the removal of the corresponding element. The truss elements are divided into 21 groups, considering symmetry around the vertical middle plane, following the approach of Deb and Gulati (2001).

Additionally, the truss is subjected to three nonstructural lumped masses, each weighing 500 kg, placed at nodes 2, 3, and 4. To satisfy the multi-load condition, two different loading conditions are considered, as depicted in Fig. 5.17.

This section demonstrates the TTO of the 39-bar truss through the utilization of proposed algorithms with 21 discrete size and topology variables. Table 5.36 presents the performance of various algorithms in TS optimization of the 39-bar truss with discrete cross-sections in 100 independent runs. It was observed that the best run resulted in the minimum mass value while satisfying all the stated structural constraints. The mean and standard deviation (SD) values of structural mass are provided to statistically assess the performance of different evolutionary methods.

Table 5.35 Design parameters of the 39-bar truss

Size variables: $X_i = G_i$, $i = 1, 2, \ldots, 21$	
Multiple load conditions:	
Condition 1 : $F_1 = 100$ kN, $F_2 = 0$ kN	Condition 2 : $F_1 = 0$ kN, $F_2 = 50$ kN
Lumped masses on nodes 2, 3, and 4 are 500 kg	
Stress and displacement constraints: $\sigma_i^{\max} = 172.43$ MPa and $\delta_{x_j/y_j}^{\max} = 50$ mm (0.05 m) (for node, $j = 2$, 3, and 4 along x- and y-axes)	
Natural frequency constraints: $f_1 \geq 4$ Hz	
The discrete cross-sectional areas: $[A^{\min}, A^{\max}] = [-50, 50]$ cm^2 = ($[-50, 50] \times 10^{-4}$ m^2) in increments of 1 cm^2 and critical area = 1 cm^2	
Material properties: $E = 69$ GPa and $\rho = 2740$kg/m^3	

The best mass results achieved by ALO, DA, GWO, MVO, SCA, WOA, HTS, TLBO, PVS, and SOS algorithms are 285.0344, 357.9742, 244.0373, 260.3312, 361.8304, 479.1845, 267.9763, 285.5877, 246.0694, and 256.3215 kg, respectively. It is evident from the results that the GWO algorithm produced the least structural mass.

The performance of various metaheuristic methods in the TS optimization of the 39-bar truss with discrete cross-sections is evaluated based on the average mass and its SD. The results presented in Table 5.36 indicate that the best run is obtained with the algorithm that provides the minimum mass value while satisfying all the stated structural constraints. The mean and SD values of structural mass are provided and will be utilized to statistically evaluate the performance of the different evolutionary methods.

It can be seen that the ALO, DA, GWO, MVO, SCA, WOA, HTS, TLBO, PVS, and SOS algorithms have a mass mean value of 466.5458, 582.9839, 319.2007, 394.4476, 423.4067, 847.9312, 376.8107, 375.7604, 305.2689, and 310.4174 kg, respectively, with corresponding SD values of 103.8478, 125.8816, 49.6891, 415.4209, 41.5999, 192.9485, 83.3802, 66.1047, 28.8670, and 42.3426. The PVS algorithm performs the best in achieving the lowest mass mean and SD values. Therefore, the PVS algorithm is considered the better algorithm for both convergence rate and consistency for this design problem.

The study found that the use of topology optimization resulted in trusses that were significantly lighter than those obtained through size optimization approaches. Specifically, the best solutions obtained through topology optimization resulted in a truss that was 53.3% lighter compared to the truss obtained through size optimization. This implies that the use of topology optimization can lead to more efficient and cost-effective truss designs, as they require less material while still maintaining their structural integrity. This has important implications for the construction industry, as it could result in lower costs, faster construction times, and reduced environmental impact. It also highlights the importance of considering multiple optimization approaches to obtain the best possible design solution.

Table 5.36 Truss topology optimization of the 39-bar truss (discrete section)

Element no.	ALO	DA	GWO	MVO	SCA	WOA	HTS	TLBO	PVS	SOS
1	—	2	1	2	6	6	—	1	1	2
2	15	13	13	13	13	14	13	13	13	16
3	3	—	—	—	—	12	4	3	—	—
4	19	36	19	19	—	16	19	20	19	19
5	23	23	23	23	26	23	23	28	23	23
6	—	—	—	—	9	32	2	—	—	—
7	3	—	—	1	—	—	2	6	—	—
8	3	6	2	2	—	8	—	3	2	2
9	5	1	—	1	—	—	3	—	—	—
10	5	14	7	7	17	—	7	5	7	7
11	31	33	27	27	—	31	27	31	27	27
12	—	7	—	—	—	23	—	—	—	—
13	5	3	—	—	4	—	—	5	—	—
14	—	2	4	4	—	—	4	—	4	4
15	—	—	—	—	—	3	—	—	—	—
16	—	—	—	—	43	—	—	—	—	—
17	2	—	—	1	2	—	—	—	—	—
18	—	—	—	—	—	—	1	1	1	2
19	—	3	1	—	7	5	—	—	—	—
21	—	—	—	—	—	29	—	—	—	—
Mass (kg)	285.0344	357.9742	244.0373	260.3312	361.8304	479.1845	267.9763	285.5877	246.0694	256.3215
σ^{max} (MPa)	166.6667	166.6667	166.6667	166.6667	166.6667	166.6667	166.6667	166.6667	166.6667	166.6667
δ^{max} (mm)	35.8811	35.8811	35.8811	35.8811	35.8811	35.8811	35.8811	35.8811	35.8811	35.8811
f_1 (Hz)	4.0016	10.7076	5.7497	9.2814	12.0014	4.4786	4.0121	5.7035	5.7507	7.9931
Mean	466.5458	582.9839	319.2007	394.4476	423.4067	847.9312	376.8107	375.7604	305.2689	310.4174
SD	103.8478	125.8816	49.6891	415.4209	41.5999	192.9485	83.3802	66.1047	28.8670	42.3426

5.5.8 *Topology Optimization of the 45-Bar Truss with Discrete Cross-Sections*

This section focuses on the size optimization of a 45-bar truss using 31 size and topology variables. The truss, referred to as the 45-bar truss, is the ninth benchmark problem chosen to analyze a complex structure under stress, buckling, displacement, and fundamental frequency constraints, considering discrete design variables.

Figure 5.18 depicts the ground structure of the truss, along with multiple loads and the boundary condition. Initially proposed by Deb and Gulati (2001) with static constraints, this problem has not been previously examined with simultaneous consideration of both static and dynamic constraints. To address this, the problem has been modified to incorporate the effects of multi-load conditions, lumped masses, natural frequency constraints, and discrete sections.

The discrete cross-sectional areas (G_i, where $i = 1, 2, \ldots, 31$) are allowed to vary within the range of $[-50, 50]$ cm^2 in increments of 1 cm^2, where a zero or negative value indicates the removal of the corresponding element. Detailed design considerations, including design variables, multi-load conditions, lumped masses, constraints, and material properties, are presented in Table 5.37. The truss elements are divided into 14 groups, based on symmetry around the vertical middle plane, with elements numbered 1–14 and 20–33 treated as symmetric groups, while the remaining elements are treated individually. Consequently, the number of size variables is reduced to 31.

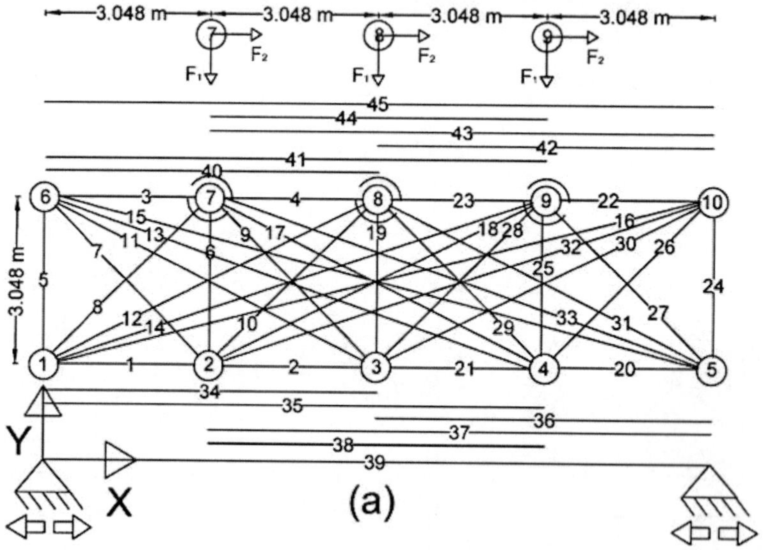

Fig. 5.18 The 45-bar truss

Table 5.37 Design parameters of the 45-bar truss

Size variables: $X_i = G_i$, $i = 1, 2, \ldots, 31$; $G =$ Group number Shape variables: $-2.5\ \text{m} \leq \xi_{x_j/y_j} \leq 2.5\ \text{m}$, for $x_1 = -x_5$ and $y_1 = y_5$ with respect to their original coordinates	
Multiple load conditions:	
Condition 1 : $F_1 = 44.537$ kN	Condition 2 : $F_2 = 4.4537$ kN
Stress and displacement constraints: $\sigma_i^{max} = 172.43$ MPa and $\delta_{x_j/y_j}^{max} = 5$ mm (0.005 m) (for node, $j = 7$, 8, and 9 along x- and y-axes)	
Lumped masses on nodes 7, 8, and 9 are 500 kg	
Natural frequency constraints: $f_1 \geq 15$ Hz	
The discrete cross-sectional areas: $[A^{\min}, A^{\max}] = [-50, 50]\ \text{cm}^2 = ([-50, 50] \times 10^{-4}\ \text{m}^2)$ in increments of 1 cm^2 and critical area = 1 cm^2	
Material properties: $E = 69$ GPa and $\rho = 2740 \text{kg/m}^3$	

Additionally, the truss is subjected to three nonstructural lumped masses, each weighing 500 kg, placed at nodes 7, 8, and 9. To satisfy the multi-load condition, two different loading conditions are considered, as illustrated in Fig. 5.18.

This section presents the optimization of the 45-bar truss for TTO with 31 size variables. Table 5.38 shows the results obtained in the TTO of the truss with discrete cross-sections for various algorithms in 100 independent runs. The best run achieved the minimum mass value while meeting all the specified structural constraints. The mean and SD values of the structural mass are provided, and they will be used to evaluate the statistical performance of the different evolutionary methods.

Among the algorithms tested, ALO, DA, GWO, MVO, SCA, WOA, HTS, TLBO, PVS, and SOS reported the best results, with mass mean values of 153.5543, 260.0420, 162.8157, 153.5543, 208.4753, 228.4355, 177.0674, 181.6670, 153.5543, and 153.5543 kg, respectively. It can be observed that ALO, MVO, PVS, and SOS resulted in the least structural mass.

To evaluate the true statistical performance of the metaheuristic methods, the average mass and its standard deviation (SD) were considered. Based on the available results, it can be concluded that ALO, DA, GWO, MVO, SCA, WOA, HTS, TLBO, PVS, and SOS algorithms have the mean mass values of 315.0151, 415.7895, 206.4008, 229.4466, 282.3139, 575.6332, 280.4644, 271.7525, 200.1187, and 183.7706 kg, respectively, with corresponding SD values of 112.7127, 96.7682, 41.4644, 67.3163, 40.9546, 188.2182, 61.6846, 65.4455, 32.5791, and 35.0708.

It is observed that the SOS algorithm has the best mean mass value, while PVS performs best in terms of SD. Therefore, the PVS and SOS algorithms are the better algorithms for both convergence rate and consistency for this design problem.

The findings of the study demonstrate that topology optimization is a significantly more efficient method of designing lightweight trusses than size optimization approaches. The truss designs achieved through topology optimization were much

Table 5.38 Truss topology optimization of the 45-bar truss (discrete section)

Group no.	ALO	DA	GWO	MVO	SCA	WOA	HTS	TLBO	PVS	SOS
1	—	—	—	—	—	3	—	—	—	—
3	—	—	—	—	—	—	1	—	—	—
4	11	22	11	11	15	13	11	11	11	11
5	—	13	—	—	5	—	2	2	—	—
6	—	5	—	—	—	—	—	1	—	—
7	—	3	—	—	—	—	—	1	—	—
8	18	18	18	18	20	29	18	18	18	18
10	—	4	—	—	—	11	—	1	—	—
12	25	26	25	25	29	25	28	25	25	25
15	—	—	1	—	—	—	—	—	—	—
23	—	—	—	—	—	7	—	—	—	—
26	—	—	1	—	—	—	—	—	—	—
29	—	14	—	—	1	—	—	—	—	—
31	—	—	—	—	5	—	—	—	—	—
Mass (kg)	153.5543	260.0420	162.8157	153.5543	208.4753	228.4355	177.0674	181.6670	153.5543	153.5543
σ^{max} (MPa)	110.4713	110.4713	110.4713	110.4713	110.4713	110.4713	110.4713	110.4713	110.4713	110.4713
δ^{max} (mm)	4.0666	4.0666	4.0666	4.0666	4.0666	4.0666	4.0666	4.0666	4.0666	4.0666
f_1 (Hz)	16.0369	17.9090	15.9929	16.0369	17.5171	16.9766	16.4661	15.9398	16.0369	16.0369
Mean	315.0151	415.7895	206.4008	229.4466	282.3139	575.6332	280.4644	271.7525	200.1187	183.7706
SD	112.7127	96.7682	41.4644	67.3163	40.9546	188.2182	61.6846	65.4455	32.5791	35.0708

lighter than those achieved through size optimization. The specific result of the study showed that the best solutions obtained through topology optimization led to a truss that was 62.5% lighter compared to the truss obtained through size optimization.

The implications of these results are significant, as they suggest that topology optimization should be the preferred design approach when aiming to achieve substantial weight reduction in trusses. This approach allows designers to create optimized truss structures that are both strong and lightweight, without sacrificing the integrity or stability of the overall structure. Furthermore, this can lead to improved efficiency and cost savings in materials and manufacturing processes, making it an attractive option for industries where lightweight structures are important such as aerospace, automotive, and civil engineering.

5.5.9 *Topology Optimization of the 25-Bar 3D Truss with Discrete Cross-Sections*

This section focuses on the size and topology optimization of a 25-bar truss using eight variables. The truss, referred to as the 25-bar truss, is the ninth benchmark problem chosen to analyze a space truss under stress, buckling, displacement, and fundamental frequency constraints, with discrete design variables.

The initial layout of the space truss is shown in Fig. 5.19. Previous studies by Deb and Gulati (2001) and Ahrari et al. (2014) have investigated this problem with static constraints, but the examination of static and dynamic constraints simultaneously has not been explored. To address this, the problem has been modified to include the effects of multi-load conditions, lumped masses, natural frequency constraints, and discrete area.

The discrete cross-sectional areas (G_i, where $i = 1, 2, \ldots, 8$) are allowed to vary within the range of $[-50, 50]$ cm^2 in increments of 1 cm^2, where a zero or negative value indicates the removal of the corresponding element. Detailed design considerations, including design variables, load conditions, lumped masses, constraints, and material properties, are presented in Table 5.39. Due to the structural symmetry of the truss, element grouping is adopted, as indicated in the table. As a result, the truss only requires eight size variables, which are combined with topology variables as discussed earlier.

In addition, the truss is subjected to two nonstructural lumped masses, each weighing 500 kg, placed at nodes 1 and 2. Two different loading conditions are considered, as depicted in the table.

This section demonstrates the TTO of a 25-bar truss through the utilization of proposed algorithms with eight discrete size and topology variables. Table 5.40 displays the outcomes of optimizing the 25-bar truss with discrete cross-sections using various algorithms in 100 independent runs. The most favorable run achieved

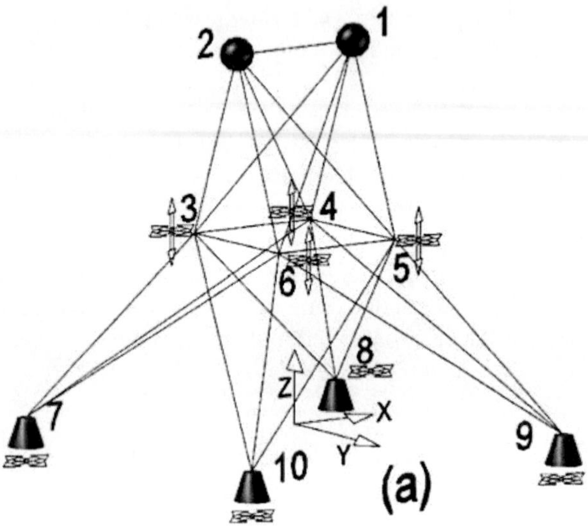

Fig. 5.19 The 25-bar 3D truss: (**a**) ground structure and (**b**) optimal structure

the minimum mass value while satisfying all structural constraints. The average and standard deviation values of the structural mass are also included, which will be utilized to statistically evaluate the effectiveness of the different evolutionary techniques. The ALO, DA, GWO, MVO, HTS, TLBO, PVS, and SOS algorithms all produced the same optimal result, which is a mass of 388.9853 kg for the truss.

In order to evaluate the performance of the various metaheuristic methods, the mean mass and its standard deviation are taken into consideration. Based on the available results, it can be concluded that the ALO, DA, GWO, MVO, SCA, WOA, HTS, TLBO, PVS, and SOS algorithms have mean mass values of 413.5271, 440.2761, 2.00E+06, 2.00E+06, 444.8937, 505.0638, 420.4687, 390.5915, 388.9853, and 389.0535 kg, respectively, with corresponding standard deviations of 24.9636, 65.2397, 1.41E+07, 1.41E+07, 16.9089, 112.2325, 106.3302, 4.1207, 0.0000, and 0.3373. As a result, the PVS algorithm is the better choice for both convergence rate and consistency in solving this particular design problem.

The results of the study indicate that the topology optimization approach is more effective in designing lighter trusses when compared to size optimization approaches. To be specific, the top-performing solutions achieved through topology optimization led to a truss that was 11.9% lighter than the truss that resulted from size optimization. This finding suggests that topology optimization may be a superior design approach to achieve a significant reduction in truss weight.

Table 5.39 Design parameters of the 25-bar 3D truss

Size variables: $X_i = G_i$, $i = 1, 2, \ldots, 8$; $G =$ Group number				
Multiple load conditions:				
Condition 1:	$F_{y_1} = 89.074$ kN	$F_{z_1} = -22.25$ kN	$F_{y_2} = -89.074$ kN	$F_{z_2} = -22.25$ kN
Condition 2:	$F_{x_1} = 4.454$ kN	$F_{y_1} = -44.537$ kN	$F_{z_1} = -44.537$ kN	$F_{y_2} = -44.537$ kN
	$F_{z_2} = -44.537$ kN	$F_{x_3} = 2.227$ kN	$F_{x_6} = 2.672$ kN	
Lumped masses at nodes 1 and 2 are 500 kg				
Stress and displacement constraints: $\sigma_i^{max} = 172.375$ MPa and $\delta_{x_j/y_j/z_j}^{max} = 8.9$ mm (0.0089 m) (for node, $j = 1, 2, \ldots, 6$ along x-, y-, and z-axes)				
Natural frequency constraints: $f_1 \geq 15$ Hz				
The discrete cross-sectional areas: $[A^{min}, A^{max}] = [-50, 50]$ cm^2 = ($[-50, 50] \times 10^{-4}$ m^2) in increments of 1 cm^2 and critical area = 1 cm^2				
Material properties: $E = 69$ GPa and $\rho = 2740$kg/m^3				
Element grouping				
Group number	Element number (end nodes)		Group number	Element number (end nodes)
G_1	1(1,2)		G_5	12(3,4), 13(5,6)
G_2	2(1,4), 3(2,3), 4(1,5), 5(2,6)		G_6	14(3,10), 15(6,7), 16(4,9), 17(5,8)
G_3	6(2,5), 7(2,4), 8(1,3), 9(1,6)		G_7	18(3,8), 19(4,7), 20(6,9), 21(5,10)
G_4	10(3,6), 11(4,5)		G_8	22(3,7), 23(4,8), 24(5,9), 25(6,10)

Table 5.40 Truss topology optimization of the 25-bar 3D truss (discrete section)

Element group no	ALO	DA	GWO	MVO	SCA	WOA	HTS	TLBO	PVS	SOS
1	—	—	—	—	8	—	—	—	—	—
2	16	16	16	16	19	16	16	16	16	16
3	16	16	16	16	16	17	16	16	16	16
4	—	—	—	—	2	—	—	—	—	—
6	13	13	13	13	13	13	13	13	13	13
7	20	20	20	20	22	20	20	20	20	20
8	18	18	18	18	18	18	18	18	18	18
Mass (kg)	388.9853	388.9853	388.9853	388.9853	416.2335	391.9585	388.9853	388.9853	388.9853	388.9853
σ^{max} (MPa)	54.8088	54.8088	54.8088	54.8088	54.8088	54.8088	54.8088	54.8088	54.8088	54.8088
δ^{max} (mm)	7.5228	7.5228	7.5228	7.5228	7.5228	7.5228	7.5228	7.5228	7.5228	7.5228
f_1 (Hz)	15.5647	15.5647	15.5647	15.5647	15.8640	15.6349	15.5647	15.5647	15.5647	15.5647
Mean	413.5271	440.2761	2.00E+06	2.00E+06	444.8937	505.0638	420.4687	390.5915	388.9853	389.0535
SD	24.9636	65.2397	1.41E+07	1.41E+07	16.9089	112.2325	106.3302	4.1207	0.0000	0.3373

5.5.10 *Topology Optimization of the 39-Bar 3D Truss with Discrete Cross-Sections*

This section aims to compare the performance of various optimization algorithms in the size optimization of a 39-bar 3D truss with discrete sections. The ground structure of this truss, consisting of 39 elements, is illustrated in Fig. 5.20. To account for symmetry about the *x*–*z* and *y*–*z* planes, the element cross-sectional areas are grouped into 11 sets.

Previous studies by Deb and Gulati (2001) and Luh and Lin (2008) have investigated this truss with static constraints. However, the examination of both static and dynamic constraints simultaneously has not been conducted. To bridge this gap, the problem is modified in this study to incorporate multi-load conditions, lumped masses, natural frequency constraints, and discrete cross-sectional areas.

In Fig. 5.20, a constant lumped mass of 500 kg is attached to the top nodes (nodes 1 and 2) of the truss. The discrete cross-sectional areas (G_i, where $i = 1, 2, \ldots, 11$) are allowed to vary within the range of $[-50, 50]$ cm^2 in increments of 1 cm^2, where a zero or negative value indicates the removal of the corresponding element. The design parameters, including material properties, loading conditions, design variable bounds, constraints, and other relevant details, are provided in Table 5.41. This comprehensive table presents all the necessary information for the truss optimization problem.

This section demonstrates the TTO of the 39-bar 3D truss through the utilization of proposed algorithms with 11 discrete size and topology variables. Table 5.42 displays the results obtained from optimizing the 39-bar 3D truss with discrete cross-sections using various algorithms in 100 independent runs. The best run achieved the minimum mass value while satisfying all the specified structural constraints. The

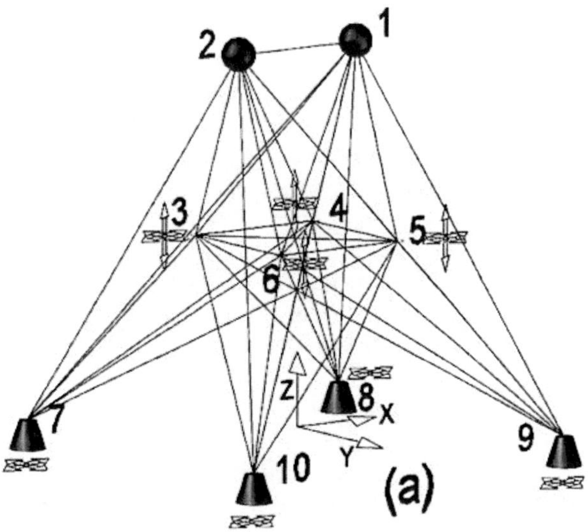

Fig. 5.20 The 39-bar 3D truss

Table 5.41 Design parameters of the 39-bar 3D truss

Size variables: $X_i = G_i$, $i = 1, 2, \ldots, 11$; $G =$ Group number			
Load condition:	$F_{z_1} = -222.685$ kN		$F_{z_2} = -222.685$ kN
Lumped masses at nodes 1 and 2 are 500 kg			
Stress and displacement constraints: $\sigma_i^{max} = 172.375$ MPa and $\delta_{x_j/y_j/z_j}^{max} = 8.9$ mm (0.0089 m) (for node, $j = 1, 2, \ldots, 6$ along x-, y-, and z-axes)			
Natural frequency constraints: $f_1 \geq 15$ Hz			
The discrete cross-sectional areas: $[A^{\min}, A^{\max}] = [-50, 50]\ \text{cm}^2 = ([-50, 50] \times 10^{-4}\ \text{m}^2)$ in increments of 1 cm^2 and critical area = 1 cm^2			
Material properties: $E = 69$ GPa and $\rho = 2740 \text{kg/m}^3$			
Element grouping			
Group number	Element number (end nodes)	Group number	Element number (end nodes)
G_1	1(1,2)	G_7	22(3,7), 23(4,8), 24(5,9), 25(6,10)
G_2	2(1,4), 3(2,3), 4(1,5), 5(2,6)	G_8	26(5,7), 27(6,8), 28(3,9), 29(4,10)
G_3	6(2,5), 7(2,4), 8(1,3), 9(1,6)	G_9	30(3,5), 31(4,6)
G_4	10(3,6), 11(4,5), 12(3,4), 13(5,6)	G_{10}	32(1,7), 33(1,10), 34(2,9), 35(2,8)
G_5	14(3,10), 15(6,7), 16(4,9), 17(5,8)	G_{11}	36(2,7), 37(2,10), 38(1,8), 39(1,9)
G_6	18(3,8), 19(4,7), 20(6,9), 21(5,10)		

average and standard deviation values of the structural mass are also provided, and will be used to statistically evaluate the effectiveness of the different evolutionary methods. The ALO, DA, GWO, MVO, SCA, WOA, HTS, TLBO, PVS, and SOS algorithms produced optimal results with masses of 378.6970, 380.6247, 360.5287, 360.5287, 370.6456, 391.5591, 373.2687, 360.5287, 360.5287, and 360.5287 kg, respectively. It is apparent that the GWO, MVO, TLBO, PVS, and SOS algorithms all achieved the lowest structural mass with an identical solution.

To evaluate the effectiveness of the various metaheuristic methods, the average mass and its standard deviation were computed. Based on the available results, it can be concluded that the ALO, DA, GWO, MVO, SCA, WOA, HTS, TLBO, PVS, and SOS algorithms produced average masses of 456.8172, 495.2505, 421.6816, 422.0433, 432.5396, 533.8161, 410.9040, 420.9139, 399.2657, and 401.6172 kg, respectively, with corresponding standard deviations of 55.1381, 63.5488, 51.8031, 35.4973, 34.1757, 98.6847, 29.3295, 33.4433, 19.3726, and 25.2316. Thus, the PVS algorithm outperformed the other methods in terms of achieving the lowest average mass and standard deviation.

To comprehensively evaluate the performance of the metaheuristic methods, both the average mass and its standard deviation are analyzed. The results indicate that the ALO, DA, GWO, MVO, SCA, WOA, HTS, TLBO, PVS, and SOS algorithms

Table 5.42 Truss topology optimization of the 39-bar 3D truss (discrete section)

Element group no	ALO	DA	GWO	MVO	SCA	WOA	HTS	TLBO	PVS	SOS
1	9	10	10	10	21	—	10	10	10	10
2	18	2	3	3	4	13	3	3	3	3
3	13	6	4	4	3	5	4	4	4	4
4	—	—	—	—	—	5	—	—	—	—
5	3	2	—	—	—	—	2	—	—	—
6	—	5	2	2	2	—	4	2	2	2
7	24	—	2	2	3	15	—	2	2	2
8	19	—	—	—	—	—	—	—	—	—
10	—	41	41	41	41	33	41	41	41	41
Mass (kg)	378.6970	380.6247	360.5287	360.5287	370.6456	391.5591	373.2687	360.5287	360.5287	360.5287
σ^{max} (MPa)	89.0350	71.7249	67.8339	67.8339	32.6971	34.1317	71.7124	67.8339	67.8339	67.8339
δ^{max} (mm)	6.1562	3.4004	3.4206	3.4206	3.2690	3.0375	3.4006	3.4206	3.4206	3.4206
f_1 (Hz)	15.0917	15.0615	15.2310	15.2310	15.6693	21.7213	15.0012	15.2310	15.2310	15.2310
Mean	456.8172	495.2505	421.6816	422.0433	432.5396	533.8161	410.9040	420.9139	399.2657	401.6172
SD	55.1381	63.5488	51.8031	35.4973	34.1757	98.6847	29.3295	33.4433	19.3726	25.2316

achieved a mean mass of 456.8172, 495.2505, 421.6816, 422.0433, 432.5396, 533.8161, 410.9040, 420.9139, 399.2657, and 401.6172 kg, respectively, with corresponding standard deviations of 55.1381, 63.5488, 51.8031, 35.4973, 34.1757, 98.6847, 29.3295, 33.4433, 19.3726, and 25.2316. Notably, the PVS algorithm outperforms the others in achieving the lowest mean mass and the best standard deviation.

This findings of the study indicate that the topology optimization approach is a more effective method for designing lightweight trusses compared to size optimization approaches. Specifically, the optimal solutions obtained through topology optimization resulted in a truss that was 40.7% lighter compared to the truss obtained through size optimization. These results suggest that topology optimization should be considered as the preferred design approach to achieve significant weight reduction in truss structures.

5.5.11 A Comprehensive Analysis

Table 5.43 presents a comprehensive analysis of the best results obtained from ten benchmark problems using the proposed algorithms in 100 independent runs. The table provides detailed information on the minimum mass and offers a comprehensive evaluation of overall, average, and individual ranks, as well as the best and worst results for each algorithm.

Upon analyzing the results in Table 5.43, it was revealed that both the PVS and SOS algorithms achieved the minimum mass and outperformed all other algorithms in terms of overall rank. Following closely were the HTS, TLBO, and GWO algorithms with the second, third, and fourth-best performances, respectively.

Furthermore, the PVS and SOS algorithms demonstrated seven best solutions, two second-best solutions, and one third-best solution, without any worse solutions.

However, the WOA algorithm presented the worst solutions for six problems, highlighting the significance of selecting the most appropriate algorithm for specific problems to achieve optimal results.

The comprehensive analysis of Table 5.43 provides valuable insights into the performance of the proposed algorithms for ten benchmark problems, which can be beneficial for researchers and practitioners in selecting the most suitable algorithm for similar problems and enhancing the efficiency of the optimization process.

Table 5.44 presents the mean mass achieved by the proposed algorithms for ten benchmark problems in 100 independent runs, similar to Table 5.43. It includes overall rank, average rank, individual ranks, as well as the best and worst results for each algorithm.

Upon analyzing the results in Table 5.44, it was observed that the PVS algorithm outperformed all other algorithms in terms of overall rank, obtaining the highest mean mass. It was followed by the SOS, TLBO, and HTS algorithms, respectively.

Table 5.43 Minimum mass obtained using TTO with discrete sections

	ALO	DA	GWO	MVO	SCA	WOA	HTS	TLBO	PVS	SOS
10-bar truss	**426.8**	510.7	**426.8**	**426.8**	464.0	602.4	**426.8**	**426.8**	**426.8**	**426.8**
14-bar truss	**369.6**	459.8	**369.6**	**369.6**	390.7	414.5	**369.6**	**369.6**	**369.6**	**369.6**
15-bar truss	105.5	124.1	**98.0**	**98.0**	124.4	130.1	98.2	98.2	**98.0**	**98.0**
24-bar truss	148.0	188.6	122.8	120.8	148.5	207.0	137.8	121.8	121.2	**120.1**
20-bar truss	157.1	202.6	162.5	155.3	195.0	282.4	157.1	157.1	**154.8**	155.3
72-bar 3D truss	459.3	618.6	**447.1**	458.8	519.1	562.8	450.4	452.1	450.4	450.4
39-bar truss	285.0	358.0	**244.0**	260.3	361.8	479.2	268.0	285.6	246.1	256.3
45-bar truss	**153.6**	260.0	162.8	**153.6**	208.5	228.4	177.1	181.7	**153.6**	**153.6**
25-bar 3D truss	**389.0**	**389.0**	**389.0**	**389.0**	416.2	392.0	**389.0**	**389.0**	**389.0**	**389.0**
39-bar 3D truss	378.7	380.6	**360.5**	**360.5**	370.6	391.6	373.3	**360.5**	**360.5**	**360.5**
Overall rank	7	9	4	3	8	10	6	5	1	1
Count best	4	1	7	6	0	0	3	4	7	7
Count 2nd best	0	0	0	2	0	0	1	0	2	2
Count 3nd best	0	0	0	0	0	0	0	0	1	1
Count worst	0	3	0	0	1	6	0	0	0	0

Note: Bold indicates the best solution

Furthermore, the PVS algorithm demonstrated seven best solutions and three second-best solutions, without any worse solutions. Similarly, the SOS algorithm provided three best solutions and seven second-best solutions, with no worse solutions. On the other hand, the TLBO algorithm had three third-best solutions and no worse solutions.

In contrast, the WOA algorithm exhibited the highest number of worst solutions for six benchmark problems. These results suggest that the performance of the WOA, DA, MVO, and ALO algorithms may not be optimal for certain types of problems.

Table 5.44 Mean mass obtained using TTO with discrete sections

	ALO	DA	GWO	MVO	SCA	WOA	HTS	TLBO	PVS	SOS
10-bar truss	547.0	725.3	496.5	560.3	541.9	944.6	521.0	484.5	466.6	**464.9**
14-bar truss	681.9	772.0	505.8	517.0	449.7	1332.4	556.8	497.3	**419.9**	425.8
15-bar truss	151.8	1.20E +07	6.00E +06	2.40E +07	1.48E +02	252.2	145.6	1.22E +02	**104.1**	106.1
24-bar truss	207.8	261.4	149.6	164.2	186.3	371.5	177.1	181.1	**138.9**	141.2
20-bar truss	263.8	2.01E +06	4.00E +06	1.20E +07	258.3	486.8	4.0E +06	497.0	**171.3**	172.8
72-bar 3D truss	722.4	8.0E +06	4.0E +06	1.00E +07	638.8	1247.9	670.8	481.2	470.4	**459.6**
39-bar truss	466.5	583.0	319.2	394.4	423.4	847.9	376.8	375.8	**305.3**	310.4
45-bar truss	315.0	415.8	206.4	229.4	282.3	575.6	280.5	271.8	200.1	**183.8**
25-bar 3D truss	413.5	440.3	2.00E +06	2.00E +06	444.9	505.1	420.5	390.6	**389.0**	389.1
39-bar 3D truss	456.8	495.3	421.7	422.0	432.5	533.8	410.9	420.9	**399.3**	401.6
Overall rank	7	9	5	8	5	10	4	3	1	2
Count best	0	0	0	0	0	0	0	0	7	3
Count 2nd best	0	0	0	0	0	0	0	0	3	7
Count 3nd best	0	0	3	0	2	0	1	4	0	0
Count worst	0	0	0	3	0	6	0	0	0	0

Note: Bold indicates the best solution

References

Achtziger, W. (1999a). Local stability of trusses in the context of topology optimization. Part I: Exact modelling. *Structural Optimization, 17*(4), 235–246. https://doi.org/10.1007/s001580050056

Achtziger, W. (1999b). Local stability of trusses in the context of topology optimization. Part II: A numerical approach. *Structural Optimization, 17*(4), 247–258. https://doi.org/10.1007/s001580050056

Ahrari, A., Atai, A. A., & Deb, K. (2014). Simultaneous topology, shape and size optimization of truss structures by fully stressed design based on evolution strategy. *Engineering Optimization, 47*(8), 37–41. https://doi.org/10.1080/0305215X.2014.947972

Bai, Y., De Klerk, E., Pasechnik, D., & Sotirov, R. (2009). Exploiting group symmetry in truss topology optimization. *Optimization and Engineering, 10*(3), 331–349. https://doi.org/10.1007/s11081-008-9050-6

Bennage, W. A., & Dhingra, A. K. (1995). Optimization of truss topology using Tabu search. *International Journal for Numerical Methods in Engineering, 38*(23), 4035–4052. https://doi.org/10.1002/nme.1620382308

Cheng, M. Y., & Prayogo, D. (2014). Symbiotic organisms search: A new metaheuristic optimization algorithm. *Computers and Structures, 139*, 98–112. https://doi.org/10.1016/j.compstruc.2014.03.007

Deb, K., & Gulati, S. (2001). Design of Truss-Structures for minimum weight using genetic algorithms. *Finite Elements in Analysis and Design, 37*(5), 447–465. https://doi.org/10.1016/S0168-874X(00)00057-3

Gholizadeh, S., & Barzegar, A. (2013). Shape optimization of structures for frequency constraints by sequential harmony search algorithm. *Engineering Optimization, 45*(6), 627–646. https://doi.org/10.1080/0305215X.2012.704028

Gonçalves, M. S., Lopez, R. H., & Miguel, L. F. F. (2015). Search group algorithm: A new Metaheuristic method for the optimization of truss structures. *Computers & Structures, 153*, 165–184. https://doi.org/10.1016/j.compstruc.2015.03.003

Hajirasouliha, I., Pilakoutas, K., & Moghaddam, H. (2011). Topology optimization for the seismic design of truss-like structures. *Computers and Structures, 89*(7–8), 702–711. https://doi.org/10.1016/j.compstruc.2011.02.003

Jin, P., & De-yu, W. (2006). Topology optimization of truss structure with fundamental frequency and frequency domain dynamic response constraints. *Acta Mechanica Solida Sinica, 19*(3), 231–240. The Chinese Society of Theoretical and Applied Mechanics. https://doi.org/10.1007/s10338-006-0628-2

Kaveh, A., & Zolghadr, A. (2013). Topology optimization of trusses considering static and dynamic constraints using the CSS. *Applied Soft Computing Journal, 13*(5), 2727–2734. https://doi.org/10.1016/j.asoc.2012.11.014

Kaveh, A., & Zolghadr, A. (2014). Democratic PSO for truss layout and size optimization with frequency constraints. *Computers and Structures, 130*, 10–21. https://doi.org/10.1016/j.compstruc.2013.09.002

Kumar, S., Tejani, G. G., & Mirjalili, S. (2019). Modified symbiotic organisms search for structural optimization. *Engineering Computations, 35*, 1269–1296. https://doi.org/10.1007/s00366-018-0662-y

Kumar, S., Tejani, G. G., Pholdee, N., & Bureerat, S. (2020). Improved metaheuristics through migration based search and an acceptance probability for truss optimization. *Asian Journal of Civil Engineering*. https://doi.org/10.1007/s42107-020-00271-x

Kumar, S., Tejani, G. G., Pholdee, N., & Bureerat, S. (2022). Performance enhancement of metaheuristics through random mutation and simulated annealing-based selection for concurrent topology and sizing optimization of truss structures. *Soft Computing, 26*, 5661–5683. https://doi.org/10.1007/s00500-022-06930-2

Li, L., & Liu, F. (2011). Group search optimization for applications in structural design. *Adaptation, Learning, and Optimization (Bd. 9)*. https://doi.org/10.1007/978-3-642-20536-1

Mela, K. (2014). Resolving issues with member buckling in truss topology optimization using a mixed variable approach. *Structural and Multidisciplinary Optimization, 50*(6), 1037–1049. https://doi.org/10.1007/s00158-014-1095-x

Mirjalili, S. (2015). The ant lion optimizer. *Advances in Engineering Software, 83*, 80–98. https://doi.org/10.1016/j.advengsoft.2015.01.010

Mirjalili, S. (2016a). Dragonfly algorithm: A new meta-heuristic optimization technique for solving single-objective, discrete, and multi-objective problems. *Neural Computing and Applications, 27*(4), 1053–1073. https://doi.org/10.1007/s00521-015-1920-1

Mirjalili, S. (2016b). SCA: A sine cosine algorithm for solving optimization problems. *Knowledge-Based Systems*. https://doi.org/10.1016/j.knosys.2015.12.022

Mirjalili, S., & Lewis, A. (2016). The whale optimization algorithm. *Advances in Engineering Software, 95*, 51–67. https://doi.org/10.1016/j.advengsoft.2016.01.008

Mirjalili, S., Mirjalili, S. M., & Lewis, A. (2014). Grey wolf optimizer. *Advances in Engineering Software, 69*, 46–61. https://doi.org/10.1016/j.advengsoft.2013.12.007

Mirjalili, S., Mirjalili, S. M., & Hatamlou, A. (2016). Multi-verse optimizer: A nature-inspired algorithm for global optimization. *Neural Computing and Applications, 27*(2), 495–513. https://doi.org/10.1007/s00521-015-1870-7

Nakamura, T., & Ohsaki, M. (1992). A natural generator of optimum topology of plane trusses for specified fundamental-frequency. *Computer Methods in Applied Mechanics and Engineering, 94*(1), 113–129. https://doi.org/10.1016/0045-7825(92)90159-h

Noilublao, N., & Bureerat, S. (2011). Simultaneous topology, shape and sizing optimisation of a three-dimensional slender truss tower using multiobjective evolutionary algorithms. *Computers and Structures, 89*(23–24), 2531–2538. https://doi.org/10.1016/j.compstruc.2011.08.010

Ohsaki, M., Fujisawa, K., Katoh, N., & Kanno, Y. (1999). Semi-definite programming for topology optimization of trusses under multiple eigenvalue constraints. *Computer Methods in Applied Mechanics and Engineering, 180*(1–2), 203–217. https://doi.org/10.1016/S0045-7825(99)00056-0

Patel, V. K., & Savsani, V. J. (2015). Heat transfer search (HTS): A novel optimization algorithm. *Information Sciences, 324*, 217–246. https://doi.org/10.1016/j.ins.2015.06.044

Pholdee, N., & Bureerat, S. (2014). Comparative performance of meta-heuristic algorithms for mass minimisation of trusses with dynamic constraints. *Advances in Engineering Software, 75*, 1–13. https://doi.org/10.1016/j.advengsoft.2014.04.005

Rao, R. V., Savsani, V. J., & Vakharia, D. P. (2011). Teaching–learning-based optimization: A novel method for constrained mechanical design optimization problems. *Computer-Aided Design, 43*(3), 303–315. https://doi.org/10.1016/j.cad.2010.12.015

Rao, R. V., Savsani, V. J., & Vakharia, D. P. (2012). Teaching–learning-based optimization: An optimization method for continuous non-linear large scale problems. *Information Sciences, 183*(1), 1–15. https://doi.org/10.1016/j.ins.2011.08.006

Richardson, J. N., Bouillard, P., Adriaenssens, S., Bouillard, P., & Filomeno, R. (2012). Multiobjective topology optimization of truss structures with kinematic stability repair. *Structural and Multidisciplinary Optimization, 46*, 513–532.

Ringertz, U. L. F. T. (1986). A branch and bound algorithm for topology optimization of truss structures. *Engineering Optimization, 10*, 111–124. https://doi.org/10.1080/03052158608902532

Rozvany, G. I. N. (1996). Difficulties in truss topology optimization with stress and local buckling constraints. *Structural Optimization, 11*(2), 134–136. https://doi.org/10.1007/BF01376857

Savsani, P. V., & Savsani, V. J. (2015). Passing Vehicle Search (PVS): A novel metaheuristic algorithm. *Applied Mathematical Modelling, 40*(5–6), 3951–3978. https://doi.org/10.1016/j.apm.2015.10.040

Savsani, V. J., Tejani, G. G., & Patel, V. K. (2016). Truss topology optimization with static and dynamic constraints using modified subpopulation teaching–learning-based optimization. *Engineering Optimization, 48*, 1–17. https://doi.org/10.1080/0305215X.2016.1150468

Savsani, V. J., Tejani, G. G., Patel, V. K., & Savsani, P. (2017). Modified meta-heuristics using random mutation for truss topology optimization with static and dynamic constraints. *Journal of Computational Design and Engineering, 4*, 106–130. https://doi.org/10.1080/0305215X.2016.1150468

Singh, P., Kottath, R., & Tejani, G. G. (2022). Ameliorated follow the leader: Algorithm and application to truss design problem. *Structure, 42*, 181–204. https://doi.org/10.1016/j.istruc.2022.05.105

Smith, O. (1996). An interactive system for truss topology design. *Advances in Engineering Software, 27*(96), 167–178. https://doi.org/10.1016/0965-9978(96)00010-5

Stolpe, M., & Svanberg, K. (2003). A note on stress-constrained truss topology optimization. *Structural and Multidisciplinary Optimization, 25*, 62–64.

Tejani, G. G. (2017). *Investigation of advanced metaheuristic techniques for simultaneous size, shape, and topology optimization of truss structures*. PhD thesis, Pandit Deendayal Petroleum University, India.

Tejani, G. G., Savsani, V. J., & Patel, V. K. (2016a). Modified sub-population teaching-learning-based optimization for Design of Truss Structures with natural frequency constraints. *Mechanics Based Design of Structures and Machines, 44*, 495–513. https://doi.org/10.1080/15397734.2015.1124023

Tejani, G. G., Savsani, V. J., & Patel, V. K. (2016b). Adaptive symbiotic organisms search (SOS) algorithm for structural design optimization. *Journal of Computational Design and Engineering, 3*, 226–249. https://doi.org/10.1016/j.jcde.2016.02.003

Tejani, G. G., Savsani, V. J., & Patel, V. K. (2017a). Modified sub-population based heat transfer search algorithm for structural optimization. *The International Journal of Applied Metaheuristic Computing, 8*, 1–23. https://doi.org/10.4018/IJAMC.2017070101

Tejani, G. G., Savsani, V. J., Patel, V. K., & Bureerat, S. (2017b). Topology, shape, and size optimization of truss structures using modified teaching-learning based optimization. *Advances in Computational Design, 2*, 313–331. https://doi.org/10.12989/acd.2017.2.4.313

Tejani, G. G., Savsani, V. J., Patel, V. K., & Mirjalili, S. (2018c). An improved heat transfer search algorithm for unconstrained optimization problems. *Journal of Computational Design and Engineering*. https://doi.org/10.1016/j.jcde.2018.04.003

Tejani, G. G., Savsani, V. J., Patel, V. K., & Mirjalili, S. (2018d). Truss optimization with natural frequency bounds using improved symbiotic organisms search. *Knowledge-Based Systems, 143*, 162–178. https://doi.org/10.1016/j.knosys.2017.12.012

Tejani, G. G., Savsani, V. J., Patel, V. K., & Savsani, P. V. (2018a). Size, shape, and topology optimization of planar and space trusses using mutation-based improved metaheuristics. *Journal of Computational Design and Engineering, 5*, 198–214. https://doi.org/10.1016/j.jcde.2017.10.001

Tejani, G. G., Savsani, V. J., Bureerat, S., & Patel, V. K. (2018b). Topology and size optimization of trusses with static and dynamic bounds by modified symbiotic organisms search. *Journal of Computing in Civil Engineering, 32*, 1–11. https://doi.org/10.1061/(ASCE)CP.1943-5487.0000741

Xu, B., Jiang, J., Tong, W., & Wu, K. (2003). Topology group concept for truss topology optimization with frequency constraints. *Journal of Sound and Vibration, 261*(5), 911–925. https://doi.org/10.1016/S0022-460X(02)01021-0

Zhou, M. (1996). Difficulties in truss topology optimization with stress and local buckling constraints. *Structural Optimization, 11*, 134–136.

Chapter 6
Topology, Shape, and Size Optimization

6.1 Introduction

The optimization of trusses can be categorized into three distinct types: size optimization, shape optimization, and topology optimization. Size optimization aims to determine the best element cross-sectional areas, while shape optimization aims to identify the optimal positions of specific nodes within the truss. On the other hand, topology optimization deals with the addition and removal of elements, making it more complex as it involves exploring various topologies to identify the optimal one. The topology, shape, and size (TSS) variables have varying effects on both the objective function and constraints, which makes simultaneous optimization of TSS a challenging task for optimization algorithms.

Using all three types of variables (topology, shape, and size) simultaneously can lead to very complex problems due to the mixed nature of design variables, including discrete, continuous, and binary variables. However, it is the best approach to achieve a lighter structure through TSS optimization. Many researchers have investigated this approach, including Rajan (1995), who applied it to 10-bar and 14-node trusses. McKeown (1998) optimized cantilever trusses using growing-complexity methods, while Rajeev and Krishnamoorthy (1997) solved the 10-bar truss problem using both single-phase and two-phase methods. Bojczuk and Mróz (1998) used biological growth models to present cantilever and simply supported trusses, and Shrestha and Ghaboussi (1999), Shea and Cagan (1999), Bojczuk and Mróz (1999), and Soh and Yang (2000) employed physical design space methodology, shape grammars, a heuristic algorithm, and ground structure method, respectively.

Deb and Gulati (2001) optimized 11-bar, 45-bar, 39-bar two-tier, 25-bar, and 39-bar trusses using a single-stage optimization method. Hasançebi and Erbatur (2001, 2002) designed a 224-bar space truss pyramid structure. Tang et al. (2005) used an improved GA (Genetic Algorithm), while Luh and Lin (2008) used a two-stage optimization method. Rahami et al. (2008) combined energy and force methods with GA. Wu and Tseng (2010) used AMPDE, and Luh and Lin (2011),

V. Savsani et al., *Truss Optimization*, https://doi.org/10.1007/978-3-031-49295-2_6

Noilublao and Bureerat (2011), Kaveh and Laknejadi (2013), Miguel et al. (2013), Cerveira et al. (2013), Noilublao and Bureerat (2013), Kaveh and Ahmadi (2014), Ahrari et al. (2014), Gonçalves et al. (2015), Kaveh and Mahdavi (2015), Ahrari and Deb (2016), Souza et al. (2016), Mortazavi and Toğan (2016), Tejani (2017), Tejani et al. (2017), Tejani et al. (2018), and Kumar et al. (2022) also presented applications on TSS optimization.

The optimal design of trusses that can withstand dynamic behavior has been a challenging area of study and has received much attention from researchers. The natural frequencies of a truss play a crucial role in enhancing its dynamic behavior. To prevent resonance with external excitations, certain restrictions on the natural frequencies of the truss are necessary. However, the truss should also be lightweight, which creates a conflict with the frequency constraints and makes truss optimization more complex. Simultaneous TSS optimization with multiple natural frequency constraints further complicates the problem and often leads to divergence. To address this issue, an efficient metaheuristic optimization approach is required. Although several researchers have applied various metaheuristics in this area, simultaneous TSS optimization with multiple natural frequency constraints has not yet been adequately addressed.

The performance of the algorithms is evaluated by modifying ten truss benchmark problems. These problems include three truss benchmarks that require simultaneous TSS optimization with continuous and discrete cross-sectional areas and subjected to multiload conditions. They are subject to static and dynamic constraints, as considered by Savsani et al. (2016), Savsani et al. (2017). Additionally, seven benchmark problems that involve static constraints are considered from Hajela and Lee (1995), Deb and Gulati (2001), Richardson et al. (2012), Miguel et al. (2013), Tejani et al. (2019), and Kumar et al. (2022). These benchmarks are proposed for the TSS optimization with static and dynamic constraints.

6.2 Problem Formulation

The primary objective of a TSS optimization problem is to minimize the weight or mass of a truss structure while satisfying specific design requirements and constraints. These constraints encompass displacement limits, stress limits, buckling instability limits, natural frequency requirements, and kinematic stability criteria.

Furthermore, if the truss incorporates nodal and element masses, they are also taken into account in the optimization objective function. The objective function serves as the mathematical representation of the optimization problem, aiming to minimize the total mass of the truss structure.

To summarize, a typical TSS optimization problem involves identifying the optimal design of a truss structure that fulfills specified constraints while minimizing the overall weight or mass of the structure, considering both nodal and element masses, if applicable. The optimization problem can be formulated as follows:

$$\text{Find}, X = \{A_1, A_2, \ldots, A_m, \xi_1, \xi_2, \ldots, \xi_n\} \tag{6.1}$$

to minimize, mass of truss,

$$F(X) = \sum\nolimits_{i=1}^{m} B_i A_i \rho_i L_i + \sum\nolimits_{j=1}^{n} b_j$$

Subject to:
Behavior constraints:

$g_1(X)$: Stress constraints, $|B_i \sigma_i| - \sigma_i^{\max} \leq 0$
$g_2(X)$: Displacement constraints, $\left|\delta_{x_j/y_j/z_j}\right| - \delta_{x_j/y_j/z_j}^{\max} \leq 0$
$g_3(X)$: Euler buckling constraints, $\left|B_i \sigma_i^{\text{comp}}\right| - \sigma_i^{cr} \leq 0$, where $\sigma_i^{cr} = \frac{k_i A_i E_i}{L_i^2}$
$g_4(X)$: $f_r - f_r^{\min} \geq 0$ for some natural frequencies.

Side constraints:

Cross − sectional area constraints, $A_i^{\min} \leq A_i \leq A_i^{\max}$
Shape constraints, $\xi_{x_j/y_j/z_j}^{\min} \leq \xi_{x_j/y_j/z_j} \leq \xi_{x_j/y_j/z_j}^{\max}$
Check on kinematic stability
Check on validity of structure
where, $i = 1, 2, \ldots, m; j = 1, 2, \ldots, n$

where B_i is a binary element, $B_i = 0$ (*if* $A_i <$ Critical area) and $B_i = 1$ (*if* $A_i \geq$ Critical area) for deleting and retaining the *i*th element, respectively). $A_i, \rho_i, L_i, E_i, \sigma_i$, and σ_i^{cr} stand for cross-sectional area, mass density, element length, Young's modulus, stress, and critical buckling stress on the element "*i*," respectively. $\delta_{x_j/y_j/z_j}, \xi_{x_j/y_j/z_j}$, and b_j are values of nodal displacement, nodal position, and mass values of node "j," respectively, where x, y, and z present x, y, and z axis, respectively. f_r is structural natural frequent obtained from solving structural free vibration analysis of the rth mode. Superscripts "max," "min," and "comp" denote maximum allowable limit, minimum allowable limit, and compressive stress, respectively. k_i is the Euler buckling coefficient calculated from elements' cross-sections.

The penalty function technique is employed to manage the constraints. This technique penalizes violated solutions by adding positive penalty terms to their objective function values, with the aim of minimizing them. The computation of the penalty value can be done by following the methodology described by Deb and Gulati (2001) and Kaveh and Zolghadr.

$$\text{Penalized } F(X) = F(X) * F_{\text{penalty}} \tag{6.2}$$

where

$$F_{\text{penalty}} = (1 + \varepsilon_1 * \mathrm{C})^{\varepsilon_2}, \mathrm{C} = \sum\nolimits_{i=1}^{q} \mathrm{C_i}, \mathrm{C_i} = \left|1 - \frac{\mathrm{p_i}}{\mathrm{p_i^*}}\right| \tag{6.3}$$

where p_i is the level of constraint violation having the bound as p_i^*. The parameter q is a number of active constraints. The variables ε_1 and ε_2 are predetermined by a user. In this study, the values of both ε_1 and ε_2 are set as 1.5, which were obtained from experimenting their effect on the balance of the exploitation–exploration balance.

In the topology optimization, the lower and upper limits of the design variables assumed to be $X^{lower} = -A^{\max}$ and $X^{\text{upper}} = A^{\max}$, respectively. The critical area is $A^{\min}$, which is applied to remove elements from the ground structure. Truss design problems involve constraints related to displacement, stress, natural frequency, and buckling, and the trusses are subjected to multiple load cases. To handle topology optimization, the search space is approximately converted into two times the limits of the design variables. This study introduces benchmark problems that involve simultaneous TSS optimization and discrete cross-sectional areas, which have not been experimentally investigated in the literature. To ensure a fair assessment, the same parameters are used for all algorithms. The proposed algorithms are used to solve each test problem 100 times, considering the stochastic nature of metaheuristics. A population size of 50 is used for all problems, and each run is evaluated 20,000 times. The results obtained from the proposed algorithms are compared with those obtained using their basic algorithms. All the problems are coded and executed using MATLAB R2013a (Ferreira, 2009).

Ten advanced metaheuristic techniques developed after 2011, including the dragonfly algorithm (DA) (Mirjalili, 2016a), multiverse optimizer (MVO) (Mirjalili et al., 2016), sine cosine algorithm (SCA) (Mirjalili, 2016b), whale optimization algorithm (WOA) (Mirjalili & Lewis, 2016), ant lion optimizer (ALO) (Mirjalili, 2015), heat transfer search (HTS) (Patel & Savsani, 2015), passing vehicle search (PVS) (Savsani & Savsani, 2015), symbiotic organisms search (SOS) (Cheng & Prayogo, 2014), gray wolf optimizer (GWO) (Mirjalili et al., 2014), and the teaching–learning-based optimization (TLBO) (Rao et al., 2011, 2012) algorithm, have been recommended to solve truss design problems. These metaheuristics employ unique search mechanisms, and their effectiveness across various engineering applications is challenging to predict. Among them, the TLBO algorithm has proven to be effective and has shown a positive impact on several engineering optimization problems.

This study examines the effectiveness of ten basic metaheuristic methods (DA, MVO, SCA, WOA, ALO, HTS, PVS, SOS, GWO, and TLBO) for the optimization of ten distinct trusses under multiple loading conditions. The objective function is the overall mass, while the constraints include stress, displacement, buckling, and frequency. The TSS optimization is performed using a single-stage approach, ground structure method, and restructuring of finite element analysis model. The results, including Tables 6.1, 6.2, 6.3, 6.4, 6.5, 6.6, 6.7, 6.8, 6.9 and 6.10, are presented, and the TSS optimization problems with continuous and discrete sections are described in subsequent sections.

Table 6.1 Design parameters of the 10-bar truss

Size variables: $X_i = A_i$, $i = 1, 2, \ldots, 10$ Shape variables: $4.572\ \text{m} \leq \xi_{y_1}, \xi_{y_3}, \xi_{y_5} \leq 25.4\ \text{m}$ (0.0254 m = 1 in.)	
Multiple load conditions:	
Condition 1 : $F_1 = 44.537$ kN, $F_2 = 44.537$ kN	Condition 2 : $F_1 = 44.537$ kN, $F_2 = 0$ kN
Lumped masses on nodes 2 and 4 are 500 kg	
Stress and displacement constraints: $\sigma_i^{max} = 172.43$ MPa and $\delta_{y_5\&y_6}^{max} = 100$ mm (0.1 m)	
Natural frequency constraint: $f_1 \geq 4$ Hz	
The continuous cross-sectional areas: $[A^{min}, A^{max}] = [-100, 100]\ \text{cm}^2$ ($[-100, 100] \times 10^{-4}\ \text{m}^2$) and critical area = 1 cm^2	
Material properties: $E = 69$ GPa and $\rho = 2740 \text{kg/m}^3$	

6.3 Results and Discussion on TSS Optimization with Continuous Sections

This section utilizes ten different benchmark problems to assess the effectiveness of the algorithms under consideration for continuous sections. The benchmark problems include 10-bar, 14-bar, 15-bar, 24-bar, 20-bar, 72-bar 3D, 39-bar, 45-bar, 25-bar 3D, and 39-bar 3D trusses, and their specifications can be found in subsequent sections. To account for the stochastic nature of metaheuristics, the proposed algorithms are applied to address each truss problem through 100 independent runs. The population size remains fixed at 50 for all problems, while each run is subjected to a total of 20,000 function evaluations (FE). The main objective is to evaluate the performance of the selected algorithms in solving the truss TSS optimization problems with continuous.

6.3.1 TSS Optimization of the 10-Bar Truss with Continuous Cross-Sections

Figure 6.1 presents the ground structure of the first benchmark truss, along with the load and boundary condition. This truss is characterized by ten size variables and three shape variables, which are utilized for TSS optimization. Several researchers, including Hajela and Lee (1995), Deb and Gulati (2001), Richardson et al. (2012), and Miguel et al. (2013), have investigated this truss for static constraints such as elemental stress and nodal displacement.

Previous studies on this truss problem primarily considered either static constraints (stress, buckling, and displacement) or dynamic constraints (natural frequencies), but the simultaneous consideration of both static and dynamic constraints has not been explored. In this study, the problem is modified to incorporate multi-load conditions, lumped masses, natural frequency constraints, and continuous cross-sectional areas.

Table 6.2 Topology, shape, and size optimization of the 10-bar truss (continuous section)

Element no.	ALO	DA	GWO	MVO	SCA	WOA	HTS	TLBO	PVS	SOS
1	3.8656	3.6427	3.3270	3.2469	7.6450	10.8426	3.2975	3.1858	3.1865	3.4551
3	47.5853	44.2867	38.2557	39.1485	47.9234	49.2277	38.3954	38.1743	38.1881	43.1261
4	38.6378	35.9439	31.2546	32.1421	51.6075	45.3404	31.3200	31.1707	31.1924	34.3608
5	—	9.5621	—	—	2.5872	—	1.0718	—	—	—
7	2.9729	2.8547	2.7534	2.7732	—	3.0589	2.7492	2.7455	2.7488	2.8461
8	10.5487	—	1.0247	1.0124	1.6671	16.5923	—	1.0008	1.0026	1.0393
9	6.9658	11.2236	3.2264	3.2921	7.1830	46.8434	3.2634	3.1836	3.1893	3.7728
y_1	313.8266	1000.0000	344.6631	953.9126	502.1078	256.8345	259.2422	913.3089	214.2489	835.0061
y_3	325.3588	380.7569	498.1437	477.8483	270.4292	290.9146	495.9119	499.9938	499.6419	421.2851
y_5	631.6692	761.7975	999.2292	966.5953	814.1472	644.4207	991.6593	999.8631	999.7202	859.8263
Mass (kg)	327.9360	322.3126	251.9777	255.1631	341.6681	520.0671	251.5316	250.6464	250.7881	269.8056
σ^{max} (MPa)	172.4299	172.4300	171.9313	172.2799	172.1432	166.7757	172.4045	172.4144	172.2925	170.1206
δ^{max} (mm)	74.4907	62.4259	96.1073	97.3965	86.6705	27.5747	95.7351	98.4922	98.4524	91.3613
f_1 (Hz)	5.2708	5.4091	4.6903	4.6573	4.9260	6.0635	4.6964	4.6342	4.6351	4.8006
Mean	508.4372	1036.3612	288.8143	331.4609	467.6564	984.9895	330.3556	331.6600	293.3991	352.5748
SD	135.5049	2841.2123	35.2280	90.8912	81.7379	300.9220	98.7174	97.6583	23.3491	45.2386

Table 6.3 Design parameters of the 14-bar truss

Size variables: $X_i = A_i$, $i = 1, 2, \ldots, 14$ Shape variables: $4.572\ \text{m} \leq \xi_{y_1}, \xi_{y_3}, \xi_{y_5} \leq 25.4\ \text{m}$ (0.0254 m = 1 in.)	
Multiple load conditions:	
Condition 1 : $F_1 = 44.537$ kN, $F_2 = 44.537$ kN	Condition 2 : $F_1 = 44.537$ kN, $F_2 = 0$ kN
Lumped mass on node 2 and 4 are 500 kg	
Stress and displacement constraints: $\sigma_i^{max} = 172.43$ MPa and $\delta_{y_5 \& y_6}^{max} = 100$ mm (0.1 m)	
Natural frequency constraint: $f_1 \geq 4$ Hz	
The continuous cross-sectional areas: $[A^{min}, A^{max}] = [-100, 100]$ cm^2 ($[1, 100] \times 10^{-4}$ m^2) and critical area = 1 cm^2	
Material properties: $E = 69$ GPa and $\rho = 2740 \text{kg/m}^3$	

For the continuous cross-sectional areas (A_i, where $i = 1, 2, \ldots, 10$) in this problem, the range is [−100, 100] cm^2, where a zero or negative value indicates removal of the element. Therefore, the continuous design variables can take values within the range of [−100, 100] cm^2, where a zero or negative value indicates the removal of the corresponding element. Table 6.1 provides comprehensive details regarding material properties, design variables, loading conditions, lumped masses, and other important data related to the test problem.

- The 10-bar truss is considered as a demonstration of the truss optimization procedure. The step-by-step procedure for implementing the size, shape, and topology optimization with a sample calculation is outlined below:

Step 1: Begin by defining the ground structure of the truss. This involves generating nodes over a predefined design domain and assigning all possible element connections, as well as material properties, loading, and boundary conditions.

- Figure 6.1 illustrates the ground structure of the 10-bar truss, showcasing the loading and boundary conditions, while Table 6.1 provides the material properties.

Step 2: Proceed to the optimization algorithm, where the objective function, population size, design variables, constraints, algorithm controlling parameters, and termination criterion are defined.

- The objective function and constraints of the 10-bar truss is:

$$\text{Find}, X = \{A_1, A_2, \ \ldots, A_{10}, \xi_{y1}, \xi_{y2}, \xi_{y3}\} \tag{6.4}$$

to minimize, mass of truss,

Table 6.4 Topology, shape, and size optimization of the 14-bar truss (continuous section)

Element no.	ALO	DA	GWO	MVO	SCA	WOA	HTS	TLBO	PVS	SOS
3	45.7003	53.6807	38.1960	38.2449	42.9467	76.0821	38.1822	38.1728	38.1728	38.1743
4	37.3218	42.7862	31.1719	31.2473	31.7006	45.4819	31.1775	31.1680	31.1680	31.1683
7	2.9065	2.8360	2.7481	2.7589	3.0926	3.5769	2.7462	2.7452	2.7452	2.7454
13	3.7116	3.4867	3.1935	3.1943	3.2591	4.9711	3.1854	3.1827	3.1827	3.1831
y_1	991.5244	373.9157	378.3156	411.2015	311.7742	852.4519	226.2308	205.8453	957.4820	1000.0000
y_3	362.1051	426.7650	276.5463	576.1026	931.4316	242.1280	643.5626	229.2188	1000.0000	701.9050
y_5	697.7033	794.0036	1000.0000	999.9002	1000.0000	472.7792	999.5201	1000.0000	1000.0000	999.9984
Mass (kg)	269.7873	304.9099	241.5124	241.9070	257.8501	369.1658	241.3917	241.3306	241.3308	241.3399
σ^{max} (MPa)	172.4300	167.5322	172.4300	172.4232	169.7572	166.5933	166.0107	172.2440	172.4300	170.6019
δ^{max} (mm)	98.5632	95.8232	98.5632	98.5565	96.9502	93.9025	78.7295	98.2386	94.6596	94.6187
f_1 (Hz)	4.6260	4.6905	4.6257	4.6259	4.6646	4.7308	5.1448	4.6331	4.7005	4.7176
Mean	587.3450	742.8925	352.7163	380.7120	381.4843	1238.0447	390.8655	377.8990	298.6982	316.1151
SD	219.3285	268.9675	108.2662	116.8030	101.0490	510.9640	116.1907	145.7468	77.4458	91.5552

Table 6.5 Design parameters of the 15-bar truss

Size variables: $X_i = A_i$, $i = 1, 2, \ldots, 15$ Shape variables: $2.54\ \text{m} \le \xi_{x_5} = \xi_{x_6} \le 3.556\ \text{m}$; $5.588\ \text{m} \le \xi_{x_3} = \xi_{x_4} \le 6.604\ \text{m}$; $1.27\ \text{m} \le \xi_{y_1} \le 2.286\ \text{m}$; $0.508\ \text{m} \le \xi_{y_2} \le 1.524\ \text{m}$; $2.54\ \text{m} \le \xi_{y_3} \le 3.556\ \text{m}$; $-0.508\ \text{m} \le \xi_{y_4} \le 0.508\ \text{m}$; $2.54\ \text{m} \le \xi_{y_5} \le 3.556\ \text{m}$; $-0.508\ \text{m} \le \xi_{y_6} \le 0.508\ \text{m}$ (1 in. = 0.0254 m)	
Multiple load conditions:	
Condition 1 : $F_1 = -44.537$ kN, $F_2 = 0$ kN	Condition 2 : $F_1 = 0$ kN, $F_2 = 4.4537$ kN
Lumped mass on node 2 is 500 kg	
Stress and displacement constraints: $\sigma_i^{max} = 172.43$ MPa and $\delta_{x_2\&y_2}^{max} = 100$ mm (0.1 m)	
Natural frequency constraint: $f_1 \ge 4$ Hz	
The continuous cross-sectional areas: $[A^{min}, A^{max}] = [-50, 50]\ \text{cm}^2$ ($[-50, 50] \times 10^{-4}\ \text{m}^2$) and critical area = 1 cm^2	
Material properties: $E = 69$ GPa and $\rho = 2740\text{kg/m}^3$	

$$F(X) = \sum_{i=1}^{10} B_i X_i \rho_i L_i + \sum b_{2\&4}\ B_i = \begin{cases} 0, \textit{if } X_i < \text{Critical area} \\ 1, \textit{if } X_i \ge \text{Critical area} \end{cases}$$

Subject to:

Behavior constraints:

$g_1(X)$: Stress constraints, $|\sigma_i| - 172.43 \le 0$, *in* MPa
$g_2(X)$: Displacement constraints, $|\delta_j| - 100 \le 0$, in mm
$g_3(X)$: Euler buckling constraints, $\left|\sigma_i^{comp}\right| - \sigma_i^{cr} \le 0$, where $\sigma_i^{cr} = \frac{k_i A_i E_i}{L_i^2}$
$g_4(X)$: $f_1 \ge 4$, *in Hz.*

Side constraints:

Cross − sectional area constraints, $-100 \le A_i \le 100\ \text{cm}^2$ and critical area = 1 cm^2
Shape constraints, $4.572 \le \xi_{y_j} \le 25.4$
where $i = 1, 2, \ldots, 10$; $j = 1, 2, \ldots, 6$

- Assume population size = 05
- Design variables:
  Size/topology variables: The continuous cross-sectional areas: $[A^{min}, A^{max}] = [-100, 100]\ \text{cm}^2 = ([1, 100] \times 10^{-4}\ \text{m}^2)$ and critical area = 1 cm^2.
  Shape variables: $4.572\ \text{m} \le \xi_{y_1}, \xi_{y_3}, \xi_{y_5} \le 25.4\ \text{m}$
- Termination criterion is presented in Eqs. 6.2 and 6.3; assume function evaluations (FE) ≤ 10

Table 6.6 Topology, shape, and size optimization of the 15-bar truss (continuous section)

Element no.	ALO	DA	GWO	MVO	SCA	WOA	HTS	TLBO	PVS	SOS
1	8.5797	5.8700	6.4661	8.1481	10.2084	11.7350	7.1860	6.2508	7.2413	6.0623
2	7.4263	5.2900	6.2745	7.7024	5.7307	10.8204	7.0178	6.3215	6.4692	5.7831
3	—	—	—	—	—	4.3712	—	—	—	—
4	9.1016	21.2353	9.9100	8.9972	12.5487	11.9928	9.6727	9.3727	10.2649	9.8772
5	11.2219	12.3050	11.6802	8.9672	11.6732	10.7959	10.2010	11.2942	10.1295	9.5951
6	12.8858	12.4554	11.7069	10.5955	12.2711	13.9013	11.3296	13.0231	11.2791	12.5328
7	—	—	—	3.6273	9.6205	—	2.6214	2.8938	2.4301	1.2659
8	—	3.1268	2.2147	1.6416	7.4455	2.8310	2.7155	1.0030	1.0159	1.4169
9	—	—	—	—	—	4.1070	—	—	—	—
10	7.3688	7.7411	3.2877	—	—	14.7094	3.8278	—	—	5.6883
11	4.6136	—	1.0844	1.5204	7.3689	5.6354	—	1.7906	1.5687	—
12	1.3745	1.0006	—	—	9.0787	7.5793	—	—	—	—
13	1.1025	3.4288	1.1828	1.5215	—	—	2.1738	2.1729	1.5900	1.3629
14	8.3593	16.9368	6.5876	9.7007	14.1549	7.3444	8.5716	6.7640	7.6456	8.7676
$x_5 = x_6$	100.3157	140.0000	108.7296	116.7164	134.9062	100.0000	118.4680	102.7638	119.2849	114.1061
$x_3 = x_4$	220.0131	260.0000	234.3858	230.5385	229.0185	220.0000	233.8120	222.0296	234.2014	221.3683
y_1	50.3300	90.0000	83.5646	59.9784	90.0000	50.0000	74.4645	52.7946	75.8358	50.4329
y_2	20.0175	20.2277	34.5842	51.2764	31.3326	20.0000	44.9235	25.7932	41.0090	39.2445
y_3	100.0035	140.0000	100.0191	100.0000	132.4761	100.0000	106.8431	100.0048	100.0048	100.0226
y_4	−1.9129	16.0665	19.9903	12.9799	0.3363	3.2624	18.1125	11.9634	19.9639	19.9586
y_5	100.5213	140.0000	128.9214	111.4757	112.9822	100.0000	105.8313	112.8361	114.2559	117.7118
y_6	−15.9542	5.5812	9.6336	5.3265	−0.0029	−6.0171	12.9643	3.4540	10.1144	13.4138
Mass (kg)	101.2324	121.2250	87.4399	88.4479	128.3670	134.0853	91.2521	87.8468	86.0658	89.9069
σ^{max} (MPa)	70.5285	35.6599	37.9409	34.7218	28.5480	31.2906	31.1365	37.8345	32.5260	35.3009
δ^{max} (mm)	10.0000	7.4095	9.8996	9.8270	5.5478	5.9965	9.8594	9.9982	9.9988	8.9360
f_1 (Hz)	4.6832	5.3554	4.7055	4.6961	6.2163	6.0028	4.7096	4.6654	4.6652	4.9456
Mean	142.3164	2.00E+07	4.00E+06	2.40E+07	163.7009	240.5608	123.6173	106.6641	96.5587	100.7155
SD	26.527	4.02E+07	1.97E+07	4.29E+07	16.463	62.150	26.721	18.861	12.077	9.011

Table 6.7 Design parameters of the 24-bar truss

Size variables: $X_i = A_i$, $i = 1, 2, \ldots, 24$ Shape variables: $-2\ \text{m} \le \xi_{x_7}, \xi_{x_8} \le 2\ \text{m}$, with respect to their original coordinates	
Multiple load conditions:	
Condition 1 : $F_1 = 50$ kN, $F_2 = 0$ kN	Condition 2 : $F_1 = 0$ kN, $F_2 = 50$ kN
Lumped mass on node 3 is 500 kg	
Stress and displacement constraints: $\sigma_i^{\max} = 172.43$ MPa and $\delta^{\max}_{y_5 \& y_6} = 10$ mm $\left(10^{-2}\ \text{m}\right)$	
Natural frequency constraint: $f_1 \ge 30$ Hz	
The continuous cross-sectional areas: $[A^{\min}, A^{\max}] = [-40, 40]\ \text{cm}^2$ ($[-40, 40] \times 10^{-4}\ \text{m}^2$) and critical area $= 1\ \text{cm}^2$	
Material properties: $E = 69$ GPa and $\rho = 2740\text{kg/m}^3$	

Step 3: Initiate a randomly generated set of topology variables (i.e., population) within its upper and lower bounds.

- $X(k) = \{A_1, A_2, \ldots, A_{10}, \xi_{y1}, \xi_{y2}, \xi_{y3}\}$
- $X(1)$ = [0.006294, −0.007124, −0.002394, −0.003245, 0.003115, 0.005155, 0.004121, 0.006469, −0.001459, −0.009476, 10.3211, 14.9519, 20.2194]
- $X(2)$ = [0.001292, −0.000970, 0.001802, 0.000771, 0.010000, 0.006568, 0.004341, 0.007007, 0.002048, 0.002150, 14.9124, 25.4000, 6.2190]
- $X(3)$ = [0.01000, 0.00385, 0.00250, 0.01000, 0.00247, 0.00664, −0.01000, 0.01000, 0.00788, 0.00953, 19.6125, 23.8296, 22.2703]
- $X(4)$ = [0.00974, −0.00150, 0.00913, 0.00783, 0.00914, 0.00885, −0.00273, 0.00196, 0.00519, 0.00534, 8.0089, 25.3463, 25.4000]
- $X(5)$ = [0.00899, −0.00624, 0.00895, 0.00837, 0.00953, 0.00933, −0.00587, 0.01000, 0.00905, 0.00798, 4.5720, 17.7417, 22.2701]

Step 4: Go to truss configurations: Generate trusses according to topology variables.

	18.288	ξ_{y1}
	18.288	0
Node coordinates =	9.144	ξ_{y2}
	9.144	0
	0	ξ_{y3}
	0	0

Table 6.8 Topology, shape, and size optimization of the 24-bar truss (continuous section)

Element no.	ALO	DA	GWO	MVO	SCA	WOA	HTS	TLBO	PVS	SOS
1	—	—	—	—	—	15.1411	—	—	—	—
6	—	—	—	—	—	13.8415	—	—	—	—
7	22.8610	18.6099	12.0772	12.3398	21.9352	27.5997	14.1924	—	12.1318	13.2593
8	2.9001	3.0385	2.9236	3.0530	3.2152	4.6740	3.6255	1.0650	2.9375	3.0166
9	—	3.6461	—	—	—	—	—	—	—	—
10	16.3249	—	—	—	16.7646	11.2267	—	—	—	—
12	18.8344	12.8410	2.7995	2.6048	21.4465	13.2370	4.1713	3.5593	2.6855	3.1031
13	—	22.6487	21.1766	21.3658	—	—	20.0579	29.9588	21.2637	21.7307
14	—	—	—	—	—	6.7709	—	4.8516	—	—
15	3.3941	4.0659	3.3300	3.2298	5.4376	8.2248	5.8301	3.2804	3.2440	3.3612
16	19.3667	26.5767	24.1469	23.8488	23.1129	20.5341	25.0408	23.5999	24.0071	23.9878
17	—	15.2745	—	—	—	15.0786	—	—	—	—
19	—	1.0012	—	—	—	23.3523	—	—	—	—
21	—	—	—	—	—	12.4706	—	—	—	—
22	—	—	1.8438	2.2592	2.5438	2.6383	4.8699	1.8262	1.1640	5.8944
23	1.0000	8.0750	1.7730	1.0440	5.3872	1.0662	2.0507	—	—	1.1922
24	2.2971	18.1983	—	—	—	—	—	1.0382	1.0586	—
x_7	−0.6320	0.4551	1.9938	1.8716	−0.1059	−0.7668	1.1891	1.9724	1.9750	1.9257
x_8	−1.7811	−1.6243	−0.2638	−0.4139	−0.6846	−0.7602	−0.1515	−0.7440	−0.3493	−0.4078
Mass (kg)	121.2085	166.0446	109.9633	109.6253	137.3382	227.7992	121.5718	113.5510	108.9194	112.0321
σ^{max} (MPa)	172.4053	164.5525	171.0218	171.6115	155.5093	120.6098	137.9115	152.4224	171.2834	165.7483
δ^{max} (mm)	10.0000	8.5717	9.7033	9.9969	8.6395	4.6423	5.9986	9.9950	9.9381	9.6384
f_1 (Hz)	30.0000	30.0000	30.2667	30.0009	30.1403	30.0000	31.2929	30.9609	30.0006	30.0638
Mean	191.7292	279.9674	130.6985	138.4959	188.2545	379.5975	181.7925	155.8492	125.3713	133.1890
SD	43.5056	67.6708	22.8625	28.0486	22.3881	86.4417	45.5741	34.0201	13.0455	14.7211

Table 6.9 Design parameters of the 20-bar truss

Size variables: $X_i = A_i$, $i = 1, 2, \ldots, 20$ Shape variables: $-1\ \text{m} \le \xi_{y_1}, \xi_{y_2}, \xi_{y_3}, \xi_{y_7}, \xi_{y_8}, \xi_{y_9} \le 1\ \text{m}$, with respect to their original coordinates	
Multiple load conditions:	
Condition 1 : $F_1 = 500$ kN, $F_2 = 0$ kN	Condition 2 : $F_1 = 0$ kN, $F_2 = 500$ kN
No lumped mass	
Stress and displacement constraints: $\sigma_i^{\max} = 172.43$ MPa and $\delta_{y_4}^{\max} = 60$ mm $\left(6 \times 10^{-2}\ \text{m}\right)$	
Natural frequency constraints: $f_1 \ge 60$ Hz and $f_2 \ge 100$ Hz	
The continuous cross-sectional areas: $[A^{\min}, A^{\max}] = [1, 100]\ \text{cm}^2$ ($[1, 100] \times 10^{-4}\ \text{m}^2$) and critical area $= 1\ \text{cm}^2$	
Material properties: $E = 69$ GPa and $\rho = 2740 \text{kg/m}^3$	

	3	5
	1	3
	4	6
	2	4
Node connectivities =	3	4
	1	2
	4	5
	3	6
	2	3
	1	4

Number Elements = 10
Number Nodes = 6
E = 69e9; % E; modulus of elasticity in Pa
ro = 2740; % E; material density in kg/m^3
% computation of B_i & Active Design variables

```
B(1)=[ 1, 0, 0, 0, 1, 1, 1, 1, 0, 0]
```

- $X(1)$= [0.006294, 0, 0, 0, 0.003115, 0.005155, 0.004121, 0.006469, 0, 0, 10.3211, 14.9519, 20.2194]

```
B(2)=[1, 0, 1, 1, 1, 1, 1, 1, 1, 1]
```

- $X(2)$= [0.001292, 0, 0.001802, 0.000771, 0.010000, 0.006568, 0.004341, 0.007007, 0.002048, 0.002150, 14.9124, 25.4000, 6.2190]

```
B(3)=[1, 1, 1, 1, 1, 1, 0, 1, 1, 1]
```

- $X(3)$= [0.01000, 0.00385, 0.00250, 0.01000, 0.00247, 0.00664, 0, 0.01000, 0.00788, 0.00953, 19.6125, 23.8296, 22.2703]

Table 6.10 Topology, shape, and size optimization of the 20-bar truss (continuous section)

Element no.	ALO	DA	GWO	MVO	SCA	WOA	HTS	TLBO	PVS	SOS
1	—	38.7483	25.2362	29.4537	41.4499	32.6850	25.2143	29.5069	28.5844	31.4888
3	—	—	—	—	—	—	9.4743	—	—	—
4	35.8408	—	7.4188	—	23.3209	38.6570	—	4.9046	—	—
5	29.0855	25.7900	29.9211	30.0667	80.5353	15.9210	30.9048	27.5683	30.9877	30.2727
7	—	—	—	—	—	—	2.2918	—	—	—
8	—	12.8408	5.6085	2.8056	11.1507	31.5045	6.0018	2.6911	3.0812	3.2613
11	18.4056	17.3088	4.1142	2.0829	6.6782	29.1404	6.0283	7.2189	1.2558	1.8979
12	—	15.0638	—	—	—	—	—	—	—	—
13	32.3091	20.6375	8.2950	2.4968	—	18.4970	3.7717	—	2.0116	1.7085
15	—	27.6975	26.4503	29.5660	32.6302	40.2960	27.9655	28.1136	28.8040	32.9338
17	—	—	—	—	—	—	3.0358	—	—	—
18	16.1779	46.5064	—	2.4781	19.3442	30.7071	—	6.6177	5.3495	9.5186
19	—	11.1983	—	—	—	—	—	—	—	—
20	46.4124	36.6352	31.9591	30.0459	33.8279	26.9473	32.8589	30.1669	31.3899	32.8142
y_1	−0.6555	0.4660	0.3024	0.5371	0.4839	0.6268	0.1558	0.8399	0.2898	0.1763
y_2	−0.3617	−0.2559	−0.7293	−0.7273	−0.6104	−0.0030	−0.8022	−0.6922	−0.8301	−0.8776
y_3	0.5093	0.1150	0.1239	0.1676	0.0186	−0.0048	−0.8529	−0.9797	0.5768	−0.4904
y_7	−0.1297	0.1253	−0.0208	0.5569	−0.0959	−0.0112	0.4284	0.2871	0.4455	−1.0000
y_8	−0.1838	0.3279	0.9302	0.9415	1.0000	−0.0150	0.6221	0.8025	0.8366	0.9999
y_9	−0.9814	0.4931	0.1178	−0.1234	−0.2180	0.3932	−0.5108	−0.1132	−0.3199	0.0695
Mass (kg)	147.7196	198.1419	114.4491	111.3688	184.7497	202.6807	125.3653	115.9569	113.3089	118.7586
σ^{max} (MPa)	172.4296	162.0703	172.4097	172.2997	158.7348	158.3728	170.7284	172.3834	172.3780	171.6484
δ^{max} (mm)	27.2861	19.6591	25.1058	22.3069	15.7510	18.1955	21.4525	21.6816	21.9572	23.3610
f_1 (Hz)	105.2877	101.0385	64.3657	60.2220	62.8617	98.8315	74.7036	60.0268	61.3796	60.4659
f_2 (Hz)	143.7056	164.8804	190.1232	139.1216	190.3068	177.6616	137.3770	149.1572	125.3771	150.7022
Mean	247.6606	2.01E+06	1.00E+06	2.10E+07	251.5663	421.7790	2784.7374	378.5922	144.8406	153.8023
SD	45.62379	1.41E+07	1.00E+07	4.09E+07	37.40319	117.9694	4417.197	1093.809	18.43191	22.81974

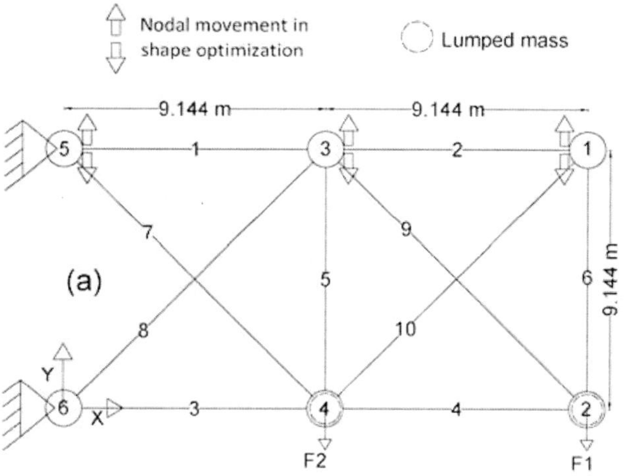

Fig. 6.1 The 10-bar truss

B(4) = [1, 0, 1, 1, 1, 1, 0, 1, 1, 1]

- *X*(4)= [0.00974, 0, 0.00913, 0.00783, 0.00914, 0.00885, 0, 0.00196, 0.00519, 0.00534, 8.0089, 25.3463, 25.4000]

B(5) = [1, 0, 1, 1, 1, 1, 0, 1, 1, 1]

- *X*(5)= [0.00899, 0, 0.00895, 0.00837, 0.00953, 0.00933, 0, 0.01000, 0.00905, 0.00798, 4.5720, 17.7417, 22.2701]

Step 5: Identify invalid trusses, which are those that lack connections to unchangeable nodes (i.e., nodes subjected to load or support), as discussed in Sect. 2.2. Evaluate the degree of freedom of valid trusses as per Eq. 2.11. If a truss is invalid or its degree of freedom is a positive number, assign a large penalty to the objective function and proceed to Step 12.

```
% population 1
DOF(1) = 3 & eigan_value = not applicable % Truss is invalid
fit_fpenalty = 100000000 kg => go to Step 9.
% population 2
DOF(2) = 0 & & eigan_value (2) > 0 % Truss is valid
% population 3
DOF(3) = 0 & & eigan_value (2) > 0 % Truss is valid
% population 4
DOF(4) = 0 & & eigan_value (2) > 0 % Truss is valid
% population 5
DOF(5) = 0 & & eigan_value (2) > 0% Truss is valid
```

Step 6: If the truss is a valid structure, proceed to finite element analysis. Compute the mass matrix, force vector, displacement vector, stresses vector, and natural frequency vector using finite element analysis.

In the context of a 2D truss, the truss bar is considered a line element with simply supported ends. It is assumed to have a length of L, cross-sectional area of A, modulus of elasticity of E, and mass density of ρ. The bar element experiences forces and displacements.

Compute the stiffness matrix using Eq. 2.2 for each element sequentially and subsequently integrating it into the global stiffness matrix.

```
GDof = 2 x Number Nodes = 12; % GDof: total number of degrees of freedom
% computation of the system stiffness matrix
```

Compute the mass matrix using Eq. 2.6 for each element sequentially and subsequently integrating it into the global mass matrix.

Compute the displacement vectors and stresses vectors using Eq. 2.5.

Compute the buckling vectors and stresses vectors using Eq. 5.4.

Compute the natural frequency vector using Eq. 2.7.

Step 7: Compute natural frequencies, element stresses, nodal displacements, and Euler buckling of the truss using finite element analysis.

```
% Result of population 1
% element stress:
sigma1(1) = Not applicable as tuss is invalid
sigma2(1) = Not applicable as tuss is invalid
% Euler buckling stress:
buckling_stress (1) = Not applicable as tuss is invalid
% Natural frequency:
frequency (1) = Not applicable as tuss is invalid
% nodal displacements:
displacements1 (1) = Not applicable as tuss is invalid
displacements2 (1) = Not applicable as tuss is invalid

% Result of population 2
% element stress:
sigma1(2) = [34001796, 0, 107425192, 20791991, 7621448, 0, 49465980,
1211052, 23114096, 0]
sigma2(2) = [26171640, 0, 73501426, 20791991, 7917585, 0, 32434478,
623473, 23114096, 0]
% Euler buckling stress:
buckling_stress (2) = [789897, 0, 5948201, 2545447, 4278009, 8151919,
9796689, 2653755, 775574, 1939295]
% Natural frequency:
frequency (2) = [3.0261, 11.5258, 12.3257, 18.9844, 30.9464, 37.5107,
```

```
62.5138, 99.5178]
% nodal displacements:
displacements1 (2) = [76.8703, 90.8934, 16.9916, 90.8934, 103.6964,
37.8343, 14.2362, 35.0287]
displacements2 (2) = [61.6307, 67.3284, 12.4959, 67.3284, 74.2746,
26.4796, 9.7405, 23.5650]

% Result of population 3
% element stress:
sigma1(3) = [5565157, 1272936, 6833699, 2153908, 14145157, 1745037,
0, 10542708, 7626745, 1105362]
sigma2(3) = [5565157, 1272936, 6833699, 2153908, 14145157, 1745037,
0, 10542708, 7626745, 1105362]
% Euler buckling stress:
buckling_stress (3) = [5565157, 1272936, 6833699, 2153908, 14145157,
1745037, 0, 10542708, 7626745, 1105362]
% Natural frequency:
frequency (3) = [9.2125, 15.2350, 24.8650, 43.5574, 54.7471, 65.4958,
74.2018, 78.0000]
% nodal displacements:
displacements1 (3) = [0.7420, 9.3601, 1.1911, 8.8641, 1.5743, 4.7812,
0.9056, 9.6663]
displacements2 (3) = [0.1536, 4.6254, 1.0764, 5.0935, 0.9346, 2.5133,
0.9056, 3.7706]

% Result of population 4
% element stress:
sigma1(4) = [4940491, 0, 1759926, 2051758, 4874391, 0, 0, 48086638,
9115432, 0]
sigma2(4) = [3293661, 0, 1759926, 2051758, 0, 0, 0, 24017823, 9115432,
0]
% Euler buckling stress:
buckling_stress (4) = [32136882, 0, 30135949, 25849553, 3925369,
38078390, 0, 746207, 1974493, 9967333]
% Natural frequency:
frequency (4) = [6.0620, 18.7791, 26.4088, 55.8260, 59.6083, 65.1171,
78.0967, 152.3956]
% nodal displacements:
displacements1 (4) = [1.8421, 24.3165, 0.5051, 24.3165, 0.5364,
20.1565, 0.2332, 21.9471]
displacements2 (4) = [3.3600, 14.2095, 0.5051, 14.2095, 0.3771,
10.1070, 0.2332, 10.1070]
```

```
% Result of population 5
% element stress:
sigma1(5) = [6812731, 0, 2563312, 2740833, 4673253, 0, 0, 6964422,
5537983, 0]
sigma2(5) = [4541821, 0, 2563312, 2740833, 0, 0, 0, 2972806, 5537983, 0]
% Euler buckling stress:
buckling_stress (5) = [23820131, 0, 29559479, 27644948, 8356410,
123135280, 0, 6928040, 6268066, 21084140]
% Natural frequency:
frequency (5) = [15.3057, 25.0928, 29.3091, 50.1544, 67.0770, 78.0267,
112.1191, 204.5012]
% nodal displacements:
displacements1 (5) = [0.1421, 4.4324, 0.7029, 4.4324, 0.0015, 2.2672,
0.3397, 3.4688]
displacements2 (5) = [0.7981, 3.3540, 0.7029, 3.3540, 0.2154, 1.0785,
0.3397, 1.0785]
```

Step 8: Check for constraint violations in the penalty function. If there is a violation, assign a penalty as per Eqs. 6.2 and 6.3, otherwise compute the total mass of the truss.

```
% Result of population 1
fit (1) = Not applicable as tuss is invalid
fpenalty (1) = Not applicable as tuss is invalid
fit_fpenalty (1) = 100000000 kg

% Result of population 2
fit (2) = 2038.4078 Kg
fpenalty (2) = 705.7591
fit_fpenalty (2) = 1438624.8067 Kg
% Result of population 3
fit (3) = 3037.5866
fpenalty (3) = 17.1805
fit_fpenalty (3) = 52187.1537 kg
% Result of population 4
fit (4) = 2233.7483
fpenalty (4) = 1763.0743
fit_fpenalty (4) = 3938264.2130 kg
% Result of population 5
fit (5) = 2560.8690
fpenalty (5) = 3.9715
fit_fpenalty (5) = 10170.6043 kg
```

Step 9: Assign functional values and proceed to the optimization algorithm

```
% Obtain functional values
F(X(1)) = fit_fpenalty (1) = 100000000 kg
F(X(2)) = fit_fpenalty (2) = 1438624.8067 Kg
F(X(3)) = fit_fpenalty (3) = 52187.1537 kg
F(X(4)) = fit_fpenalty (4) = 3938264.2130 kg
F(X(5)) = fit_fpenalty (5) = 10170.6043 kg
% Generate new set of variables (i.e., population) within its upper and
lower bounds as per the optimization algorithm rules
% iteration 1
```

- $X(k) = \{A_1, A_2, \ldots, A_{10}\}$
- $X(1)$ = [0.01000, 0.00586, 0.00000, 0.01000, 0.01000, 0.01000, −0.00301, 0.00908, 0.00228, 0.00988, 5.4026, 9.6620, 4.5720]
- $X(2)$ = [0.00891, −0.00355, 0.00491, 0.00582, 0.01000, 0.01000, 0.00023, 0.00296, 0.00678, 0.00141, 13.2120, 7.3312, 4.8331]
- $X(3)$ = [0.00908, −0.00702, 0.00504, 0.00692, 0.00080, 0.00774, −0.00581, 0.00570, 0.00798, 0.01000, 14.0855, 25.4000, 25.4000]
- $X(4)$ = [0.00550, −0.00792, 0.00780, 0.00522, 0.00705, 0.00362, −0.00176, 0.00914, 0.00427, 0.01000, 4.5720, 12.0700, 16.7601]
- $X(5)$ = [0.00345, −0.00140, 0.00347, 0.00653, 0.00415, 0.00731, −0.00459, 0.01000, 0.00815, 0.00878, 4.5720, 12.7977, 4.5720]

```
% functional values after iteration 1
F(X(1)) = fit_fpenalty (1) = 107104366627468 Kg
F(X(2)) =fit_fpenalty (2) = 255405.7751 kg
F(X(3)) =fit_fpenalty (3) = 341840.4126 Kg
F(X(4)) =fit_fpenalty (4) = 1626.3552 Kg
F(X(5)) =fit_fpenalty (5) = 94012.1344 Kg
% End of iteration 1
% Select best solotions for the next iteration as per the algorithm.
F(X(1)) = fit_fpenalty (1) = 1626.3552 Kg
F(X(2)) =fit_fpenalty (2) = 10170.6043 kg
F(X(3)) =fit_fpenalty (3) = 94012.1344 Kg
F(X(4)) =fit_fpenalty (4) = 255405.7751 kg
F(X(5)) =fit_fpenalty (5) = 341840.4126 Kg
```

Step 10: Check the termination criteria. If not satisfied, generate new trusses (i.e., solutions) according to the algorithm and proceed to Step 4.

```
>> FE = 10 % function evaluations
```

Step 11: Output: The best solution.

```
X= [0.00550, -0.00792, 0.00780, 0.00522, 0.00705, 0.00362, -0.00176,
0.00914, 0.00427, 0.01000, 4.5720, 12.0700, 16.7601]
F(X) = 1626.3552 Kg
```

The step-by-step procedure for implementing the 10-bar size, shape, and topology optimization in Matlab is presented below:

Step 1:

```
>>  n1=[2 6 10]; % elements connected with node 1
>>  n2=[4 6 9]; % elements connected with node 2
>>  n3=[1 2 5 8 9]; % elements connected with node 3
>>  n4=[3 4 5 7 10]; % elements connected with node 4
>>  n5=[1 7]; % elements connected with node 5
>>  n6=[3 8]; % elements connected with node 6
```

Step 2:

```
>>  fit= sum(((A).*L))* ro + nodal mass; % Objective function
>>  fpenalty= (1+e1*sum(Constraint violation))^e2; % Penalty function
>>  fit_fpenalty =fit*fpenalty; % Penalized objective function
>>  OPTIONS.popsize = 5; % Total population size
>>  OPTIONS.numVar = 13; % Number of variables
>>  PTIONS.maxFE = 10; % Termination criterion
>>  ll= [-100*ones(1,10)/10000 4.572 4.572 4.572];; % Lower bound
>>  ul= [100*ones(1,10)/10000 25.4 25.4 25.4]; % Upper bound
```

Step 3:

```
>>  for popindex = 1 : OPTIONS.popsize
>>  for k = 1 : OPTIONS.numVar
>>  chrom(k) = (ll(k)+ (ul(k) - ll(k)) * rand);
>>  end
>>  Population(popindex).chrom = chrom;
>>  end
```

Step 4:

```
% generation of coordinates and connectivities
>>  nodeCoordinates=[9.144*2 ξy1;9.144*2 0;9.144 ξy2;9.144 0 ;0 ξy3;0 0];
% nodal Coordinates in m
>>      elementNodes=[3 5;1 3;4 6;2 4;3 4;1 2;4 5;3 6;2 3;1 4]; %
connections at elements
```

```
>>   numberElements=size(elementNodes,1); % number of Elements
>>   numberNodes=size(nodeCoordinates,1); % number of nodes
>>   xx=nodeCoordinates(:,1);
>>   yy=nodeCoordinates(:,2);
>>   E = 69e9; % E; modulus of elasticity in Pa
>>   ro = 2740; % E; material density in kg/m^3
% computation of B_i
>> B=x;
>>   B(B>=1e-4)=1;
>>   B(B<1e-4)=0; % 0 = absent node, 1 = present
```

Step 5:

```
% basic nodes for supports and load
>>   node_topo=[max((B(n1))),max((B(n2))),max((B(n3))),max((B
(n4))), max((B(n5))),max((B(n6)))]; % 0 = absent node, 1 = present
>>   DOF=2*sum(node_topo)-sum(B)-4; % DOF=2n-m-nl, Grubler's
criterion n=node, m=link, nl=4=DOF lost at suppot
>>   if
>>   sum(B(n1))==0||sum(B(n2))==0||sum(B(n3))==0||sum(B(n4))
==0||sum(B(n5)
)==0||sum(B(n6))==0
>>   fit=1e9;
>>   elseif DOF>0.1
>>   fit=1e8; % Go to step 8
>>   else
>>   ...% Go to step 6
>>   end
```

Step 6:

```
% computation of the system stiffness matrix
>>   GDof=2*numberNodes; % GDof: total number of degrees of freedom
% calculation of the system stiffness matrix
>>   stiffness=zeros(GDof);
>>   for e=1:numberElements
% elementDof: element degrees of freedom (Dof)
>>   indice=elementNodes(e,:) ;
>>elementDof=[indice(1)*2-1 indice(1)*2 indice(2)*2-1 indice(2)*2] ;
>>   xa=xx(indice(2))-xx(indice(1));
>>   ya=yy(indice(2))-yy(indice(1));
>>   L(e)=sqrt(xa*xa+ya*ya); % L: length of bar
>>   C=xa/L(e);
>>   S=ya/L(e);
```

```
>>  k1=E*(B(e)*A(e))/L(e)*[C*C C*S -C*C -C*S; ...
C*S S*S -C*S -S*S;-C*C -C*S C*C C*S;-C*S -S*S C*S S*S];
>>  stiffness(elementDof,elementDof)= ... stiffness(elementDof,
elementDof)+k1;
>>  end

% mass : mass matrix
>>  mass=zeros(GDof);
>>  for e=1:numberElements
% elementDof: element degrees of freedom (Dof)
>>  indice=elementNodes(e,:) ;
>>  elementDof=[indice(1)*2-1 indice(1)*2 indice(2)*2-1 indice(2)
*2];
>>  xa=xx(indice(2))-xx(indice(1));
>>  ya=yy(indice(2))-yy(indice(1));
>>  L(e)=sqrt(xa*xa+ya*ya); % L: length of bar
>>  m1=(ro*(A(e))*L(e)/6)*[2 0 1 0;0 2 0 1;1 0 2 0;0 1 0 2];
>>  mass(elementDof,elementDof)= ... mass(elementDof,elementDof)
+m1;
>>  end
% force : force vector
>>  force=zeros(GDof,1); % force vector
% applied load Condition 1
>>  force 1 (4)=-44.537e3; % in N; Condition 1
>>  force 1 (8)=-44.537e3; % in N; Condition 1
% applied load Condition 2
>>  force 2 (4)=-44.537e3; % in N; Condition 2

% displacements: displacement vector
>>  prescribedDof=(9:12); % boundary conditions; Restricted DOF
>>  activeDof=setdiff([1:GDof]', [prescribedDof]); % active DOF
>>  U1=stiffness(activeDof,activeDof)\force1(activeDof);
>>  displacements1=zeros(GDof,1);
>>  displacements1(activeDof)=U1; % global displacements vector1
>>  U2=stiffness(activeDof,activeDof)\force2(activeDof);
>>  displacements2=zeros(GDof,1);
>>  displacements2(activeDof)=U2; % global displacements vector2

% stress: stress vector
>>  for e=1:numberElements
>>  indice=elementNodes(e,:);
>>elementDof=[indice(1)*2-1 indice(1)*2 indice(2)*2-1 indice(2)*2] ;
>>  xa=xx(indice(2))-xx(indice(1));
>>  ya=yy(indice(2))-yy(indice(1));
>>  L=sqrt(xa*xa+ya*ya); % L: length of bar
```

```
>>   C=xa/L;
>>   S=ya/L;
>>   sigma1(e)=E/L*[-C -S C S]*displacements1(elementDof);
>>   sigma2(e)=E/L*[-C -S C S]*displacements2(elementDof);
>>   end

% Euler buckling stress: buckling stress vector
>>   buckling_stress=(4*E*B).*(A./(L.^2));

% Natural frequency: Natural frequency vector
>>   aa=zeros(12,1);aa(3:4)=500;aa(7:8)=500; % lumped masses
>>   mass1=diag(aa); % lumped mass matrix
>>   mass=mass+mass1; % mass matrix
>>   f=eig(stiffness(activeDof,activeDof),mass(activeDof,
activeDof));
>>   f=realsqrt(f);
>>   frequency =f/(2*pi); % Natural frequency vector
```

Step 7:

```
>>   disp('sigma1')
>>   disp('sigma2')
>>   disp('buckling_stress')
>>   disp('frequency')
>>   disp('displacements1')
>>   disp('displacements2')
```

Step 8:

```
>>   fit= sum(((A).*L.*B))* ro + nodal mass; % total mass
>>   fpenalty=(1+e1*sum(Constraint violation))^e2; % penalty function
>>   fit_fpenalty =fit*fpenalty;
```

Step 9:

```
% iteration 1
% Generate new set of size variables (i.e., population) within its upper
and lower bounds as per the optimization algorithm rules
End of iteration 1
% Select best solotions for the next iteration as per the algorithm.
```

Step 10:

```
>>   while FE ≤ maxFE
>>   ...Algorithms code
>>   end
```

Step 11: Output: The best solution.

```
>>   disp('X')
>>   disp('fit_fpenalty')
```

In this section, we evaluate the proposed algorithms' performance in optimizing the 10-bar truss with ten continuous size and three continuous shape variables. The results of the TSS optimization of the 10-bar truss with continuous cross-sections for the algorithms in 100 independent runs are presented in Table 6.2.

We identified the best run, which produced the minimum mass value while fulfilling all the stated structural constraints. To assess the performance of the various evolutionary methods statistically, we present the mean and standard deviation (SD) values of structural mass.

Among the evaluated algorithms, the ALO, DA, GWO, MVO, SCA, WOA, HTS, TLBO, PVS, and SOS algorithms produced the best results, with mass values of 327.9360, 322.3126, 251.9777, 255.1631, 341.6681, 520.0671, 251.5316, 250.6464, 250.7881, and 269.8056 kg, respectively.

It is worth noting that the TLBO algorithm resulted in the least structural mass, demonstrating its efficiency in optimizing the 10-bar truss. Overall, the results of this evaluation demonstrate that the proposed algorithms can effectively optimize the 10-bar truss, providing valuable insights for future TSS optimization studies.

This section presents the evaluation of the proposed algorithms on optimizing the 10-bar truss with 10 continuous size and three continuous shape variables. The obtained results in TSS optimization of the 10-bar truss with continuous cross-sections for the algorithms in 100 independent runs are shown in Table 6.2. The best run was identified, which provided the minimum mass value while fulfilling all the stated structural constraints. The mean and standard deviation (SD) values of structural mass are presented to assess the performance of the various evolutionary methods statistically. The ALO, DA, GWO, MVO, SCA, WOA, HTS, TLBO, PVS, and SOS algorithms produced the best results with mass values of 327.9360, 322.3126, 251.9777, 255.1631, 341.6681, 520.0671, 251.5316, 250.6464, 250.7881, and 269.8056 kg, respectively. It is evident that the TLBO algorithm resulted in the least structural mass.

To evaluate the performance of the metaheuristic methods, the average mass and its standard deviation were taken into consideration. Based on the obtained results, it can be concluded that the ALO, DA, GWO, MVO, SCA, WOA, HTS, TLBO, PVS,

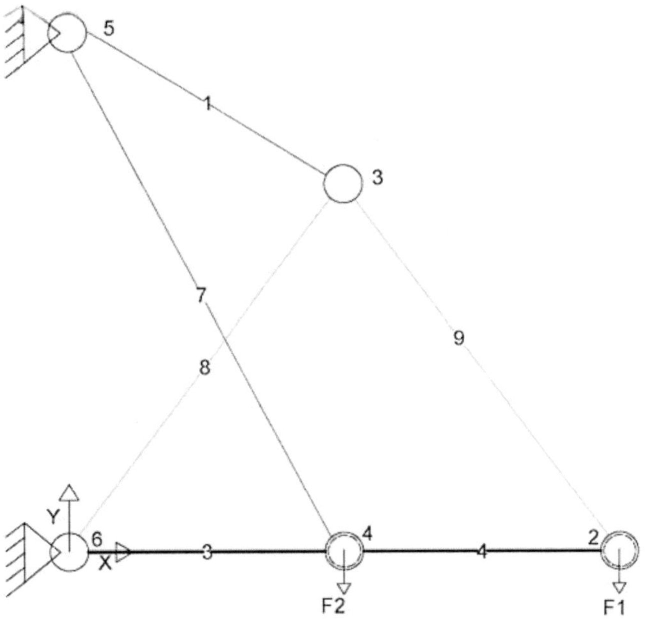

Fig. 6.2 Optimal structure of the 10-bar truss

and SOS algorithms achieved mean mass values of 508.4372, 1036.3612, 288.8143, 331.4609, 467.6564, 984.9895, 330.3556, 331.6600, 293.3991, and 352.5748 kg, respectively, with standard deviations of 135.5049, 2841.2123, 35.2280, 90.8912, 81.7379, 300.9220, 98.7174, 97.6583, 23.3491, and 45.2386, respectively. The GWO algorithm performed the best in achieving the lowest mean mass value, while the PVS algorithm resulted in the best standard deviation value. Therefore, for this specific design problem, the GWO and PVS algorithms were identified as the better algorithms in terms of both convergence rate and consistency.

The results of the study reveal that the TSS optimization approach leads to significantly lighter trusses compared to topology optimization and size optimization approaches. Specifically, the best solutions obtained through TSS optimization resulted in a truss that was 54.8% lighter compared to the truss obtained through size optimization and 40.8% lighter compared to the truss obtained through topology optimization.

In Fig. 6.2, the optimal structure of the 10-bar truss is depicted. Notably, upon observation, it becomes apparent that this optimal configuration requires only six out of the ten available elements. This finding highlights the efficiency and effectiveness of the selected design, as it achieves the desired structural integrity while utilizing a reduced number of components.

This significant improvement in weight reduction demonstrates the effectiveness of the TSS approach in optimizing truss designs for minimum mass, while ensuring structural performance and practical manufacturing constraints. By simultaneously considering the TSS optimization, the TSS approach can produce more efficient truss designs with a reduced weight compared to other optimization approaches.

These findings have important implications for engineering design, as the reduction in weight of a truss can lead to significant cost savings in manufacturing, transportation, and installation. Therefore, TSS optimization can be a valuable tool for engineers and designers seeking to create optimized truss designs that are both structurally sound and cost-effective.

6.3.2 *TSS Optimization of the 14-Bar Truss with Continuous Cross-Sections*

Figure 6.3 illustrates the ground structure of the second benchmark truss, known as the 14-bar truss, along with the load and boundary conditions. This truss problem has been optimized in previous studies such as Tang et al. (2005), Rahami et al. (2008), and Miguel et al. (2013) for TSS (topology, shape, and size) optimization using 14 continuous cross-sectional areas and eight continuous shape variables under static constraints.

However, this problem has not been addressed with the simultaneous consideration of static and dynamic constraints. The optimization process in previous studies focused solely on static constraints.

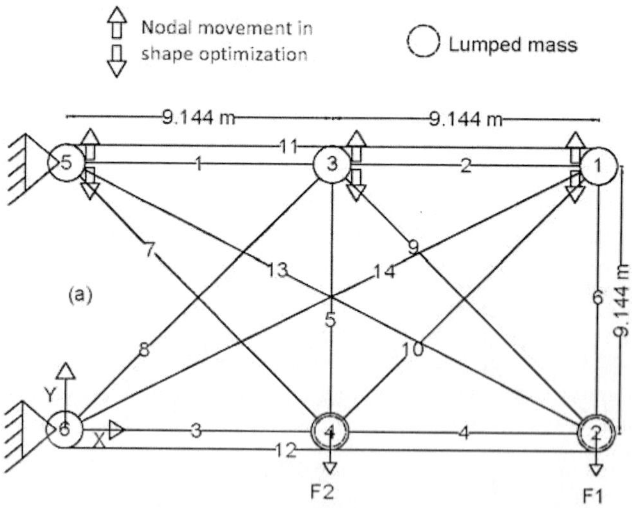

Fig. 6.3 The 14-bar truss

Table 6.3 provides detailed information about the design variables, load conditions, design constraints, and other relevant data specific to the test problem.

This section evaluates the effectiveness of the proposed algorithms in optimizing the 14-bar truss, with 14 continuous size and three continuous shape variables. The results obtained from 100 independent runs of TSS optimization of the 14-bar truss with continuous cross-sections for the algorithms are presented in Table 6.4. The best run is determined by finding the minimum mass value while ensuring that all stated structural constraints are met. The mean and SD values of structural mass are calculated to statistically evaluate the performance of the various evolutionary methods. The ALO, DA, GWO, MVO, SCA, WOA, HTS, TLBO, PVS, and SOS algorithms produced the best results, with a mass mean value of 266.7873, 304.9099, 241.5124, 241.9070, 257.8501, 369.1658, 241.3917, 241.3306, 241.3308, and 241.3399 kg, respectively. It is observed that the TLBO, SOS, and PVS algorithms have the lowest structural mass.

In order to evaluate the performance of metaheuristic methods accurately, we consider their average mass and standard deviation. Based on the available results, we can conclude that the ALO, DA, GWO, MVO, SCA, WOA, HTS, TLBO, PVS, and SOS algorithms have mean mass values of 587.3450, 742.8925, 352.7163, 380.7120, 381.4843, 1238.0447, 390.8655, 377.8990, 298.6982, and 316.1151 kg, respectively. These algorithms also have standard deviations of 219.3285, 268.9675, 108.2662, 116.8030, 101.0490, 510.9640, 116.1907, 145.7468, 77.4458, and 91.5552 kg, respectively.

Based on these findings, it can be concluded that the PVS algorithm performs the best in achieving both the highest mean mass and the lowest standard deviation. Therefore, the PVS algorithm is considered the best algorithm for this design problem, as it has the best convergence rate and consistency.

The results of the study indicate that TSS optimization produces significantly lighter truss designs compared to both size optimization and topology optimization approaches. Specifically, the best solutions obtained through TSS optimization resulted in a truss that was 67.3% lighter compared to the truss obtained through size optimization and 33.6% lighter compared to the truss obtained through topology optimization.

Fig. 6.4 illustrates the ideal arrangement of the 14-bar truss. Upon careful examination of the diagram, it becomes evident that this optimal arrangement necessitates the utilization of merely 4 elements out of the total 14. This discovery emphasizes the efficiency and effectiveness of the chosen design, as it successfully achieves the desired structural integrity while minimizing the number of components required.

The reduction in weight achieved through TSS optimization is particularly noteworthy as it can lead to substantial cost savings in manufacturing, transportation, and installation. By simultaneously considering TSS optimization, the TSS approach can produce more efficient truss designs that are both structurally sound and cost-effective.

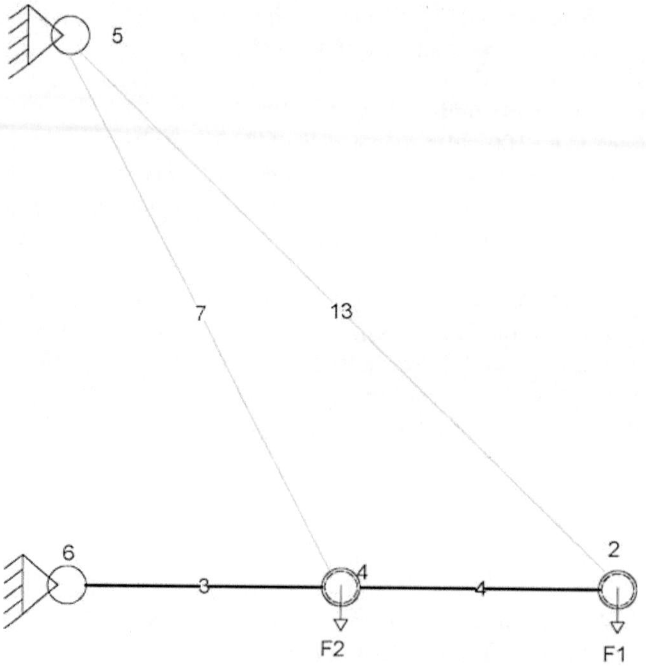

Fig. 6.4 Optimal structure of the 14-bar truss

Overall, these findings highlight the effectiveness of the TSS approach in optimizing truss designs for minimum mass, while accounting for structural performance and practical manufacturing constraints. Thus, TSS optimization can be a valuable tool for engineers and designers looking to develop optimized truss designs that are both lightweight and efficient.

6.3.3 *TSS Optimization of the 15-Bar Truss with Continuous Cross-Sections*

The third benchmark problem involves the exploration of the 15-bar planar truss to examine the impact of both continuous size variables. The ground structure of this truss can be seen in Fig. 6.3, while detailed information about the design variables, including continuous size variables, as well as other relevant data, is provided in Table 6.5.

This problem is based on previous works by Rahami et al. (2008) and Ahrari et al. (2014). Earlier studies focused on investigating this problem under static constraints such as stress, buckling, and displacement. However, the simultaneous consideration of both static and dynamic constraints, particularly natural frequency constraints, has not been examined.

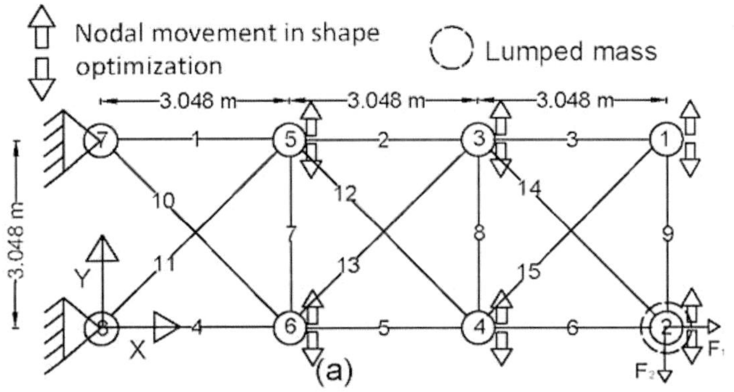

Fig. 6.5 The 15-bar truss

To address this gap, the problem is modified in this study to incorporate the effects of multi-load conditions, lumped masses, natural frequency constraints, and continuous cross-sectional areas. Design considerations and material properties are also provided in the table for a comprehensive understanding of the problem (Fig. 6.5).

Table 6.6 presents the results obtained from the TSS optimization of a 15-bar truss with 15 continuous cross-sections and eight shape variables using various evolutionary algorithms in 100 independent runs. The best run is identified as the one that achieves the minimum mass value while satisfying all the stated structural constraints. The mean and standard deviation (SD) values of the structural mass are provided to assess the performance of the different evolutionary methods statistically.

The results reveal that the ALO, DA, GWO, MVO, SCA, WOA, HTS, TLBO, PVS, and SOS algorithms produce the best results with minimum mass values of 101.2324, 121.2250, 87.4399, 88.4479, 128.3670, 134.0853, 91.2521, 87.8468, 86.0658, and 89.9069 kg, respectively. From the results, it is evident that the PVS algorithm outperforms the others, giving the least structural mass, with GWO and TLBO algorithms being the second and third-best, respectively.

In order to evaluate the effectiveness of the metaheuristic methods, we consider the average mass and its standard deviation (SD). Based on the available results, we can conclude that the ALO, DA, GWO, MVO, SCA, WOA, HTS, TLBO, PVS, and SOS algorithms yield mean mass values of 142.3164, 2.00E+07, 4.00E+06, 2.40E +07, 163.7009, 240.5608, 123.6173, 106.6641, 96.5587, and 100.7155 kg, respectively. The corresponding SD values are 26.527, 4.02E+07, 1.97E+07, 4.29E+07, 16.463, 62.150, 26.721, 18.861, 12.077, and 9.011 kg, respectively.

Based on the results, we can observe that the PVS algorithm performs the best in terms of achieving the minimum mean mass value, while the SOS algorithm outperforms the others in terms of achieving the minimum SD value.

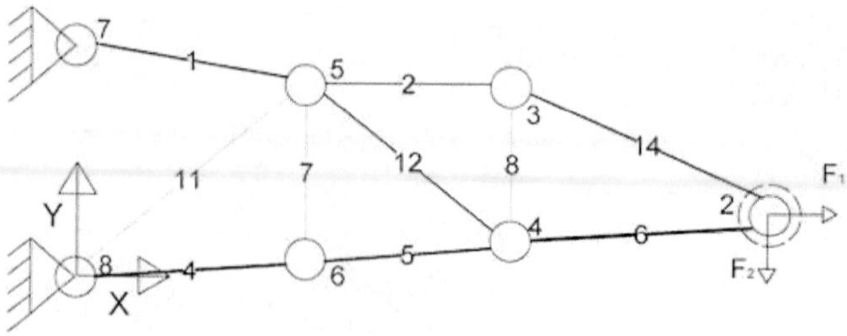

Fig. 6.6 Optimal structure of the 15-bar truss

The optimal arrangement of the 15-bar truss is visually represented in Fig. 6.6. A close examination of the diagram reveals that this ideal configuration only requires the utilization of 10 elements out of the total 15 available. This significant finding highlights the efficiency and effectiveness of the chosen design, as it successfully accomplishes the desired level of structural integrity while minimizing the overall number of required components.

Furthermore, the TSS optimization approach yields a truss that is 23.6% lighter than that obtained through size optimization and 8.1% lighter than that obtained through topology optimization for the best solutions in each case. These results clearly indicate that the TSS optimization approach outperforms the topology optimization and traditional size optimization approaches in designing lighter trusses.

6.3.4 *TSS Optimization of the 24-Bar Truss with Continuous Cross-Sections*

Figure 6.4 displays the ground structure of the fourth benchmark problem, which is the 24-bar truss. This truss problem was initially investigated by using a Branch and Bound algorithm for topology optimization, with a focus on static constraints. However, this study aims to investigate the truss with continuous design variables as well.

The truss is subjected to two external load cases, as specified in Table 6.7. In this problem, the cross-sectional areas of the 24 bars are treated as continuous size variables, while the horizontal movement of nodes 7 and 8 is considered as the shape variables. The table also provides information on nodal masses, material properties, and other required data for a comprehensive understanding of the problem. Additionally, a nonstructural lumped mass of 500 kg is applied at node 3 of the truss (Fig. 6.7).

In this section, we present the optimization results for TSS of the 24-bar truss with 24 continuous cross-sections and two shape variables. The algorithms were run for 100 independent runs and the results are reported in Table 6.8. The best run, which

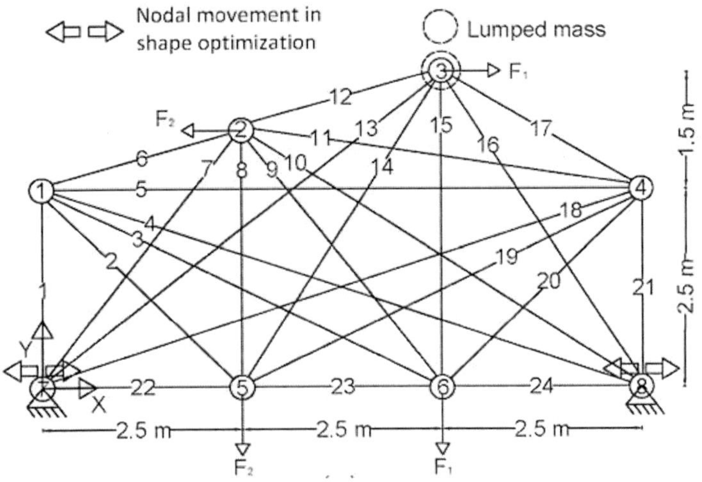

Fig. 6.7 The 24-bar truss

meets all design constraints and has the least mass value, is also shown in the table. Moreover, the mean and standard deviation (SD) of the structural mass are provided and will be used for statistical comparison of the various metaheuristics.

The best results obtained by ALO, DA, GWO, MVO, SCA, WOA, HTS, TLBO, PVS, and SOS algorithms are 121.2085, 166.0446, 109.9633, 109.6253, 137.3382, 227.7992, 121.5718, 113.5510, 108.9194, and 112.0321 kg, respectively. It is worth noting that the PVS algorithm yields the least mass value, while the MVO and GWO algorithms are the second and third best, respectively. However, the difference between the two best mass values is insignificant.

In order to assess the true performance of the metaheuristics, the average mass and its standard deviation will be used. The results table shows that the ALO, DA, GWO, MVO, SCA, WOA, HTS, TLBO, PVS, and SOS algorithms have a mean mass value of 191.7292, 279.9674, 130.6985, 138.4959, 188.2545, 379.5975, 181.7925, 155.8492, 125.3713, and 133.1890 kg, respectively, with the corresponding standard deviations of 43.5056, 67.6708, 22.8625, 28.0486, 22.3881, 86.4417, 45.5741, 34.0201, 13.0455, and 14.7211.

Based on the available results, it can be concluded that the PVS algorithm, with a mean mass value of 125.3713 kg and a standard deviation of 13.0455, is significantly better than the other approaches. Therefore, the PVS algorithm is the better algorithm for both convergence rate and consistency for this design problem.

Figure 6.8 visually presents the optimal arrangement of the 24-bar truss. Upon careful observation of the diagram, it becomes apparent that this ideal configuration necessitates the utilization of just 8 elements out of the total available 24. This

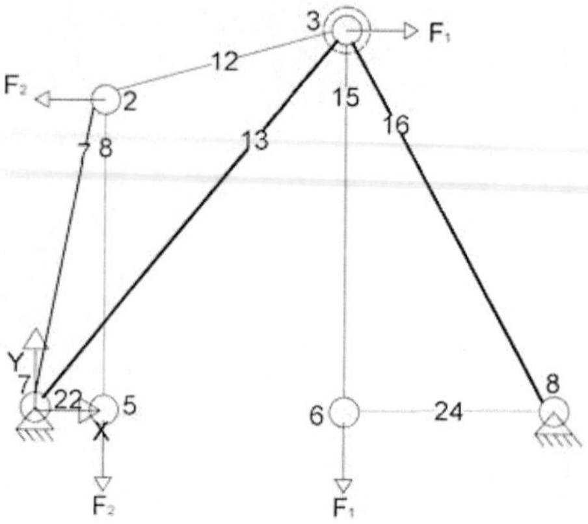

Fig. 6.8 Optimal structure of the 24-bar truss

notable discovery underscores the efficiency and effectiveness of the selected design, as it effectively achieves the desired level of structural integrity while minimizing the overall number of components required.

Additionally, the consideration of simultaneously TSS optimization results in a 45.2 and 8.6% reduction in the weight of the truss compared to the best solutions obtained through size optimization and topology optimization, respectively. This indicates that TSS optimization performs significantly better than topology optimization and size optimization in designing lighter trusses.

6.3.5 TSS Optimization of the 20-Bar Truss with Continuous Cross-Sections

Figure 6.9 illustrates the ground structure of the 20-bar truss for the benchmark problem. Initially, Xu et al. (2003) formulated this problem for topology and size optimization, which was later tested by Kaveh and Zolghadr (2013) for continuous optimization.

Table 6.9 provides comprehensive details regarding the design variables, loading conditions, structural safety constraints, and other relevant data specific to the test problem. It is worth noting that no lumped masses are added to the truss for this particular problem.

In this section, proposed algorithms for optimizing the 20-bar truss with 20 size and six shape variables are examined. The results obtained from TSS optimization of the truss with continuous cross-sections using these algorithms in 100 independent runs are presented in Table 6.10. The best run achieved the minimum mass value while satisfying all of the specified structural constraints. The mean and standard deviation values of the structural mass are provided, which will be utilized to statistically evaluate the performance of the different evolutionary methods. The ALO, DA, GWO, MVO, SCA, WOA, HTS, TLBO, PVS, and SOS algorithms

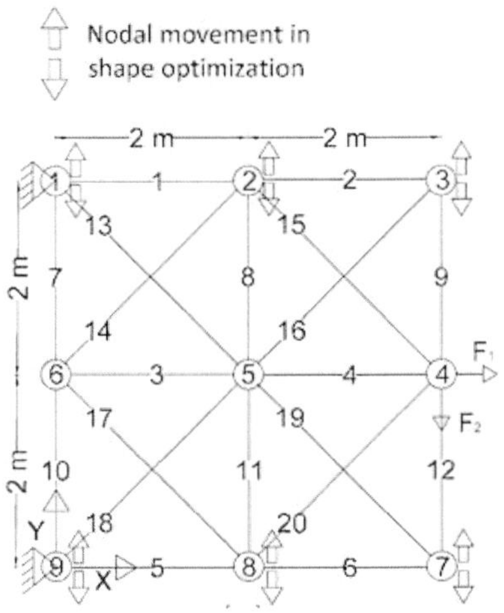

Fig. 6.9 The 20-bar truss

yielded the best results of 147.7196, 198.1419, 114.4491, 111.3688, 184.7497, 202.6807, 125.3653, 115.9569, 113.3089, and 118.7586 kg, respectively. It can be observed that the MVO algorithm produced the least structural mass, with the PVS and GWO algorithms closely following in second and third place, respectively.

To evaluate the true performance of the metaheuristic methods, the average mass and its standard deviation are taken into consideration. Based on the available results, it can be concluded that the ALO, DA, GWO, MVO, SCA, WOA, HTS, TLBO, PVS, and SOS algorithms have mean mass values of 247.6606, 2.01E+06, 1.00E+06, 2.10E+07, 251.5663, 421.7790, 2784.7374, 378.5922, 144.8406, and 153.8023 kg, respectively, with the corresponding standard deviations of 45.62379, 1.41E+07, 1.00E+07, 4.09E+07, 37.40319, 117.9694, 4417.197, 1093.809, 18.43191, and 22.81974. The PVS algorithm performs the best, achieving the best mean mass and standard deviation values. Therefore, the PVS and MVO algorithms are the actual better algorithms for both convergence rate and consistency for this design problem.

The optimal arrangement of the 20-bar truss is depicted in Fig. 6.10, providing a visual representation of the ideal configuration. Upon a meticulous examination of the diagram, it becomes evident that this optimal arrangement requires the utilization of only 8 elements out of the total 20 available. This significant discovery emphasizes the efficiency and effectiveness of the chosen design, as it successfully achieves the desired level of structural integrity while minimizing the overall number of components required.

In addition, when the TSS optimization is considered simultaneously, it results in a reduction of 45.7 and 26.1% in the weight of the truss compared to topology optimization and size optimization, respectively, for the best solutions in this case. Therefore, TSS optimization is significantly more effective than TS optimization in designing lighter trusses.

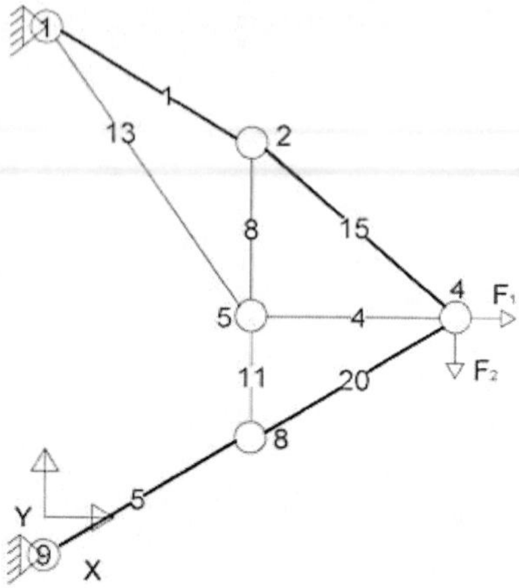

Fig. 6.10 Optimal structure of the 20-bar truss

6.3.6 TSS Optimization of the 72-Bar 3D Truss with Continuous Cross-Sections

The ground structure of the sixth test problem, the 72-bar truss, is depicted in Fig. 6.11. This truss is a 3D structure that was originally used by Kaveh and Zolghadr (2013) for continuous optimization. In this study, the problem is further investigated for continuous optimization.

Table 6.11 provides comprehensive information regarding material properties, design variables, loading conditions, lumped mass, and other relevant data specific to the test problem. This table includes all the necessary details to understand and analyze the truss optimization problem.

The elements are clustered into 16 groups (i.e., G_1 (A_1–A_4), G_2 (A_5–A_{12}), G_3 (A_{13}–A_{16}), G_4 (A_{17}–A_{18}), G_5 (A_{19}–A_{22}), G_6 (A_{23}–A_{30}), G_7 (A_{31}–A_{34}), G_8 (A_{35}–A_{36}), G_9 (A_{37}–A_{40}), G_{10} (A_{41}–A_{48}), G_{11} (A_{49}–A_{52}), G_{12} (A_{53}–A_{54}), G_{13} (A_{55}–A_{58}), G_{14} (A_{59}–A_{66}), G_{15} (A_{67}–A_{70}), and G_{16} (A_{71}–A_{72})) by considering structural symmetry as per Kaveh and Zolghadr (2013). The truss has four nonstructural lumped masses of 2270 kg added to all the four top nodes (node numbers 1–4).

Table 6.12 presents the results obtained from TSS optimization of the 72-bar truss with continuous cross-sections using various algorithms in 100 independent runs. The best run achieved the minimum mass value while satisfying all of the specified structural constraints. The mean and standard deviation values of the structural mass are provided and will be utilized to statistically evaluate the performance of the different evolutionary methods. The ALO, DA, GWO, MVO, SCA, WOA, HTS, TLBO, PVS, and SOS algorithms yielded the best results of 463.4872, 567.9175, 439.5694, 462.4715, 507.3739, 638.5970, 443.3485, 435.4902, 439.6855, and 444.2760 kg, respectively. It can be observed that the TLBO algorithm produced the least structural mass.

Fig. 6.11 The 72-bar 3D truss

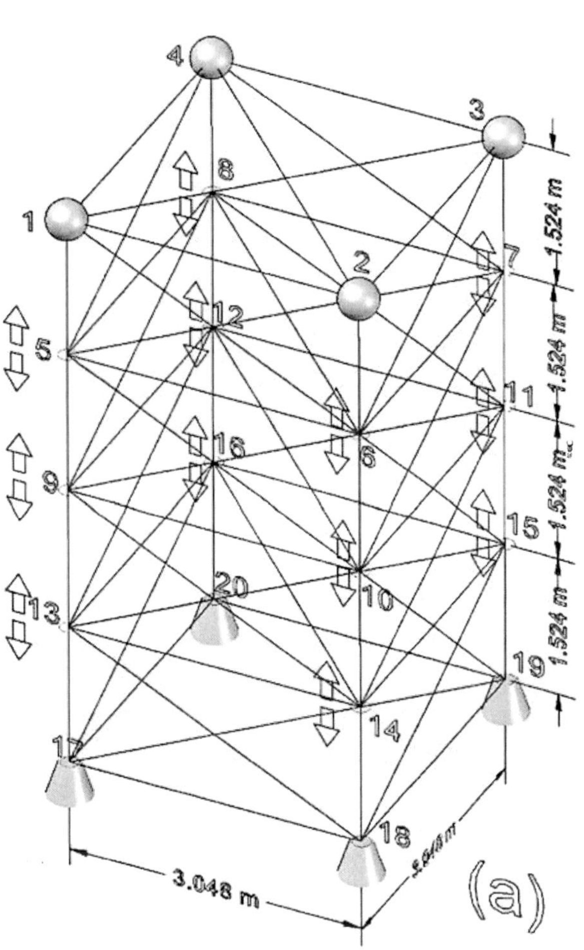

Table 6.11 Design parameters of the 72-bar truss

Size variables: G_i, $i = 1, 2, \ldots, 16$; $G =$ Group number Shape variables: $-1\ \text{m} \le \xi_{y_5}, \xi_{y_9}, \xi_{y_{13}} \le 1\ \text{m}$, with respect to their original coordinates				
Multiple load conditions:				
Condition 1:	$F_{1x} = 22.25$ kN	$F_{1y} = 22.25$ kN		$F_{1z} = -22.25$ kN
Condition 2:	$F_{1z} = -22.25$ kN	$F_{2z} = -22.25$ kN	$F_{3z} = -22.25$ kN	$F_{4z} = -22.25$ kN
Lumped masses at nodes 1, 2, 3, and 4 are 2270 kg				
Stress and displacement constraints: $\sigma_i^{\max} = 172.375$ MPa and $\delta_{x_j/y_j}^{\max} = 6.35$ mm $\left(6.35 \times 10^{-3}\ \text{m}\right)$ (for node, $j = 1, 2, 3$, and 4 along x- and y-axes)				
Natural frequency constraints: $f_1 \ge 4$ Hz and $f_3 \ge 6$ Hz				
The continuous cross-sectional areas: $[A^{\min}, A^{\max}] = [-30, 30]$ cm^2($[-30, 30] \times 10^{-4}$ m^2) and critical area $= 1$ cm^2				
Material properties: $E = 68.95$ GPa and $\rho = 2767.99$kg/m^3				

Table 6.12 Topology, shape, and size optimization of the 72-bar 3D truss (continuous section)

Element group no.	ALO	DA	GWO	MVO	SCA	WOA	HTS	TLBO	PVS	SOS
1	6.1014	10.8690	3.4047	4.6629	8.6161	4.0795	6.1127	5.0990	4.2059	7.7409
2	8.9284	8.3506	9.5811	7.9269	8.7320	21.2013	9.9863	10.2154	10.1325	10.2308
3	—	—	—	4.9010	4.9972	—	—	—	—	—
4	9.5836	9.2632	—	8.2583	13.8123	14.3463	—	—	—	—
5	7.6958	38.6901	7.2051	8.0054	12.9789	7.9498	8.2284	9.7830	7.9684	10.2607
6	8.4222	8.2630	8.2046	7.1293	7.9706	11.3178	8.5155	7.1585	7.0028	8.0919
7	—	3.6546	2.5134	—	—	—	—	—	2.5745	—
8	—	—	4.8290	—	—	—	3.1332	3.4278	5.0462	3.9232
9	13.1806	24.5455	12.4544	10.4671	17.8471	6.2547	12.9122	14.5708	12.0646	11.6126
10	7.9991	8.1146	7.6761	9.2415	6.7603	13.4284	7.7691	7.5565	7.5511	7.6575
11	4.2580	6.7882	—	—	—	—	—	—	—	—
13	17.5589	14.9562	18.0221	17.5918	13.0643	23.4268	14.2587	13.8273	16.8478	13.9811
14	8.0474	8.0970	7.5098	8.9129	11.9146	5.4515	8.8644	7.9806	8.1840	7.5450
15	—	—	—	—	—	12.5492	—	1.6610	—	1.8322
y_{13}	−0.4935	0.0842	−0.0556	0.1297	0.0152	0.1633	0.0958	0.1908	0.1325	0.4686
y_9	0.0033	−0.0757	0.0188	0.7753	0.3790	−0.4883	0.0490	0.0341	0.1298	0.3321
y_5	−0.4852	0.3741	0.3675	0.0142	−0.0623	0.1309	0.1429	0.0378	0.0768	0.0614
Mass (kg)	463.4872	567.9175	439.5694	462.4715	507.3739	638.5970	443.3485	435.4902	439.6855	444.2760
σ^{max} (MPa)	56.3198	31.9232	65.3505	43.6274	25.0203	54.4914	104.8607	100.8942	65.6660	94.8008
δ^{max} (mm)	4.2472	3.4649	4.6671	3.2614	3.1426	3.5143	6.0644	6.0993	4.8955	5.9278
f_1 (Hz)	4.0000	4.3738	4.0001	4.0097	4.1088	4.0000	4.0018	4.0002	4.0008	4.0064
f_3 (Hz)	6.0000	6.0000	6.0000	6.0014	6.0908	6.5180	6.0010	6.0029	6.0003	6.0216
Mean	730.6399	1.00E+06	6.00E+06	1.50E+07	652.5153	1434.7452	603.1818	482.8751	460.7335	467.3752
SD	211.2282	1.00E+07	2.39E+07	3.59E+07	70.5941	558.5113	274.5793	92.7977	39.1570	23.3611

After analyzing the average mass and standard deviation values of the different algorithms used to optimize the 72-bar truss with continuous cross-sections, it can be concluded that the ALO, DA, GWO, MVO, SCA, WOA, HTS, TLBO, PVS, and SOS algorithms have a mass mean value of 730.6399, 1.00E+06, 6.00E+06, 1.50E +07, 652.5153, 1434.7452, 603.1818, 482.8751, 460.7335, and 211.2282 kg, respectively, with the corresponding SD values of 1.00E+07, 2.39E+07, 3.59E +07, 70.5941, 558.5113, 274.5793, 92.7977, 39.1570, and 23.3611. It can be observed that the PVS algorithm performs best in terms of the mass mean value, while the SOS algorithm performs best in terms of the standard deviation value. Thus, the PVS and SOS algorithms are the actual best algorithms for achieving both convergence rate and consistency for this design problem.

The optimal arrangement of the 72-bar 3D truss is depicted in Fig. 6.12, providing a visual representation of the ideal configuration. Upon a meticulous examination of the diagram, it becomes evident that this optimal arrangement requires the utilization

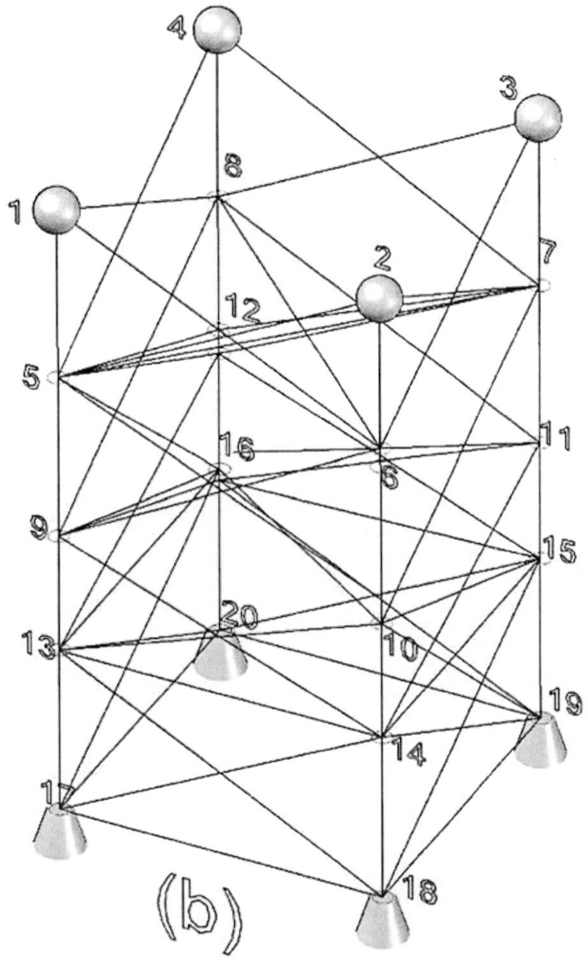

Fig. 6.12 Optimal structure of the 72-bar 3D truss

of only 10 groups of elements out of the total 15 available. This significant discovery emphasizes the efficiency and effectiveness of the chosen design, as it successfully achieves the desired level of structural integrity while minimizing the overall number of components required.

Figure 6.12 visually illustrates the optimal arrangement of the 72-bar 3D truss, offering a clear representation of the ideal configuration. Upon careful scrutiny of the diagram, it becomes apparent that this optimal arrangement necessitates the utilization of only 10 groups of elements out of the total 16 available. This noteworthy finding highlights the efficiency and effectiveness of the chosen design, as it successfully achieves the desired level of structural integrity while minimizing the overall number of components required.

The TSS optimization approach yields a 7.5 and 1.5% reduction in weight compared to size optimization and topology optimization, respectively, for the best solutions in each case. This indicates that TSS optimization is significantly more effective than size optimization and topology optimization in achieving lighter truss designs.

6.3.7 *TSS Optimization of the 39-Bar Truss with Continuous Cross-Sections*

The seventh benchmark problem involves the investigation of a two-tier 39-bar planar truss, as shown in Fig. 6.13. The purpose of this problem is to study the behavior of a large-sized truss under stress, buckling, displacement, and fundamental

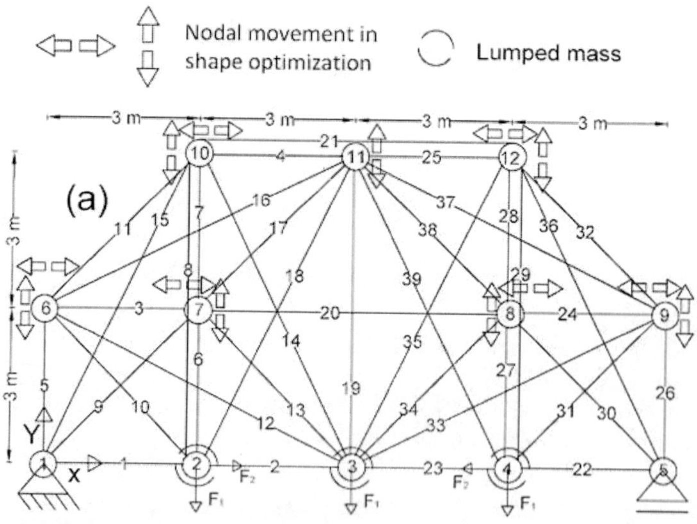

Fig. 6.13 The 39-bar truss

Table 6.13 Design parameters of the 39-bar truss

Size variables: $X_i = G_i$, $i = 1, 2, \ldots, 21$ Shape variables: $-2.5\ \text{m} \leq \xi_{x_j/y_j} \leq 2.5$ m, for $x_6 = -x_9, y_6 = y_9, x_7 = -x_8, y_7 = y_8, x_{10} = -x_{12}$, $y_{10} = y_{12}, y_{11}$ with respect to their original coordinates	
Multiple load conditions:	
Condition 1 : $F_1 = 100$ kN, $F_2 = 0$ kN	Condition 2 : $F_1 = 0$ kN, $F_2 = 50$ kN
Lumped masses on nodes 2, 3, and 4 are 500 kg	
Stress and displacement constraints: $\sigma_i^{\max} = 172.43$ MPa and $\delta_{x_j/y_j}^{\max} = 50$ mm (0.05 m) (for node, $j = 2, 3$, and 4 along x- and y-axes)	
Natural frequency constraints: $f_1 \geq 4$ Hz	
The continuous cross-sectional areas: $[A^{\min}, A^{\max}] = [-50, 50]\ \text{cm}^2$ ($[-50, 50] \times 10^{-4}\ \text{m}^2$) and critical area $= 1\ \text{cm}^2$	
Material properties: $E = 69$ GPa and $\rho = 2740\text{kg/m}^3$	

frequency constraints, considering both continuous design variables. Initial work on this problem was conducted by Deb and Gulati (2001), followed by subsequent studies by researchers such as Luh and Lin (2008), Wu and Tseng (2010), and Ahrari et al. (2014), primarily focusing on static constraints.

However, this problem has not been explored with the simultaneous consideration of both static and dynamic constraints. To address this gap, the problem has been modified in this study to incorporate the effects of multi-load conditions, lumped masses, and natural frequency constraints. Detailed information regarding design variables, multi-load conditions, lumped masses, constraints, and material properties can be found in Table 6.13.

The truss elements are grouped into 21 groups, taking into account the symmetry about the vertical middle plane, similar to the approach used by Deb and Gulati (2001). Furthermore, the truss is subjected to three nonstructural lumped masses of 500 kg each, placed at nodes 2, 3, and 4. To satisfy the multi-load condition, two different loading conditions are considered, as depicted in Fig. 6.13.

The objective of this problem is to optimize the TSS of a two-tire truss using 21 continuous size and seven shape variables. The shape variables for nodes 6–12 are restricted to $-2.5\ \text{m} \leq \xi(x, y) \leq 2.5$ m with respect to their original coordinates, and the following equations are used for shape variables: $x_6 = -x_9, y_6 = -y_9, x_7 = -x_8, y_7 = -y_8, x_{10} = -x_{12}, y_{10} = -y_{12}, y_{11}$. Table 6.14 presents the results obtained from TSS optimization of the 39-bar truss with continuous cross-sections for various algorithms in 100 independent runs. The best run was chosen based on the minimum mass value while satisfying all the structural constraints. The mean and standard deviation (SD) values of the structural mass are provided to statistically evaluate the performance of different evolutionary methods. The results indicate that the ALO, DA, GWO, MVO, SCA, WOA, HTS, TLBO, PVS, and SOS algorithms produced optimum weight designs of 227.1870, 356.1997, 199.2802, 189.2037, 294.2668, 482.7424, 209.1827, 193.7194, 186.7655, and 196.4616 kg, respectively. The PVS algorithm exhibited better results than other proposed algorithms without violating any constraints.

Table 6.14 Topology, shape, and size optimization of the 39-bar truss (continuous section)

Element no.	ALO	DA	GWO	MVO	SCA	WOA	HTS	TLBO	PVS	SOS
1	4.2491	7.0471	3.3022	6.6025	8.6892	17.4432	10.8598	6.8333	5.0393	6.0072
2	28.3355	16.2137	12.8184	12.7715	14.0414	13.5743	15.1313	12.7826	12.9448	13.2294
3	—	6.8422	17.3050	20.6213	—	—	25.2153	20.7936	—	—
4	—	—	—	—	—	—	—	—	19.8611	—
5	25.0591				—	26.5937	26.0482	21.4623	20.4590	22.4272
6	—	—	—	—	5.8774	—	2.7291	1.1643	—	—
7	—	3.3692	—	—	—	1.9389	—	—	—	—
8	—	—	—	—	—	2.8715	—	—	2.7727	1.1746
9	—	—	1.0027	1.0042	37.5528	—	—	—	—	—
10	7.0108	6.9734	7.4667	5.8235	—	6.8517	7.6584	6.1340	4.1790	6.2191
11	22.4959	—	—	—	—	—	—	—	19.6200	25.6989
12	1.0000	—	—	—	—	6.9082	—	—	—	—
13	—	10.8118	—	—	4.4155	33.8798	4.5780	3.5718	—	—
14	3.1223	—	—	—	—	4.1890	—	—	1.0493	3.5604
15	—	1.0000	—	—	—	—	—	—	—	—
16	—	50.0000	—	—	—	48.4276	—	—	—	—
17	—	31.2468	25.9338	22.9187	36.3248	—	—	—	—	—
18	—	—	1.6665	1.6580	15.8859	5.7471	—	—	—	—
19	—	7.9947	5.9010	6.5921	—		—	—	4.7285	—
20	—	—	2.8441	—	—	—	31.8586	35.7093	—	—
21	29.6409	—	—	—	—	—	—	—	—	25.3965
x_6	1.2549	1.4300	0.7976	1.4617	−0.6675	−0.4623	1.8091	1.5840	1.2851	1.4783
y_6	−0.4308	−0.5932	−0.8461	−1.0906	0.0082	0.3699	−1.2293	−0.9540	−0.7694	−0.6737
x_7	−1.2074	1.7325	−0.1162	0.6809	−0.0174	−0.0283	1.3443	0.9438	0.3418	2.0472
y_7	−2.3706	−1.2274	0.0345	−0.1472	0.0691	−2.0273	−0.6098	−0.1104	−2.5000	−1.8310
x_{10}	1.1243	−1.3951	−0.0128	0.4221	0.0001	0.5319	0.5258	−2.2067	0.3520	1.4436

y_{10}	−2.4973	0.4866	−0.1181	1.8743	−1.8955	−2.4773	−1.8126	−0.8520	−2.2333	−2.4994
y_{11}	−1.2662	−2.5000	−1.7112	−2.1895	−1.8394	−0.6073	0.9752	−1.8843	−1.4822	−1.3536
Mass (kg)	227.1870	356.1997	199.2802	189.2037	294.2668	482.7424	209.1827	193.7194	186.7655	196.4616
σ^{max} (MPa)	172.4298	171.2143	169.6981	171.4139	167.7623	156.5321	152.1072	172.4206	171.4927	171.3046
δ^{max} (mm)	22.8734	21.6728	31.4097	33.9297	34.6280	40.8035	41.7354	37.6389	29.7428	30.9583
f_1 (Hz)	6.9413	10.0874	9.3194	10.1178	11.4579	10.1952	10.5372	10.3610	10.2673	10.6761
Mean	364.9732	578.4476	256.3261	255.4384	445.5555	922.3355	338.6629	312.5435	246.5263	255.6555
SD	76.87001	147.5279	33.59341	46.75581	78.67888	226.5646	75.3663	75.6727	31.35422	32.65645

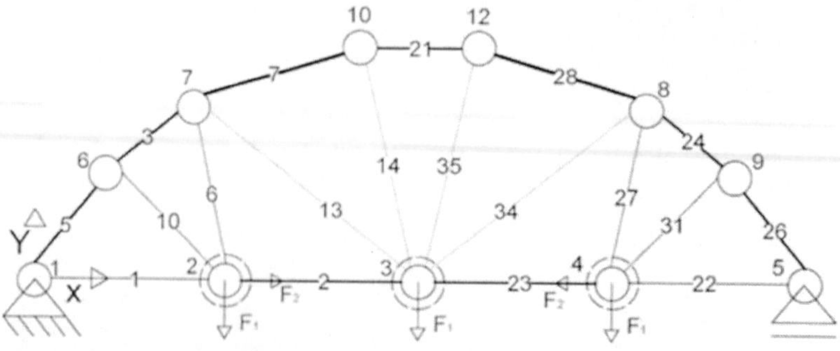

Fig. 6.14 Optimal structure of the 39-bar truss

To evaluate the true performance of the metaheuristic methods, the average mass and its standard deviation (SD) are considered. Based on the available results, it can be concluded that the ALO, DA, GWO, MVO, SCA, WOA, HTS, TLBO, PVS, and SOS algorithms have mean mass values of 364.9732, 578.4476, 256.3261, 255.4384, 445.5555, 922.3355, 338.6629, 312.5435, 246.5263, and 255.6555 kg, respectively, with the corresponding SD values of 76.87001, 147.5279, 33.59341, 46.75581, 78.67888, 226.5646, 75.3663, 75.6727, 31.35422, and 32.65645. The PVS algorithm performs the best in terms of achieving the lowest mean mass and SD values. Therefore, the PVS algorithm is the most suitable algorithm for this design problem, considering both convergence rate and consistency.

Figure 6.14 presents the optimal arrangement of the 39-bar truss, offering a visual depiction of the ideal configuration. Upon a careful examination of the diagram, it becomes apparent that this optimal arrangement requires the utilization of only 19 elements out of the total 39 available. This notable discovery underscores the efficiency and effectiveness of the chosen design, as it successfully achieves the desired level of structural integrity while minimizing the overall number of components required.

Furthermore, the TSS optimization approach results in a truss that is 63.9 and 20.7% lighter compared to the size optimization and topology optimization, respectively, for the best solutions in each case. Therefore, the TSS approach outperforms both size optimization and topology optimization significantly when it comes to designing lighter trusses.

It is important to note that the TSS optimization approach outperforms the traditional methods of size and topology optimization in terms of reducing the weight of trusses. Specifically, the simultaneous consideration of TSS optimization results in significantly lighter trusses compared to size optimization and topology optimization, as demonstrated by the percentage differences in weight reduction in each case. This indicates that the TSS approach can lead to more optimal designs and therefore potentially significant cost savings in material usage and manufacturing.

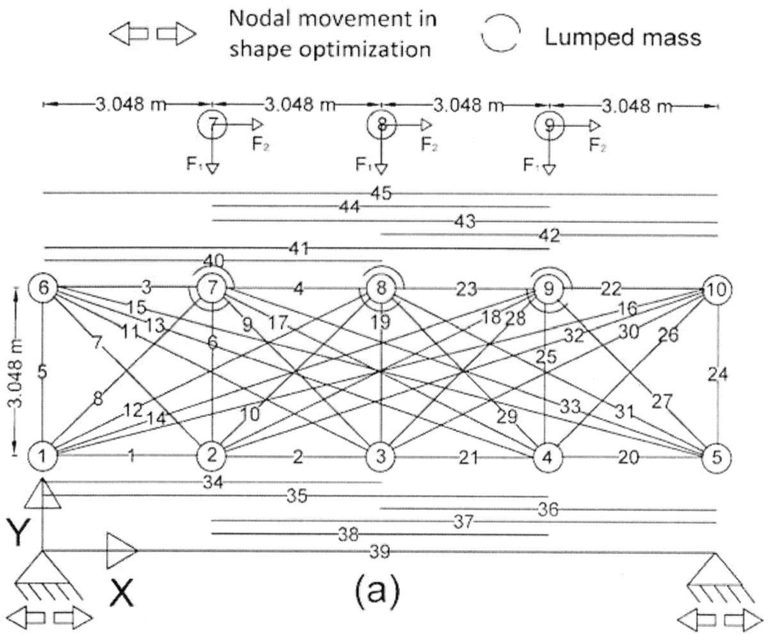

Fig. 6.15 The 45-bar truss

6.3.8 *TSS Optimization of the 45-Bar Truss with Continuous Cross-Sections*

The ninth benchmark problem focuses on the investigation of the 45-bar truss, which represents a complex structure subjected to stress, buckling, displacement, and fundamental frequency constraints, taking into account continuous design variables. The ground structure of this truss, along with multiple loads and the boundary condition, is depicted in Fig. 6.15. The problem was initially introduced by Deb and Gulati (2001) and examined under static constraints. However, it has not been explored with the simultaneous consideration of static and dynamic constraints.

To address this gap, the problem has been modified to incorporate the effects of multi-load conditions, lumped masses, natural frequency constraints, and continuous sections. Detailed design considerations, including design variables, multi-load conditions, lumped masses, constraints, and material properties, are provided in Table 6.15. In this problem, the 28 elements (element numbers 1–14 and 20–33) are grouped into 14 groups based on their symmetry about the vertical middle plane, while the remaining elements are treated as distinct. As a result, the number of size variables is reduced to 31.

Additionally, the truss is subjected to three nonstructural lumped masses of 500 kg each, placed at nodes 7, 8, and 9. To fulfill the multi-load condition, two different loading conditions are considered, as illustrated in Fig. 6.15.

Table 6.15 Design parameters of the 45-bar truss

Size variables: $X_i = G_i$, $i = 1, 2, \ldots, 31$; $G =$ Group number Shape variables: $-2.5\ \text{m} \le \xi_{x_j/y_j} \le 2.5\ \text{m}$, for $x_1 = -x_5$ and $y_1 = y_5$ with respect to their original coordinates	
Multiple load conditions:	
Condition 1 : $F_1 = 44.537$ kN	Condition 2 : $F_2 = 4.4537$ kN
Stress and displacement constraints: $\sigma_i^{max} = 172.43$ MPa and $\delta_{x_j/y_j}^{max} = 5$ mm (0.005 m) (for node, $j = 7, 8$, and 9 along x- and y-axes)	
Lumped masses on nodes 7, 8, and 9 are 500 kg	
Natural frequency constraints: $f_1 \ge 15$ Hz	
The continuous cross-sectional areas: $[A^{min}, A^{max}] = [-50, 50]\ \text{cm}^2$ ($[-50, 50] \times 10^{-4}\ \text{m}^2$) and critical area = 1 cm^2	
Material properties: $E = 69$ GPa and $\rho = 2740\text{kg/m}^3$	

The aim of this section is to optimize the 45-bar truss for TSS, using 31 size variables and two shape variables. The obtained results are presented in Table 6.16, which shows the outcomes of TSS optimization of the 45-bar truss with continuous cross-sections for the algorithms in 100 independent runs. The best run among all these runs is selected based on the minimum mass value while fulfilling all the structural constraints. The ALO, DA, GWO, MVO, SCA, WOA, HTS, TLBO, PVS, and SOS algorithms report the best results of 71.1636, 117.2089, 69.5285, 69.5394, 106.0708, 165.3516, 87.6375, 69.5202, 69.5445, and 69.5223 kg, respectively. Among the algorithms, the best results in terms of the structural mass are achieved by the TLBO, PVS, and SOS algorithms with the mass values of 69.5202, 69.5223, and 69.5445 kg, respectively. The TLBO algorithm gives the lowest structural mass, whereas PVS provides the second-best results.

To obtain a better understanding of the metaheuristic methods' actual performance, the average mass and its SD are taken into consideration. The results reveal that among the ALO, DA, GWO, MVO, SCA, WOA, HTS, TLBO, PVS, and SOS algorithms, the mass mean values are 170.0811, 280.7560, 97.4722, 110.6992, 217.5403, 514.2350, 176.2751, 154.1027, 108.3578, and 95.1853 kg, respectively, with the corresponding SD values of 60.8680, 73.1218, 24.0147, 38.5800, 51.2218, 162.4577, 42.8382, 55.8915, 24.6438, and 27.5470. Based on the results, it is concluded that the SOS algorithm and the GWO algorithm perform best in terms of achieving the lowest SD and mass mean values, respectively. Therefore, the actual better algorithms in terms of convergence rate and consistency for this design problem are the SOS, GWO, and PVS algorithms.

The ideal configuration of the 45-bar truss is visually illustrated in Fig. 6.16, providing a clear depiction of the optimal arrangement. Upon a thorough examination of the diagram, it becomes evident that this ideal configuration only requires the utilization of 6 elements out of the total 45 available. This significant finding emphasizes the efficiency and effectiveness of the selected design, as it successfully achieves the desired level of structural integrity while minimizing the overall number of components required.

Table 6.16 Topology, shape, and size optimization of the 45-bar truss (continuous section)

Group no.	ALO	DA	GWO	MVO	SCA	WOA	HTS	TLBO	PVS	SOS
1	—	—	—	—	—	18.6017	—	—	—	—
4	6.0731	8.5890	6.8939	6.9583	8.6188	16.7999	6.6958	6.9167	7.1154	6.9684
6	—	1.1073	—	—	—	11.3821	—	—	—	—
7	—	—	—	—	—	2.3370	—	—	—	—
8	7.8677	7.0889	6.4466	6.3755	28.7301	11.6535	8.1605	6.4114	6.1507	6.3478
10	—	1.3698	—	—	—	—	—	—	—	—
11	—	—	—	—	—	1.0000	—	—	—	—
12	14.7707	16.2132	15.5404	15.6200	22.0700	18.6089	18.0651	15.5722	15.8116	15.6278
15	—	—	—	—	—	1.1813	—	—	—	—
19	—	—	—	—	—	—	4.9608	—	—	—
20	—	—	—	—	—	—	1.8225	—	—	—
22	—	—	—	—	—	2.5341	—	—	—	—
23	—	—	—	—	—	1.2560	—	—	—	—
30	—	20.0987	—	—	—	—	—	—	—	—
x_1	75.0000	75.0000	75.0000	75.0000	68.4773	74.9993	74.0958	75.0000	74.9997	75.0000
y_1	29.4342	38.2332	45.1699	46.1897	45.1521	12.8266	36.7096	45.5927	48.5912	46.3284
Mass (kg)	71.1636	117.2089	69.5285	69.5394	106.0708	165.3516	87.6375	69.5202	69.5445	69.5223
σ^{max} (MPa)	60.0550	67.7775	75.9290	77.0275	29.6986	39.7585	58.9710	76.4472	80.5177	77.3988
δ^{max} (mm)	3.2963	3.5921	4.4621	4.5682	3.5728	2.0516	3.1699	4.5057	4.8435	4.5840
f_1 (Hz)	19.0717	20.7829	19.7440	19.7727	22.0464	23.2427	20.4189	19.7529	19.8134	19.7735
Mean	170.0811	280.7560	97.4722	110.6992	217.5403	514.2350	176.2751	154.1027	108.3578	95.1853
SD	60.8680	73.1218	24.0147	38.5800	51.2218	162.4577	42.8382	55.8915	24.6438	27.5470

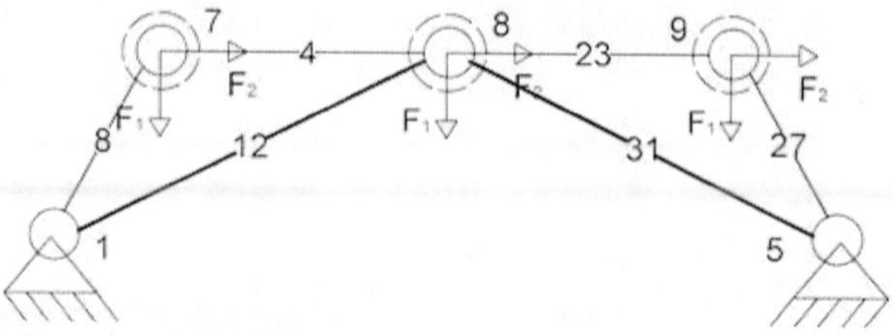

Fig. 6.16 Optimal structure of the 45-bar truss

Moreover, the simultaneous TSS optimization leads to a significant reduction in the weight of trusses compared to the individual optimization of topology and size. Specifically, the TSS approach results in a reduction of 82.9 and 53.0% in weight compared to topology optimization and size optimization, respectively, for the best solutions of each case. This means that by simultaneously considering TSS optimization, the TSS approach is able to find more efficient truss designs with a much lower weight. These results highlight the importance of considering multiple optimization criteria in engineering design, as optimizing for a single criterion may not always result in the most optimal design.

6.3.9 *TSS Optimization of the 25-Bar 3D Truss with Continuous Cross-Sections*

The ninth benchmark problem focuses on the investigation of the 25-bar truss, which represents a space truss subjected to stress, buckling, displacement, and fundamental frequency constraints, considering both continuous design variables. The initial layout of this space truss is depicted in Fig. 6.17. Previous studies by Deb and Gulati (2001) and Ahrari et al. (2014) have examined this problem under static constraints but have not explored it with the simultaneous consideration of static and dynamic constraints.

To address this research gap, the problem has been modified to incorporate the effects of multi-load conditions, lumped masses, natural frequency constraints, and continuous areas. Detailed design considerations, including design variables, load conditions, lumped masses, constraints, and material properties, are provided in Table 6.17. In this problem, element grouping is adopted due to structural symmetry, as indicated in the table. As a result, the truss has only eight size variables, while topology variables are incorporated with them, as discussed earlier.

Furthermore, the truss is subjected to two nonstructural lumped masses of 500 kg each, placed at nodes 1 and 2. Two different loading conditions are considered to fulfill the multi-load condition, as described in the table.

Table 6.18 presents the results of optimizing the 25-bar truss with eight continuous size and five continuous shape variables using various algorithms in 100 independent runs. The best run is identified as the one that yields the minimum mass value while satisfying all the stated structural constraints. ALO, DA, GWO, MVO,

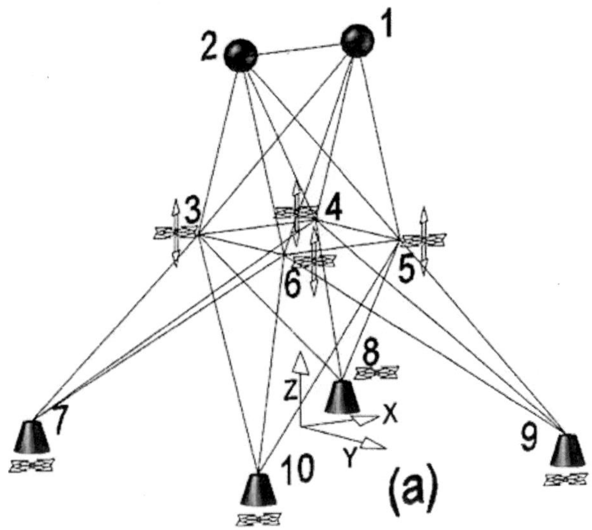

Fig. 6.17 The 25-bar 3D truss

SCA, WOA, HTS, TLBO, PVS, and SOS algorithms yield the best results of 259.3264, 300.0249, 260.9474, 255.8244, 323.8298, 351.9796, 255.8321, 253.6679, 256.9905, and 256.1837 kg, respectively. Among these algorithms, TLBO yields the least structural mass.

To assess the performance of these metaheuristic methods, the mean and SD values of structural mass are used. The average mass and its SD are computed for each algorithm, which reveals that ALO, DA, GWO, MVO, SCA, WOA, HTS, TLBO, PVS, and SOS algorithms have the mass mean value of 310.8660, 403.6285, 268.8140, 1.00E+06, 376.9654, 530.9819, 282.4127, 268.6501, 261.0151, and 264.8201 kg, respectively. The SD values for these algorithms are 39.9370, 88.0694, 6.4869, 1.00E+07, 23.4726, 101.0103, 67.2626, 43.2474, 2.4259, and 5.3726, respectively. Among these algorithms, the PVS algorithm performs best in terms of achieving the best mass mean and SD values. Therefore, the PVS algorithm is the best algorithm for both convergence rate and consistency for this design problem.

Figure 6.18 showcases the optimal arrangement of the 25-bar 3D truss, offering a visual representation of the ideal configuration. A careful examination of the diagram reveals that this optimal arrangement necessitates the utilization of only five groups of elements out of the total eight available. This notable discovery underscores the efficiency and effectiveness of the chosen design, as it successfully achieves the desired level of structural integrity while minimizing the overall number of components required.

The TSS optimization approach also outperforms the individual size and topology optimizations in terms of designing lighter trusses. Specifically, considering TSS optimization leads to 33.9 and 33.2% lighter trusses compared to the size optimization and topology optimization, respectively, for the best solutions of each

Table 6.17 Design parameters of the 25-bar 3D truss

Size variables: $X_i = G_i$, $i = 1, 2, \ldots, 8$; $G =$ Group number
Shape variables: $0.508\ \text{m} \le \xi_{x_4} \le 1.524\ \text{m}$, where $x_4 = x_5 = -x_3 = -x_6$
$1.016\ \text{m} \le \xi_{y_3} \le 2.032\ \text{m}$, where $y_3 = y_4 = -y_5 = -y_6$
$2.286\ \text{m} \le \xi_{z_3} \le 3.302\ \text{m}$, where $z_3 = z_4 = z_5 = z_6$
$1.016\ \text{m} \le \xi_{x_8} \le 2.032\ \text{m}$, where $x_8 = x_9 = -x_7 = -x_{10}$
$2.54\ \text{m} \le \xi_{y_7} \le 3.556\ \text{m}$, where $y_7 = y_8 = -y_9 = -y_{10}$ (1 in. = 0.0254 m)

Multiple load conditions:

Condition 1:	$F_{y_1} = 89.074$ kN	$F_{z_1} = -22.25$ kN	$F_{y_2} = -89.074$ kN	$F_{z_2} = -22.25$ kN
Condition 2:	$F_{x_1} = 4.454$ kN	$F_{y_1} = -44.537$ kN	$F_{z_1} = -44.537$ kN	$F_{y_2} = -44.537$ kN
	$F_{z_2} = -44.537$ kN	$F_{x_3} = 2.227$ kN	$F_{x_6} = 2.672$ kN	

Lumped masses at nodes 1 and 2 are 500 kg

Stress and displacement constraints: $\sigma_i^{\max} = 172.375$ MPa and $\delta_{x_j/y_j/z_j}^{max} = 8.9$ mm (0.0089 m) (for node, $j = 1, 2, \ldots, 6$ along x-, y-, and z-axes)

Natural frequency constraints: $f_1 \ge 15$ Hz

The continuous cross-sectional areas: $[A^{\min}, A^{\max}] = [-50, 50]\ \text{cm}^2$ ($[-50, 50] \times 10^{-4}\ \text{m}^2$) and critical area = 1 cm^2

Material properties: $E = 69$ GPa and $\rho = 2740 \text{kg/m}^3$

Element grouping

Group number	Element number (end nodes)	Group number	Element number (end nodes)
G_1	1(1,2)	G_5	12(3,4), 13(5,6)
G_2	2(1,4), 3(2,3), 4(1,5), 5(2,6)	G_6	14(3,10), 15(6,7), 16(4,9), 17(5,8)
G_3	6(2,5), 7(2,4), 8(1,3), 9(1,6)	G_7	18(3,8), 19(4,7), 20(6,9), 21(5,10)
G_4	10(3,6), 11(4,5)	G_8	22(3,7), 23(4,8), 24(5,9), 25(6,10)

Table 6.18 Topology, shape, and size optimization of the 25-bar 3D truss (continuous section)

Element group no.	ALO	DA	GWO	MVO	SCA	WOA	HTS	TLBO	PVS	SOS
1	—	9.4288	—	—	—	21.4651	—	—	—	—
2	12.5542	13.7405	13.4501	13.8406	17.7956	20.1650	13.1901	13.2714	13.8226	13.0402
3	14.7620	15.1186	13.2842	14.2384	12.8418	15.9140	13.1585	13.0567	14.7957	13.2421
6	1.0000	4.5518	2.4152	1.0054	4.4461	9.0392	1.0747	1.0015	1.0003	1.0251
7	11.7404	11.9162	8.3151	8.4069	9.2277	12.8368	9.0766	8.1592	10.6493	9.0955
8	18.0337	16.5466	19.6084	19.1852	23.9079	19.7907	19.7455	20.0898	16.9548	20.0435
x_4	50.3811	60.0000	59.9641	59.8638	58.8888	44.2366	58.0494	59.9921	53.7066	57.7642
x_8	56.1079	80.0000	59.9436	59.0504	67.6798	55.9812	58.0849	59.8378	57.3240	57.7502
y_4	44.9832	58.4234	42.7964	43.4400	46.4520	60.1697	41.5078	40.6004	48.6107	40.0075
y_8	101.9500	109.5539	100.8604	100.1150	126.4509	100.0000	101.0043	100.7731	101.3649	100.0000
z_4	112.2373	108.4092	117.6467	113.0692	130.0000	90.0000	118.1743	119.8635	104.6814	119.9257
Mass (kg)	259.3264	300.0249	260.9474	255.8244	323.8298	351.9796	255.8321	253.6679	256.9905	256.1837
σ^{max} (MPa)	55.5443	45.8233	61.9699	58.7650	56.6252	43.1822	63.6664	64.7226	54.0457	64.3529
δ^{max} (mm)	6.0745	4.6759	7.6632	7.8808	4.6936	4.0327	7.6704	8.0019	6.6108	7.7558
f_1 (Hz)	15.0000	17.9585	15.0571	15.0131	16.5323	15.0795	15.0253	15.0000	15.0000	15.0138
Mean	310.8660	403.6285	268.8140	1.00E+06	376.9654	530.9819	282.4127	268.6501	261.0151	264.8201
SD	39.9370	88.0694	6.4869	1.00E+07	23.4726	101.0103	67.2626	43.2474	2.4259	5.3726

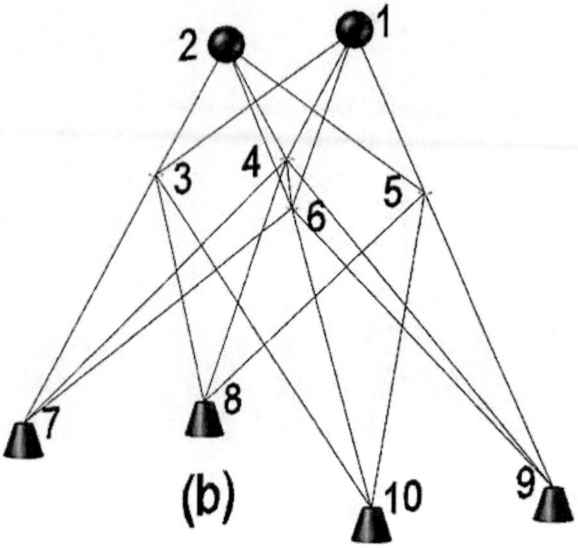

Fig. 6.18 Optimal structure of the 25-bar 3D truss

case. This indicates that considering both size and shape variables simultaneously can result in significant improvements in weight reduction compared to optimizing them separately. Therefore, TSS optimization can be a more effective strategy to achieve optimal designs with improved performance and reduced weight.

6.3.10 TSS Optimization of the 39-Bar 3D Truss with Continuous Cross-Sections

The 39-element 3D truss is the focus of the benchmark problem presented in Fig. 6.19. The element cross-sectional areas are grouped into 11 sets, considering the symmetry about the *x–z* and *y–z* planes. Previous studies by Deb and Gulati (2001) and Luh and Lin (2008) have investigated this truss under static constraints but have not examined it with simultaneous static and dynamic constraints.

To address this research gap, the problem has been modified to incorporate the effects of multi-load conditions, lumped masses, natural frequency constraints, and continuous areas. Additionally, a constant lumped mass of 500 kg is attached to the top nodes (nodes 1 and 2) of the truss, as shown in Fig. 6.19. The design parameters, including material properties, loading conditions, design variable bounds, constraints, and other relevant details, are presented in Table 6.19.

This comprehensive table provides all the necessary information for the optimization of the truss problem, considering the aforementioned modifications. It serves as a reference for conducting further analysis and investigations on the truss structure.

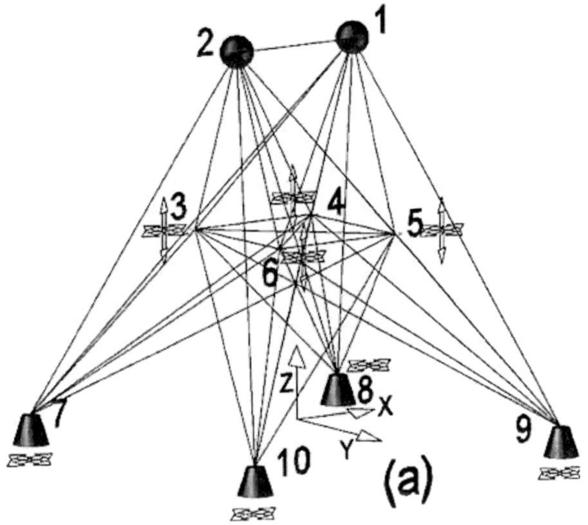

Fig. 6.19 The 39-bar 3D truss

Table 6.19 Design parameters of the 39-bar 3D truss

Size variables: $X_i = G_i$, $i = 1, 2, \ldots, 11$; $G =$ Group number Shape variables: $0.508\ \text{m} \leq \xi_{x_4} \leq 1.524\ \text{m}$, where, $x_4 = x_5 = -x_3 = -x_6$ $1.016\ \text{m} \leq \xi_{y_3} \leq 2.032\ \text{m}$, where, $y_3 = y_4 = -y_5 = -y_6$ $2.286\ \text{m} \leq \xi_{z_3} \leq 3.302\ \text{m}$, where, $z_3 = z_4 = z_5 = z_6$ $1.016\ \text{m} \leq \xi_{x_8} \leq 2.032\ \text{m}$, where, $x_8 = x_9 = -x_7 = -x_{10}$ $2.54\ \text{m} \leq \xi_{y_7} \leq 3.556\ \text{m}$, where, $y_7 = y_8 = -y_9 = -y_{10}$ (1 in. = 0.0254 m)			
Load condition:	$F_{z_1} = -222.685$ kN		$F_{z_2} = -222.685$ kN
Lumped masses at nodes 1 and 2 are 500 kg			
Stress and displacement constraints: $\sigma_i^{max} = 172.375$ MPa and $\delta_{x_j/y_j/z_j}^{max} = 8.9$ mm (0.0089 m) (for node, $j = 1, 2, \ldots, 6$ along x-, y-, and z-axes)			
Natural frequency constraints: $f_1 \geq 15$ Hz			
The continuous cross-sectional areas: $[A^{min}, A^{max}] = [-50, 50]$ cm^2 ($[-50, 50] \times 10^{-4}$ m^2) and critical area = 1 cm^2			
Material properties: $E = 69$ GPa and $\rho = 2740$kg/m^3			
Element grouping			
Group number	Element number (end nodes)	Group number	Element number (end nodes)
G_1	1(1,2)	G_7	22(3,7), 23(4,8), 24(5,9), 25(6,10)
G_2	2(1,4), 3(2,3), 4(1,5), 5(2,6)	G_8	26(5,7), 27(6,8), 28(3,9), 29(4,10)
G_3	6(2,5), 7(2,4), 8(1,3), 9(1,6)	G_9	30(3,5), 31(4,6)
G_4	10(3,6), 11(4,5), 12(3,4), 13(5,6)	G_{10}	32(1,7), 33(1,10), 34(2,9), 35(2,8)
G_5	14(3,10), 15(6,7), 16(4,9), 17(5,8)	G_{11}	36(2,7), 37(2,10), 38(1,8), 39(1,9)
G_6	18(3,8), 19(4,7), 20(6,9), 21(5,10)		

In this section, we investigate the effects of the proposed algorithms on TSS optimization by using continuous elemental cross-sections. Table 6.20 displays the results obtained from TSS optimization of the 39-bar 3D truss with continuous cross-sections for the algorithms in 100 independent runs. It was found that the best run produced the minimum mass value while meeting all the structural constraints. The mean and SD values of the structural mass are given and will be used to assess the performance of the various evolutionary methods statistically. The results reported by the ALO, DA, GWO, MVO, SCA, WOA, HTS, TLBO, PVS, and SOS algorithms that produced the best structural mass values are 229.5686, 273.9633, 222.5339, 221.0164, 269.3925, 337.1842, 221.2232, 222.0958, 220.2474, and 222.7984 kg, respectively. It can be observed that the PVS algorithm produced the least structural mass, with MVO and HTS algorithms being a close second and third, respectively.

To truly assess the performance of the metaheuristic methods, we consider the average mass and its SD. Based on the available results, we conclude that the ALO, DA, GWO, MVO, SCA, WOA, HTS, TLBO, PVS, and SOS algorithms have a mass mean value of 385.0075, 462.0064, 337.2098, 336.7691, 411.3055, 539.1184, 332.3905, 353.0946, 309.0815, and 312.3422 kg, respectively, with an SD of 63.3188, 81.8506, 57.3521, 54.2240, 47.0537, 116.0023, 53.8613, 61.0993, 53.8415, and 58.0386, respectively. The PVS algorithm performs the best to achieve the lowest mass mean, while the SCA algorithm gives the best SD. Therefore, the better algorithms for both convergence rate and consistency for this design problem are the PVS, MVO, HTS, and SCA algorithms.

The ideal configuration of the 39-bar 3D truss is visually presented in Fig. 6.20, providing a clear representation of the optimal arrangement. Upon careful examination of the diagram, it becomes apparent that this optimal configuration requires the utilization of only 5 groups of elements out of the total 11 available. This significant discovery highlights the efficiency and effectiveness of the chosen design, as it successfully achieves the desired level of structural integrity while minimizing the overall number of components required.

The TSS optimization approach results in a significant improvement in designing lighter trusses when compared to size optimization and topology optimization. Specifically, the simultaneous consideration of TSS optimization gives a 62.9 and 37.6% lighter truss compared to the best solutions obtained from size optimization and topology optimization, respectively, for each case. This highlights the effectiveness of TSS optimization in reducing the weight of trusses.

6.3.11 A Comprehensive Analysis

It seems that the PVS algorithm consistently performs well in minimizing the structural mass for all the truss design problems considered in this section. However, it is important to note that different algorithms may perform differently depending on the specific design problem and constraints. Therefore, it is recommended to test

Table 6.20 Topology, shape, and size optimization of the 39-bar 3D truss (continuous section)

Element group no.	ALO	DA	GWO	MVO	SCA	WOA	HTS	TLBO	PVS	SOS
1	—	10.2733	10.2547	—	26.4731	—	4.3699	—	—	—
2	8.7456	8.9711	5.8334	8.4545	6.8309	12.6635	7.5905	8.7882	8.1615	9.8325
3	17.3681	20.2442	17.5074	15.8651	17.7405	18.6059	16.4513	15.7798	16.6204	16.8887
5	1.0000	6.2400	1.1557	—	2.0072	—	1.0179	—	1.0031	1.0136
6	4.8003	4.4427	3.0510	2.4708	5.6894	12.9933	2.3959	2.1952	3.1762	1.7674
7	20.8191	19.6145	21.3952	22.1753	25.7626	19.5347	22.1643	22.3612	21.1821	21.2822
8	—	—	—	1.0046	—	11.0473	—	1.0456	—	—
x_4	45.1256	56.4556	54.2982	47.9620	60.0000	41.5944	50.0406	48.6457	46.8753	50.5541
x_8	63.5014	67.6901	75.4770	70.9353	70.0842	53.4741	74.1396	73.1013	68.2147	75.8820
y_4	47.7550	54.9701	46.5061	44.4849	43.2036	58.1155	44.0731	44.1127	45.7888	47.6513
y_8	100.0019	108.5611	100.0316	100.0000	100.0000	100.0000	100.0868	100.1790	100.1713	100.0000
z_4	105.4951	96.0673	108.4791	113.5635	114.0075	108.3353	112.0987	113.4801	108.8864	107.2956
Mass (kg)	229.5686	273.9633	222.5339	221.0164	269.3925	337.1842	221.2232	222.0958	220.2474	222.7984
σ^{max} (MPa)	65.9288	68.5764	69.0604	70.7271	75.3477	67.4166	69.9583	70.5893	67.7620	81.1893
δ^{max} (mm)	5.6553	6.6097	6.1717	5.9099	7.0091	6.5497	5.8246	5.8557	5.7653	5.9028
f_1 (Hz)	15.0000	15.0001	15.0266	15.0575	15.9924	15.0000	15.0529	15.0392	15.0034	15.0405
Mean	385.0075	462.0064	337.2098	336.7691	411.3055	539.1184	332.3905	353.0946	309.0815	312.3422
SD	63.3188	81.8506	57.3521	54.2240	47.0537	116.0023	53.8613	61.0993	53.8415	58.0386

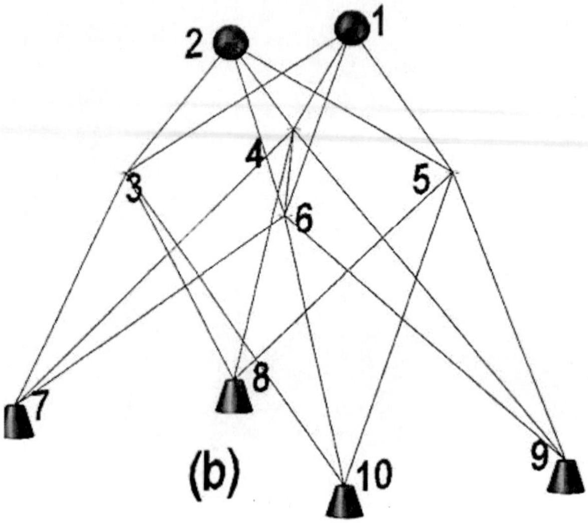

Fig. 6.20 Optimal structure of the 39-bar 3D truss

multiple algorithms and compare their results to determine the most suitable algorithm for a specific design problem. Additionally, further analysis such as sensitivity analysis and robustness analysis may be conducted to validate the performance of the chosen algorithm.

Table 6.21 presents a comprehensive analysis of the best results obtained from ten benchmark problems using proposed algorithms in 100 independent runs. The table showcases minimum mass and offers a detailed evaluation of overall, average, and individual ranks, as well as the best and worst results for each algorithm.

Upon reviewing the results in Table 6.21, it was observed that the PVS algorithm outperformed all other algorithms in terms of overall rank to obtain minimum mass. The TLBO, MVO, and SOS algorithms followed with the second, third, and fourth-best performances, respectively.

Moreover, the PVS algorithm provided the highest number of best solutions, including six best solutions, two second-best solutions, and one third-best solution. Notably, the algorithm did not have any worse solutions. Similarly, the TLBO algorithm offered five best solutions, two third-best solutions, and no worse solutions. On the other hand, the WOA algorithm presented the worst solutions for all problems. This observation highlights the importance of selecting the right algorithm for specific problems to achieve optimal results.

Overall, the analysis of Table 6.22 provides valuable insights into the performance of proposed algorithms for ten benchmark problems. These findings can assist researchers and practitioners in selecting the most suitable algorithm for similar problems and improving the optimization process's efficiency.

Table 6.21 Minimum mass obtained using TSS optimization with continuous sections

	DA	MVO	SCA	WOA	ALO	HTS	PVS	SOS	GWO	TLBO
10-bar truss	322.3	255.2	341.7	520.1	327.9	251.5	250.8	269.8	252	**250.6**
14-bar truss	304.9	241.9	257.9	369.2	269.8	241.4	**241.3**	**241.3**	241.5	**241.3**
15-bar truss	121.2	88.4	128.4	134.1	101.2	91.3	**86.1**	89.9	87.4	87.8
24-bar truss	166	109.6	137.3	227.8	121.2	121.6	**108.9**	112	110	113.6
20-bar truss	198.1	**111.4**	184.7	202.7	147.7	125.4	113.3	118.8	114.4	116
72-bar 3D truss	567.9	462.5	507.4	638.6	463.5	443.3	439.7	444.3	439.6	**435.5**
39-bar truss	356.2	189.2	294.3	482.7	227.2	209.2	**186.8**	196.5	199.3	193.7
45-bar truss	117.2	**69.5**	106.1	165.4	71.2	87.6	**69.5**	**69.5**	**69.5**	**69.5**
25-bar 3D truss	300	255.8	323.8	352	259.3	255.8	257	256.2	260.9	253.7
39-bar 3D truss	274	221	269.4	337.2	229.6	221.2	**220.2**	222.8	222.5	222.1
Overall rank	9	3	8	10	7	6	1	5	4	2
Count best	0	2	0	0	0	0	6	2	1	5
Count second best	0	4	0	0	0	1	2	0	2	0
Count 3nd best	0	0	0	0	0	2	1	0	2	2
Count worst	0	0	0	10	0	0	0	0	0	0

Note: Bold indicates the best solution

Table 6.22 showcases the mean mass obtained for ten benchmark problems using the proposed algorithms in 100 independent runs. Similar to previous table, this table provides an overall rank, average rank, individual ranks, best and worst results for each algorithm.

Upon reviewing the results in Table 6.22, it was observed that the PVS algorithm is the best-performing algorithm in terms of overall rank to obtain mean mass. It was followed by the SOS, TLBO, and GWO algorithms, respectively.

Furthermore, the PVS algorithm presented eight best solutions and one second-best solution without any worse solution count. In contrast, the SOS algorithm provided one best count, six second-best counts, and did not have any worse solution count.

In contrast, the WOA, MVO, and DA algorithms provided five, four, and one worst solutions, respectively. These results suggest that the performance of these algorithms may be suboptimal for certain types of problems.

Table 6.22 Mean mass obtained using TSS optimization with continuous sections

	DA	MVO	SCA	WOA	ALO	HTS	PVS	SOS	GWO	TLBO
10-bar truss	1036.4	331.5	467.7	985.0	508.4	330.4	293.4	352.6	**288.8**	331.7
14-bar truss	742.9	380.7	381.5	1238.0	587.3	390.9	**298.7**	316.1	352.7	377.9
15-bar truss	INF	INF	163.7	240.6	142.3	123.6	**96.6**	100.7	INF	106.7
24-bar truss	280.0	138.5	188.3	379.6	191.7	181.8	**125.4**	133.2	130.7	155.8
20-bar truss	INF	INF	251.6	421.8	247.7	2784.7	**144.8**	153.8	INF	378.6
72-bar 3D truss	INF	INF	652.5	1434.7	730.6	603.2	**460.7**	467.4	INF	482.9
39-bar truss	578.4	255.4	445.6	922.3	365.0	338.7	**246.5**	255.7	256.3	312.5
45-bar truss	280.8	110.7	217.5	514.2	170.1	176.3	108.4	**95.2**	97.5	154.1
25-bar 3D truss	403.6	INF	377.0	531.0	310.9	282.4	**261.0**	264.8	268.8	268.7
39-bar 3D truss	462.0	336.8	411.3	539.1	385.0	332.4	**309.1**	312.3	337.2	353.1
Overall rank	10	6	8	9	7	5	1	2	4	3
Count best	0	0	0	0	0	0	8	1	1	0
Count 2nd best	0	1	0	0	0	0	1	6	2	0
Count 3nd best	0	0	0	0	1	2	1	2	1	3
Count worst	1	4	0	5	0	0	0	0	0	0

Note: Bold indicates the best solution and INF indicates the infeasible solution

The analysis of Table 6.22 provides important insights into the efficiency of the proposed algorithms for ten benchmark problems. These findings can help researchers and practitioners to select the most suitable algorithm for their specific problem to achieve optimal results.

In addition to evaluating the minimum and mean mass obtained by optimization algorithms, it is also important to assess their success rate and computational effort. This book conducted experiments to obtain these metrics for the algorithms used to solve TTO problems.

Evaluating the success rate of an algorithm is crucial as it indicates how often the algorithm can successfully solve a problem. This metric is calculated by dividing the number of successful runs by the total number of runs. Furthermore, evaluating the computational effort required by an algorithm is important to ensure that it is practical and efficient. This metric can be calculated by measuring the execution time or the number of function evaluations required by the algorithm.

In this book, experiments were carried out to evaluate the success rate and computational effort required by the algorithms for solving TTO problems. These experiments provide a more comprehensive understanding of the performance of the proposed algorithms and their practicality for real-world applications.

By considering all these metrics together, the proposed algorithms can be compared and evaluated comprehensively to select the most suitable algorithm for a specific optimization problem.

The success rate of an optimization algorithm is a crucial metric that indicates the percentage of times an algorithm can successfully solve a problem. In this book, success rate has been defined as the number of times an algorithm reaches the best mean functional value among all applied algorithms in 100 independent runs. If an algorithm fails to reach the best mean functional value, it can be assumed that the algorithm is not beneficial for solving such problems.

Table 6.23 presents the success rate of the algorithms for TTO problems in 100 independent runs. As shown in Table 6.22, the GWO algorithm offers the best mean mass of 288.8 kg among ten algorithms. The success rate of the GWO

Table 6.23 The success rate of TSS optimization with continuous sections

	DA	MVO	SCA	WOA	ALO	HTS	PVS	SOS	GWO	TLBO
10-bar truss	0	32	0	0	0	48	44	5	64	47
14-bar truss	0	30	3	0	2	24	58	50	46	36
15-bar truss	0	3	0	0	0	0	49	4	18	23
24-bar truss	0	32	0	0	2	1	50	29	43	13
20-bar truss	0	22	0	0	0	2	40	22	17	9
72-bar 3D truss	0	0	0	0	0	9	41	18	55	36
39-bar truss	0	35	0	0	1	2	30	23	17	3
45-bar truss	0	36	0	0	13	1	30	53	47	12
25-bar 3D truss	0	8	0	0	0	19	13	6	0	31
39-bar 3D truss	2	16	1	0	7	19	38	37	19	17
Total count	2	214	4	0	25	125	393	247	326	227

algorithm is 64, as illustrated in Table 6.23, indicating that the best functional value of the GWO algorithm reached 64 times the best mean mass of 288.8 kg out of 100 independent runs.

The results in Table 6.23 indicate that the PVS algorithm is the best-performing algorithm in terms of overall success rate, followed by the GWO and SOS algorithms, respectively. It is important to note that the WOA algorithm failed to reach the best mean mass for all problems within the predefined experimental conditions and produced a 0% success rate.

These findings provide important insights into the performance of the algorithms for TTO problems. The success rate of an algorithm is a critical metric to consider when selecting the most suitable algorithm for a specific problem. By analyzing the success rate along with other performance metrics such as mean mass and computational effort, researchers and practitioners can choose the most effective algorithm for their specific optimization problem.

FE count is a crucial criterion for comparing optimization algorithms along with statistical analyses such as standard deviation, Friedman rank test, and so on. This is because many metaheuristics have multiple stages in their optimization process, such as the TLBO algorithm with two phases (teacher phase and student phase), the GWO algorithm with one phase (searching and hunting), and the SOS, HTS, and PVS algorithms with three phases each. Although the number of iterations or generations may be the same for all algorithms, it does not reflect the true computational effort required by the algorithm. Therefore, using FE provides a more meaningful comparison since it represents the number of times the fitness function was called.

Table 6.24 presents the comparative results of each algorithm with respect to the mean number of FE used to reach the best mean mass value with a 25% error of 100 independent runs for each of the TTO problems. The results indicate that the PVS, SOS, and GWO algorithms are computationally more efficient than the other comparative algorithms. Additionally, the PVS algorithm outperforms the other comparative algorithms. On the other hand, the DA, SCA, and WOA algorithms failed to reach the best mean mass value with a 25% error for all the problems under the predefined experimental conditions and, hence, formed a 0% success rate.

The mean mass and standard deviation of mass obtained from ten basic metaheuristics are illustrated in Figs. 6.21, 6.22, 6.23, 6.24, 6.25, 6.26, 6.27, 6.28, 6.29 and 6.30. Figure 6.21 indicates that, for the 10-bar truss problem, the GWO and PVS algorithms have superior mean and SD of mass compared to other algorithms. Furthermore, there is a 72.1% difference between the best mean mass obtained by the GWO algorithm and the worst mean mass obtained by the DA algorithm. The comparative results for the 14-bar truss problem are presented in Fig. 6.22, where the PVS algorithm outperforms the others. Additionally, the percentage difference between the best mean mass achieved by the PVS algorithm and the worst mean mass achieved by the WOA algorithm is 77.5%.

Table 6.24 The mean number of FE required to reach the best mean optimum value with 25% error in TSS with continuous sections

	DA	MVO	SCA	WOA	ALO	HTS	PVS	SOS	GWO	TLBO
10-bar truss	—	15600	—	—	—	12600	6000	12000	4000	10300
14-bar truss	—	—	—	—	—	—	10800	13400	—	—
15-bar truss	—	—	—	—	—	—	8800	12200	—	12800
24-bar truss	—	12000	—	—	—	—	10200	12000	5000	—
20-bar truss	—	—	—	—	—	—	12200	15000	—	—
72-bar 3D truss	—	—	—	—	—	—	6200	7400	—	17700
39-bar truss	—	14100	—	—	—	—	13500	14500	9400	—
45-bar truss	—	13900	—	—	—	—	15100	13400	5400	—
25-bar 3D truss	—	—	—	—	6600	7300	3100	4200	5800	4100
39-bar 3D truss	—	17100	—	—	—	14100	7400	8200	15600	—

— Indicates that algorithm is failed to reach best mean optimum value with 25% error of that problem

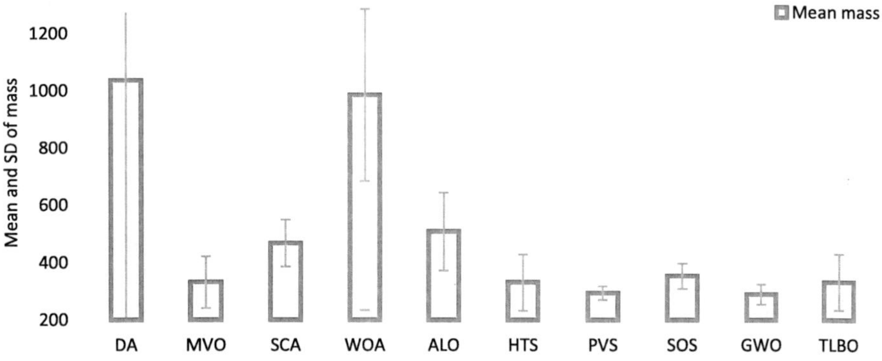

Fig. 6.21 The 10-bar truss: TSS with continuous sections

Figure 6.23 demonstrates that the PVS and SOS algorithms perform better for the 15-bar truss problem. Additionally, the percentage difference between the best mean mass obtained from the PVS algorithm and the worst mean mass obtained from the MVO algorithm is 99.9%. For the 24-bar truss problem, Fig. 6.24 shows that the

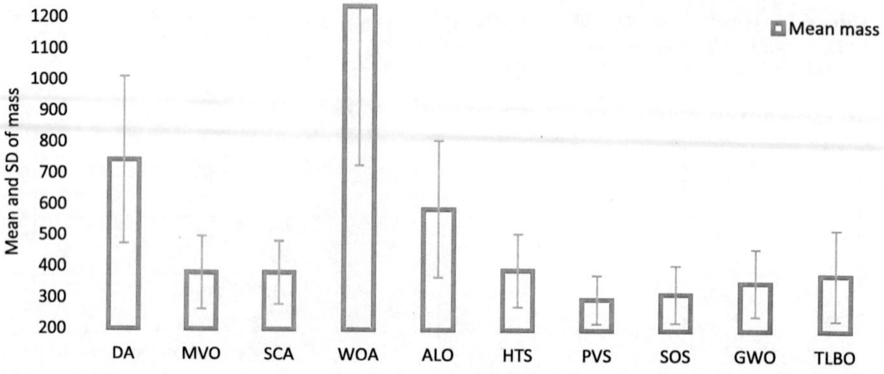

Fig. 6.22 The 14-bar truss: TSS with continuous sections

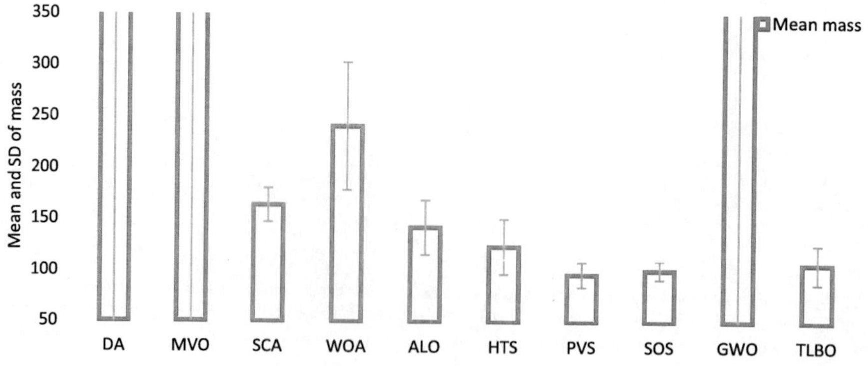

Fig. 6.23 The 15-bar truss: TSS with continuous sections

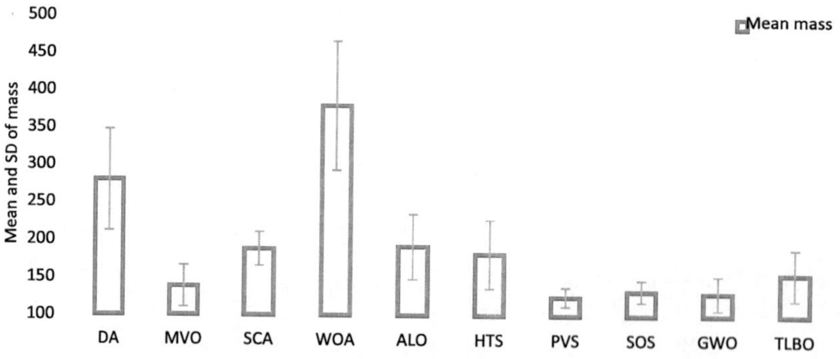

Fig. 6.24 The 24-bar truss: TSS with continuous sections

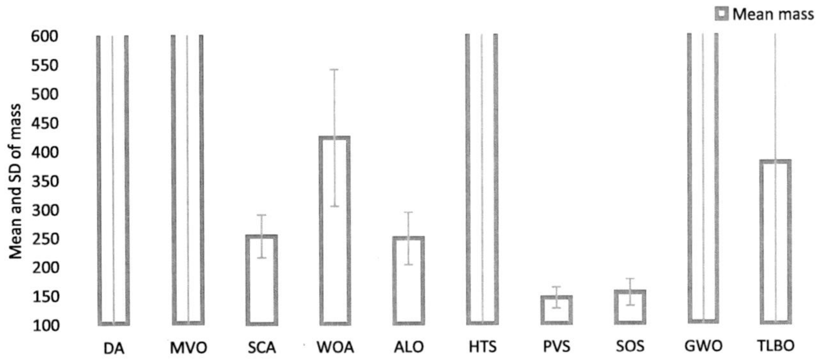

Fig. 6.25 The 20-bar truss: TSS with continuous sections

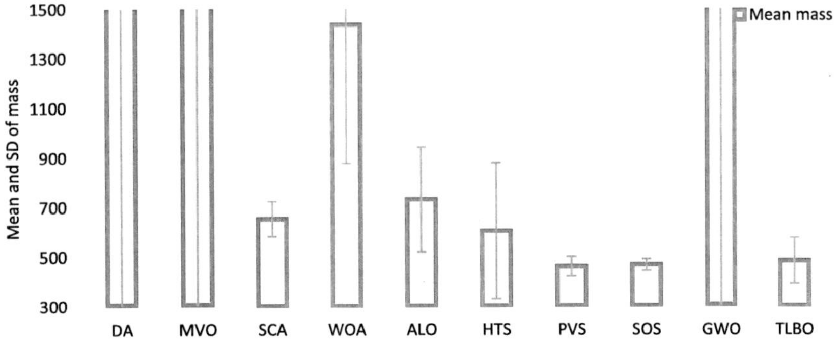

Fig. 6.26 The 72-bar 3D truss: TSS with continuous sections

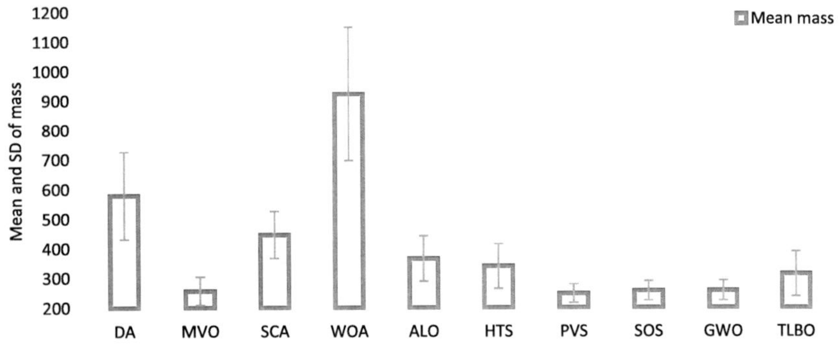

Fig. 6.27 The 39-bar truss: TSS with continuous sections

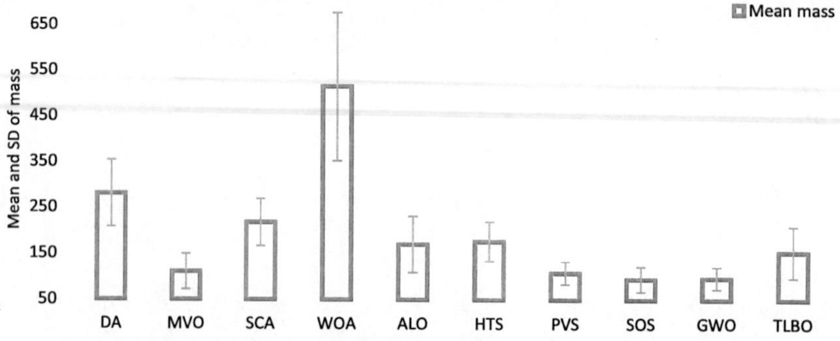

Fig. 6.28 The 45-bar truss: TSS with continuous sections

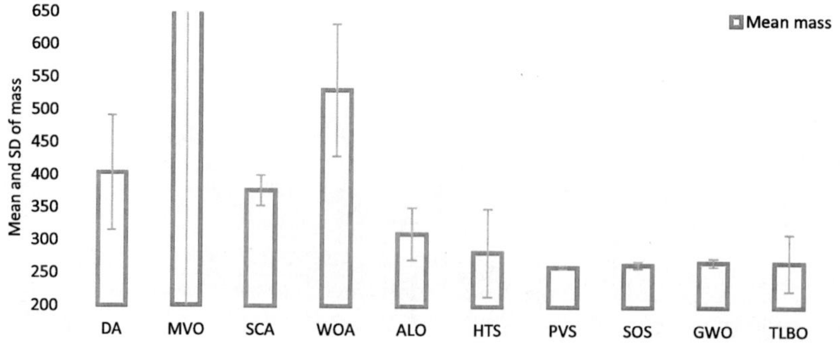

Fig. 6.29 The 25-bar 3D truss: TSS with continuous sections

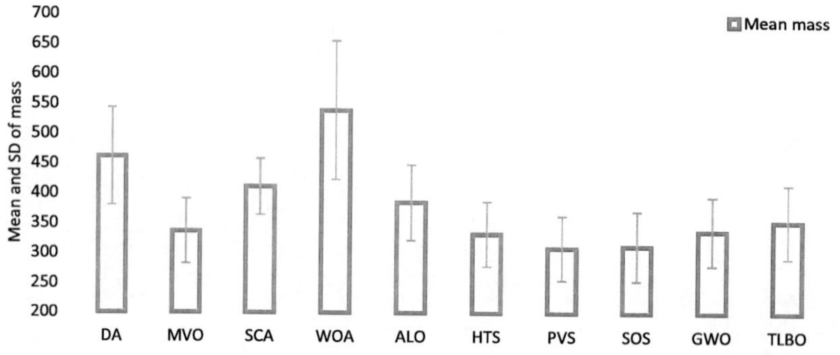

Fig. 6.30 The 39-bar 3D truss: TSS with continuous sections

Table 6.25 The Friedman rank test of mean mass and success rate obtained using TTO for continuous sections

	DA	MVO	SCA	WOA	ALO	HTS	PVS	SOS	GWO	TLBO
Friedman rank of mean mass	9	6	8	10	7	5	1	2	3	4
Friedman rank of success rate	9	4	8	10	7	6	1	2	3	5

PVS and GWO algorithms perform better, with a percentage difference of 64% between the best mean mass obtained from the PVS algorithm and the worst mean mass obtained from the WOA algorithm. Figure 6.25 presents the results for the 20-bar truss problem, where the PVS and SOS algorithms perform better, and the percentage difference between the best mean mass obtained from the PVS algorithm and the worst mean mass obtained from the MVO algorithm is 99.9%.

Figure 6.26 shows that the PVS and SOS algorithms perform better for the 72-bar 3D truss problem, with a percentage difference of 99.9% between the best mean mass obtained from the PVS algorithm and the worst mean mass obtained from the MVO algorithm. For the 39-bar truss problem, Fig. 6.27 indicates that the PVS, MVO, and SOS algorithms perform better, and the percentage difference between the best mean mass obtained from the PVS algorithm and the worst mean mass obtained from the WOA algorithm is 73.3%. Figure 6.28 demonstrates that the SOS and GWO algorithms perform better for the 45-bar truss problem, with a percentage difference of 81.5% between the best mean mass obtained from the SOS algorithm and the worst mean mass obtained from the WOA algorithm.

Figure 6.29 shows that the PVS, SOS, and TLBO algorithms perform better for the 25-bar 3D truss problem, with a percentage difference of 99.9% between the best mean mass obtained from the PVS algorithm and the worst mean mass obtained from the MVO algorithm. For the 39-bar 3D truss problem, Fig. 6.30 demonstrates that the PVS algorithm performs better, and the percentage difference between the best mean mass obtained from the PVS algorithm and the worst mean mass obtained from the MVO algorithm is 46.4%. These results indicate that the choice of algorithm is an important factor, as the percentage differences between the best and worst results range from 46.4 to 99.9% (Table 6.25).

6.3.12 *The Friedman Rank Test*

From the obtained results, it can be understood that the PVS, SOS, and GWO algorithms exhibit better performance compared to the other algorithms. However, it is important to conduct statistical tests to rank all algorithms based on the obtained results from the proposed method. Draa (2015) suggests the use of the Wilcoxon

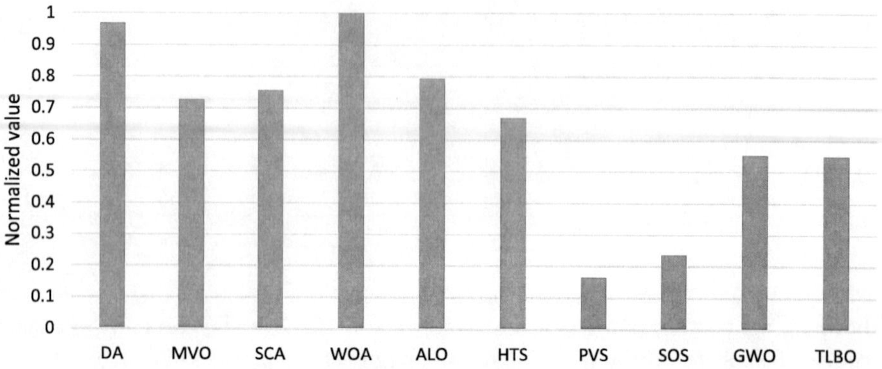

Fig. 6.31 The Friedman rank test of mean mass obtained using TSS for continuous sections

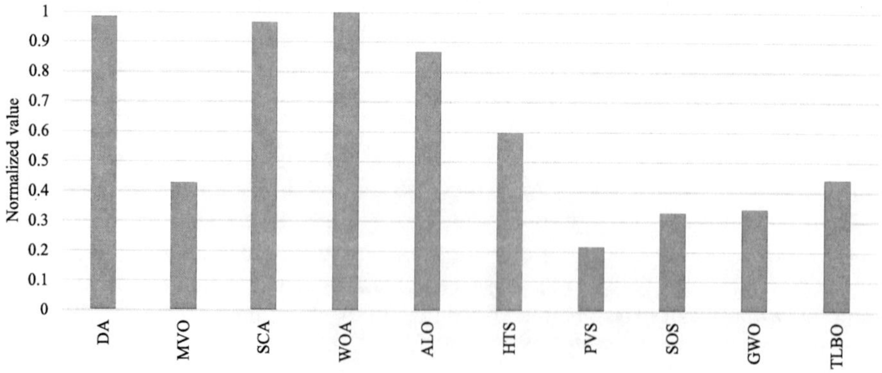

Fig. 6.32 The Friedman rank test of success rate of TSS for continuous sections

signed-rank test as an alternate method, but it only compares two related sample algorithms. On the other hand, the Friedman rank test compares all algorithms simultaneously and provides individual algorithm rank. Hence, the Friedman rank test is performed on the mean mass and success rate of the ALO, DA, GWO, MVO, SCA, WOA, HTS, TLBO, PVS, and SOS algorithms for the continuous sections test problems. Table 6.31 presents the results of the Friedman rank test, which are normalized based on the best value obtained. The algorithms are then ranked based on the normalized value.

The graphical representation of the normalized values of mean mass and success rate is shown in Figs. 6.31 and 6.32, respectively. The results indicate that the PVS algorithm stands first, followed by the SOS and GWO algorithms in terms of the best mean mass and success rate of algorithm for continuous sections. Therefore, based on the performance assessment, the PVS algorithm can be considered the best method.

6.4 Results and Discussion on TSS Optimization with Discrete Sections

In this section, we aim to evaluate the performance of the algorithms with discrete sections, and to accomplish this objective, we use ten benchmark problems. These problems include the 10-bar truss, 14-bar truss, 15-bar truss, 24-bar truss, 20-bar truss, 72-bar 3D truss, 39-bar truss, 45-bar truss, 25-bar 3D truss, and 39-bar 3D truss. In order to accommodate the stochastic nature of metaheuristics, the suggested algorithms are implemented to tackle each truss problem using 100 independent runs. The population size remains constant at 50 for all problems, with each run undergoing a total of 20,000 function evaluations (FE). By assessing the performance of the algorithms on these problems, we can gain insights into their effectiveness and potential for practical applications.

6.4.1 TSS Optimization of the 10-Bar Truss with Discrete Cross-Sections

Figure 6.33 showcases the ground structure of the initial benchmark truss, along with the load and boundary condition. This particular truss has ten size variables and three shape variables, which are utilized for TSS optimization. Several researchers, such as Hajela and Lee (1995), Deb and Gulati (2001), Richardson et al. (2012), and Miguel et al. (2013), have examined this truss for static constraints such as elemental stress and nodal displacement. It has also been investigated for size and shape

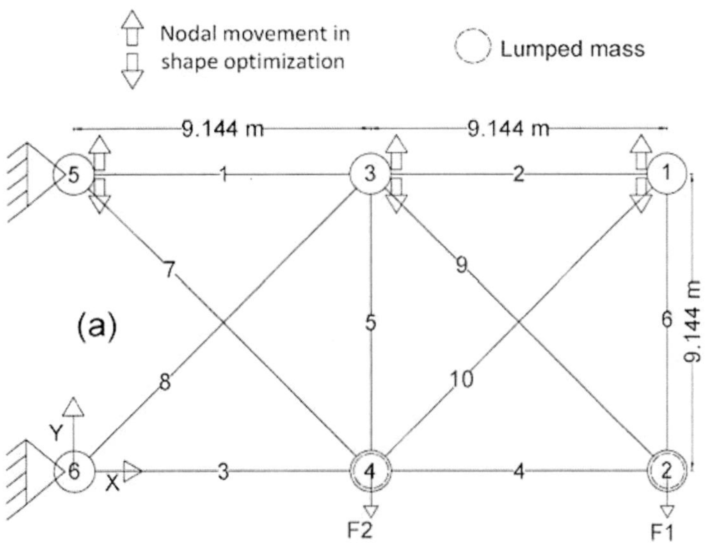

Fig. 6.33 The 10-bar truss

Table 6.26 Design parameters of the 10-bar truss

Size variables: $X_i = A_i$, $i = 1, 2, \ldots, 10$ Shape variables: $4.572 \text{ m} \leq \xi_{y_1}, \xi_{y_3}, \xi_{y_5} \leq 25.4 \text{ m}$ (0.0254 m = 1 in.)	
Multiple load conditions:	
Condition 1 : $F_1 = 44.537$ kN, $F_2 = 44.537$ kN	Condition 2 : $F_1 = 44.537$ kN, $F_2 = 0$ kN
Lumped masses on node 2 and 4 are 500 kg	
Stress and displacement constraints: $\sigma_i^{max} = 172.43$ MPa and $\delta_{y_5\&y_6}^{max} = 100$ mm (0.1 m)	
Natural frequency constraint: $f_1 \geq 4$ Hz	
The discrete cross-sectional areas: $[A^{min}, A^{max}] = [-100, 100]$ cm^2 ($[-100, 100] \times 10^{-4}$ m^2) in increments of 1 cm^2 and critical area = 1 cm^2	
Material properties: $E = 69$ GPa and $\rho = 2740$kg/m^3	

optimization by researchers such as Wei et al. (2005, 2011), Gomes (2011), Kaveh and Zolghadr (2011, 2012), Miguel and Miguel (2012), and Farshchin et al. (2016) with a focus on dynamic constraints, particularly natural frequencies.

Previous studies on this truss problem primarily focused on either static constraints (stress, buckling, and displacement) or dynamic constraints (natural frequencies). However, the simultaneous consideration of both static and dynamic constraints has not been explored. In this study, the problem has been modified to include multiple load conditions, lumped masses, natural frequency constraints, and discrete cross-sectional areas.

Regarding the discrete cross-sectional areas (A_i, where $i = 1, 2, \ldots, 10$) in this problem, the range is $[-100, 100]$ cm^2, where a zero or negative value indicates the removal of the element. Therefore, the discrete design variables can take integer values within the range of $[-100, 100]$ cm^2 in increments of 1 cm^2, where a zero or negative value indicates the removal of the corresponding element. For comprehensive details regarding material properties, design variables, loading conditions, lumped masses, and other important data related to the test problem, refer to Table 6.26.

This section provides an analysis of the performance of proposed algorithms in optimizing a 10-bar truss with ten discrete size and three continuous shape variables. The study involves testing the algorithms using TSS optimization of the 10-bar truss with discrete cross-sections. Table 6.27 summarizes the results obtained from 100 independent runs of the algorithms. The best run was selected based on the lowest mass value achieved while satisfying all the specified structural constraints.

To evaluate the performance of the different evolutionary methods, the mean and standard deviation (SD) values of the structural mass are reported. It is worth noting that these values will be used to make statistical assessments of the methods' effectiveness. The ALO, DA, GWO, MVO, SCA, WOA, HTS, TLBO, PVS, and SOS algorithms all report mass values, with TLBO producing the least structural mass at 261.7825 kg. The other algorithms achieved mass values of 316.1291, 383.5209, 261.7839, 261.7845, 338.3707, 534.0751, 261.8120, 261.7875, and 269.2916 kg, respectively.

Table 6.27 Topology, shape, and size optimization of the 10-bar truss (discrete section)

Element no.	ALO	DA	GWO	MVO	SCA	WOA	HTS	TLBO	PVS	SOS
1	8	4	4	4	5	12	4	4	4	4
2	2	1	—	—	—	—	—	—	—	—
3	47	42	39	39	51	58	39	39	39	42
4	40	42	32	32	41	55	32	32	32	34
5	—	—	—	—	—	3	—	—	—	—
6	1	24	—	—	—	—	—	—	—	—
7	3	3	3	3	3	—	3	3	3	3
8	1	13	1	1	2	42	1	1	1	1
9	4	4	4	4	8	24	4	4	4	4
y_1	439.5494	217.3761	347.2492	693.4442	196.8038	235.3143	184.6006	187.6487	1000.0000	712.9579
y_3	329.8540	452.8258	479.0401	479.0512	459.3476	181.2670	478.9051	479.0489	479.1002	420.9547
y_5	659.5285	877.8912	957.9283	957.9419	947.1421	440.2006	958.6194	957.8929	958.0078	843.5230
Mass (kg)	316.1291	383.5209	261.7839	261.7845	338.3707	534.0751	261.8120	261.7825	261.7875	269.2916
σ^{max} (MPa)	169.1329	169.1329	169.1329	169.1329	169.1329	169.1329	169.1329	169.1329	169.1329	169.1329
δ^{max} (mm)	69.0242	69.0242	69.0242	69.0242	69.0242	69.0242	69.0242	69.0242	69.0242	69.0242
f_1 (Hz)	5.4464	4.9855	5.1536	5.1536	5.4995	6.0683	5.1512	5.1537	5.1537	5.0812
Mean	518.9694	691.8571	289.1947	316.8198	473.7440	1035.8224	320.9270	366.8739	304.2437	341.4147
SD	106.9298	213.4684	18.8231	70.6820	85.9940	332.2037	67.6816	95.5701	29.0133	37.7534

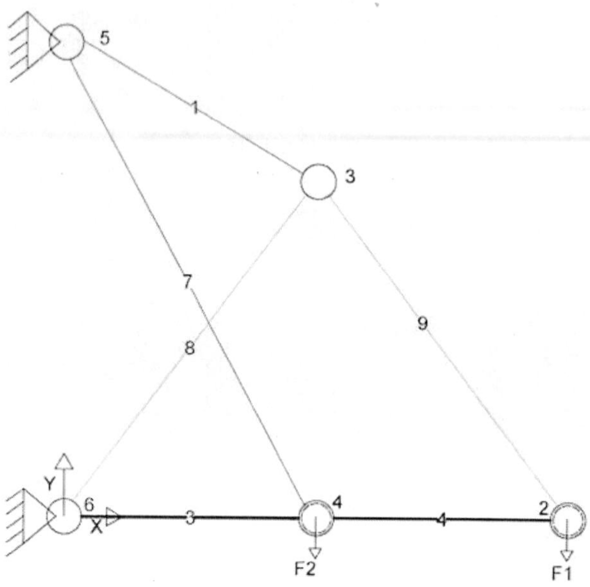

Fig. 6.34 Optimal structure of the 10-bar truss

In conclusion, the results show that TLBO algorithm is the most effective in optimizing the 10-bar truss, producing the lowest structural mass while satisfying all the specified constraints.

To evaluate the performance of metaheuristic methods in optimizing the given design problem, the average mass and its standard deviation (SD) are considered. The available results show that the ALO, DA, GWO, MVO, SCA, WOA, HTS, TLBO, PVS, and SOS algorithms have mass mean values of 518.9694, 691.8571, 289.1947, 316.8198, 473.7440, 1035.8224, 320.9270, 366.8739, 304.2437, and 341.4147 kg, respectively, with the corresponding SD values of 106.9298, 213.4684, 18.8231, 70.6820, 85.9940, 332.2037, 67.6816, 95.5701, 29.0133, and 37.7534.

Based on these results, it can be concluded that the GWO algorithm performs the best, achieving the lowest mass mean and SD values. Therefore, the TLBO and GWO algorithms are the better algorithms for both convergence rate and consistency for this design problem.

Figure 6.34 illustrates the optimal structure of the 10-bar truss. It is worth noting that upon examination, it becomes evident that this optimal configuration utilizes only six out of the ten available elements. This discovery emphasizes the efficiency and effectiveness of the chosen design, as it successfully achieves the desired structural integrity while employing a reduced number of components.

To elaborate further, truss optimization is an important aspect of structural engineering to achieve optimal designs that can withstand applied loads while being as lightweight as possible. Traditionally, size optimization and topology optimization have been used to obtain the optimal truss designs. However, the

TSS optimization approach offers a more comprehensive solution by considering both size and shape optimization simultaneously.

In addition, when TSS optimization is considered simultaneously, the resulting truss is 67.3% lighter than that produced by size optimization and 33.6% lighter than that produced by topology optimization, considering the best solutions for each case. Therefore, TSS optimization outperforms size optimization and topology optimization significantly in terms of designing lighter trusses.

6.4.2 TSS Optimization of the 14-Bar Truss with Discrete Cross-Sections

Figure 6.35 depicts the ground structure of the second benchmark truss, which is commonly referred to as the 14-bar truss. It also showcases the load and boundary conditions associated with it. Previous studies, including those conducted by Tang et al. (2005), Rahami et al. (2008), and Miguel et al. (2013), have optimized this truss problem using TSS. These studies focused on static and dynamic constraints and utilized 14 discrete cross-sectional areas and eight continuous shape variables.

However, the consideration of static and dynamic constraints simultaneously has not been explored for this problem. Previous optimization processes concentrated solely on static constraints.

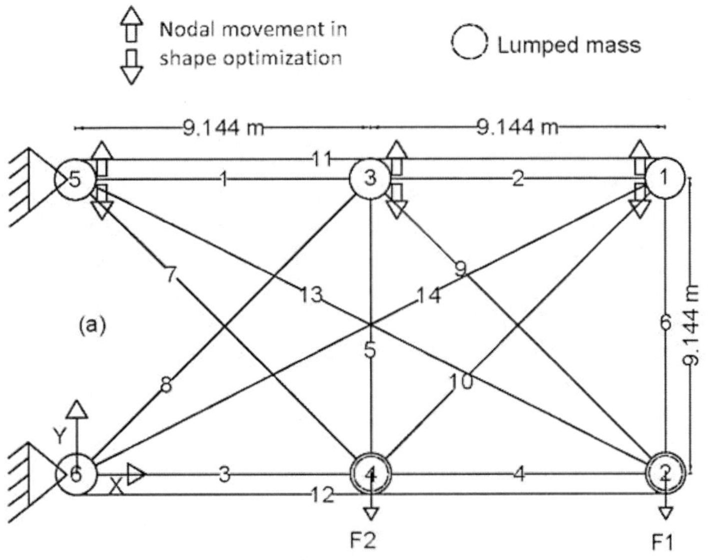

Fig. 6.35 The 14-bar truss

Table 6.28 Design parameters of the 14-bar truss

Size variables: $X_i = A_i$, $i = 1, 2, \ldots, 14$ Shape variables: $4.572\ \text{m} \leq \xi_{y_1}, \xi_{y_3}, \xi_{y_5} \leq 25.4\ \text{m}$ (0.0254 m = 1 in.)	
Multiple load conditions:	
Condition 1 : $F_1 = 44.537$ kN, $F_2 = 44.537$ kN	Condition 2 : $F_1 = 44.537$ kN, $F_2 = 0$ kN
Lumped mass on node 2 and 4 are 500 kg	
Stress and displacement constraints: $\sigma_i^{\max} = 172.43$ MPa and $\delta_{y_5 \& y_6}^{\max} = 100$ mm (0.1 m)	
Natural frequency constraint: $f_1 \geq 4$ Hz	
The discrete cross-sectional areas: $[A^{\min}, A^{\max}] = [-100, 100]$ cm^2 ($[1, 100] \times 10^{-4}$ m^2) in increments of 1 cm^2 and critical area = 1 cm^2	
Material properties: $E = 69$ GPa and $\rho = 2740$kg/m^3	

For comprehensive information about the design variables, load conditions, design constraints, and other pertinent data specific to the test problem, refer to Table 6.28.

In this section, we focused on optimizing the 14-bar truss using 14 discrete size variables and three continuous shape variables in the context of TS. Table 6.29 presents the results obtained from the TSS optimization of the 14-bar truss with discrete cross-sections for different evolutionary algorithms in 100 independent runs. The best run resulted in the minimum mass value while fulfilling all the specified structural constraints. The mean and SD values of structural mass are also provided, which will be used to statistically evaluate the performance of the different evolutionary methods.

The top-performing results for the ALO, DA, GWO, MVO, SCA, WOA, HTS, TLBO, PVS, and SOS algorithms are 274.4180, 282.3526, 252.6178, 252.6178, 259.7888, 354.1471, 252.6177, 252.6177, 252.6177, and 252.6177 kg, respectively. It is evident from the results that the HTS, TLBO, PVS, and SOS algorithms produced identical solutions with the lowest structural mass.

To accurately evaluate the performance of the different metaheuristic methods, we need to consider the average mass and its standard deviation (SD). Based on the available results, we can conclude that the ALO, DA, GWO, MVO, SCA, WOA, HTS, TLBO, PVS, and SOS algorithms have mass mean values of 625.7964, 663.8041, 382.6427, 380.4641, 376.3344, 1109.2196, 362.5218, 367.5719, 282.5537, and 326.5344 kg, respectively, along with the corresponding SD values of 252.8503, 239.5409, 120.9032, 109.8131, 109.5882, 469.3832, 117.7051, 105.6575, 57.8193, and 91.0214.

After analyzing these results, we can conclude that the PVS algorithm performs the best in achieving the lowest mass mean and SD values, making it the most efficient algorithm for this design problem in terms of both convergence rate and consistency. Therefore, the PVS algorithm is the recommended choice for this problem.

Table 6.29 Topology, shape, and size optimization of the 14-bar truss (discrete section)

Element no.	ALO	DA	GWO	MVO	SCA	WOA	HTS	TLBO	PVS	SOS
3	46	48	39	39	40	45	39	39	39	39
4	38	40	32	32	34	—	32	32	32	32
6	—	—	—	—	—	3	—	—	—	—
7	3	3	3	3	3	4	3	3	3	3
10	—	—	—	—	—	38	—	—	—	—
11	—	—	—	—	—	4	—	—	—	—
12	—	—	—	—	—	11	—	—	—	—
13	4	4	4	4	4	—	4	4	4	4
y_1	522.2430	200.6726	520.0898	901.2361	357.2108	180.0000	793.3899	793.3899	793.3899	703.7322
y_3	689.6401	196.3882	788.5464	554.6178	503.3601	849.8682	618.0673	618.0673	618.0673	889.7881
y_5	688.6403	632.4491	958.0321	958.0327	949.7604	804.7174	958.0294	958.0294	958.0294	958.0294
Mass (kg)	274.4180	282.3526	252.6178	252.6178	259.7888	354.1471	252.6177	252.6177	252.6177	252.6177
σ^{max} (MPa)	167.5186	167.5186	167.5186	167.5186	167.5186	167.5186	167.5186	167.5186	167.5186	167.5186
δ^{max} (mm)	89.2767	89.2767	89.2767	89.2767	89.2767	89.2767	89.2767	89.2767	89.2767	89.2767
f_1 (Hz)	4.8411	4.7019	5.1430	5.1430	5.1345	5.4221	5.1430	5.1430	5.1430	5.1430
Mean	625.7964	663.8041	382.6427	380.4641	376.3344	1109.2196	362.5218	367.5719	282.5537	326.5344
SD	252.8503	239.5409	120.9032	109.8131	109.5882	469.3832	117.7051	105.6575	57.8193	91.0214

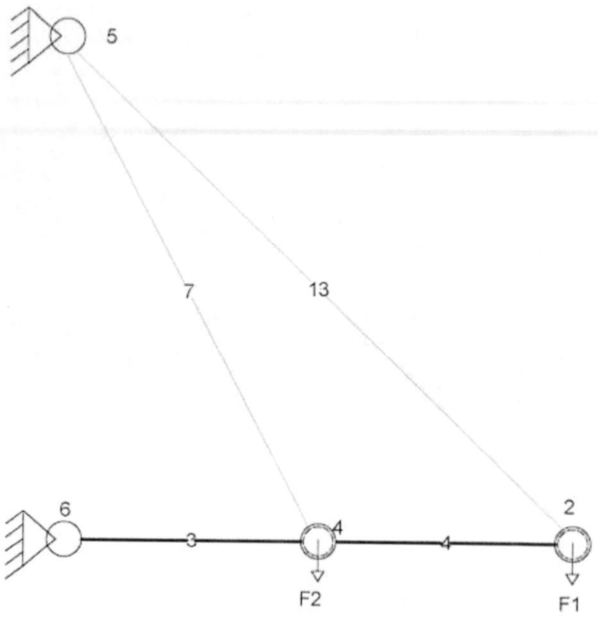

Fig. 6.36 Optimal structure of the 14-bar truss

The TSS optimization technique is a powerful tool for designing lighter trusses. When compared to other optimization methods such as size optimization and topology optimization, TSS optimization performs significantly better in achieving a lightweight design.

The ideal configuration of the 14-bar truss is depicted in Fig. 6.36. A close examination reveals that this optimal arrangement only requires 4 out of the total 14 elements. This finding highlights the efficiency and effectiveness of the selected design, as it achieves the desired structural integrity while minimizing the number of necessary components.

In fact, the results of a recent study showed that when considering TSS optimization, the resulting truss is 66.3% lighter than when using size optimization and 31.7% lighter than when using topology optimization. These numbers are based on the best solutions for each optimization case, and they clearly indicate the superiority of TSS optimization in achieving lightweight truss designs.

Overall, the simultaneous consideration of topological, size, and shape optimization allows for the creation of highly efficient truss structures that are lighter and more robust compared to traditional designs. TSS optimization is a promising technique that can be utilized in a variety of industries, from aerospace engineering to construction, to create stronger, lighter, and more efficient structures.

6.4.3 *TSS Optimization of the 15-Bar Truss with Discrete Cross-Sections*

The third benchmark problem involves analyzing the 15-bar planar truss to assess the influence of discrete size variables. Figure 6.37 displays the ground structure of this truss, while Table 6.30 provides detailed information regarding the design variables, including discrete size variables and other pertinent data.

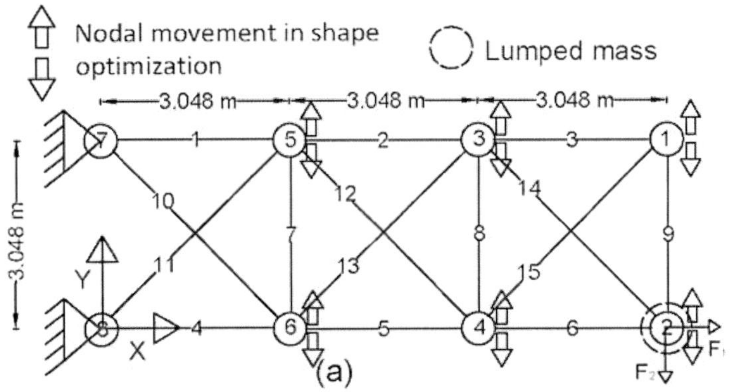

Fig. 6.37 The 15-bar truss

Table 6.30 Design parameters of the 15-bar truss

Size variables: $X_i = A_i$, $i = 1, 2, \ldots, 15$ Shape variables: $2.54\ \text{m} \leq \xi_{x_5} = \xi_{x_6} \leq 3.556\ \text{m}$; $5.588\ \text{m} \leq \xi_{x_3} = \xi_{x_4} \leq 6.604\ \text{m}$; $1.27\ \text{m} \leq \xi_{y_1} \leq 2.286\ \text{m}$; $0.508\ \text{m} \leq \xi_{y_2} \leq 1.524\ \text{m}$; $2.54\ \text{m} \leq \xi_{y_3} \leq 3.556\ \text{m}$; $-0.508\ \text{m} \leq \xi_{y_4} \leq 0.508\ \text{m}$; $2.54\ \text{m} \leq \xi_{y_5} \leq 3.556\ \text{m}$; $-0.508\ \text{m} \leq \xi_{y_6} \leq 0.508\ \text{m}$ (1 in. = 0.0254 m)	
Multiple load conditions:	
Condition 1 : $F_1 = -44.537$ kN, $F_2 = 0$ kN	Condition 2 : $F_1 = 0$ kN, $F_2 = 4.4537$ kN
Lumped mass on node 2 is 500 kg	
Stress and displacement constraints: $\sigma_i^{max} = 172.43$ MPa and $\delta_{x_2\&y_2}^{max} = 100$ mm (0.1 m)	
Natural frequency constraint: $f_1 \geq 4$ Hz	
The discrete cross-sectional areas: $[A^{min}, A^{max}] = [-50, 50]\ \text{cm}^2$ ($[-50, 50] \times 10^{-4}\ \text{m}^2$) in increments of 1 cm^2 and critical area = 1 cm^2	
Material properties: $E = 69$ GPa and $\rho = 2740 \text{kg/m}^3$	

This problem builds upon the previous works of Rahami et al. (2008) and Ahrari et al. (2014). Previous studies primarily focused on investigating the problem under static constraints, such as stress, buckling, and displacement. However, the simultaneous consideration of both static and dynamic constraints, particularly natural frequency constraints, has not been explored.

To address this research gap, the current study modifies the problem by incorporating the effects of multi-load conditions, lumped masses, natural frequency constraints, and discrete cross-sectional areas. The table also includes information about design considerations and material properties, providing a comprehensive understanding of the problem.

Table 6.31 presents the results obtained from TSS optimization of the 15-bar truss with 15 discrete cross-sections and eight shape variables using various algorithms in 100 independent runs. The best run was found to provide the minimum mass value while satisfying all the specified structural constraints. The mean and SD values of the structural mass are provided and will be used to evaluate the statistical performance of the different evolutionary methods.

Among the ALO, DA, GWO, MVO, SCA, WOA, HTS, TLBO, PVS, and SOS algorithms, the best results were achieved by PVS, with a structural mass of 88.2623 kg, followed by HTS, TLBO, and SOS, with masses of 94.2802, 92.5738, and 90.5705 kg, respectively. On the other hand, ALO, DA, GWO, MVO, and SCA algorithms reported higher masses of 103.7078, 131.8855, 91.2259, 88.9083, and 129.3851 kg, respectively.

It can be inferred that PVS algorithm provides the lowest structural mass, which is highly desirable for this design problem.

The results obtained in the TSS optimization of the 15-bar truss using 15 discrete cross-sections and eight shape variables for various algorithms in 100 independent runs are shown in Table 6.31. The best run is identified based on the minimum mass value while meeting all the structural constraints. The mean and SD values of the structural mass are provided to evaluate the performance of the evolutionary methods statistically. The algorithms that produced the best results are ALO, DA, GWO, MVO, SCA, WOA, HTS, TLBO, PVS, and SOS with mass mean values of 155.0170, 1.60E+07, 4.00E+06, 1.40E+07, 162.1220, 239.5319, 124.9370, 113.4083, 96.1872, and 103.9591 kg, respectively. The SD values for these algorithms are 29.0231, 3.70E+07, 1.98E+07, 3.51E+07, 21.5410, 71.1037, 27.5577, 16.0965, 7.2452, and 12.5840, respectively. The algorithm that produced the least structural mass is PVS. Therefore, based on the mass mean and SD values, PVS is the best algorithm in terms of convergence rate and consistency for this design problem.

Additionally, the simultaneous consideration of TSS optimization leads to even more significant improvements in truss design. In comparison to size optimization and topology optimization, the resulting truss is 26.4 and 9.9% lighter, respectively, when using TSS optimization for the best solutions in each case.

Figure 6.38 visually presents the optimal configuration of the 15-bar truss. Upon careful observation of the diagram, it becomes evident that this ideal arrangement only necessitates the use of 10 out of the total 15 elements. This noteworthy

Table 6.31 Topology, shape, and size optimization of the 15-bar truss (discrete section)

Element no.	ALO	DA	GWO	MVO	SCA	WOA	HTS	TLBO	PVS	SOS
1	9	9	8	8	12	9	6	6	7	6
2	8	29	5	5	6	12	8	7	7	6
4	16	11	12	12	16	10	9	10	10	10
5	12	10	9	10	14	12	11	11	12	12
6	13	13	13	12	14	14	12	13	11	14
7	4	2	—	—	7	11	—	3	—	3
8	9	2	1	1	10	—	3	2	1	2
10	—	10	—	—	—	5	5	—	4	—
11	4	—	2	1	4	—	—	1	1	2
12	—	8	5	5	—	4	1	—	—	—
13	1	—	2	1	4	1	3	5	1	2
14	7	10	7	7	14	10	9	8	7	7
$x_5 = x_6$	102.0681	140.0000	122.0652	129.1202	140.0000	100.0000	105.5801	100.1648	109.0038	100.0136
$x_3 = x_4$	225.5975	260.0000	221.4051	230.5372	220.0000	220.0000	227.6567	220.0053	239.7283	220.0006
y_1	51.4365	90.0000	55.1521	87.7970	57.9095	58.9843	87.4481	88.3463	81.8763	50.0000
y_2	20.2151	54.9698	29.3196	33.1319	39.7828	20.0000	38.4805	20.0161	38.0797	24.2930
y_3	100.0000	133.8059	100.0000	100.0000	140.0000	100.0000	100.0000	100.0004	100.0000	100.0000
y_4	10.5242	6.1926	14.5660	17.0889	0.8399	4.3166	8.9202	−0.0663	19.8644	9.7632
y_5	105.6286	134.0360	100.0000	100.0000	122.3193	100.0000	113.1153	100.0000	119.0706	114.4047
y_6	−2.4324	6.7514	7.3338	8.9276	5.2085	−6.2890	7.5471	9.5509	12.2797	2.1338
Mass (kg)	103.7078	131.8855	91.2259	88.9083	129.3851	112.5226	94.2802	92.5738	88.2623	90.5705
σ^{max} (MPa)	58.5937	58.5937	58.5937	58.5937	58.5937	58.5937	58.5937	58.5937	58.5937	58.5937
δ^{max} (mm)	9.9993	9.9993	9.9993	9.9993	9.9993	9.9993	9.9993	9.9993	9.9993	9.9993
f_1 (Hz)	4.6441	5.5360	4.8356	4.7644	5.2292	4.7365	5.0215	4.7576	4.6788	4.6667
Mean	155.0170	1.60E+07	4.00E+06	1.40E+07	162.1220	239.5319	124.9370	113.4083	96.1872	103.9591
SD	29.0231	3.70E+07	1.98E+07	3.51E+07	21.5410	71.1037	27.5577	16.0965	7.2452	12.5840

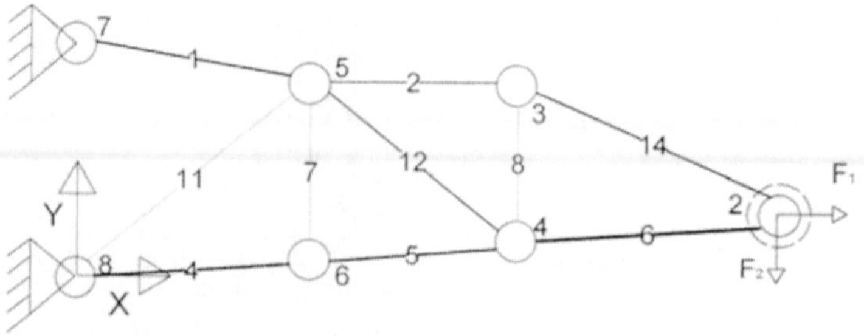

Fig. 6.38 Optimal structure of the 15-bar truss

discovery emphasizes the efficiency and effectiveness of the selected design, as it effectively achieves the desired level of structural integrity while minimizing the overall number of components required.

These findings demonstrate the impressive performance of TSS optimization in achieving lightweight truss designs, which is of great importance in various industries such as aerospace and civil engineering. The ability to design lighter and stronger truss structures can lead to significant cost savings, reduced environmental impact, and improved safety and performance.

In summary, the use of TSS optimization provides a valuable tool for achieving optimal truss designs that are both lightweight and robust. This technique can be utilized in a wide range of applications to create more efficient structures that meet the demands of modern engineering.

6.4.4 TSS Optimization of the 24-Bar Truss with Discrete Cross-Sections

Figure 6.39 illustrates the ground structure of the fourth benchmark problem, which is referred to as the 24-bar truss. Initially, Ringertz (1986) investigated this truss problem using a Branch and Bound algorithm to optimize its topology, primarily focusing on static constraints. Subsequently, Xu et al. (2003) employed a 1-D search method to test the truss, and Kaveh and Zolghadr (2013) examined the search performance of charged system search and particle swarm optimization algorithms, considering continuous design variables. However, the objective of this study is to investigate the truss with discrete design variables.

In this problem, the truss is subjected to two external load cases, which are specified in Table 6.32. The cross-sectional areas of the 24 bars are treated as discrete size variables, while the horizontal movement of nodes 7 and 8 is considered as the shape variables. The table provides additional information such as nodal masses, material properties, and other necessary data for a comprehensive understanding of the problem. Furthermore, a nonstructural lumped mass of 500 kg is applied at node 3 of the truss.

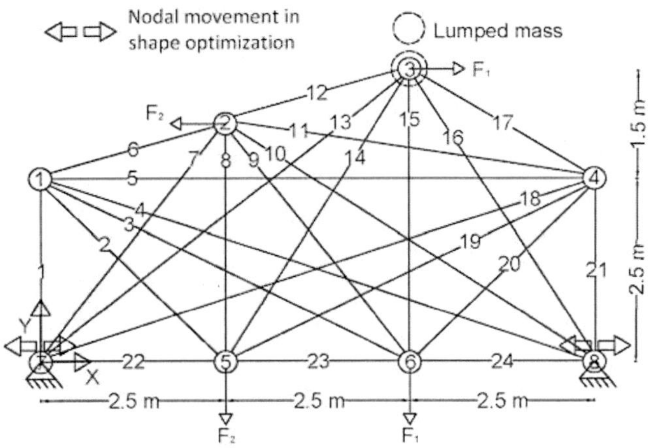

Fig. 6.39 The 24-bar truss

Table 6.32 Design parameters of the 24-bar truss

Size variables: $X_i = A_i$, $i = 1, 2, \ldots, 24$ Shape variables: $-2\text{ m} \le \xi_{x_7}, \xi_{x_8} \le 2$ m, with respect to their original coordinates	
Multiple load conditions:	
Condition 1 : $F_1 = 50$ kN, $F_2 = 0$ kN	Condition 2 : $F_1 = 0$ kN, $F_2 = 50$ kN
Lumped mass on node 3 is 500 kg	
Stress and displacement constraints: $\sigma_i^{\max} = 172.43$ MPa and $\delta_{y_5 \& y_6}^{\max} = 10$ mm $\left(10^{-2}\text{ m}\right)$	
Natural frequency constraint: $f_1 \ge 30$ Hz	
The discrete cross-sectional areas: $[A^{\min}, A^{\max}] = [-40, 40]$ cm^2 ($[-40, 40] \times 10^{-4}$ m^2) in increments of 1 cm^2 and critical area = 1 cm^2	
Material properties: $E = 69$ GPa and $\rho = 2740$kg/m^3	

In this section, we discuss the TSS optimization of a 24-bar truss with 24 discrete size and two continuous shape variables. The results obtained for various algorithms in 100 independent runs are presented in Table 6.33.

It is observed that the best run is giving the minimum mass value while satisfying all the imposed structural constraints. The mean and standard deviation (SD) values of the structural mass are provided, and they are used to evaluate the performance of different evolutionary methods statistically.

The best results reported by ALO, DA, GWO, MVO, SCA, WOA, HTS, TLBO, PVS, and SOS algorithms are 136.4515, 163.1389, 112.9946, 111.4551, 141.6025, 217.6452, 123.1010, 123.3892, 112.2889, and 112.4878 kg, respectively. It can be observed that the MVO algorithm provides the least structural mass, with the PVS algorithm being a close second. The two mass values can be considered similar.

Table 6.33 Topology, shape, and size optimization of the 24-bar truss (discrete section)

Element no.	ALO	DA	GWO	MVO	SCA	WOA	HTS	TLBO	PVS	SOS
1	—	3	—	—	—	—	—	—	—	—
4	—	—	—	—	—	2	—	—	—	—
5	—	—	—	—	—	16	—	—	—	—
6	—	7	—	—	—	5	—	—	—	—
7	18	21	13	13	22	32	13	18	13	14
8	11	3	3	3	8	15	3	9	3	3
10	3	—	—	—	12	12	—	3	—	—
12	—	4	3	3	16	23	6	—	3	3
13	18	25	21	22	—	—	22	20	21	22
14	—	—	1	—	7	10	—	—	—	—
15	4	6	4	4	10	4	5	4	4	4
16	25	27	24	24	22	26	25	25	25	24
17	7	—	—	—	—	—	—	—	—	—
18	—	1	—	—	—	—	—	—	—	—
19	1	—	—	—	—	—	—	—	—	—
20	1	1	—	—	—	—	—	—	—	—
21	—	—	—	—	—	1	—	—	—	—
22	—	8	—	1	4	—	8	4	1	1
23	—	9	—	—	—	7	—	—	—	1
24	1	6	2	1	2	—	8	1	1	—
y_7	2.0000	2.0000	1.8200	2.0000	−0.0153	−0.9895	1.5333	1.9281	2.0000	1.9958
y_8	0.0870	0.2338	−0.2878	−0.4059	−0.6386	−1.9941	−0.5891	−0.3284	−0.1378	−0.4053
Mass (kg)	136.4515	163.1389	112.9946	111.4551	141.6025	217.6452	123.1010	123.3892	112.2889	112.4878
σ^{max} (MPa)	152.3732	152.3732	152.3732	152.3732	152.3732	152.3732	152.3732	152.3732	152.3732	152.3732
δ^{max} (mm)	8.1956	8.1956	8.1956	8.1956	8.1956	8.1956	8.1956	8.1956	8.1956	8.1956
f_1 (Hz)	30.0310	30.4120	30.2479	30.4431	31.1488	30.0036	30.3957	30.4055	30.0806	30.0025
Mean	194.0142	265.6511	136.5462	133.7365	187.4308	391.6002	180.1949	174.1376	131.2473	130.6953
SD	50.5668	51.3444	14.4927	18.4005	23.4563	95.4333	40.4768	43.2435	10.0399	14.5219

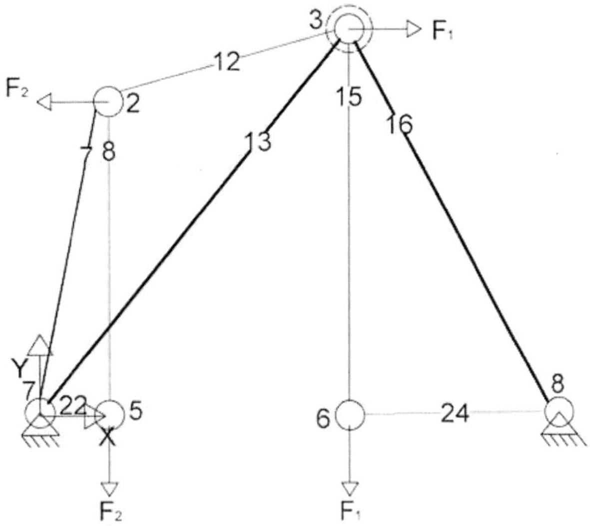

Fig. 6.40 Optimal structure of the 24-bar truss

In order to accurately evaluate the performance of various metaheuristic methods in optimizing the 24-bar truss with discrete cross-sections and continuous shape variables, the average mass and standard deviation are taken into consideration. Based on the results obtained from 100 independent runs, it can be concluded that the ALO, DA, GWO, MVO, SCA, WOA, HTS, TLBO, PVS, and SOS algorithms have average mass values of 194.0142, 265.6511, 136.5462, 133.7365, 187.4308, 391.6002, 180.1949, 174.1376, 131.2473, and 130.6953 kg, respectively, with the corresponding standard deviations of 50.5668, 51.3444, 14.4927, 18.4005, 23.4563, 95.4333, 40.4768, 43.2435, 10.0399, and 14.5219.

Upon analyzing these results, it can be inferred that the SOS algorithm has the best average mass value, while the PVS algorithm achieves the best standard deviation value. However, it is important to note that the MVO algorithm closely follows the PVS algorithm in terms of achieving the least structural mass. Therefore, for this particular design problem, the MVO, SOS, and PVS algorithms can be considered the better algorithms for both convergence rate and consistency.

The optimal arrangement of the 24-bar truss is visually depicted in Fig. 6.40. By closely examining the diagram, it becomes evident that this optimal configuration only utilizes 8 out of the available 24 elements. This significant finding highlights the efficiency and effectiveness of the chosen design, as it successfully achieves the desired level of structural integrity while minimizing the overall number of required components.

Furthermore, when TSS optimization is considered alongside size and topology optimization, it results in even greater improvements in truss design. Specifically, the resulting truss is 45.1% lighter compared to size optimization and 7.2% lighter compared to topology optimization for the best solutions in each case.

This significant improvement in truss weight demonstrates the effectiveness of TSS optimization in creating lighter and more efficient truss structures. By taking into account both topological and geometric design considerations, TSS optimization can help engineers achieve the optimal balance between strength, weight, and cost.

In summary, the use of TSS optimization is a powerful tool for designing lightweight truss structures. Its ability to outperform traditional optimization techniques such as size and topology optimization makes it an important consideration in various industries, including aerospace, civil engineering, and automotive. By leveraging the benefits of TSS optimization, engineers can create more efficient and sustainable structures that meet the evolving demands of modern engineering.

6.4.5 *TSS Optimization of the 20-Bar Truss with Discrete Cross-Sections*

The ground structure of the benchmark problem, known as the 20-bar truss, is depicted in Fig. 6.41. This problem was first formulated by Xu et al. (2003) for topology and size optimization and subsequently tested by Kaveh and Zolghadr (2013) for continuous optimization.

Table 6.34 offers a comprehensive overview of the design variables, loading conditions, structural safety constraints, and other pertinent data related to the test problem. It is important to highlight that, unlike other cases, no lumped masses are applied to the truss in this particular problem.

This particular section of the study is dedicated to optimizing the 20-bar truss, using a combination of 20 discrete size and six continuous shape variables. To

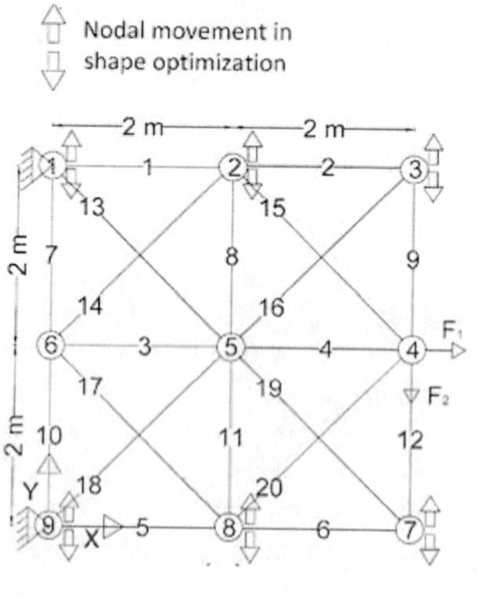

Fig. 6.41 The 20-bar truss

Table 6.34 Design parameters of the 20-bar truss

Size variables: $X_i = A_i$, $i = 1, 2, \ldots, 20$ Shape variables: $-1\ \text{m} \leq \xi_{y_1}, \xi_{y_2}, \xi_{y_3}, \xi_{y_7}, \xi_{y_8}, \xi_{y_9} \leq 1$ m, with respect to their original coordinates	
Multiple load conditions:	
Condition 1 : $F_1 = 500$ kN, $F_2 = 0$ kN	Condition 2 : $F_1 = 0$ kN, $F_2 = 500$ kN
No lumped mass	
Stress and displacement constraints: $\sigma_i^{max} = 172.43$ MPa and $\delta_{y_4}^{max} = 60$ mm $\left(6 \times 10^{-2}\ \text{m}\right)$	
Natural frequency constraints: $f_1 \geq 60$ Hz and $f_2 \geq 100$ Hz	
The discrete cross-sectional areas: $[A^{min}, A^{max}] = [1, 100]$ cm^2 ([1, 100] $\times$ 10^{-4} m^2) in increments of 1 cm^2 and critical area = 1 cm^2	
Material properties: $E = 69$ GPa and $\rho = 2740$kg/m^3	

evaluate the effectiveness of the optimization process, the results obtained from the truss optimization with discrete cross-sections are presented in Table 6.35. This table showcases the outcomes of 100 independent runs of the TSS algorithm.

Upon analyzing the table, it was found that the best run resulted in achieving the minimum mass value for the optimized truss while still fulfilling all the prescribed structural constraints. In addition, the mean and standard deviation (SD) values of the structural mass are given in the table, which can be utilized to assess the performance of various evolutionary methods statistically.

The results section of Table 6.35 displays the best outcomes achieved by the ALO, DA, GWO, MVO, SCA, WOA, HTS, TLBO, PVS, and SOS algorithms. The minimum structural mass values achieved by these algorithms are 131.6950, 219.3218, 117.7873, 116.9796, 192.7776, 271.3296, 135.8179, 126.6655, 116.6868, and 128.7385 kg, respectively. It can be seen that among all the algorithms used, the PVS algorithm resulted in the lowest structural mass value.

Overall, these results provide a comprehensive understanding of the effectiveness of various optimization algorithms in reducing the structural mass of the 20-bar truss, while still satisfying all the given structural constraints.

To assess the performance of the metaheuristic methods accurately, the average mass and its standard deviation (SD) are considered. Based on the available results, it can be concluded that the ALO, DA, GWO, MVO, SCA, WOA, HTS, TLBO, PVS, and SOS algorithms achieved mean mass values of 239.2611, 8.01E+06, 2.00E+06, 1.02E+07, 249.9738, 421.0046, 2128.1980, 476.4928, 147.1198, and 156.7586 kg, respectively. Additionally, the SD values for these algorithms are 44.2000, 2.74E+07, 1.41E+07, 3.06E+07, 38.4641, 81.6203, 5630.2628, 1411.6822, 16.8557, and 19.0723 kg, respectively.

It is worth noting that the PVS algorithm achieved the best mean and SD values for the mass, making it the best algorithm in terms of both convergence rate and consistency for this design problem. Therefore, it can be concluded that the PVS algorithm is the most effective optimization algorithm for reducing the structural mass of the 20-bar truss while still satisfying all the prescribed structural constraints.

Table 6.35 Topology, shape, and size optimization of the 20-bar truss (discrete section)

Element no.	ALO	DA	GWO	MVO	SCA	WOA	HTS	TLBO	PVS	SOS
1	29	17	32	23	69	33	29	27	30	30
2	—	—	—	—	—	28	—	—	—	—
3	—	2	—	—	—	—	—	3	—	—
4	—		2	5	−61	24	17	3	1	—
5	28	45	32	32	30	42	21	30	31	36
7	—	36	—	—	—	1	—	—	—	—
8	9	19	4	11	21	32	9	4	3	4
9	—	25	—	—	—	—	—	—	—	—
10	—	—	—	—	—	34	—	6	2	—
11	15	26	1	2	6	19	7	—	1	4
12	—	—	1	—	—	—	—	—	—	—
13	14	16	2	7	10	11	1	2	—	7
14	—	11	—	1	—	33	—	3	2	—
15	32	22	32	26	33	22	31	27	30	36
16	—	1	—	—	—	11	—	—	—	—
17	—	—	—	—	—	15	—	4	1	—
18	21	35	6	—	27	28	26	—	—	3
19	—	—	1	—	—	—	—	—	—	—
20	31	34	32	31	71	35	26	32	31	39
y_1	−0.2370	−0.2309	0.0018	0.2743	0.1812	−0.7333	1.0000	0.8906	0.4098	−0.0090
y_2	−0.9310	−0.0837	−0.9914	−0.5764	−0.8751	−0.7333	−0.3409	−0.5072	−0.7893	−1.0000
y_3	0.9350	−1.0000	−0.0391	0.2349	−0.0528	−0.7333	1.0000	−0.8248	0.6118	−0.0582
y_7	−0.9376	−0.3305	−0.8283	0.6551	−0.7749	0.1858	−0.9618	0.9927	0.2640	−0.6293
y_8	0.9871	−0.0245	0.9905	0.9465	0.8479	0.1775	0.7508	0.7929	0.9436	0.8009
y_9	0.7297	−0.3709	0.0109	−0.1666	0.0433	−0.7333	0.2132	−0.4201	−0.1095	−0.4023
Mass (kg)	131.6950	219.3218	117.7873	116.9796	192.7776	271.3296	135.8179	126.6655	116.6868	128.7385

σ^{max} (MPa)	659.9911	659.9911	659.9911	659.9911	659.9911	659.9911	659.9911	659.9911	659.9911	659.9911
δ^{max} (mm)	28.0732	28.0732	28.0732	28.0732	28.0732	28.0732	28.0732	28.0732	28.0732	28.0732
f_1 (Hz)	112.1823	104.9429	63.4397	62.6739	97.0670	106.9277	74.6674	60.6878	63.1932	69.2463
f_2 (Hz)	222.3253	149.4072	120.1095	160.9463	183.5850	145.8060	177.1716	102.8726	110.7973	167.7071
Mean	239.2611	8.01E+06	2.00E+06	1.02E+07	249.9738	421.0046	2128.1980	476.4928	147.1198	156.7586
SD	44.2000	2.74E+07	1.41E+07	3.06E+07	38.4641	81.6203	5630.2628	1411.6822	16.8557	19.0723

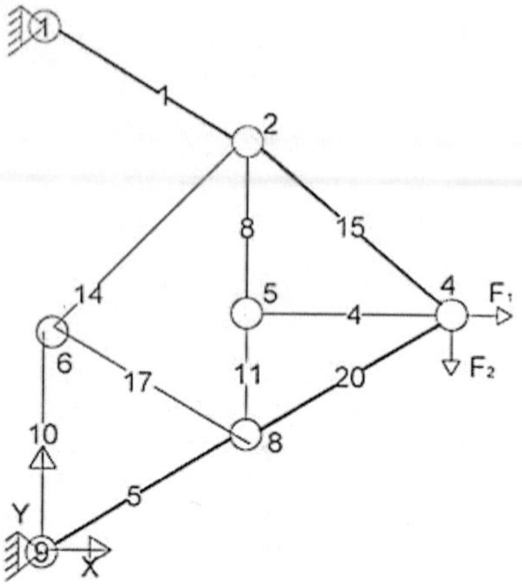

Fig. 6.42 Optimal structure of the 20-bar truss

Figure 6.42 illustrates the optimal arrangement of the 20-bar truss, visually presenting the ideal configuration. Through careful examination of the diagram, it becomes apparent that this optimal arrangement utilizes only 10 out of the 20 available elements. This noteworthy discovery highlights the efficiency and effectiveness of the selected design, as it effectively achieves the desired level of structural integrity while minimizing the overall number of components required.

Additionally, the simultaneous consideration of TSS optimization leads to even more significant improvements in truss design. In comparison to size optimization and topology optimization, the resulting truss is 43.7 and 24.6% lighter, respectively, when using TSS optimization for the best solutions in each case.

These impressive results highlight the effectiveness of TSS optimization in achieving lightweight and efficient truss designs. By considering both the topology and geometric features of the structure, TSS optimization can identify the most optimal design configuration for achieving the desired performance goals while minimizing weight.

Overall, the use of TSS optimization is a valuable tool in a wide range of applications, including aerospace, civil engineering, and automotive, where the demand for lightweight and efficient structures is crucial. By leveraging the benefits of TSS optimization, engineers can create truss designs that are not only lighter but also more robust and cost-effective.

6.4.6 *TSS Optimization of the 72-Bar 3D Truss with Discrete Cross-Sections*

Figure 6.43 presents the ground structure of the sixth test problem, which is the 72-bar truss. This truss is a 3D structure that was initially utilized by Kaveh and Zolghadr (2013) for continuous optimization. However, in this study, the problem is further investigated for discrete optimization as well.

Fig. 6.43 The 72-bar 3D truss

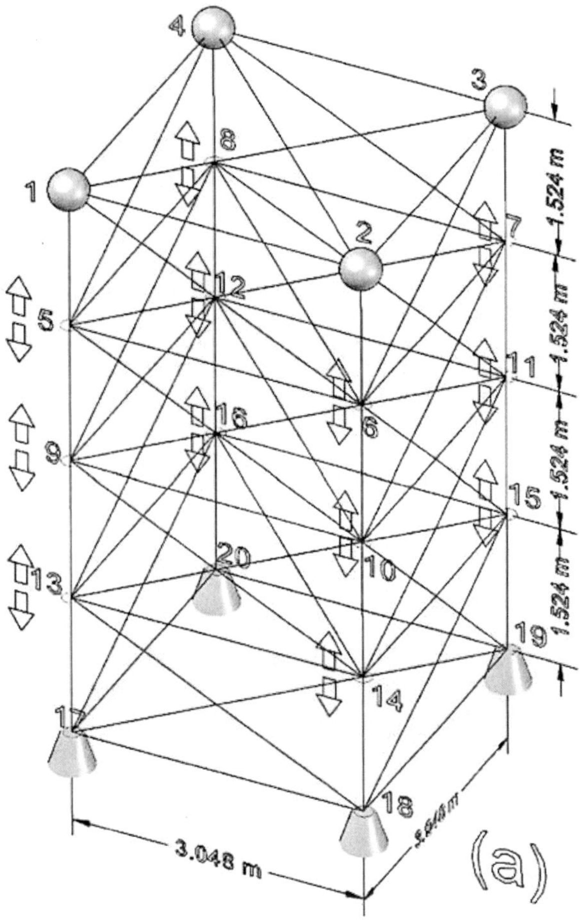

Table 6.36 provides comprehensive information regarding material properties, design variables, loading conditions, lumped mass, and other relevant data specific to the test problem. This table includes all the necessary details required to understand and analyze the truss optimization problem.

In order to maintain structural symmetry, the elements are grouped into 16 clusters, denoted as G_1 (A_1–A_4), G_2 (A_5–A_{12}), G_3 (A_{13}–A_{16}), G_4 (A_{17}–A_{18}), G_5 (A_{19}–A_{22}), G_6 (A_{23}–A_{30}), G_7 (A_{31}–A_{34}), G_8 (A_{35}–A_{36}), G_9 (A_{37}–A_{40}), G_{10} (A_{41}–A_{48}), G_{11} (A_{49}–A_{52}), G_{12} (A_{53}–A_{54}), G_{13} (A_{55}–A_{58}), G_{14} (A_{59}–A_{66}), G_{15} (A_{67}–A_{70}), and G_{16} (A_{71}–A_{72}), as per Kaveh and Zolghadr (2013).

Additionally, the truss includes four nonstructural lumped masses of 2270 kg each, which are added to the top nodes numbered 1–4.

Table 6.36 Design parameters of the 72-bar truss

Size variables: G_i, $i = 1, 2, \ldots, 16$; $G =$ Group number Shape variables: $-1\ \text{m} \leq \xi_{y_5}, \xi_{y_9}, \xi_{y_{13}} \leq 1$ m, with respect to their original coordinates				
Multiple load conditions:				
Condition 1:	$F_{1x} = 22.25$ kN	$F_{1y} = 22.25$ kN		$F_{1z} = -22.25$ kN
Condition 2:	$F_{1z} = -22.25$ kN	$F_{2z} = -22.25$ kN	$F_{3z} = -22.25$ kN	$F_{4z} = -22.25$ kN
Lumped masses at nodes 1, 2, 3, and 4 are 2270 kg				
Stress and displacement constraints: $\sigma_i^{max} = 172.375$ MPa and $\delta_{x_j/y_j}^{max} = 6.35$ mm $(6.35 \times 10^{-3}\ \text{m})$ (for node, $j = 1, 2, 3$, and 4 along x- and y-axes)				
Natural frequency constraints: $f_1 \geq 4$ Hz and $f_3 \geq 6$ Hz				
The discrete cross-sectional areas: $[A^{min}, A^{max}] = [-30, 30]$ cm^2($[-30, 30] \times 10^{-4}$ m^2) in increments of 1 cm^2 and critical area = 1 cm^2				
Material properties: $E = 68.95$ GPa and $\rho = 2767.99$kg/m^3				

Table 6.37 presents the results of TSS optimization of the 72-bar truss using discrete cross-sections for different algorithms in 100 independent runs. The best run resulted in the minimum mass value while satisfying all the prescribed structural constraints. The mean and SD values of the structural mass are provided to assess the performance of various evolutionary methods statistically.

Among the ALO, DA, GWO, MVO, SCA, WOA, HTS, TLBO, PVS, and SOS algorithms, the best results were reported as 472.4064, 533.8369, 446.7584, 461.1806, 533.5511, 598.5806, 450.7343, 439.8038, 446.2105, and 447.7331 kg, respectively. The TLBO algorithm gave the least structural mass, whereas the second-best result was observed for the PVS algorithm. These findings suggest that the TLBO algorithm is the most effective optimization algorithm for reducing the structural mass of the 72-bar truss while still satisfying all the prescribed structural constraints, followed closely by the PVS algorithm.

After analyzing the results obtained from TSS optimization of the 72-bar truss using discrete cross-sections with various metaheuristic methods, it can be concluded that the average mass and its SD are important indicators to assess the true performance of these methods.

Among the ALO, DA, GWO, MVO, SCA, WOA, HTS, TLBO, PVS, and SOS algorithms, the mass mean values were 744.6466, 4.00E+06, 469.5318, 1.00E+07, 652.9597, 1413.2644, 597.0388, 501.9354, 470.7167, and 469.5140 kg, respectively, while the SD values were 198.6668, 1.98E+07, 44.3020, 3.03E+07, 83.1197, 502.5712, 217.6537, 90.9292, 53.7416, and 24.8457, respectively.

It was observed that the SOS algorithm outperformed the other methods with the best mass mean and the best SD value. Therefore, the SOS algorithm can be considered the most effective method for optimizing the 72-bar truss while satisfying all the prescribed structural constraints.

Table 6.37 Topology, shape, and size optimization of the 72-bar 3D truss (discrete section)

Element group no.	ALO	DA	GWO	MVO	SCA	WOA	HTS	TLBO	PVS	SOS
1	4	10	5	5	20	21	5	4	4	5
2	10	9	10	10	12	14	10	10	10	10
4	10	—	—	—	—	10	—	—	—	—
5	8	9	7	11	19	3	8	8	8	8
6	9	7	8	8	8	6	7	8	9	9
7	—	—	—	—	—	—	3	—	—	—
8	—	15	3	4	9	—	6	4	3	3
9	11	7	12	12	11	19	12	14	12	11
10	9	9	9	9	13	9	8	8	9	9
11	—	7	—	—	—	11	—	—	—	—
12	—	—	—	—	—	5	—	—	—	—
13	19	28	16	15	14	18	16	17	15	17
14	9	12	9	10	7	8	8	7	8	8
15	—	—	—	—	—	6	—	3	—	—
16	—	—	—	—	—	—	2	—	—	—
y_{13}	−0.4731	0.1193	0.0272	−0.5530	−0.3186	−0.6507	0.2567	−0.1847	0.0963	−0.1306
y_9	0.1093	1.0000	−0.0009	−0.1497	−0.2730	−0.0692	0.3421	−0.2565	0.0481	−0.2020
y_5	0.2375	1.0000	0.2395	0.2106	0.0031	−0.6963	0.1400	0.1207	0.2502	0.2305
Mass (kg)	472.4064	533.8369	446.7584	461.1806	533.5511	598.5806	450.7343	439.8038	446.2105	447.7331
σ^{max} (MPa)	85.4425	85.4425	85.4425	85.4425	85.4425	85.4425	85.4425	85.4425	85.4425	85.4425
δ^{max} (mm)	4.8190	4.8190	4.8190	4.8190	4.8190	4.8190	4.8190	4.8190	4.8190	4.8190
f_1 (Hz)	4.0035	4.0889	4.0018	4.0251	4.1341	4.0002	4.0113	4.0000	4.0000	4.0007
f_3 (Hz)	6.0092	6.1197	6.0002	6.0059	6.2629	6.1142	6.0080	6.0015	6.0211	6.0004
Mean	744.6466	4.00E+06	469.5318	1.00E+07	652.9597	1413.2644	597.0388	501.9354	470.7167	469.5140
SD	198.6668	1.98E+07	44.3020	3.03E+07	83.1197	502.5712	217.6537	90.9292	53.7416	24.8457

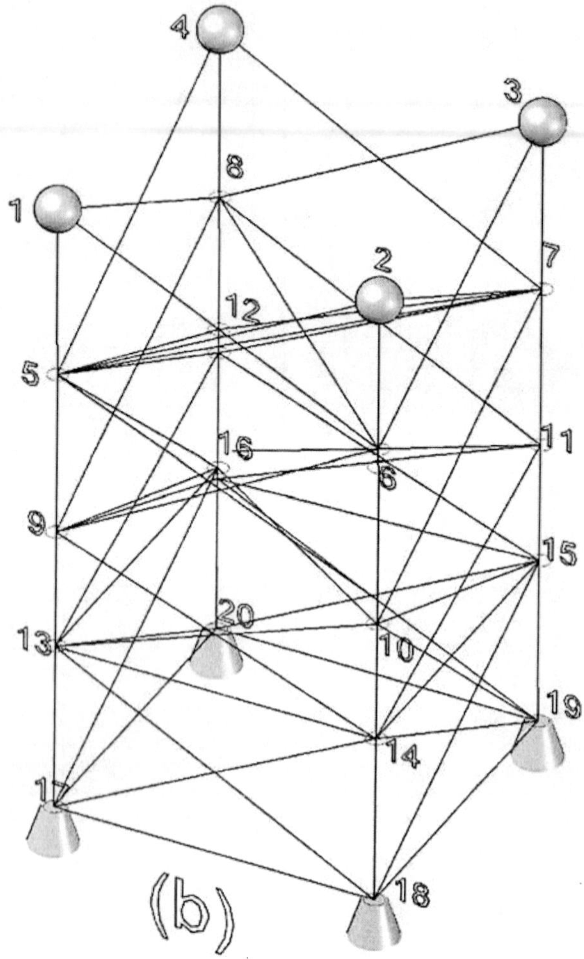

Fig. 6.44 Optimal structure of the 72-bar 3D truss

Moreover, the PVS and TLBO algorithms also performed well, with close mass mean values and SD values, and can be considered as the second and third-best algorithms, respectively, for this design problem.

Figure 6.44 illustrates the optimal arrangement of the 72-bar 3D truss, offering a visual representation of the ideal configuration. Upon careful examination of the diagram, it becomes evident that this optimal arrangement utilizes only 9 groups of elements out of the available 15. This significant finding highlights the efficiency and effectiveness of the selected design, as it successfully achieves the desired level of structural integrity while minimizing the overall number of components required.

Furthermore, when TSS optimization is considered in isolation and in combination with other optimization methods, it is still effective in achieving lighter truss designs. Specifically, the simultaneous consideration of TSS optimization results in a truss that is 8.3% lighter compared to using size optimization alone, and 1.6% lighter compared to using topology optimization for the best solutions in each case.

These results indicate that TSS optimization is a valuable tool for achieving optimal truss designs, even when used in combination with other optimization methods. By considering both the topology and geometric features of the structure, TSS optimization can identify the most optimal design configuration for achieving the desired performance goals while minimizing weight.

6.4.7 *TSS Optimization of the 39-Bar Truss with Discrete Cross-Sections*

The seventh benchmark problem focuses on the investigation of a two-tier 39-bar planar truss, as displayed in Fig. 6.45. The objective of this problem is to analyze the behavior of a large-sized truss under stress, buckling, displacement, and fundamental frequency constraints while considering discrete design variables. Initial work on this problem was conducted by Deb and Gulati (2001), followed by subsequent studies by Luh and Lin (2008), Wu and Tseng (2010), and Ahrari et al. (2014), with a primary emphasis on static constraints.

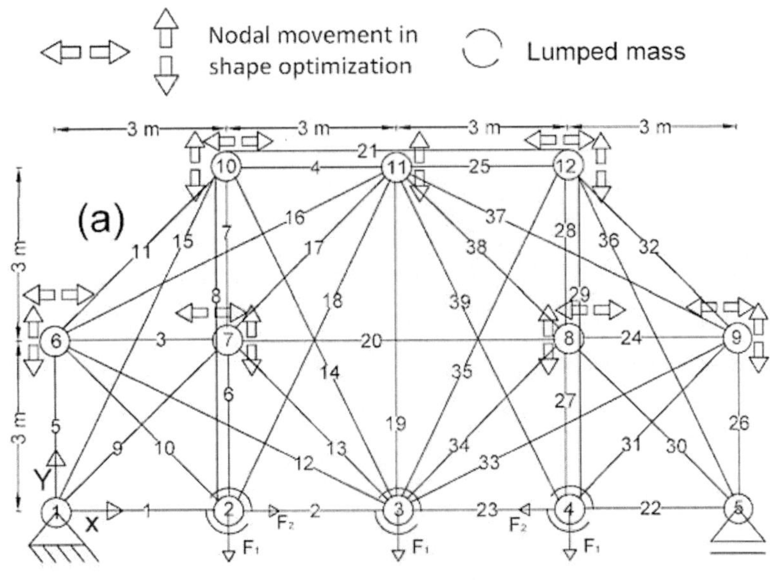

Fig. 6.45 The 39-bar truss

Table 6.38 Design parameters of the 39-bar truss

Size variables: $X_i = G_i$, $i = 1, 2, \ldots, 21$ Shape variables: $-2.5\ \text{m} \leq \xi_{x_j/y_j} \leq 2.5\ \text{m}$, for $x_6 = -x_9$, $y_6 = y_9$, $x_7 = -x_8$, $y_7 = y_8$, $x_{10} = -x_{12}$, $y_{10} = y_{12}$, y_{11} with respect to their original coordinates	
Multiple load conditions:	
Condition 1 : $F_1 = 100$ kN, $F_2 = 0$ kN	Condition 2 : $F_1 = 0$ kN, $F_2 = 50$ kN
Lumped masses on nodes 2, 3, and 4 are 500 kg	
Stress and displacement constraints: $\sigma_i^{max} = 172.43$ MPa and $\delta_{x_j/y_j}^{max} = 50$ mm (0.05 m) (for node, $j = 2, 3$, and 4 along x- and y-axes)	
Natural frequency constraints: $f_1 \geq 4$ Hz	
The discrete cross-sectional areas: $[A^{min}, A^{max}] = [-50, 50]\ \text{cm}^2$ ($[-50, 50] \times 10^{-4}\ \text{m}^2$) in increments of 1 cm^2 and critical area = 1 cm^2	
Material properties: $E = 69$ GPa and $\rho = 2740\text{kg/m}^3$	

However, the exploration of this problem has not yet considered the simultaneous consideration of both static and dynamic constraints. To address this gap, this study modifies the problem to incorporate the effects of multi-load conditions, lumped masses, and natural frequency constraints. Detailed information regarding the design variables, multi-load conditions, lumped masses, constraints, and material properties can be found in Table 6.38.

The truss elements are categorized into 21 groups, taking into account the symmetry about the vertical middle plane, similar to the approach employed by Deb and Gulati (2001). Additionally, the truss is subjected to three nonstructural lumped masses of 500 kg each, positioned at nodes 2, 3, and 4. To satisfy the multi-load condition, two different loading conditions are considered, as depicted in Fig. 6.38.

In this section, the optimization of a 39-bar 3D truss with 11 discrete size and five shape variables is performed. The obtained results from Table 6.39 show the performance of various evolutionary algorithms in 100 independent runs, where the best run is selected based on the minimum mass value while satisfying all the stated structural constraints. The mean and standard deviation (SD) values of the structural mass are also provided for statistical performance assessment.

The best results reported by ALO, DA, GWO, MVO, SCA, WOA, HTS, TLBO, PVS, and SOS algorithms are 264.1801, 361.6735, 210.0713, 199.6929, 307.0478, 525.1092, 220.8201, 194.8783, 193.6027, and 202.5162 kg, respectively. These results indicate that the PVS algorithm has achieved the best performance compared to the other algorithms in terms of minimizing the structural mass.

Therefore, based on the obtained results, it can be concluded that the PVS algorithm is the most effective in optimizing the 39-bar 3D truss, resulting in the lowest mass value while satisfying all the structural constraints.

Table 6.39 Topology, shape, and size optimization of the 39-bar truss (discrete section)

Element no.	ALO	DA	GWO	MVO	SCA	WOA	HTS	TLBO	PVS	SOS
1	11	7	4	2	17	7	9	6	6	8
2	13	16	13	13	20	17	13	13	13	13
3	—	—	—	22	—	15	—	30	—	—
4	—	11	25	—	—	12	—	—	24	—
5	16	8	—	18	1	44	—	21	15	23
6	4	25	5	2	11	—	8	—	—	—
7	—	40	21	15	—	—	—	—	—	—
8	—	—	3	—	—	—	—	—	5	4
9	29	30	16	—	46	—	—	—	—	—
10	1	2	—	6	2	23	—	8	3	4
11	—	7	7	—	—	9	—	—	22	23
12	—	—	1	—	—	—	—	1	—	—
13	—	—	—	1	6	20	—	3	—	—
14	—	13	4	3	—	—	—	—	2	4
15	—	—	—	—	—	17	—	—	—	—
16	—	—	1	—	—	33	—	—	—	—
17	35	—	—	—	25	4	33	—	—	—
18	3	—	—	—	—	—	1	—	—	—
19	26	—	—	—	6	6	7	—	4	—
20	—	—	—	—	—	15	—	19	—	—
21	—	48	—	24	—	—	—	—	—	31
x_6	−0.0010	2.5000	2.5000	0.5337	−1.8266	1.1871	2.3721	1.4507	0.9922	1.7370
y_6	−1.6578	−1.2850	−0.2926	−0.6781	−1.9833	0.3949	1.8486	−0.8960	−1.4106	−0.9287
x_7	−0.5733	−1.8047	−2.1454	−0.0468	0.1665	1.2403	−0.7626	1.8914	−0.2764	−0.3052
y_7	−1.0485	−1.3068	−1.1413	1.2710	0.0459	0.5067	−0.5913	−0.0396	−1.7000	−0.3970

(continued)

Table 6.39 (continued)

Element no.	ALO	DA	GWO	MVO	SCA	WOA	HTS	TLBO	PVS	SOS
x_{10}	2.1321	0.9949	−0.0903	1.9749	−0.0016	0.7786	1.4026	1.7602	−0.0450	1.1384
y_{10}	−1.1577	−2.5000	−2.5000	−0.8205	1.2528	−0.9920	−0.7635	0.3929	−2.4998	−2.4998
y_{11}	−2.5000	−0.4724	−2.5000	1.2763	−2.1196	0.1863	−2.3387	0.4972	−1.9736	−0.2878
Mass (kg)	264.1801	361.6735	210.0713	199.6929	307.0478	525.1092	220.8201	194.8783	193.6027	202.5162
σ^{max} (MPa)	169.6265	169.6265	169.6265	169.6265	169.6265	169.6265	169.6265	169.6265	169.6265	169.6265
δ^{max} (mm)	36.1972	36.1972	36.1972	36.1972	36.1972	36.1972	36.1972	36.1972	36.1972	36.1972
f_1 (Hz)	10.4799	9.9157	9.9905	7.8969	13.5919	7.6931	8.9913	6.4360	10.8666	11.0878
Mean	399.9471	615.7221	265.1195	260.0991	444.1273	880.0261	325.8080	331.2878	253.5581	260.1438
SD	77.7952	123.0112	34.7282	37.1752	72.2827	196.9775	54.1571	94.2993	35.3661	34.7060

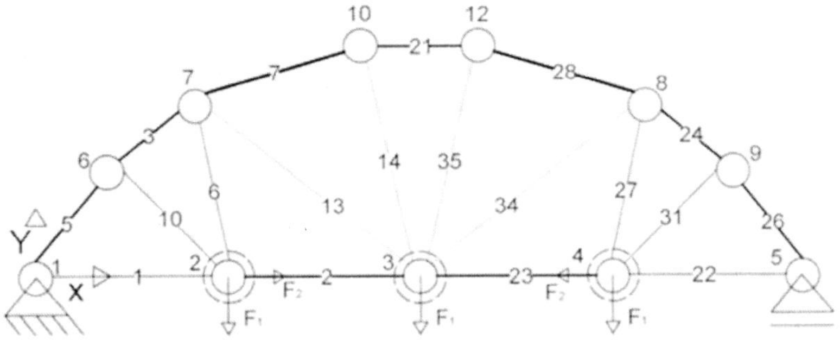

Fig. 6.46 Optimal structure of the 39-bar truss

To assess the true performance of the metaheuristic methods, the average mass and its SD are considered. With the available results, it can be concluded that the PVS algorithm performs the best to achieve the best mean mass value and SOS for SD values. The mass mean values obtained for ALO, DA, GWO, MVO, SCA, WOA, HTS, TLBO, PVS, and SOS algorithms are 399.9471, 615.7221, 265.1195, 260.0991, 444.1273, 880.0261, 325.8080, 331.2878, 253.5581, and 260.1438 kg, respectively, with the corresponding standard deviations (SD) of 77.7952, 123.0112, 34.7282, 37.1752, 72.2827, 196.9775, 54.1571, 94.2993, 35.3661, and 34.7060.

Therefore, based on the mean mass and SD values, it can be concluded that the PVS algorithm outperforms the other algorithms in achieving the best trade-off between the mass of the structure and the constraints imposed. It is, therefore, the actual better algorithm for both convergence rate and consistency for this design problem.

Figure 6.46 displays the optimal arrangement of the 39-bar truss, providing a visual representation of the ideal configuration. Upon careful examination of the diagram, it becomes evident that this optimal arrangement utilizes only 19 elements out of the available 39. This significant finding emphasizes the efficiency and effectiveness of the chosen design, as it successfully achieves the desired level of structural integrity while minimizing the overall number of components required.

Furthermore, the simultaneous consideration of TSS optimization results in significant improvements in truss design when compared to using size optimization and topology optimization alone. Specifically, the resulting truss is 63.0% lighter compared to size optimization and 20.7% lighter compared to topology optimization for the best solutions in each case.

These results demonstrate the superior performance of TSS optimization in achieving lightweight truss designs when compared to traditional optimization methods. By considering both topology and geometric design considerations, TSS optimization can help engineers achieve the optimal balance between strength, weight, and cost.

Overall, the use of TSS optimization is a powerful tool for achieving optimal truss designs that are both lightweight and robust. This technique has applications in various industries, including aerospace, civil engineering, and automotive, where lightweight and efficient structures are essential. By leveraging the benefits of TSS optimization, engineers can create truss designs that are not only lighter but also more durable, cost-effective, and sustainable.

6.4.8 TSS Optimization of the 45-Bar Truss with Discrete Cross-Sections

The ninth benchmark problem focuses on the investigation of the 45-bar truss, which represents a complex structure subjected to stress, buckling, displacement, and fundamental frequency constraints, taking into account discrete design variables. The ground structure of this truss, along with multiple loads and the boundary condition, is depicted in Fig. 6.15. The problem was initially introduced by Deb and Gulati (2001) and examined under static constraints. However, it has not been explored with the simultaneous consideration of static and dynamic constraints.

To address this gap, the problem has been modified to incorporate the effects of multi-load conditions, lumped masses, natural frequency constraints, and discrete sections. Detailed design considerations, including design variables, multi-load conditions, lumped masses, constraints, and material properties, are provided in Table 6.40. In this problem, the 28 elements (element numbers 1–14 and 20–33) are grouped into 14 groups based on their symmetry about the vertical middle plane, while the remaining elements are treated as distinct. As a result, the number of size variables is reduced to 31.

Additionally, the truss is subjected to three nonstructural lumped masses of 500 kg each, placed at nodes 7, 8, and 9. To fulfill the multi-load condition, two different loading conditions are considered, as illustrated in Fig. 6.47.

Table 6.40 Design parameters of the 45-bar truss

Size variables: $X_i = G_i$, $i = 1, 2, \ldots, 31$; $G =$ Group number Shape variables: $-2.5\text{ m} \leq \xi_{x_j/y_j} \leq 2.5\text{ m}$, for $x_1 = -x_5$ and $y_1 = y_5$ with respect to their original coordinates	
Multiple load conditions:	
Condition 1 : $F_1 = 44.537$ kN	Condition 2 : $F_2 = 4.4537$ kN
Stress and displacement constraints: $\sigma_i^{\max} = 172.43$ MPa and $\delta_{x_j/y_j}^{\max} = 5$ mm (0.005 m) (for node, $j = 7, 8$, and 9 along x- and y-axes)	
Lumped masses on nodes 7, 8, and 9 are 500 kg	
Natural frequency constraints: $f_1 \geq 15$ Hz	
The discrete cross-sectional areas: $[A^{\min}, A^{\max}] = [-50, 50]\text{ cm}^2$ ($[-50, 50] \times 10^{-4}\text{ m}^2$) in increments of 1 cm^2 and critical area $= 1\text{ cm}^2$	
Material properties: $E = 69$ GPa and $\rho = 2740\text{kg/m}^3$	

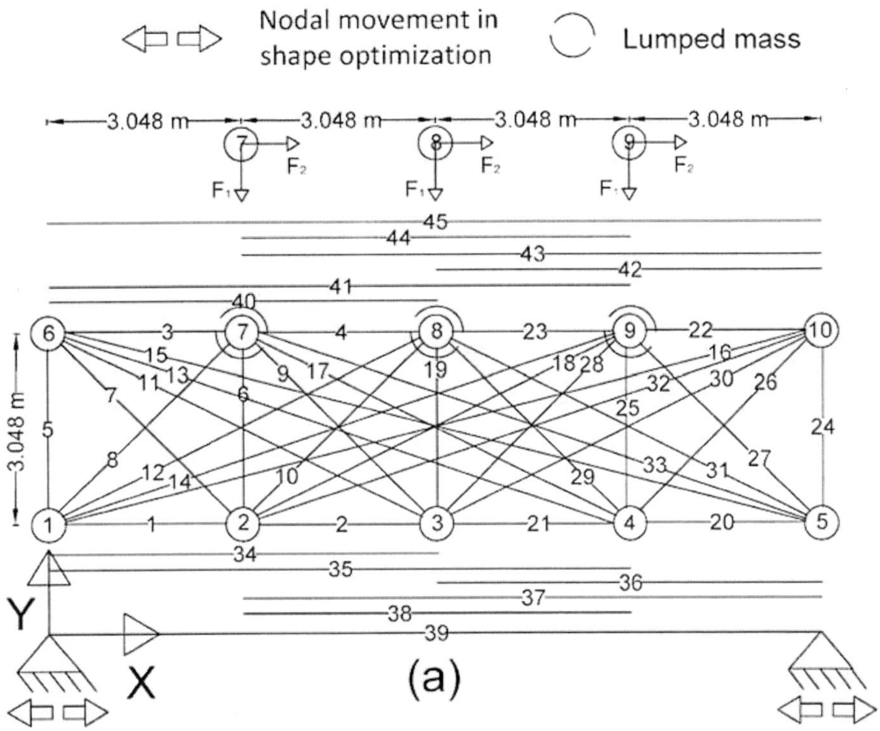

Fig. 6.47 The 45-bar truss

This section of the report delves into the optimization of the 45-bar truss for TSS, using a combination of 31 discrete size variables and two continuous shape variables. To present the results of this optimization, Table 6.41 has been included, which outlines the outcomes obtained from various algorithms in 100 independent runs.

The results show that the best run achieved the minimum mass value while also meeting all the necessary structural constraints. Additionally, the mean and standard deviation (SD) values of the structural mass are provided to help assess the performance of the different evolutionary methods statistically.

Of the algorithms used, including ALO, DA, GWO, MVO, SCA, WOA, HTS, TLBO, PVS, and SOS, the lowest structural mass values were achieved by the GWO, MVO, and SOS algorithms. Specifically, these algorithms reported mass values of 70.7108 kg, which is identical in each case.

These findings suggest that the GWO, MVO, and SOS algorithms are the most effective when it comes to optimizing the 45-bar truss for TSS and should be considered for future use. Overall, the results of this optimization are significant, as they provide valuable insights into how best to optimize this type of truss for optimal performance.

Table 6.41 Topology, shape, and size optimization of the 45-bar truss (discrete section)

Element no.	ALO	DA	GWO	MVO	SCA	WOA	HTS	TLBO	PVS	SOS
1	—	5	—	—	—	—	—	—	—	—
4	6	8	7	7	10	—	—	6	7	7
5	—	—	—	—	—	—	—	1	1	—
6	—	3	—	—	—	—	—	—	—	—
8	9	8	7	7	17	11	6	13	7	7
9	—	—	—	—	4	—	—	—	—	—
12	15	28	16	16	19	15	16	16	16	16
14	—	—	—	—	—	13	16	—	—	—
16	—	7	—	—	—	—	—	1	—	—
18	—	2	—	—	—	—	—	—	—	—
19	—	—	—	—	1	—	—	—	—	—
20	—	4	—	—	—	—	—	—	—	—
22	—	1	—	—	—	—	—	—	—	—
27	—	—	—	—	—	—	—	1	1	—
28	—	17	—	—	—	—	—	—	—	—
29	—	—	—	—	—	—	—	—	1	—
x_1	75.0000	68.9162	75.0000	75.0000	50.9410	75.0000	75.0000	71.0390	75.0000	75.0000
y_1	27.7036	38.2049	46.8842	46.8842	4.4042	−4.9158	47.8771	4.7105	46.8843	46.8843
Mass (kg)	73.0838	175.8563	70.7108	70.7108	124.7457	120.5088	111.4256	103.2090	86.1658	70.7108
σ^{max} (MPa)	85.5225	79.5272	123.5909	123.5909	47.3433	58.6181	127.2554	63.1917	123.5913	123.5912
δ^{max} (mm)	1.8869	2.3819	4.5496	4.5496	2.6177	2.1578	4.6845	2.3262	4.5496	4.5496
f_1 (Hz)	19.9706	16.5227	19.9705	19.9705	19.5249	25.5331	21.7291	18.3988	19.9027	19.9705
Mean	191.8259	337.2810	102.9085	110.4501	210.7179	443.0329	174.2819	169.8723	120.8072	101.9250
SD	84.2063	101.4243	23.1007	40.8203	43.9842	188.5357	43.9260	48.9406	21.5210	32.0840

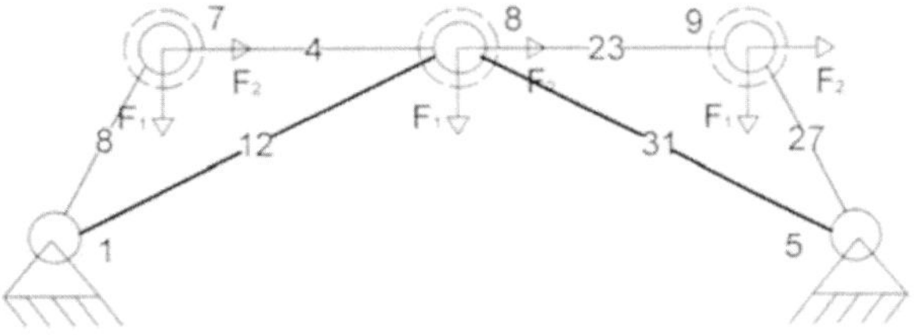

Fig. 6.48 Optimal structure of the 45-bar truss

To accurately evaluate the performance of the metaheuristic methods used in this study, it is essential to consider both the average mass value and its standard deviation. Based on the results obtained, it can be concluded that the ALO, DA, GWO, MVO, SCA, WOA, HTS, TLBO, PVS, and SOS algorithms yielded mean mass values of 191.8259, 337.2810, 102.9085, 110.4501, 210.7179, 443.0329, 174.2819, 169.8723, 120.8072, and 101.9250 kg, respectively. The standard deviation values for these algorithms were 84.2063, 101.4243, 23.1007, 40.8203, 43.9842, 188.5357, 43.9260, 48.9406, 21.5210, and 32.0840, respectively.

Based on these results, it can be concluded that the SOS algorithm performed the best in achieving the lowest mean mass value, while the PVS algorithm achieved the lowest standard deviation value. Therefore, it can be argued that the SOS and PVS algorithms are the most effective for both convergence rate and consistency in solving this design problem.

In conclusion, the findings of this study highlight the importance of considering both the mean mass value and its standard deviation in evaluating the performance of metaheuristic methods. The SOS and PVS algorithms proved to be the most effective in optimizing the 45-bar truss for TSS, and these algorithms should be considered for future use in similar design problems.

Figure 6.48 visually represents the ideal configuration of the 45-bar truss, providing a clear illustration of the optimal arrangement. Upon careful examination of the diagram, it becomes apparent that this ideal configuration utilizes only 6 elements out of the available 45. This significant discovery highlights the efficiency and effectiveness of the selected design, as it successfully achieves the desired level of structural integrity while minimizing the overall number of components required.

In addition, the simultaneous consideration of TSS optimization leads to significant improvements in truss design when compared to using size optimization and topology optimization alone. Specifically, the resulting truss is 82.7% lighter compared to size optimization and 54.0% lighter compared to topology optimization for the best solutions in each case.

These results underscore the remarkable performance of TSS optimization in achieving lightweight truss designs and its superiority over traditional optimization methods. By considering both topology and geometric design features, TSS optimization can help engineers achieve optimal truss designs that are not only lightweight but also durable and cost-effective.

Overall, the use of TSS optimization is a powerful tool for achieving optimal truss designs that meet the demands of modern engineering. This technique has a broad range of applications, including aerospace, civil engineering, and automotive industries, where weight reduction and efficiency are crucial. By leveraging the benefits of TSS optimization, engineers can create truss designs that are not only lighter but also stronger, more cost-effective, and environmentally sustainable.

6.4.9 TSS Optimization of the 25-Bar 3D Truss with Discrete Cross-Sections

The ninth benchmark problem focuses on the investigation of the 25-bar truss, which represents a space truss subjected to stress, buckling, displacement, and fundamental frequency constraints while considering discrete design variables. The initial layout of this space truss is depicted in Fig. 6.49. Previous studies conducted by Deb and Gulati (2001) and Ahrari et al. (2014) have examined this problem under static constraints, but the simultaneous consideration of static and dynamic constraints has not been explored.

To address this research gap, the problem has been modified to incorporate the effects of multi-load conditions, lumped masses, natural frequency constraints, and discrete areas. Detailed design considerations, including design variables, load conditions, lumped masses, constraints, and material properties, are provided in Table 6.42. In this problem, element grouping is adopted due to structural symmetry, as indicated in the table. As a result, the truss has only eight size variables, which include topology variables.

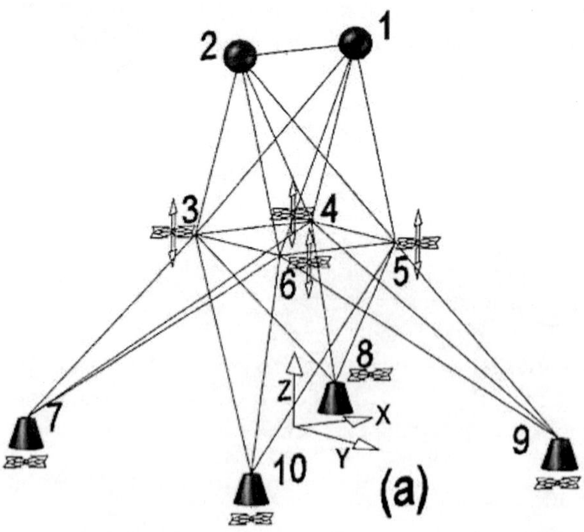

Fig. 6.49 The 25-bar 3D truss

Table 6.42 Design parameters of the 25-bar 3D truss

Size variables: $X_i = G_i$, $i = 1, 2, \ldots, 8$; G = Group number Shape variables: $0.508 \text{ m} \leq \xi_{x_4} \leq 1.524$ m, where $x_4 = x_5 = -x_3 = -x_6$ $1.016 \text{ m} \leq \xi_{y_3} \leq 2.032$ m, where, $y_3 = y_4 = -y_5 = -y_6$ $2.286 \text{ m} \leq \xi_{z_3} \leq 3.302$ m, where $z_3 = z_4 = z_5 = z_6$ $1.016 \text{ m} \leq \xi_{x_8} \leq 2.032$ m, where $x_8 = x_9 = -x_7 = -x_{10}$ $2.54 \text{ m} \leq \xi_{y_7} \leq 3.556$ m, where $y_7 = y_8 = -y_9 = -y_{10}$ (1 in. = 0.0254 m)				
Multiple load conditions:				
Condition 1:	$F_{y_1} = 89.074$ kN	$F_{z_1} = -22.25$ kN	$F_{y_2} = -89.074$ kN	$F_{z_2} = -22.25$ kN
Condition 2:	$F_{x_1} = 4.454$ kN	$F_{y_1} = -44.537$ kN	$F_{z_1} = -44.537$ kN	$F_{y_2} = -44.537$ kN
	$F_{z_2} = -44.537$ kN	$F_{x_3} = 2.227$ kN	$F_{x_6} = 2.672$ kN	
Lumped masses at nodes 1 and 2 are 500 kg				
Stress and displacement constraints: $\sigma_i^{max} = 172.375$ MPa and $\delta_{x_j/y_j/z_j}^{max} = 8.9$ mm (0.0089 m) (for node, $j = 1, 2, \ldots, 6$ along x-, y-, and z-axes)				
Natural frequency constraints: $f_1 \geq 15$ Hz				
The discrete cross-sectional areas: $[A^{min}, A^{max}] = [-50, 50]$ cm^2 ($[-50, 50] \times 10^{-4}$ m^2) in increments of 1 cm^2 and critical area = 1 cm^2				
Material properties: $E = 69$ GPa and $\rho = 2740 \text{kg/m}^3$				
Element grouping				
Group number	Element number (end nodes)		Group number	Element number (end nodes)
G_1	1(1,2)		G_5	12(3,4), 13(5,6)
G_2	2(1,4), 3(2,3), 4(1,5), 5(2,6)		G_6	14(3,10), 15(6,7), 16(4,9), 17(5,8)
G_3	6(2,5), 7(2,4), 8(1,3), 9(1,6)		G_7	18(3,8), 19(4,7), 20(6,9), 21(5,10)
G_4	10(3,6), 11(4,5)		G_8	22(3,7), 23(4,8), 24(5,9), 25(6,10)

Additionally, the truss is subjected to two nonstructural lumped masses of 500 kg each, placed at nodes 1 and 2. To satisfy the multi-load condition, two different loading conditions are considered, as described in the table.

Table 6.43 presents the results of the TSS optimization of the 25-bar truss using discrete cross-sections for various evolutionary algorithms in 100 independent runs. The best run was identified to be the one that gave the minimum mass value while meeting all the structural constraints specified. The mean and standard deviation values of the structural mass obtained from the different algorithms were also recorded for statistical analysis.

The best results were reported by the ALO, DA, GWO, MVO, SCA, WOA, HTS, TLBO, PVS, and SOS algorithms, with the corresponding mass values of 263.6642, 309.7794, 264.8168, 259.4121, 335.3444, 341.4888, 260.6916, 261.0766, 258.6078, and 259.4956 kg. Upon closer examination, it can be observed that the PVS algorithm produced the least structural mass, with MVO and SOS algorithms following closely as the second and third-best, respectively.

The available results for the TSS optimization of the 25-bar truss with discrete cross-sections indicate that the PVS algorithm has the least structural mass, with the MVO and SOS algorithms following closely as the second and third-best options, respectively. To accurately assess the performance of the metaheuristic methods, the average mass and its SD are considered. The mean mass values for the ALO, DA, GWO, MVO, SCA, WOA, HTS, TLBO, PVS, and SOS algorithms are 318.8568, 389.4518, 282.3058, 4.00E+06, 376.6004, 540.2276, 286.7519, 275.5365, 266.4168, and 269.2277 kg, respectively, with the corresponding SD values of 44.5799, 103.7340, 45.6531, 1.98E+07, 24.7689, 110.4743, 63.3336, 12.0239, 5.4263, and 5.8978. The PVS algorithm has the best mean mass and SD values among the evaluated methods, indicating that it is the better algorithm for both convergence rate and consistency for the design problem.

In conclusion, the TSS optimization of the 25-bar truss using discrete cross-sections was successfully performed using various evolutionary algorithms. The results show that the PVS algorithm is the most effective in achieving the lowest structural mass, followed closely by the MVO and SOS algorithms. These findings can be used to guide the selection of appropriate metaheuristic algorithms for similar design problems in the future.

Figure 6.50 presents the optimal arrangement of the 25-bar 3D truss, providing a visual representation of the ideal configuration. A thorough examination of the diagram reveals that this optimal arrangement utilizes only five groups of elements out of the available eight. This significant finding emphasizes the efficiency and effectiveness of the selected design, as it successfully achieves the desired level of structural integrity while minimizing the overall number of components required.

Furthermore, the simultaneous consideration of TSS optimization leads to substantial improvements in truss design when compared to using size optimization and topology optimization alone. Specifically, the resulting truss is 34.3% lighter compared to size optimization and 33.5% lighter compared to topology optimization for the best solutions in each case.

Table 6.43 Topology, shape, and size optimization of the 25-bar 3D truss (discrete section)

Element group no.	ALO	DA	GWO	MVO	SCA	WOA	HTS	TLBO	PVS	SOS
1	—	—	—	—	—	30	—	—	—	—
2	15	13	13	13	17	16	14	13	13	13
3	16	13	13	14	15	18	15	16	14	14
4	—	—	1	—	—	5	—	—	—	—
6	1	6	3	1	8	8	1	1	1	1
7	12	11	9	11	10	14	11	8	12	11
8	15	19	20	19	24	16	17	20	18	19
x_4	48.6216	60.0000	57.7988	53.2397	60.0000	42.4882	53.7911	59.2493	49.7187	52.8836
x_8	60.3130	80.0000	58.0089	55.2124	58.3532	61.8326	57.4047	60.7464	55.6360	55.2739
y_4	52.1932	65.6116	42.1075	43.5916	43.5570	52.4773	49.6793	44.9537	45.0239	42.8413
y_8	100.0000	128.6184	100.0000	100.0000	100.0000	100.0000	103.0683	107.4383	100.0000	100.0000
z_4	95.8686	118.4298	118.9654	112.7240	130.0000	93.8723	104.0099	116.5328	110.4699	114.6358
Mass (kg)	263.6642	309.7794	264.8168	259.4121	335.3444	341.4888	260.6916	261.0766	258.6078	259.4956
σ^{max} (MPa)	48.7280	48.7280	48.7280	48.7280	48.7280	48.7280	48.7280	48.7280	48.7280	48.7280
δ^{max} (mm)	5.7625	5.7625	5.7625	5.7625	5.7625	5.7625	5.7625	5.7625	5.7625	5.7625
f_1 (Hz)	15.0000	17.3273	15.1619	15.0157	15.5438	15.0022	15.0390	15.0480	15.0000	15.0079
Mean	318.8568	389.4518	282.3058	4.00E+06	376.6004	540.2276	286.7519	275.5365	266.4168	269.2277
SD	44.5799	103.7340	45.6531	1.98E+07	24.7689	110.4743	63.3336	12.0239	5.4263	5.8978

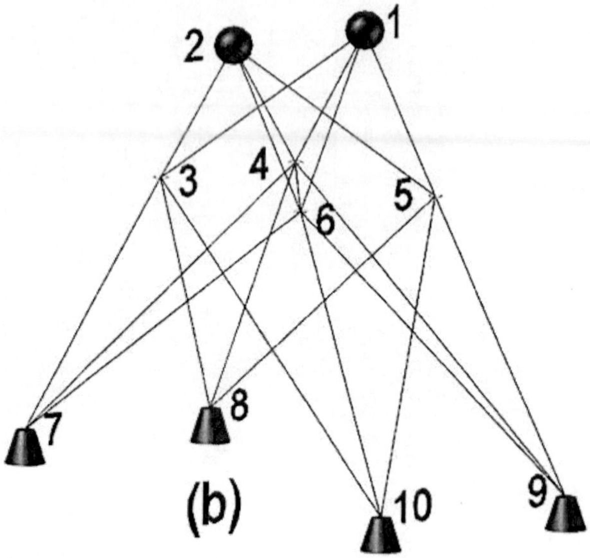

Fig. 6.50 Optimal structure of the 25-bar 3D truss

These results highlight the significant advantages of TSS optimization in achieving lightweight truss designs and its superiority over traditional optimization methods. By considering both topology and geometric design considerations, TSS optimization can help engineers achieve optimal truss designs that are both lightweight and structurally sound.

In summary, the use of TSS optimization is an effective approach for achieving optimal truss designs that meet the performance requirements of modern engineering. This technique has a wide range of applications, including aerospace, civil engineering, and automotive industries, where weight reduction and efficiency are critical. By harnessing the benefits of TSS optimization, engineers can create truss designs that are not only lighter but also more robust, cost-effective, and sustainable.

6.4.10 TSS Optimization of the 39-Bar 3D Truss with Discrete Cross-Sections

The benchmark problem presented in Fig. 6.51 focuses on the 39-element 3D truss. The element cross-sectional areas are grouped into 11 sets, taking into account the symmetry about the *x–z* and *y–z* planes. Previous studies conducted by Deb and Gulati (2001) and Luh and Lin (2008) have examined this truss under static constraints but have not explored it with simultaneous static and dynamic constraints.

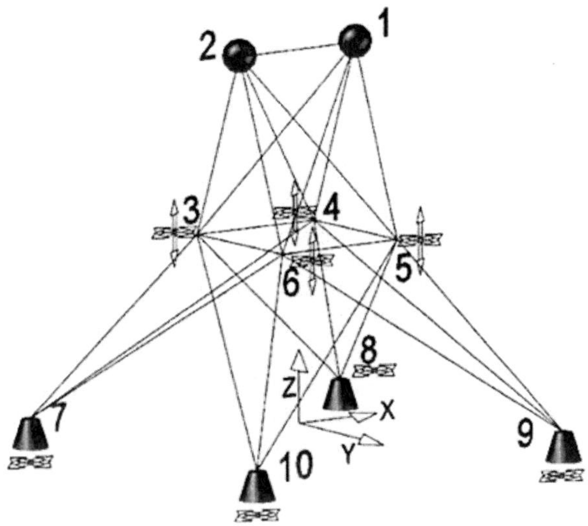

Fig. 6.51 The 39-bar 3D truss

To address this research gap, the problem has been modified to incorporate the effects of multi-load conditions, lumped masses, natural frequency constraints, and discrete areas. Additionally, a constant lumped mass of 500 kg is attached to the top nodes (nodes 1 and 2) of the truss, as depicted in Fig. 6.51. The design parameters, including material properties, loading conditions, design variable bounds, constraints, and other relevant details, are presented in Table 6.44.

This comprehensive table provides all the necessary information for optimizing the truss problem, considering the aforementioned modifications. It serves as a reference for conducting further analysis and investigations on the truss structure.

Table 6.45 displays the outcomes of TSS optimization for the 39-bar 3D truss with discrete cross-sections using various algorithms in 100 independent runs. The results reveal that the optimal run yields the lowest mass value while fulfilling all the stated structural constraints. The mean and standard deviation values of the structural mass are also presented to enable statistical evaluation of the performance of the different evolutionary methods. The ALO, DA, GWO, MVO, SCA, WOA, HTS, TLBO, PVS, and SOS algorithms produce the best results of 229.2730, 336.6299, 225.9504, 223.7883, 305.1848, 368.3861, 224.4931, 226.1253, 230.2560, and 228.6720 kg, respectively. It is observed that the MVO algorithm generates the lowest structural mass, followed closely by the HTS and GWO algorithms, which exhibit similar mass values. These three mass values can be considered interchangeable.

To evaluate the true effectiveness of the metaheuristic methods, we consider the average mass and its standard deviation. Based on the available results, it can be concluded that the ALO, DA, GWO, MVO, SCA, WOA, HTS, TLBO, PVS, and SOS algorithms have average mass values of 391.1211, 477.2727, 348.6549, 340.4151, 402.0006, 557.5680, 323.2108, 366.0633, 318.0890, and 319.2202 kg, respectively, along with the corresponding standard deviations of 62.0506, 71.6356,

Table 6.44 Design parameters of the 39-bar 3D truss

Size variables: $X_i = G_i$, $i = 1, 2, \ldots, 11$; $G =$ Group number Shape variables: $0.508\ \text{m} \le \xi_{x_4} \le 1.524\ \text{m}$, where $x_4 = x_5 = -x_3 = -x_6$ $1.016\ \text{m} \le \xi_{y_3} \le 2.032\ \text{m}$, where $y_3 = y_4 = -y_5 = -y_6$ $2.286\ \text{m} \le \xi_{z_3} \le 3.302\ \text{m}$, where $z_3 = z_4 = z_5 = z_6$ $1.016\ \text{m} \le \xi_{x_8} \le 2.032\ \text{m}$, where $x_8 = x_9 = -x_7 = -x_{10}$ $2.54\ \text{m} \le \xi_{y_7} \le 3.556\ \text{m}$, where $y_7 = y_8 = -y_9 = -y_{10}$ (1 in. = 0.0254 m)			
Load condition:	$F_{z_1} = -222.685$ kN		$F_{z_2} = -222.685$ kN
Lumped masses at nodes 1 and 2 are 500 kg			
Stress and displacement constraints: $\sigma_i^{max} = 172.375$ MPa and $\delta_{x_j/y_j/z_j}^{\max} = 8.9$ mm (0.0089 m) (for node, $j = 1, 2, \ldots, 6$ along x-, y-, and z-axes)			
Natural frequency constraints: $f_1 \ge 15$ Hz			
The discrete cross-sectional areas: $[A^{\min}, A^{\max}] = [-50, 50]\ \text{cm}^2$ ($[-50, 50] \times 10^{-4}\ \text{m}^2$) in increments of 1 cm^2 and critical area = 1 cm^2			
Material properties: $E = 69$ GPa and $\rho = 2740\text{kg/m}^3$			
Element grouping			
Group number	Element number (end nodes)	Group number	Element number (end nodes)
G_1	1(1,2)	G_7	22(3,7), 23(4,8), 24(5,9), 25(6,10)
G_2	2(1,4), 3(2,3), 4(1,5), 5(2,6)	G_8	26(5,7), 27(6,8), 28(3,9), 29(4,10)
G_3	6(2,5), 7(2,4), 8(1,3), 9(1,6)	G_9	30(3,5), 31(4,6)
G_4	10(3,6), 11(4,5), 12(3,4), 13(5,6)	G_{10}	32(1,7), 33(1,10), 34(2,9), 35(2,8)
G_5	14(3,10), 15(6,7), 16(4,9), 17(5,8)	G_{11}	36(2,7), 37(2,10), 38(1,8), 39(1,9)
G_6	18(3,8), 19(4,7), 20(6,9), 21(5,10)		

54.6911, 59.5971, 31.0578, 129.8382, 55.7307, 56.8645, 48.1850, and 60.6504 kg. The PVS algorithm is found to be the most effective in achieving the lowest average mass value, while the SCA algorithm produces the best standard deviation result.

Figure 6.52 visually presents the ideal configuration of the 39-bar 3D truss, providing a clear representation of the optimal arrangement. Upon careful examination of the diagram, it becomes apparent that this optimal configuration utilizes only 5 groups of elements out of the available 11. This significant discovery underscores the efficiency and effectiveness of the chosen design, as it successfully achieves the desired level of structural integrity while minimizing the overall number of components required.

Furthermore, the simultaneous consideration of TSS optimization leads to a substantial improvement in truss design compared to using size optimization and topology optimization alone. Specifically, the resulting truss is 63.2% lighter compared to size optimization and 37.9% lighter compared to topology optimization for the best solutions in each case.

Table 6.45 Topology, shape, and size optimization of the 39-bar 3D truss (discrete section)

Element group no.	ALO	DA	GWO	MVO	SCA	WOA	HTS	TLBO	PVS	SOS
1	4	—	10	—	—	—	6	—	—	—
2	7	14	5	9	13	—	6	9	10	8
3	18	21	19	17	20	—	17	17	17	16
4	—	6	—	—	—	—	—	—	—	—
5	1	1	1	—	5	—	1	1	1	—
6	5	8	4	3	8	—	4	3	4	4
7	21	25	21	21	24	—	22	22	21	23
8	—	—	—	1	—	—	—	—	—	1
10	—	—	—	—	—	36	—	—	—	—
11	—	—	—	—	—	17	—	—	—	—
x_4	47.6948	60.0000	55.7684	47.7489	60.0000	20.0000	50.6400	46.7498	45.4298	45.1324
x_8	65.3396	67.6320	74.1652	66.4361	68.1716	51.3134	70.2332	67.1068	63.8459	64.1519
y_4	46.9206	68.6841	47.2922	48.6828	45.4773	40.8451	45.5927	46.3299	46.5988	43.5000
y_8	100.0000	140.0000	100.0000	100.0000	100.0000	100.0000	100.7253	100.0000	100.0195	100.3707
z_4	106.0320	120.1249	105.5383	108.0181	107.4068	90.0000	111.9568	107.2476	107.8977	115.9843
Mass (kg)	229.2730	336.6299	225.9504	**223.7883**	305.1848	368.3861	224.4931	226.1253	230.2560	228.6720
σ^{max} (MPa)	64.4961	64.4961	64.4961	64.4961	64.4961	64.4961	64.4961	64.4961	64.4961	64.4961
δ^{max} (mm)	5.6891	5.6891	5.6891	5.6891	5.6891	5.6891	5.6891	5.6891	5.6891	5.6891
f_1 (Hz)	15.0000	16.1962	15.0006	15.0004	17.8097	15.1306	15.0181	15.0124	15.0013	15.0149
Mean	391.1211	477.2727	348.6549	340.4151	402.0006	557.5680	323.2108	366.0633	318.0890	319.2202
SD	62.0506	71.6356	54.6911	59.5971	31.0578	129.8382	55.7307	56.8645	48.1850	60.6504

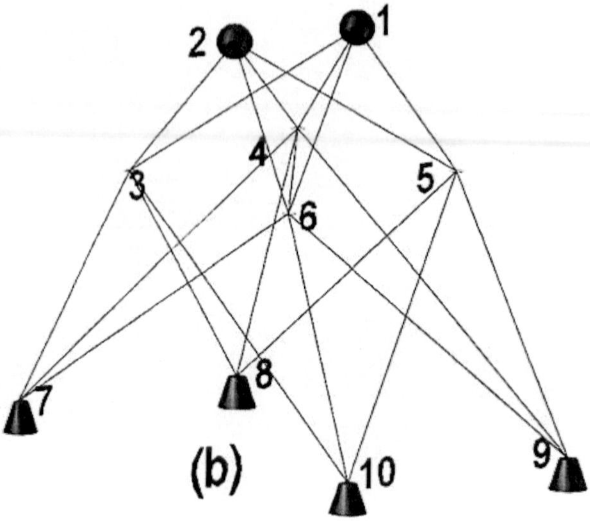

Fig. 6.52 Optimal structure of the 39-bar 3D truss

These results demonstrate the effectiveness of TSS optimization in achieving lightweight truss designs and its superiority over traditional optimization methods. By combining topology, size, and shape design considerations, TSS optimization can help engineers achieve optimal truss designs that are not only lighter but also stronger and more cost-effective.

6.4.11 A Comprehensive Analysis of the Best Results Obtained for TSS with Discrete Sections

Overall, the use of TSS optimization is a powerful tool for achieving optimal truss designs that meet the demands of modern engineering. This technique has broad applications, including in aerospace, civil engineering, and automotive industries, where weight reduction and efficiency are crucial. By leveraging the benefits of TSS optimization, engineers can create truss designs that are not only lighter but also more durable, cost-effective, and environmentally sustainable.

The study presented in Table 6.46 aimed to evaluate the performance of proposed algorithms for solving ten benchmark problems. We ran each algorithm independently for 100 times and recorded the minimum mass obtained in each run. To further analyze the results, the table was examined based on overall, average, and individual ranks, as well as the best and worst solutions obtained by each algorithm.

Upon analyzing the results, it was found that the PVS algorithm performed the best overall, as it produced the minimum mass in six out of ten problems and achieved the second-best result in the remaining four. The MVO algorithm also performed well, producing the minimum mass in five problems, the second-best

Table 6.46 Minimum mass obtained using TSS optimization with discrete sections

	ALO	DA	GWO	MVO	SCA	WOA	HTS	TLBO	PVS	SOS
10-bar truss	383.5	**261.8**	338.4	534.1	316.1	**261.8**	**261.8**	269.3	**261.8**	**261.8**
14-bar truss	274.4	282.4	**252.6**	**252.6**	259.8	354.1	**252.6**	**252.6**	**252.6**	**252.6**
15-bar truss	103.7	131.9	91.2	88.9	129.4	112.5	94.3	92.6	**88.3**	90.6
24-bar truss	136.5	163.1	113	**111.5**	141.6	217.6	123.1	123.4	112.3	112.5
20-bar truss	131.7	219.3	117.8	117	192.8	271.3	135.8	126.7	**116.7**	128.7
72-bar 3D truss	472.4	533.8	446.8	461.2	533.6	598.6	450.7	**439.8**	446.2	447.7
39-bar truss	264.2	361.7	210.1	199.7	307	525.1	220.8	194.9	**193.6**	202.5
45-bar truss	73.1	175.9	**70.7**	**70.7**	124.7	120.5	111.4	103.2	86.2	**70.7**
25-bar 3D truss	263.7	309.8	264.8	259.4	335.3	341.5	260.7	261.1	**258.6**	259.5
39-bar 3D truss	229.3	336.6	226	**223.8**	305.2	368.4	224.5	226.1	230.3	228.7
Overall rank	7	9	4	2	8	10	6	5	1	3
Count best	0	1	2	4	0	1	2	2	6	3
Count 2nd best	0	0	0	3	0	0	1	1	2	0
Count 3nd best	0	0	3	1	0	0	0	0	0	3
Count worst	0	2	0	1	0	7	0	0	0	0

Note: Bold indicates the best solution

result in three problems, and the third-best result in one problem. The SOS algorithm produce three best solutions. On the other hand, the TLBO and GWO algorithms achieved two best solutions each, without producing any worse solution.

Despite their good performance, the PVS and MVO algorithms had some limitations. They produced the average solutions for some the problems, indicating that they may not always provide the best solution. In contrast, the WOA and DA algorithms produced the worst solutions in eight and two problems, respectively.

In conclusion, Table 6.46 provides a comprehensive analysis of the proposed algorithms' performance for solving benchmark problems. The PVS algorithm was found to be the best overall, followed by the MVO algorithm, while the WOA and DA algorithms had some limitations. The results of this study could be useful for researchers and practitioners in selecting an appropriate algorithm for solving optimization problems.

Table 6.47 Mean mass obtained using TSS optimization with discrete sections

	ALO	DA	GWO	MVO	SCA	WOA	HTS	TLBO	PVS	SOS
10-bar truss	519	691.9	**289.2**	316.8	473.7	1035.8	320.9	366.9	304.2	341.4
14-bar truss	625.8	663.8	382.6	380.5	376.3	1109.2	362.5	367.6	**282.6**	326.5
15-bar truss	155	1.60E +07	4.00E +06	1.40E +07	162.1	239.5	124.9	113.4	**96.2**	104
24-bar truss	194	265.7	136.5	133.7	187.4	391.6	180.2	174.1	131.2	**130.7**
20-bar truss	239.3	8.01E +06	2.00E +06	1.02E +07	250	421	2128.2	476.5	**147.1**	156.8
72-bar 3D truss	744.6	4.00E +06	**469.5**	1.00E +07	653	1413.3	597	501.9	470.7	**469.5**
39-bar truss	399.9	615.7	265.1	260.1	444.1	880	325.8	331.3	**253.6**	260.1
45-bar truss	191.8	337.3	102.9	110.5	210.7	443	174.3	169.9	120.8	**101.9**
25-bar 3D truss	318.9	389.5	282.3	4.00E +06	376.6	540.2	286.8	275.5	**266.4**	269.2
39-bar 3D truss	391.1	477.3	348.7	340.4	402	557.6	323.2	366.1	**318.1**	319.2
Overall rank	7	10	3	6	7	9	4	4	1	2
Count best	0	0	2	0	0	0	0	0	6	3
Count 2nd best	0	0	1	1	0	0	0	0	2	6
Count 3nd best	1	0	0	3	0	0	2	2	1	0
Count worst	0	1	0	3	0	6	0	0	0	0

Note: Bold indicates the best solution

The proposed study investigated the mean mass obtained for ten benchmark problems using the proposed algorithms in 100 independent runs, which is presented in Table 6.47. The table was analyzed based on overall, average, and individual ranks, including the best and worst results of each algorithm.

The results indicated that the PVS algorithm was the best-performing algorithm overall, as it achieved the mean mass in six out of ten problems and the second-best mean mass in the two. The SOS algorithm also performed well, with three third-best counts and six second-best counts, without producing any worse solution. The GWO algorithm achieved two best counts and one second-best count without any worse solution.

On the other hand, the WOA, MVO, and DA algorithms produced the worst solutions in six, three, and one problems, respectively, indicating that they may not always provide the best solution.

In conclusion, Table 6.47 provides a detailed analysis of the proposed algorithms' performance in terms of mean mass for solving benchmark problems. The PVS algorithm was found to be the best overall, followed by the SOS and GWO algorithms. The results of this study could be useful for researchers and practitioners in selecting an appropriate algorithm for solving optimization problems.

In addition to evaluating the performance of optimization algorithms in terms of minimum mass and mean mass, the success rate and computational effort required are also important metrics. To evaluate the success rate and computational effort of the algorithms for solving TTO problems with discrete sections, experiments were conducted.

Table 6.48 presents the success rate of the algorithms for truss problems with discrete sections in 100 independent runs. It was observed that the PVS algorithm had the highest overall success rate, followed by the SOS algorithm, while the GWO and MVO algorithms shared the third position with a success rate of 131.

However, the DA and WOA algorithms failed to reach the best mean mass for all problems within the predefined experimental conditions and therefore produced a 0% success rate. This indicates that these algorithms may not be suitable for solving TTO problems with discrete sections.

In conclusion, Table 6.48 provides valuable insights into the success rate of the algorithms for solving TTO problems with discrete sections. The PVS and SOS algorithms were found to be the best performers, while the DA and WOA algorithms were not successful in producing the best mean mass for any of the problems under the given experimental conditions. These findings could be useful for researchers and practitioners in selecting an appropriate algorithm for solving TTO problems with discrete sections.

Table 6.48 The success rate of the TSS optimization with discrete sections

	ALO	DA	GWO	MVO	SCA	WOA	HTS	TLBO	PVS	SOS
10-bar truss	0	0	25	18	0	0	17	9	14	7
14-bar truss	1	0	13	18	2	0	19	13	32	23
15-bar truss	0	0	2	1	0	0	0	0	18	2
24-bar truss	0	0	12	22	0	0	2	3	21	24
20-bar truss	1	0	6	17	0	0	1	2	14	3
72-bar 3D truss	0	0	34	0	0	0	8	12	31	19
39-bar truss	0	0	8	11	0	0	2	3	18	9
45-bar truss	6	0	23	20	0	0	0	0	6	26
25-bar 3D truss	1	0	0	14	0	0	9	3	16	8
39-bar 3D truss	4	0	8	10	0	0	14	8	19	18
Total success rate	13	0	131	131	2	0	72	53	189	139

Table 6.49 The mean number of FE required to reach the best mean optimum value in the algorithms with discrete sections

	ALO	DA	GWO	MVO	SCA	WOA	HTS	TLBO	PVS	SOS
10-bar truss	—	—	4900	14500	—	—	12400	—	6700	12400
14-bar truss	—	—	—	—	—	—	—	—	9700	14600
15-bar truss	—	—	—	—	—	—	—	17600	8200	12800
24-bar truss	—	—	3900	7700	—	—	—	—	10100	10600
20-bar truss	—	—	—	—	—	—	—	—	12400	15600
72-bar 3D truss	—	—	7700	—	—	—	—	7500	6300	6800
39-bar truss	—	—	9300	13600	—	—	—	—	13100	14200
45-bar truss	—	—	4600	11400	—	—	—	—	18900	14000
25-bar 3D truss	6000	—	6800	—	—	—	7000	4400	3000	4000
39-bar 3D truss	—	—	7800	9400	—	—	9000	11900	5000	6000

Table 6.49 displays a comparison of the results obtained by ten different algorithms in terms of the mean number of function evaluations (FE) required to achieve the best mean mass value with a 25% error for each of the truss problems with discrete sections. The data presented in the table has been collected from 100 independent runs for each of the problems under consideration.

Based on the results presented in the table, it can be inferred that the PVS, SOS, and GWO algorithms are more computationally efficient than the other comparative algorithms. These algorithms have demonstrated better performance in terms of the number of function evaluations required to achieve the desired solution quality. Moreover, among these three algorithms, the PVS algorithm has shown to be the most efficient in terms of computational effort.

It is also noteworthy that the DA, SCA, and WOA algorithms have not been able to reach the best mean mass value with a 25% error for any of the problems under the predefined experimental conditions. As a result, these algorithms have formed a 0% success rate.

Overall, the results presented in Table 6.49 provide valuable insights into the performance of different algorithms for solving truss problems with discrete sections. These findings can guide future research on developing more efficient and effective optimization algorithms for solving similar engineering problems.

Figures 6.53, 6.54, 6.55, 6.56, 6.57, 6.58, 6.59, 6.60, 6.61 and 6.62 exhibit the graphical representation of the mean mass and standard deviation of the mass obtained by utilizing ten fundamental metaheuristics. As demonstrated in Fig. 6.53, the PVS, GWO, and MVO algorithms yield superior results for the mean and standard deviation of mass when compared to other comparative algorithms for the 10-bar truss problem. In addition, the percentage difference between the best mean mass (achieved by the GWO algorithm) and the worst mean mass

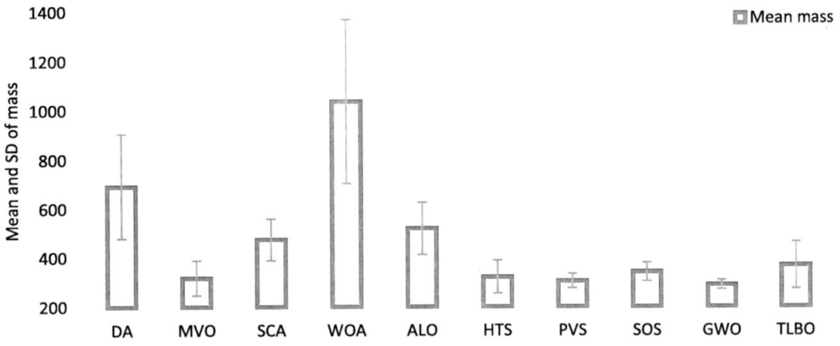

Fig. 6.53 The 10-bar truss: TSS with discrete sections

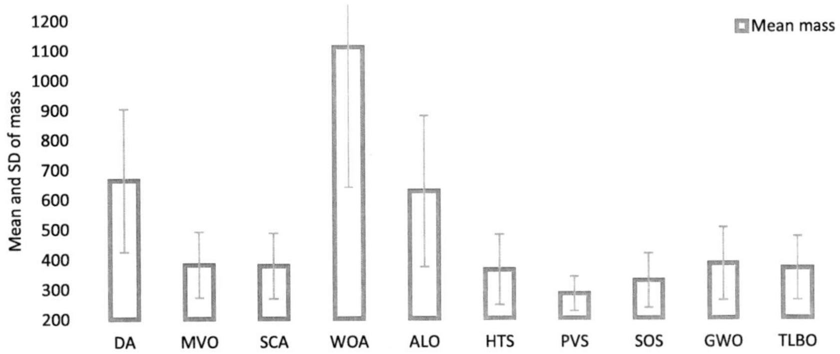

Fig. 6.54 The 14-bar truss: TSS with discrete sections

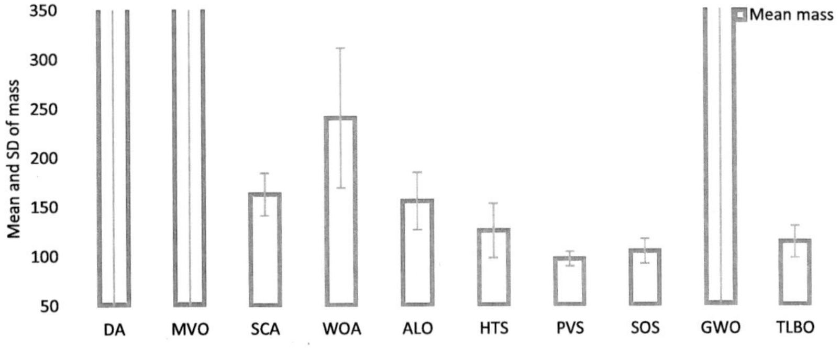

Fig. 6.55 The 15-bar truss: TSS with discrete sections

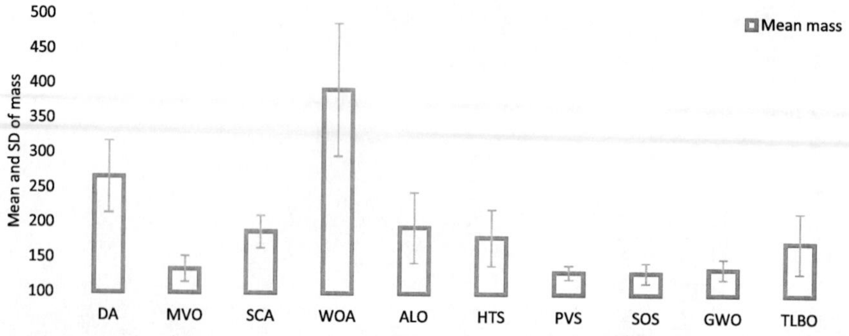

Fig. 6.56 The 24-bar truss: TSS with discrete sections

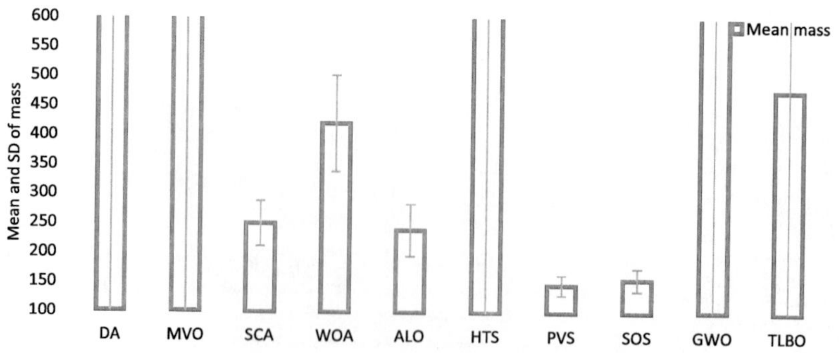

Fig. 6.57 The 20-bar truss: TSS with discrete sections

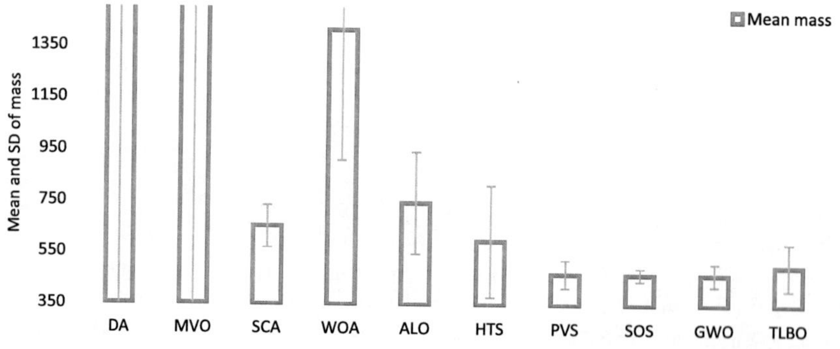

Fig. 6.58 The 72-bar 3D truss: TSS with discrete sections

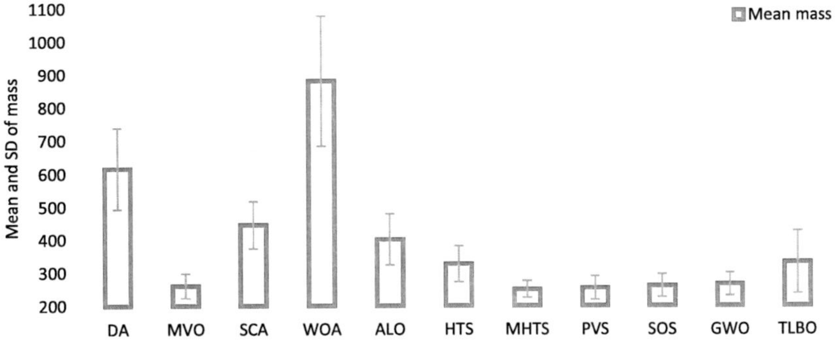

Fig. 6.59 The 39-bar truss: TSS with discrete sections

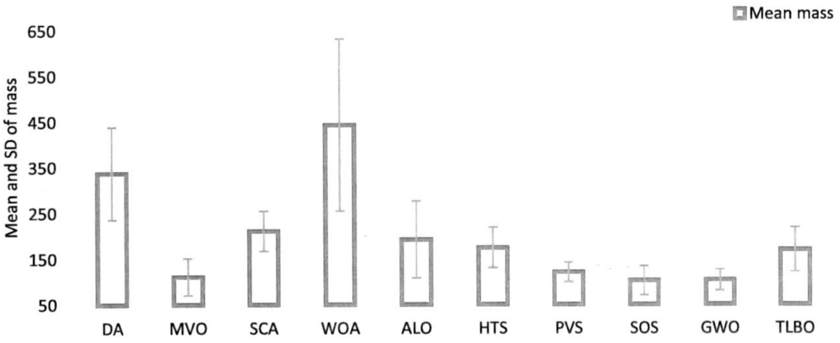

Fig. 6.60 The 45-bar truss: TSS with discrete sections

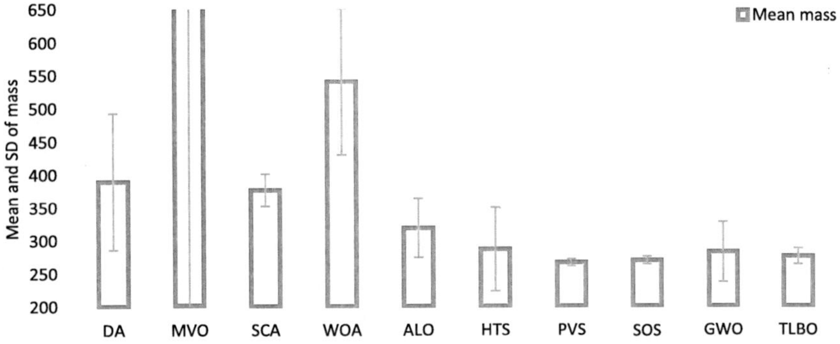

Fig. 6.61 The 25-bar 3D truss: TSS with discrete sections

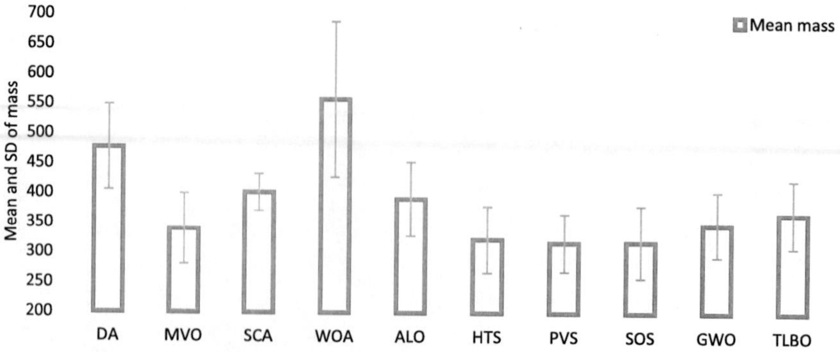

Fig. 6.62 The 39-bar 3D truss: TSS with discrete sections

(achieved by the WOA algorithm) is 72.1%. For the 14-bar truss problem, Fig. 6.54 displays the comparative results of each algorithm, indicating that the PVS, SOS, and HTS algorithms perform better. Moreover, the percentage difference between the best mean mass (achieved by the PVS algorithm) and the worst mean mass (achieved by the DA algorithm) is 74.5%.

Similarly, the graphical representation in Fig. 6.55 shows that the PVS, SOS, and TLBO algorithms produce better results for the 15-bar truss problem. Furthermore, the percentage difference between the best mean mass (obtained by the PVS algorithm) and the worst mean mass (obtained by the DA algorithm) is 99.9%. For the 24-bar truss problem, as demonstrated in Fig. 6.56, the SOS and PVS algorithms perform better, and the percentage difference between the best mean mass (obtained by the SOS algorithm) and the worst mean mass (obtained by the WOA algorithm) is 66.6%.

Moreover, the PVS, SOS, and ALO algorithms produce better results for the 20-bar truss problem, as illustrated in Fig. 6.57. Furthermore, the percentage difference between the best mean mass (obtained by the PVS algorithm) and the worst mean mass (obtained by the MVO algorithm) is 99.9%. In Fig. 6.58, the graphical representation indicates that the SOS, GWO, and PVS algorithms perform better for the 72-bar 3D truss problem. Moreover, the percentage difference between the best mean mass (achieved by the SOS and GWO algorithms) and the worst mean mass (achieved by the MVO algorithm) is 99.9%.

Furthermore, for the 39-bar truss problem, the PVS, SOS, and MVO algorithms yield better results, as presented in Fig. 6.59. Additionally, the percentage difference between the best mean mass (achieved by the PVS algorithm) and the worst mean mass (achieved by the WOA algorithm) is 71.1%. For the 45-bar truss problem, as shown in Fig. 6.60, the SOS, GWO, and PVS algorithms perform better, and the percentage difference between the best mean mass (obtained by the SOS algorithm) and the worst mean mass (obtained by the WOA algorithm) is 77%.

Moreover, the PVS, SOS, and TLBO algorithms yield better results for the 25-bar 3D truss problem, as illustrated in Fig. 6.61. In addition, the percentage difference between the best mean mass (achieved by the PVS algorithm) and the worst mean mass (achieved by the MVO algorithm) is 99.9%. For the 39-bar 3D truss problem, Fig. 6.62 demonstrates that the PVS algorithm performs better, and the percentage difference between the best mean mass (achieved by the PVS algorithm) and the worst mean mass (achieved by the WOA algorithm) is 45.9%.

In summary, the study evaluated the performance of ten basic metaheuristic algorithms for optimizing the mass of truss structures with varying numbers of bars. The graphical representations of the mean mass and standard deviation of the mass for each algorithm were presented in Figs. 6.53, 6.54, 6.55, 6.56, 6.57, 6.58, 6.59, 6.60, 6.61 and 6.62. The results showed that certain algorithms performed better than others for each truss configuration, with differences in the percentage of the best mean mass obtained compared to the worst mean mass obtained ranging from 45.9 to 99.9%. The algorithms that consistently performed well across multiple truss configurations included PVS, SOS, and GWO. These findings can help guide the selection of appropriate metaheuristic algorithms for optimizing the mass of truss structures.

6.4.12 The Friedman Rank Test

Based on the comprehensive analysis of the results, it can be deduced that the PVS, SOS, and GWO algorithms exhibit significantly better performance compared to other algorithms. However, in order to provide a more comprehensive and reliable ranking of all the algorithms based on the obtained results from the proposed method over other comparative algorithms (Draa, 2015), statistical tests such as the Friedman rank test are necessary.

The Friedman rank test is conducted on the mean mass and success rate of the ALO, DA, GWO, MVO, SCA, WOA, HTS, TLBO, PVS, and SOS algorithms. The results of the Friedman test are then normalized based on the best value obtained, which allows for a fair and unbiased comparison among the algorithms. The algorithms are ranked based on the normalized values, providing a clear understanding of their relative performance.

Table 6.50 presents the detailed results of the Friedman rank test for the discrete sections of the test problems. The table provides valuable insights into the performance of each algorithm and their rankings based on the obtained results. Additionally, Figs. 6.63 and 6.64 graphically represent the normalized values of mean mass and success rate, respectively, allowing for a visual comparison of the algorithms.

The findings reveal that the PVS algorithm consistently ranks first in terms of obtaining the best mean mass, followed by the SOS and GWO algorithms. This indicates that the PVS algorithm outperforms other algorithms in terms of achieving the optimal mean mass. Similarly, the results indicate that the PVS algorithm also ranks first in terms of success rate for discrete sections, followed by the SOS and MVO algorithms.

Table 6.50 The Friedman rank test of mean mass and success rate obtained using the algorithms with discrete sections

	DA	MVO	SCA	WOA	ALO	HTS	PVS	SOS	GWO	TLBO
Friedman rank of mean mass	9	6	7	10	8	4	1	2	3	5
Friedman rank of success rate	9	3	8	10	7	5	1	2	4	6

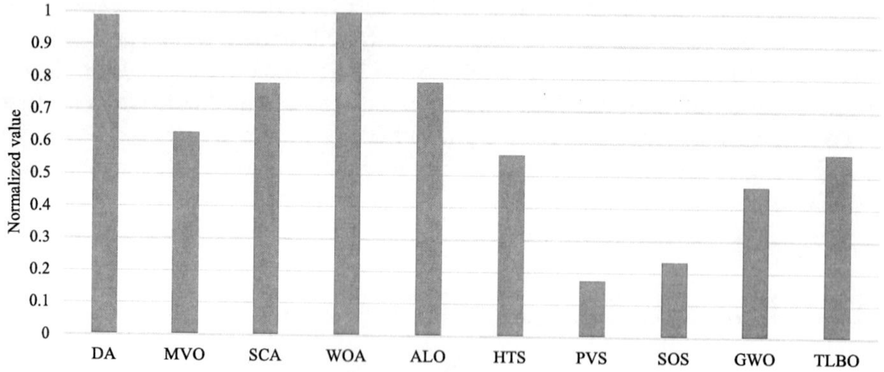

Fig. 6.63 The Friedman rank test of mean mass obtained using TSS with discrete sections

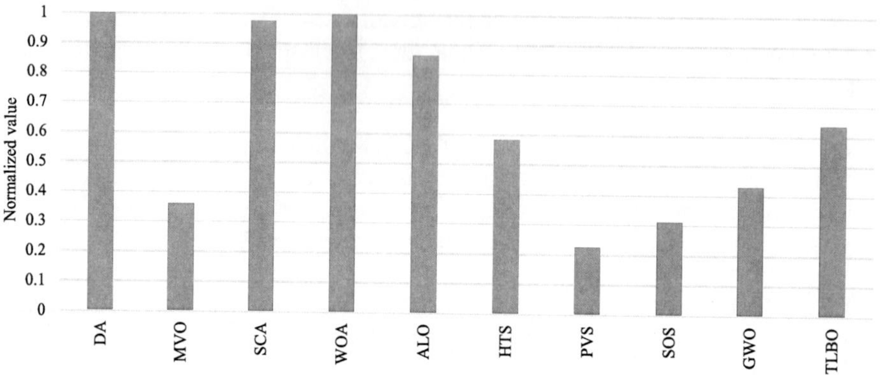

Fig. 6.64 The Friedman rank test of success rate of TSS with discrete sections

These results highlight the superior performance of the PVS, SOS, and GWO algorithms in comparison to other algorithms based on the obtained results from the proposed method. The statistical tests provide a rigorous evaluation of the algorithms, allowing for confident conclusions about their relative rankings.

References

Ahrari, A., & Deb, K. (2016). An improved fully stressed design evolution strategy for layout optimization of truss structures. *Computers & Structures, 164*(February), 127–144. https://doi.org/10.1016/j.compstruc.2015.11.009

Ahrari, A., Atai, A. A., & Deb, K. (2014). Simultaneous topology, shape and size optimization of truss structures by fully stressed design based on evolution strategy. *Engineering Optimization, 47*(8), 37–41. https://doi.org/10.1080/0305215X.2014.947972

Bojczuk, D., & Mróz, Z. (1998). Optimal design of trusses with account for topology variation. *Mechanics of Structures and Machines, 26*(1), 21–40. https://doi.org/10.1080/08905459808945418

Bojczuk, D., & Mróz, Z. (1999). Optimal topology and configuration Design of Trusses with stress and buckling constraints. *Structural Optimization, 17*, 25–35. https://doi.org/10.1007/BF01197710

Cerveira, A., Agra, A., Bastos, F., & Gromicho, J. (2013). A new branch and bound method for a discrete truss topology design problem. *Computational Optimization and Applications, 54*(1), 163–187. https://doi.org/10.1007/s10589-012-9487-6

Cheng, M. Y., & Prayogo, D. (2014). Symbiotic organisms search: A new metaheuristic optimization algorithm. *Computers and Structures, 139*, 98–112. https://doi.org/10.1016/j.compstruc.2014.03.007

Deb, K., & Gulati, S. (2001). Design of truss-structures for minimum weight using genetic algorithms. *Finite Elements in Analysis and Design, 37*(5), 447–465. https://doi.org/10.1016/S0168-874X(00)00057-3

Draa, A. (2015). On the performances of the flower pollination algorithm – Qualitative and quantitative analyses. *Applied Soft Computing, 34*, 349–371. https://doi.org/10.1016/j.asoc.2015.05.015

Ferreira, A. J. (2009). *MATLAB codes for finite element analysis: Solids and structures* (p. 157). Springer. https://doi.org/10.1007/978-1-4020-9200-8

Gonçalves, M. S., Lopez, R. H., & Miguel, L. F. F. (2015). Search group algorithm: A new metaheuristic method for the optimization of truss structures. *Computers & Structures, 153*, 165–184. https://doi.org/10.1016/j.compstruc.2015.03.003

Hajela, P., & Lee, E. (1995). Genetic algorithms in truss topological optimization. *International Journal of Solids and Structures, 32*(22), 3341–3357.

Hasançebi, O., & Erbatur, F. (2001). Layout optimization of trusses using improved GA methodologies. *Acta Mechanica, 146*(1–2), 87–107. https://doi.org/10.1007/BF01178797

Hasançebi, O., & Erbatur, F. (2002). Layout optimisation of trusses using simulated annealing. *Advances in Engineering Software, 33*, 681–696.

Kaveh, A., & Ahmadi, B. (2014). Sizing, geometry and topology optimization of trusses using force method and supervised charged system search. *Structural Engineering and Mechanics, 50*(3), 365–382. https://doi.org/10.12989/sem.2014.50.3.365

Kaveh, A., & Laknejadi, K. (2013). A hybrid evolutionary graph-based multi-objective algorithm for layout optimization of truss structures. *Acta Mechanica, 224*, 343–364. https://doi.org/10.1007/s00707-012-0754-5

Kaveh, A., & Mahdavi, V. R. (2015). Colliding bodies optimization for size and topology optimization of truss structures. *Structural Engineering and Mechanics, 53*(5), 847–865. https://doi.org/10.12989/sem.2015.53.5.847

Kumar, S., Tejani, G. G., Pholdee, N., & Bureerat, S. (2022). Performance enhancement of metaheuristics through random mutation and simulated annealing-based selection for concurrent topology and sizing optimization of truss structures. *Soft Computing, 26*, 5661–5683. https://doi.org/10.1007/s00500-022-06930-2

Luh, G., & Lin, C. (2008). Optimal design of truss structures using ant algorithm. *Structural and Multidisciplinary Optimization, 36*(4), 365–379. https://doi.org/10.1007/s00158-007-0175-6

Luh, G., & Lin, C. (2011). Optimal design of truss-structures using particle swarm optimization. *Computers & Structures, 89*(23–24), 2221–2232. https://doi.org/10.1016/j.compstruc.2011.08.013

McKeown, J. J. (1998). Growing optimal pin-jointed frames. *Structural Optimization, 15*(2), 92–100. https://doi.org/10.1007/BF01278495

Miguel, L. F. F., Lopez, R. H., & Miguel, L. F. F. F. (2013). Multimodal size, shape, and topology optimisation of truss structures using the firefly algorithm. *Advances in Engineering Software, 56*, 23–37. https://doi.org/10.1016/j.advengsoft.2012.11.006

Mirjalili, S. (2015). The ant lion optimizer. *Advances in Engineering Software, 83*, 80–98. https://doi.org/10.1016/j.advengsoft.2015.01.010

Mirjalili, S. (2016a). Dragonfly algorithm: A new meta-heuristic optimization technique for solving single-objective, discrete, and multi-objective problems. *Neural Computing and Applications, 27*(4), 1053–1073. https://doi.org/10.1007/s00521-015-1920-1

Mirjalili, S. (2016b). SCA: A sine cosine algorithm for solving optimization problems. *Knowledge-Based Systems., 96*, 120. https://doi.org/10.1016/j.knosys.2015.12.022

Mirjalili, S., & Lewis, A. (2016). The whale optimization algorithm. *Advances in Engineering Software, 95*, 51–67. https://doi.org/10.1016/j.advengsoft.2016.01.008

Mirjalili, S., Mirjalili, S. M., & Lewis, A. (2014). Grey wolf optimizer. *Advances in Engineering Software, 69*, 46–61. https://doi.org/10.1016/j.advengsoft.2013.12.007

Mirjalili, S., Mirjalili, S. M., & Hatamlou, A. (2016). Multi-verse optimizer: A nature-inspired algorithm for global optimization. *Neural Computing and Applications, 27*(2), 495–513. https://doi.org/10.1007/s00521-015-1870-7

Mortazavi, A., & Toğan, V. (2016). Simultaneous size, shape, and topology optimization of truss structures using integrated particle swarm optimizer. *Structural and Multidisciplinary Optimization., 54*, 715. https://doi.org/10.1007/s00158-016-1449-7

Noilublao, N., & Bureerat, S. (2011). Simultaneous topology, shape and sizing optimisation of a three-dimensional slender truss tower using multiobjective evolutionary algorithms. *Computers and Structures, 89*(23–24), 2531–2538. https://doi.org/10.1016/j.compstruc.2011.08.010

Noilublao, N., & Bureerat, S. (2013). Simultaneous topology, shape, and sizing optimisation of plane trusses with adaptive ground finite elements using MOEAs. *Mathematical Problems in Engineering, 2013*, 1. https://doi.org/10.1155/2013/838102

Patel, V. K., & Savsani, V. J. (2015). Heat Transfer Search (HTS): A novel optimization algorithm. *Information Sciences, 324*, 217–246. https://doi.org/10.1016/j.ins.2015.06.044

Rahami, H., Kaveh, A., & Gholipour, Y. (2008). Sizing, geometry and topology optimization of trusses via force method and genetic algorithm. *Engineering Structures, 30*, 2360–2369. https://doi.org/10.1016/j.engstruct.2008.01.012

Rajan, S. D. (1995). Sizing, shape, and topology design optimization of trusses using genetic algorithm. *Journal of Structural Engineering, 121*(10), 1480–1487.

Rajeev, S., & Krishnamoorthy, C. S. (1997). Genetic algorithms-based methodologies for design optimization of trusses. *Journal of Structural Engineering, 123*(3), 350. https://doi.org/10.1061/(ASCE)0733-9445(1997)123:3(350)

Rao, R. V., Savsani, V. J., & Vakharia, D. P. (2011). Teaching–learning-based optimization: A novel method for constrained mechanical design optimization problems. *Computer-Aided Design, 43*(3), 303–315. https://doi.org/10.1016/j.cad.2010.12.015

Rao, R. V., Savsani, V. J., & Vakharia, D. P. (2012). Teaching–learning-based optimization: An optimization method for continuous non-linear large scale problems. *Information Sciences, 183*(1), 1–15. https://doi.org/10.1016/j.ins.2011.08.006

Richardson, J. N., Bouillard, P., Adriaenssens, S., Bouillard, P., & Filomeno, R. (2012). Multiobjective topology optimization of truss structures with kinematic stability repair. *Structural and Multidisciplinary Optimization, 46*, 513–532. https://doi.org/10.1007/s00158-012-0777-5

Savsani, P. V., & Savsani, V. J. (2015). Passing Vehicle Search (PVS): A novel metaheuristic algorithm. *Applied Mathematical Modelling, 40*(5–6), 3951–3978. https://doi.org/10.1016/j.apm.2015.10.040

Savsani, V. J., Tejani, G. G., & Patel, V. K. (2016). Truss topology optimization with static and dynamic constraints using modified subpopulation teaching–learning-based optimization. *Engineering Optimization, 48*, 1–17. https://doi.org/10.1080/0305215X.2016.1150468

Savsani, V. J., Tejani, G. G., Patel, V. K., & Savsani, P. (2017). Modified meta-heuristics using random mutation for truss topology optimization with static and dynamic constraints. *Journal of Computational Design and Engineering, 4*, 106–130. https://doi.org/10.1080/0305215X.2016.1150468

Shea, K., & Cagan, J. (1999). The Design of Novel Roof Trusses with shape annealing: Assessing the ability of a computational method in aiding structural designers with varying design intent. *Design Studies, 20*(1), 3–23. https://doi.org/10.1016/S0142-694X(98)00019-2

Shrestha, S. M., & Ghaboussi, J. (1999). Evolution of optimum structural shapes using genetic algorithm. *Journal of Structural Engineering, 124*(11), 1331–1338.

Soh, C. K., & Yang, Y. (2000). Genetic programming-based approach for structural optimization. *Journal of Computing in Civil Engineering, 14*(1), 31–37. https://doi.org/10.1061/(ASCE)0887-3801(2000)14:1(31)

Souza, R. R., Miguel, L. F. F., Lopez, R. H., Miguel, L. F. F., & Torii, A. J. (2016). A procedure for the size, shape and topology optimization of transmission line tower structures. *Engineering Structures, 111*, 162–184. https://doi.org/10.1016/j.engstruct.2015.12.005

Tang, W., Tong, L., & Gu, Y. (2005). Improved genetic algorithm for design optimization of truss structures with sizing, shape and topology variables. *International Journal for Numerical Methods in Engineering, 62*, 1737–1762. https://doi.org/10.1002/nme.1244

Tejani G. G. (2017). *Investigation of advanced metaheuristic techniques for simultaneous size, shape, and topology optimization of truss structures*. PhD thesis, Pandit Deendayal Petroleum University, India

Tejani, G. G., Savsani, V. J., Patel, V. K., & Bureerat, S. (2017). Topology, shape, and size optimization of truss structures using modified teaching-learning based optimization. *Advances in Computational Design, 2*, 313–331. https://doi.org/10.12989/acd.2017.2.4.313

Tejani, G. G., Savsani, V. J., Patel, V. K., & Savsani, P. V. (2018). Size, shape, and topology optimization of planar and space trusses using mutation-based improved metaheuristics. *Journal of Computational Design and Engineering, 5*, 198–214. https://doi.org/10.1016/j.jcde.2017.10.001

Tejani, G. G., Savsani, V. J., Bureerat, S., et al. (2019). Topology optimization of truss subjected to static and dynamic constraints by integrating simulated annealing into passing vehicle search algorithms. *Engineering Computations, 35*, 1–19.

Wu, C., & Tseng, K. (2010). Truss structure optimization using adaptive multi-population differential evolution. *Structural and Multidisciplinary Optimization, 42*(4), 575–590. https://doi.org/10.1007/s00158-010-0507-9

Chapter 7
Validation

7.1 The Finite Element Analysis Process Using ANSYS

Given the availability of numerous finite element analysis programs, ANSYS is recognized as a highly efficient software. Hence, to verify the accuracy of our obtained results, we will compare them with those reported in previous chapters using ANSYS. The following steps outline the finite element analysis process of the sample problem, which involves the 24-bar truss, using ANSYS (Tejani, 2017):

Launch ANSYS

Step 1: Preprocessing: Define the problem.

Select job name and analysis type.

```
>>    /filename, 24 bar truss
>>  /Title Truss Sample Problem Case 1
>>  /prep7
```

Step 2: Enter nodal points.

```
>>  n,1,0,2.5,0 ! Node number, Nadal coordinates
>>  n,2,2.5,3.25,0
>>  n,3,5,4,0
>>  n,4,7.5,2.5,0
>>  n,5,2.5,0,0
>>  n,6,5,0,0
>>  n,7,1.792,0,0
>>  n,8,7.186189,0,0
```

Step 3: Define element type.

```
>>  et, 1, link1 ! LINK 1 is a 2D truss element
>>  ! or
>>  et, 1, link180 ! LINK 180 is a 3D truss element
>>  et,2,mass21 ! MASS 21 is a nodal element
```


V. Savsani et al., *Truss Optimization*, https://doi.org/10.1007/978-3-031-49295-2_7

Step 4: Define element cross-sectional areas.

```
>> r,1, 0.0013, , ! Real consts: cross-sectional area of element 1,
>> r,2, 0.0003, ,
>> r,3, 0.0003, ,
>> r,4, 0.0021, ,
>> r,5, 0.0004, ,
>> r,6, 0.0024, ,
>> r,7, 0.0001, ,
>> r,8, 0.0001, ,
```

Step 5: Define material property.

```
>> mp, ex, 1, 69e9 ! Material property, Modulus of Elasticity
>> mp, prxy, 1, 0.3
>> mp,dens,1,2740 ! Material density
```

Step 6: Enter element connectivity and assign their element cross-sectional areas.

```
>> real, 1 ! Section number 1
>> en,1,2,7 ! Element number 1 connected to nodes 2 and 7
>> real, 2
>> en,2,2,5
>> real, 3
>> en,3,2,3
>> real, 4
>> en,4,3,7
>> real, 5
>> en,5,3,6
>> real, 6
>> en,6,3,8
>> real, 7
>> en,7,7,5
>> real, 8
>> en,8,6,8
```

Step 7: Assign nodal elements and their mass.

```
>> R,9,500,500,0,0,0,0, ! Node number, element mass along X, Y, and Z
coordinates
```

Step 8: Define boundary conditions.

```
>> d, 7, ux, 0 ! Displacement at node 7 in X-direction is zero
>> d, 7, uy, 0 ! Displacement at node 7 in Y-direction is zero
>> d, 8, ux, 0
>> d, 8, uy, 0
```

Step 9: Define loads.

```
>> f, 2, fx, -50000 ! Force at node 2 in X-direction
>> f, 5, fy, -50000
>> /pnum, elem, 1 ! Plot element numbers
>> eplot ! Plot the elements
>> /pbc, u, , 1
finish
```

Step 10: Define the material behavior.

```
>> /solu ! Select static load solution
>> antype, static ! If static solution
>> antype,2 ! If dynamic analysis
```

Step 11: Analysis /solution: Solve the problem.

```
>> solve
>> save
>> finish
```

Step 12: Post-processing: Examine the results.

```
>> /post1
>> etable, stress, ls, 1 ! Create a table of elemental stress values
>> pretab, ! Display elemental stress values
>> prnsol,u,comp ! Display nodal displacement values
>> set,list ! Display frequency
```

The detailed discussion on validation of obtained results is presented in following subsections.

7.2 Validation of the 10-Bar Truss

The TLBO algorithm has yielded the best result for the 10-bar truss structure of topology, shape, and size optimization, demonstrating the least truss mass as compared to other algorithms. This outcome is significant as it indicates that the TLBO algorithm is a more effective approach for optimizing the truss design than other methods. The truss structure consists of ten bars that are subjected to external forces applied at nodes 2 and 4, while nodes 5 and 6 are fixed. The design parameters for the truss are presented in Chap. 6.

The obtained stress, displacement, and frequency values were computed using two programs, namely, MATLAB and ANSYS. The results obtained from the two software programs are presented in Table 7.1. From the result table, it can be observed that the MATLAB results are 100% accurate compared to the ANSYS results. This observation provides strong evidence that the results obtained using MATLAB are reliable and can be used for further analysis.

Table 7.1 The 10-bar truss

	MATLAB results	ANSYS results	Accuracy (%)	MATLAB results	ANSYS results	Accuracy (%)
	Load case 1			Load case 2		
σ_1 (N/m^2)	1.3882E+08	1.3882E+08	100	1.3882E+08	1.3882E+08	100
σ_3 (N/m^2)	−1.2744E+07	−1.2744E+07	100	−8.4957E+06	−8.4957E+06	100
σ_4 (N/m^2)	−1.0356E+07	−1.0356E+07	100	−1.0356E+07	−1.0356E+07	100
σ_5 (N/m^2)	1.6832E+05	1.6832E+05	100	1.6832E+05	1.6832E+05	100
σ_7 (N/m^2)	1.5833E+08	1.5833E+08	100	−5.9864E+04	−5.9864E+04	100
σ_9 (N/m^2)	1.3878E+08	1.3878E+08	100	1.3878E+08	1.3878E+08	100
δ_{x_2} (m)	3.0612E-03	−3.5488E-02	100	−2.4983E-03	−2.4983E-03	100
δ_{y_2} (m)	7.9087E-02	−6.4812E-02	100	−7.8692E-02	−7.8692E-02	100
f_1 (Hz)	5.1542	5.1542	100			
f_2 (Hz)	5.3782	5.3782	100			
f_3 (Hz)	19.7574	19.757	100			
f_4 (Hz)	22.5538	22.554	100			

Finally, Fig. 7.1 presents the displacement diagrams of the truss structure. These diagrams illustrate how the truss structure deforms under the influence of the external forces applied at nodes 2 and 4. The diagrams provide valuable insight into the behavior of the truss structure and can be used to optimize its design further. Overall, the findings of this study have important implications for the design and optimization of truss structures.

7.3 Validation of the 14-Bar Truss

The TLBO algorithm has yielded the least truss mass for the 14-bar truss structure of topology, shape, and size optimization, as evidenced by the best reported result. In Fig. 7.2, the finite element model indicates that the two leftmost nodes (nodes 5 and 6) of the truss are fixed, while external forces are applied at nodes 2 and 4.

Table 7.2 presents the stress, displacement, and frequency values obtained using MATLAB and ANSYS programs. The results obtained using MATLAB software are found to be 100% accurate compared to the results obtained using ANSYS software, as depicted in the result table. This observation validates the accuracy of the results obtained using MATLAB, which can be utilized for further analysis.

Additionally, Fig. 7.2 shows the displacement diagrams of the 14-bar truss structure, which display how the truss deforms under the external forces applied at nodes 2 and 4. These diagrams provide valuable insight into the behavior of the truss

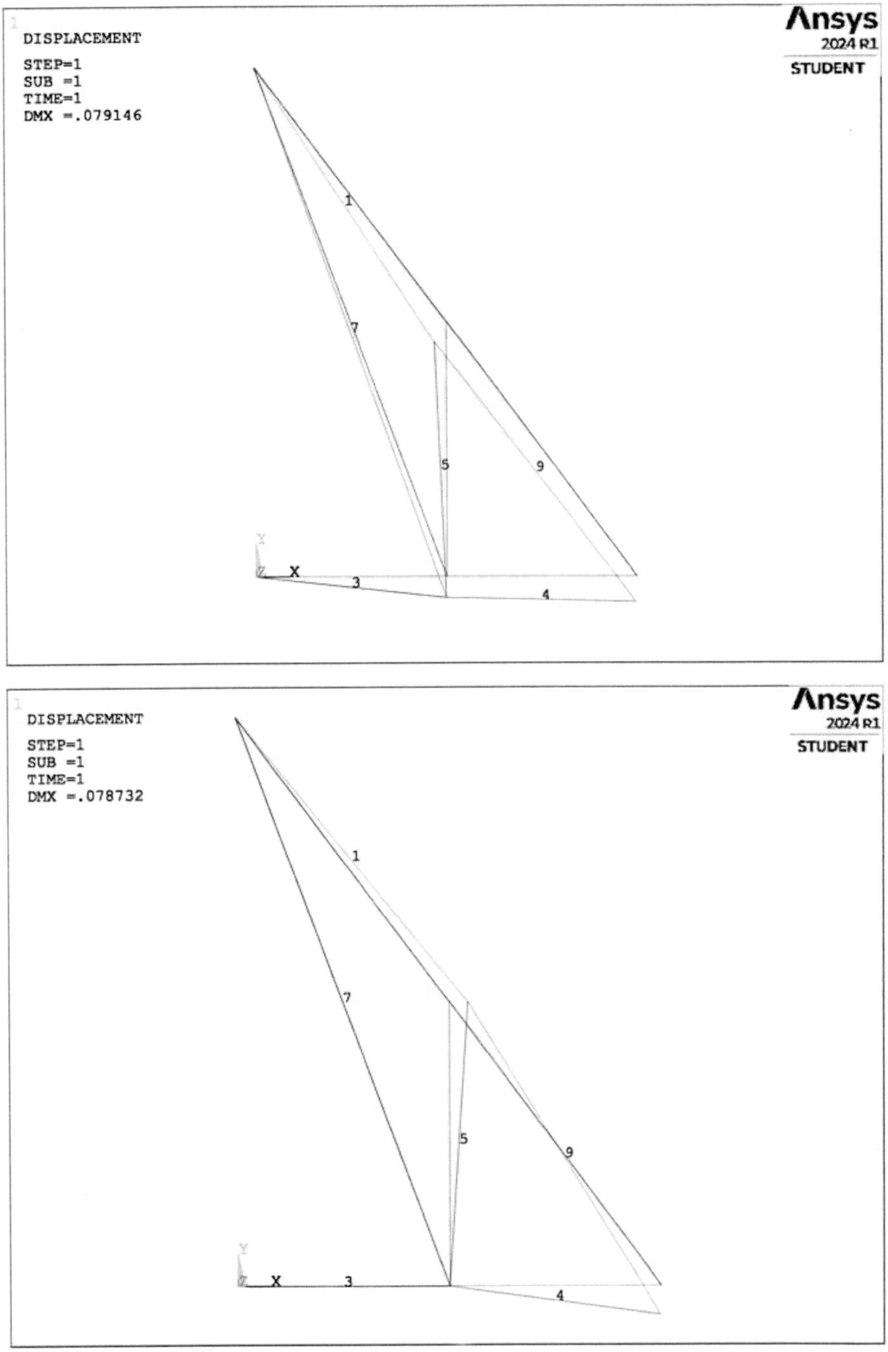

Fig. 7.1 Displacement diagrams of the 10-bar truss under multiple load conditions

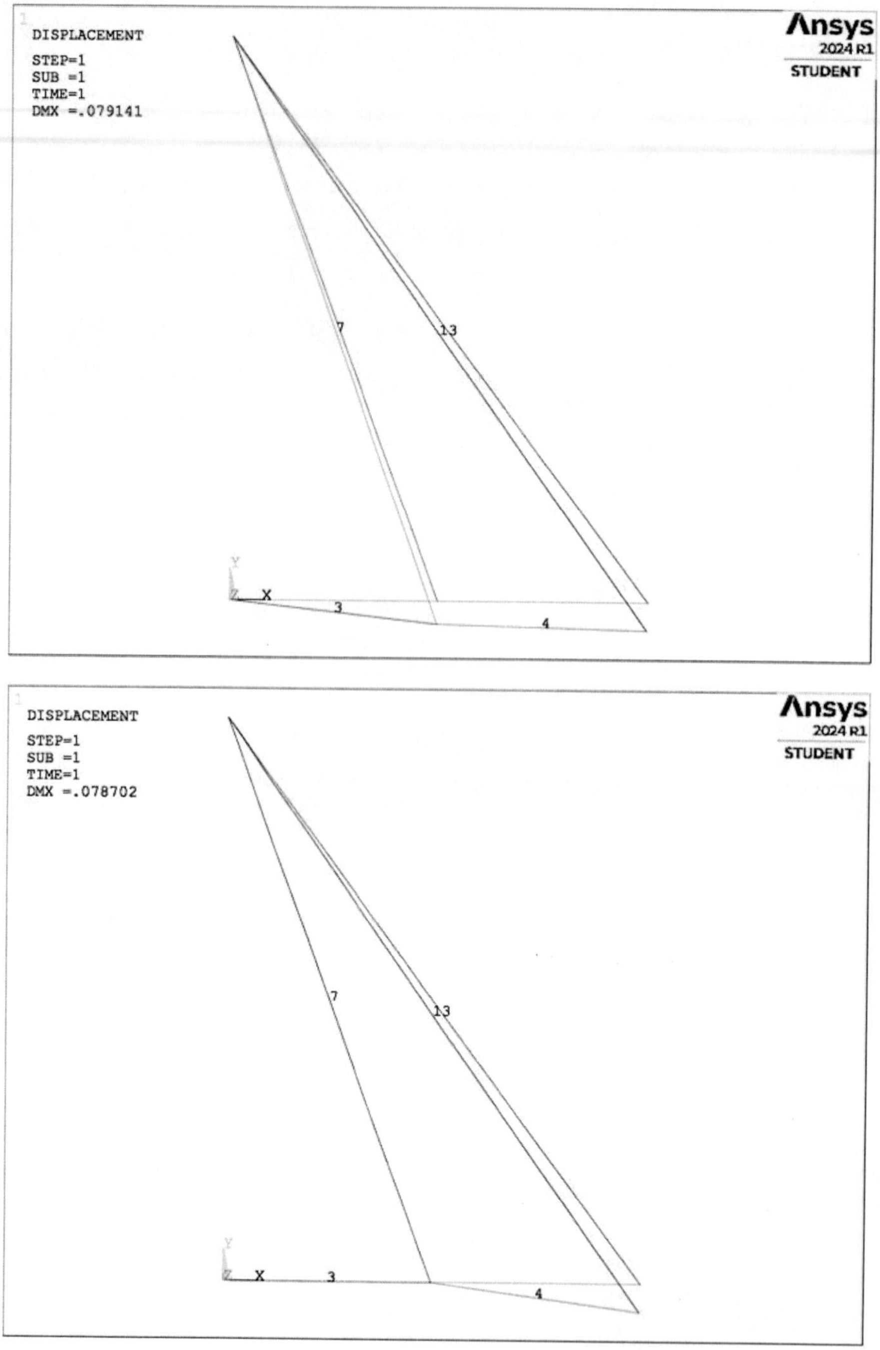

Fig. 7.2 Displacement diagrams of the 14-bar truss under multiple load conditions

Table 7.2 The 14-bar truss

	MATLAB results	ANSYS results	Accuracy (%)	MATLAB results	ANSYS results	Accuracy (%)
	Load case 1			Load case 2		
σ_3 (N/m^2)	−1.2874E+07	−1.2744E+07	100	−8.5824E+06	−8.4957E+06	100
σ_4 (N/m^2)	−1.0460E+07	−1.0354E+07	100	−1.0460E+07	−1.0354E+07	100
σ_7 (N/m^2)	1.5859E+08	1.5839E+08	100	0.0000E+00	0.0000E+00	100
σ_{13} (N/m^2)	1.3928E+08	1.3878E+08	100	1.3928E+08	1.3878E+08	100
δ_{x_2} (m)	−3.0922E-03	−3.0610E-03	100	−2.5235E-03	−2.4980E-03	100
δ_{y_2} (m)	−7.9187E-02	−7.9081E-02	100	−7.8760E-02	−7.8663E-02	100
f_1 (Hz)	5.1430	5.1454	100			
f_2 (Hz)	5.5934	5.576	100			
f_3 (Hz)	22.2499	22.243	100			
f_4 (Hz)	55.8075	55.803	100			

structure and can be used to further optimize its design. Overall, these findings have important implications for the optimization of truss structures, and the results obtained using the TLBO algorithm and MATLAB software are reliable and can be used for further analysis.

7.4 Validation of the 15-Bar Truss

According to the PVS algorithm, the least truss mass for the 15-bar truss structure of topology, shape, and size optimization has been achieved. This outcome is the best result reported for this algorithm. In Fig. 7.3, in the finite element model of the truss structure, external forces are applied at node 2, while nodes 7 and 8 are fixed.

Table 7.3 presents the stress, displacement, and frequency values obtained using MATLAB and ANSYS programs. The results obtained using MATLAB software are 100% accurate compared to the results obtained using ANSYS software, as illustrated in the result table. This observation validates the accuracy of the results obtained using MATLAB, which can be used for further analysis.

Figure 7.3 displays the displacement diagrams of the 15-bar truss structure, which show how the truss deforms under the external forces applied at node 2. These diagrams provide important insights into the behavior of the truss structure and can be used to optimize its design further.

In conclusion, the PVS algorithm has yielded the least truss mass for the 15-bar truss structure, and the results obtained using MATLAB software have been validated and can be used for further analysis. The displacement diagrams presented in Fig. 7.3 provide valuable insights into the behavior of the truss structure and can be used to optimize its design further.

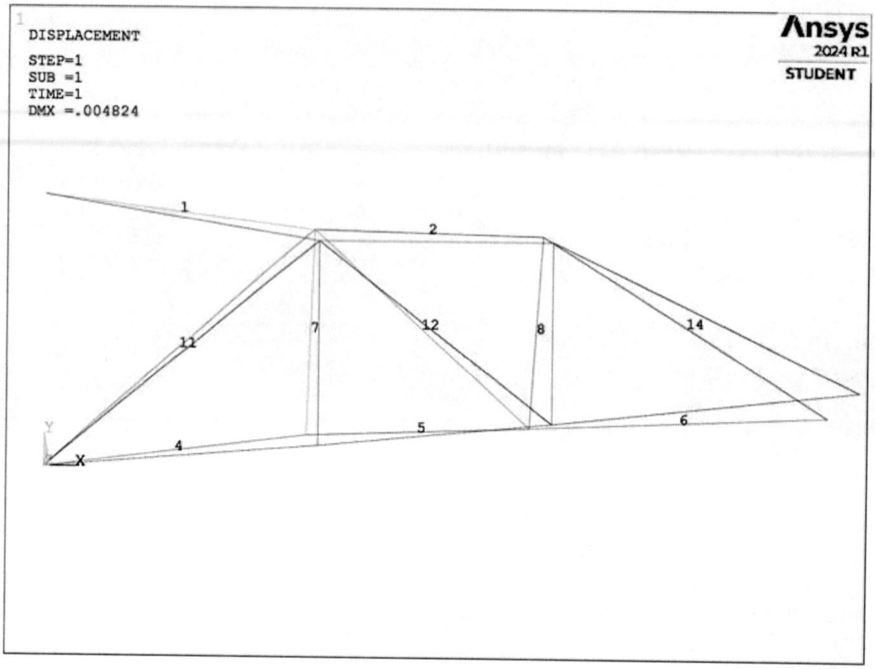

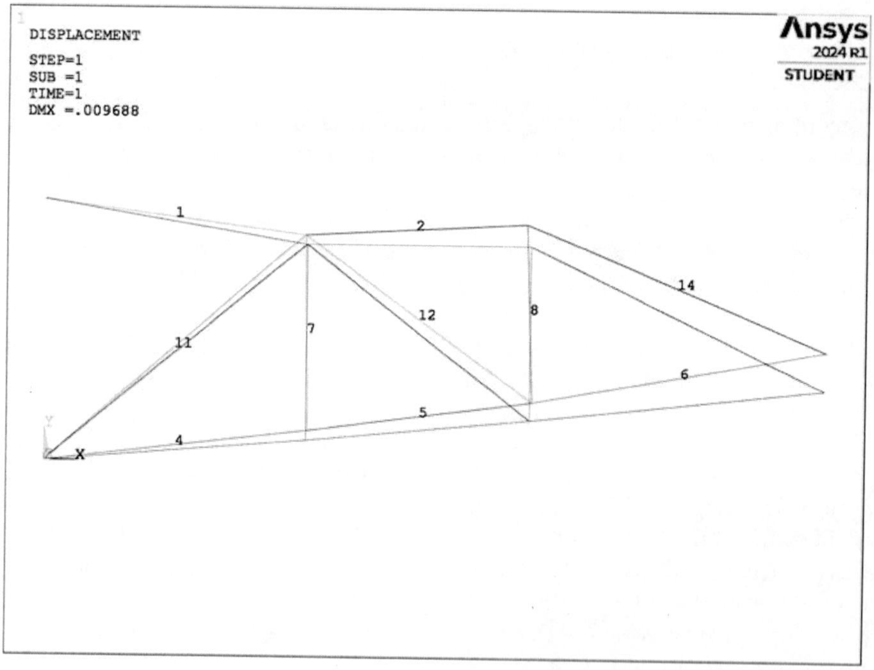

Fig. 7.3 Displacement diagrams of the15-bar truss under multiple load conditions

Table 7.3 The 15-bar truss

	MATLAB results	ANSYS results	Accuracy (%)	ATLAB results	ANSYS results	Accuracy (%)
	Load case 1			Load case 2		
σ_1 (N/m^2)	−1.6498E+07	−1.6498E+07	100	−1.6926E+07	−1.69E+07	100
σ_2 (N/m^2)	−1.6803E+07	−1.6803E+07	100	−1.5052E+07	−1.51E+07	100
σ_4 (N/m^2)	−2.9291E+07	−2.9291E+07	100	1.0755E+07	1.08E+07	100
σ_5 (N/m^2)	−3.5849E+07	−3.5849E+07	100	1.3163E+07	1.32E+07	100
σ_6 (N/m^2)	−2.7969E+07	−2.7969E+07	100	5.8252E+06	5.83E+06	100
σ_7 (N/m^2)	4.8631E+06	4.8631E+06	100	−1.7857E+06	−1.79E+06	100
σ_8 (N/m^2)	4.0340E+07	4.0340E+07	100	3.6135E+07	3.61E+07	100
σ_{11} (N/m^2)	7.7603E+06	7.7603E+06	100	2.0321E+07	2.03E+07	100
σ_{12} (N/m^2)	−1.0182E+07	−1.0182E+07	100	−1.0796E+07	−1.08E+07	100
σ_{14} (N/m^2)	−1.3314E+07	−1.3314E+07	100	−1.1926E+07	−1.19E+07	100
δ_{x_2} (m)	−3.7556E-03	−3.7556E-03	100	3.0274E-04	3.03E-04	100
δ_{y_2} (m)	−3.0274E-03	−3.0274E-03	100	9.6831E-03	9.68E-03	100
f_1 (Hz)	4.7576	4.7576				
f_2 (Hz)	24.1819	24.182				
f_3 (Hz)	37.1016	37.102				
f_4 (Hz)	119.8357	119.84				

7.5 Validation of the 24-Bar Truss

The PVS algorithm has yielded the least truss mass for the 24-bar truss structure of topology, shape, and size optimization, and this is the best result reported for this algorithm. In Fig. 7.4, in the finite element model of the truss structure, external forces are applied at nodes 2 and 3, while nodes 7 and 8 are fixed.

Table 7.4 presents the stress, displacement, and frequency values obtained using MATLAB and ANSYS programs. The results obtained using MATLAB software are 100% accurate compared to the results obtained using ANSYS software, as illustrated in the result table. This observation validates the accuracy of the results obtained using MATLAB, which can be used for further analysis.

Figure 7.4 displays the displacement diagrams of the 24-bar truss structure, which show how the truss deforms under the external forces applied at nodes 2 and 3. These diagrams provide important insights into the behavior of the truss structure and can be used to optimize its design further.

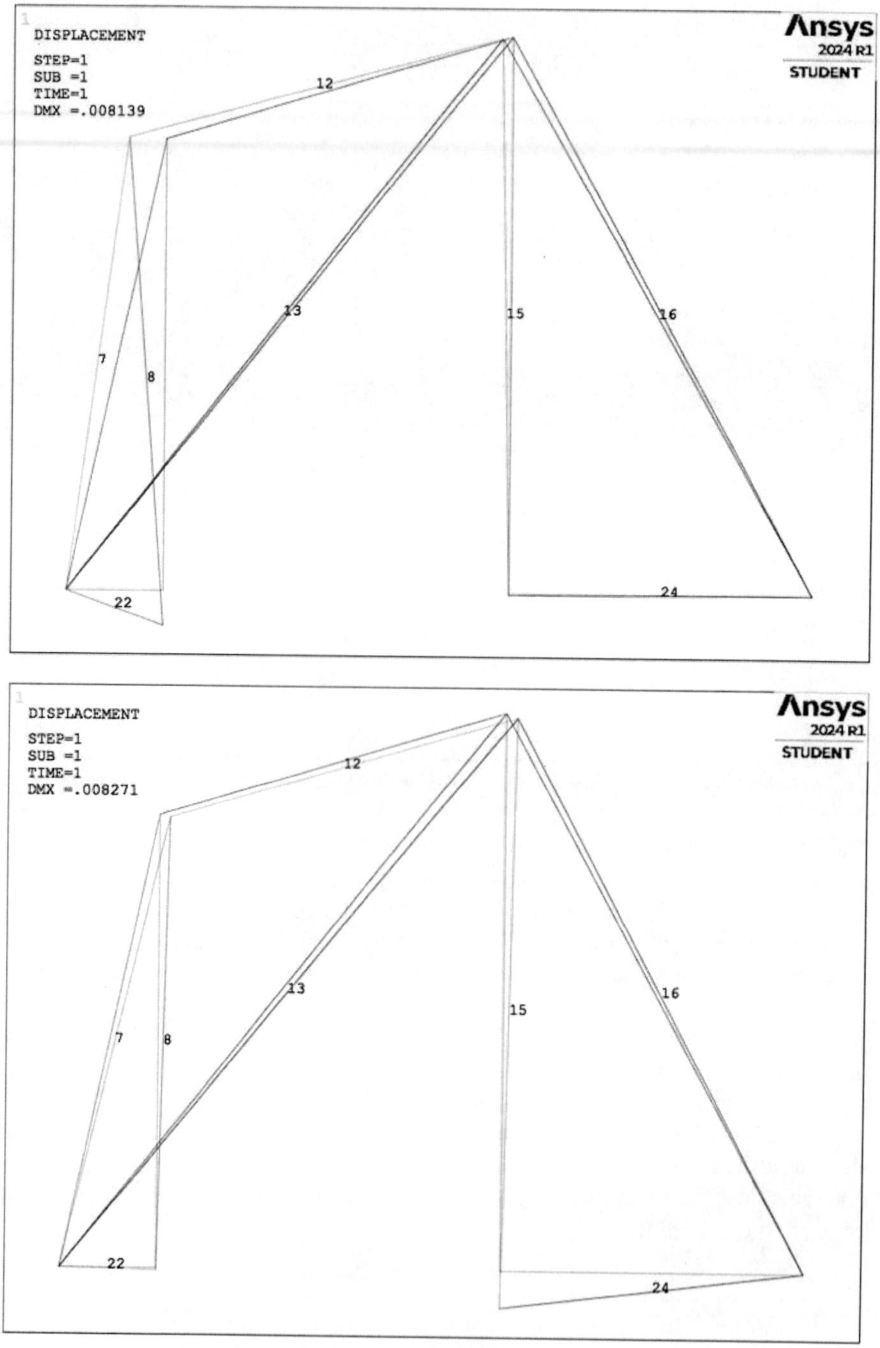

Fig. 7.4 Displacement diagrams of the 24-bar truss under multiple load conditions

Table 7.4 The 24-bar truss

	MATLAB results	ANSYS results	Accuracy (%)	MATLAB results	ANSYS results	Accuracy (%)
	Load case 1			Load case 2		
σ_7 (N/m^2)	−2.9481E+07	−2.9481E+07	100	0.0000E+00	0.0000E+00	100
σ_8 (N/m^2)	1.6667E+08	1.6667E+08	100	0.0000E+00	0.0000E+00	100
σ_{12} (N/m^2)	1.4562E+08	1.4562E+08	100	0.0000E+00	0.0000E+00	100
σ_{13} (N/m^2)	−2.2045E+07	−2.2045E+07	100	1.0263E+07	1.0263E+07	100
σ_{15} (N/m^2)	0.0000E+00	0.0000E+00	100	1.2500E+08	1.2500E+08	100
σ_{16} (N/m^2)	1.1188E+07	1.1188E+07	100	−3.1725E+07	−3.1725E+07	100
σ_{22} (N/m^2)	0.0000E+00	0.0000E+00	100	0.0000E+00	0.0000E+00	100
σ_{24} (N/m^2)	0.0000E+00	0.0000E+00	100	0.0000E+00	0.0000E+00	100
δ_{y_5} (m)	−7.5330E-03	−7.5330E-03	100	0.0000E+00	0.0000E+00	100
δ_{y_6} (m)	−7.5330E-03	−7.5330E-03	100	0.0000E+00	0.0000E+00	100
f_1 (Hz)	30.3936	30.394	100			
f_2 (Hz)	46.7040	46.704	100			
f_3 (Hz)	156.4569	156.46	100			
f_4 (Hz)	219.4590	219.46	100			

In conclusion, the PVS algorithm has yielded the least truss mass for the 24-bar truss structure, and the results obtained using MATLAB software have been validated and can be used for further analysis. The displacement diagrams presented in Fig. 7.4 provide valuable insights into the behavior of the truss structure and can be used to optimize its design further.

7.6 Validation of the 20-Bar Truss

The MVO algorithm has resulted in the least truss mass for the 20-bar truss structure of topology, shape, and size optimization, making it the best reported result for this algorithm. In Fig. 7.5, the finite element model of the truss structure in this case has nodes 1 and 9 fixed, while external forces are applied at node 4.

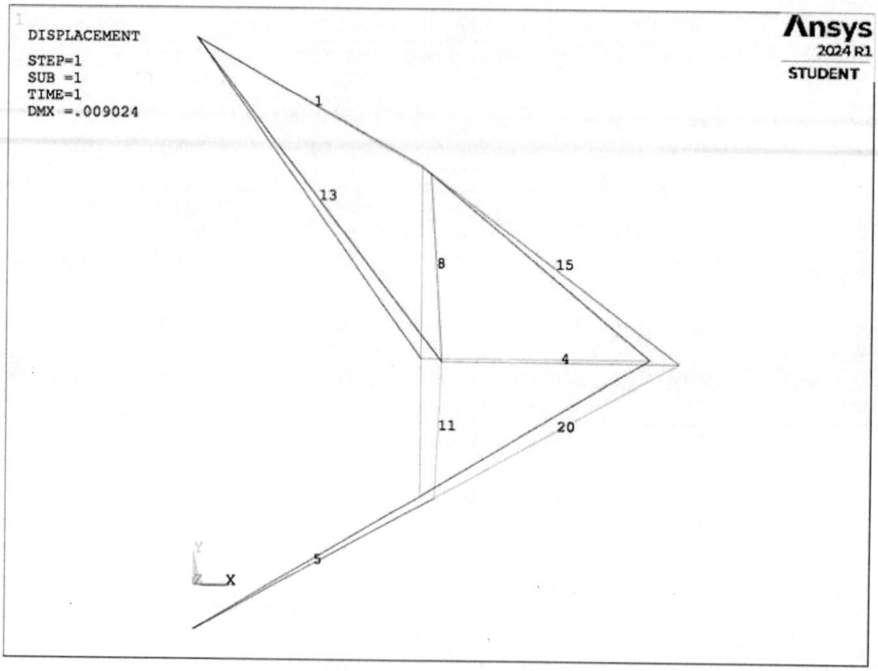

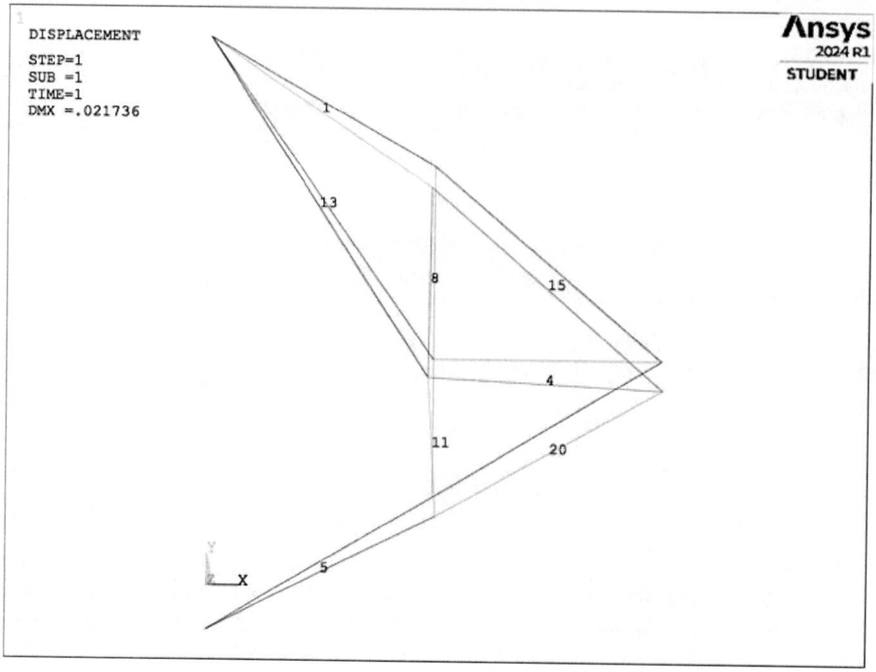

Fig. 7.5 Displacement diagrams of the 20-bar truss under multiple load conditions

Table 7.5 The 20-bar truss

	MATLAB results	ANSYS results	Accuracy (%)	MATLAB results	ANSYS results	Accuracy (%)
	Load case 1			Load case 2		
σ_1 (N/m^2)	1.0557E+08	1.0557E+08	100	1.7242E+08	1.72E+08	100
σ_4 (N/m^2)	8.9159E+07	8.9159E+07	100	1.7216E+08	1.72E+08	100
σ_5 (N/m^2)	1.0475E+08	1.0475E+08	100	−1.4856E+08	−1.49E+08	100
σ_8 (N/m^2)	−5.5202E+07	−5.5202E+07	100	−9.0159E+07	−9.02E+07	100
σ_{11} (N/m^2)	−2.4532E+07	−2.4532E+07	100	3.4791E+07	3.48E+07	100
σ_{13} (N/m^2)	8.8080E+07	8.8080E+07	100	1.7008E+08	1.70E+08	100
σ_{15} (N/m^2)	1.0556E+08	1.0556E+08	100	1.7242E+08	1.72E+08	100
σ_{20} (N/m^2)	1.0559E+08	1.0559E+08	100	−1.4975E+08	−1.50E+08	100
δ_{y_4} (m)	−1.1573E-03	−1.1573E-03	100	−2.1706E-02	-2.17E-02	100
f_1 (Hz)	61.7131	61.713	100			
f_2 (Hz)	152.5688	152.57	100			
f_3 (Hz)	200.8862	200.89	100			
f_4 (Hz)	239.6034	239.6	100			

Table 7.5 presents the obtained stress, displacement, and frequency values for the truss structure using both MATLAB and ANSYS programs. The results obtained using MATLAB software are 100% accurate compared to the results obtained using ANSYS software, as indicated in the result table. Hence, the results are validated, and they can be used for further analysis.

Figure 7.5 displays the displacement diagrams of the 20-bar truss structure, which illustrate how the truss deforms under the external forces applied at node 4. These diagrams provide valuable insights into the behavior of the truss structure and can be used to optimize its design further.

In summary, the MVO algorithm has yielded the least truss mass for the 20-bar truss structure, and the results obtained using MATLAB software are accurate and validated. The displacement diagrams presented in Fig. 7.5 provide insights into the behavior of the truss structure and can be utilized to optimize its design further.

7.7 Validation of the 72-Bar 3D Truss

The TLBO algorithm has yielded the optimal outcome for the 72-bar 3D truss of topology, shape, and size optimization, resulting in the least truss mass. The four bottom nodes, namely, nodes 17, 18, 19, and 20, of the truss have been fixed in the finite element model, while external forces have been applied at the top nodes 1, 2, 3, and 4. Table 7.6 presents the obtained stress, displacement, and frequency values

Table 7.6 The 72-bar 3D truss

	MATLAB results	ANSYS results	Accuracy (%)	MATLAB results	ANSYS results	Accuracy (%)
	Load case 1			Load case 2		
σ_1 (N/m^2)	−4.4063E+06	−4.4063E+06	100	−5.5625E+07	−5.5625E+07	100
σ_2 (N/m^2)	0.0000E+00	0.0000E+00	100	−5.5625E+07	−5.5625E+07	100
σ_3 (N/m^2)	0.0000E+00	0.0000E+00	100	−5.5625E+07	−5.5625E+07	100
σ_4 (N/m^2)	0.0000E+00	0.0000E+00	100	−5.5625E+07	−5.5625E+07	100
σ_{12} (N/m^2)	−2.4495E+07	−2.4495E+07	100	0.0000E+00	0.0000E+00	100
σ_{19} (N/m^2)	−3.7000E+06	−3.7000E+06	100	−2.5993E+07	−2.5993E+07	100
σ_{20} (N/m^2)	−1.4672E+07	−1.4672E+07	100	−2.5993E+07	−2.5993E+07	100
σ_{21} (N/m^2)	−1.4968E+06	−1.4968E+06	100	−2.5993E+07	−2.5993E+07	100
σ_{22} (N/m^2)	−1.4672E+07	−1.4672E+07	100	−2.5993E+07	−2.5993E+07	100
σ_{23} (N/m^2)	1.4141E+06	1.4141E+06	100	−1.7194E+06	−1.7194E+06	100
σ_{24} (N/m^2)	1.8154E+07	1.8154E+07	100	−1.7194E+06	−1.7194E+06	100
σ_{25} (N/m^2)	−1.4626E+07	−1.4626E+07	100	−1.7194E+06	−1.7194E+06	100
σ_{26} (N/m^2)	1.4141E+06	1.4141E+06	100	−1.7194E+06	−1.7194E+06	100
σ_{27} (N/m^2)	1.4141E+06	1.4141E+06	100	−1.7194E+06	−1.7194E+06	100
σ_{28} (N/m^2)	−1.4626E+07	−1.4626E+07	100	−1.7194E+06	−1.7194E+06	100
σ_{29} (N/m^2)	1.8154E+07	1.8154E+07	100	−1.7194E+06	−1.7194E+06	100
σ_{30} (N/m^2)	1.4141E+06	1.4141E+06	100	−1.7194E+06	−1.7194E+06	100
σ_{35} (N/m^2)	3.5099E+07	3.5099E+07	100	4.1264E+06	4.1264E+06	100
σ_{36} (N/m^2)	−3.3936E+06	−3.3936E+06	100	4.1264E+06	4.1264E+06	100
σ_{37} (N/m^2)	1.7253E+07	1.7253E+07	100	−1.6687E+07	−1.6687E+07	100
σ_{38} (N/m^2)	−6.8753E+06	−6.8753E+06	100	−1.6687E+07	−1.6687E+07	100
σ_{39} (N/m^2)	−1.6459E+07	−1.6459E+07	100	−1.6687E+07	−1.6687E+07	100
σ_{40} (N/m^2)	−6.8753E+06	−6.8753E+06	100	−1.6687E+07	−1.6687E+07	100
σ_{41} (N/m^2)	−1.7062E+07	−1.7062E+07	100	1.6160E+06	1.6160E+06	100
σ_{42} (N/m^2)	−1.3291E+06	−1.3291E+06	100	1.6160E+06	1.6160E+06	100
σ_{43} (N/m^2)	−1.3291E+06	−1.3291E+06	100	1.6160E+06	1.6160E+06	100
σ_{44} (N/m^2)	1.3746E+07	1.3746E+07	100	1.6160E+06	1.6160E+06	100
σ_{45} (N/m^2)	1.3746E+07	1.3746E+07	100	1.6160E+06	1.6160E+06	100
σ_{46} (N/m^2)	−1.3291E+06	−1.3291E+06	100	1.6160E+06	1.6160E+06	100
σ_{47} (N/m^2)	−1.3291E+06	−1.3291E+06	100	1.6160E+06	1.6160E+06	100
σ_{48} (N/m^2)	−1.7062E+07	−1.7062E+07	100	1.6160E+06	1.6160E+06	100
σ_{55} (N/m^2)	1.4868E+07	1.4868E+07	100	−1.2451E+07	−1.2451E+07	100
σ_{56} (N/m^2)	−6.5607E+06	−6.5607E+06	100	−1.2451E+07	−1.2451E+07	100
σ_{57} (N/m^2)	−1.5499E+07	−1.5499E+07	100	−1.2451E+07	−1.2451E+07	100
σ_{58} (N/m^2)	−6.5607E+06	−6.5607E+06	100	−1.2451E+07	−1.2451E+07	100
σ_{59} (N/m^2)	−3.6154E+06	−3.6154E+06	100	−1.9235E+06	−1.9235E+06	100
σ_{60} (N/m^2)	1.4115E+07	1.4115E+07	100	−1.9235E+06	−1.9235E+06	100
σ_{61} (N/m^2)	−1.2741E+07	−1.2741E+07	100	−1.9235E+06	−1.9235E+06	100
σ_{62} (N/m^2)	4.2476E+06	4.2476E+06	100	−1.9235E+06	−1.9235E+06	100

(continued)

Table 7.6 (continued)

	MATLAB results	ANSYS results	Accuracy (%)	MATLAB results	ANSYS results	Accuracy (%)
σ_{63} (N/m^2)	4.2476E+06	4.2476E+06	100	−1.9235E+06	−1.9235E+06	100
σ_{64} (N/m^2)	−1.2741E+07	−1.2741E+07	100	−1.9235E+06	−1.9235E+06	100
σ_{65} (N/m^2)	1.4115E+07	1.4115E+07	100	−1.9235E+06	−1.9235E+06	100
σ_{66} (N/m^2)	−3.6154E+06	−3.6154E+06	100	−1.9235E+06	−1.9235E+06	100
σ_{67} (N/m^2)	1.0923E+07	1.0923E+07	100	2.1851E+05	2.1851E+05	100
σ_{68} (N/m^2)	−5.8742E+06	−5.8742E+06	100	2.1851E+05	2.1851E+05	100
σ_{69} (N/m^2)	−5.8742E+06	−5.8742E+06	100	2.1851E+05	2.1851E+05	100
σ_{70} (N/m^2)	1.0923E+07	1.0923E+07	100	2.1851E+05	2.1851E+05	100
δ_{x_1} (m)	5.8769E-03	5.8769E-03	100	−4.3000E-04	−4.3000E-04	100
δ_{y_1} (m)	5.8769E-03	5.8769E-03	100	−4.3000E-04	−4.3000E-04	100
f_1 (Hz)	4.0000	3.9989	100.03			
f_2 (Hz)	4.0000	3.9989	100.03			
f_3 (Hz)	6.0015	5.989	100.21			
f_4 (Hz)	6.0570	6.0504	100.11			

through the use of MATLAB and ANSYS programs. It can be observed from the result table that all the results obtained using MATLAB software are similar to those obtained using ANSYS software. Therefore, the results have been validated, and they can be used for further analysis. The displacement diagrams of this truss are shown in Fig. 7.6.

7.8 Validation of the 39-Bar Truss

According to the PVS algorithm, the 39-bar truss of topology, shape, and size optimization can achieve the least mass with the best result. In Fig. 7.7, the left bottom node (node 1) of the truss is fixed in both directions, while the right bottom node (node 5) can only move horizontally. External forces are applied at nodes 2, 3, and 4 for two loading cases. Table 7.7 presents the obtained stress, displacement, and frequency values using both MATLAB and ANSYS programs. The result table indicates that all the results obtained using MATLAB software are 100% accurate compared to those obtained using ANSYS software, confirming their validity for further analysis. Figure 7.7 shows the displacement diagrams of this truss.

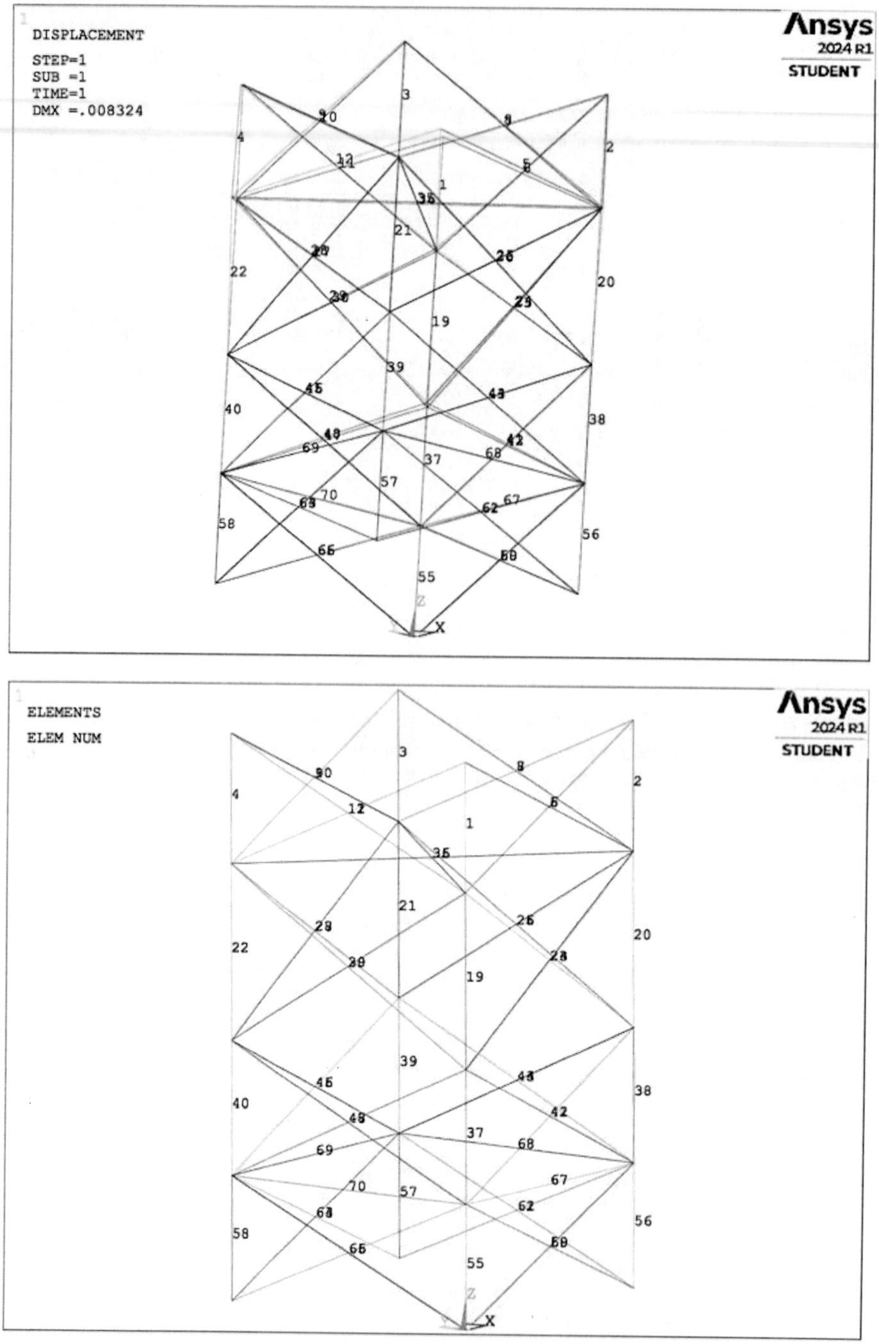

Fig. 7.6 Displacement diagrams of the 72-bar 3D truss under multiple load conditions

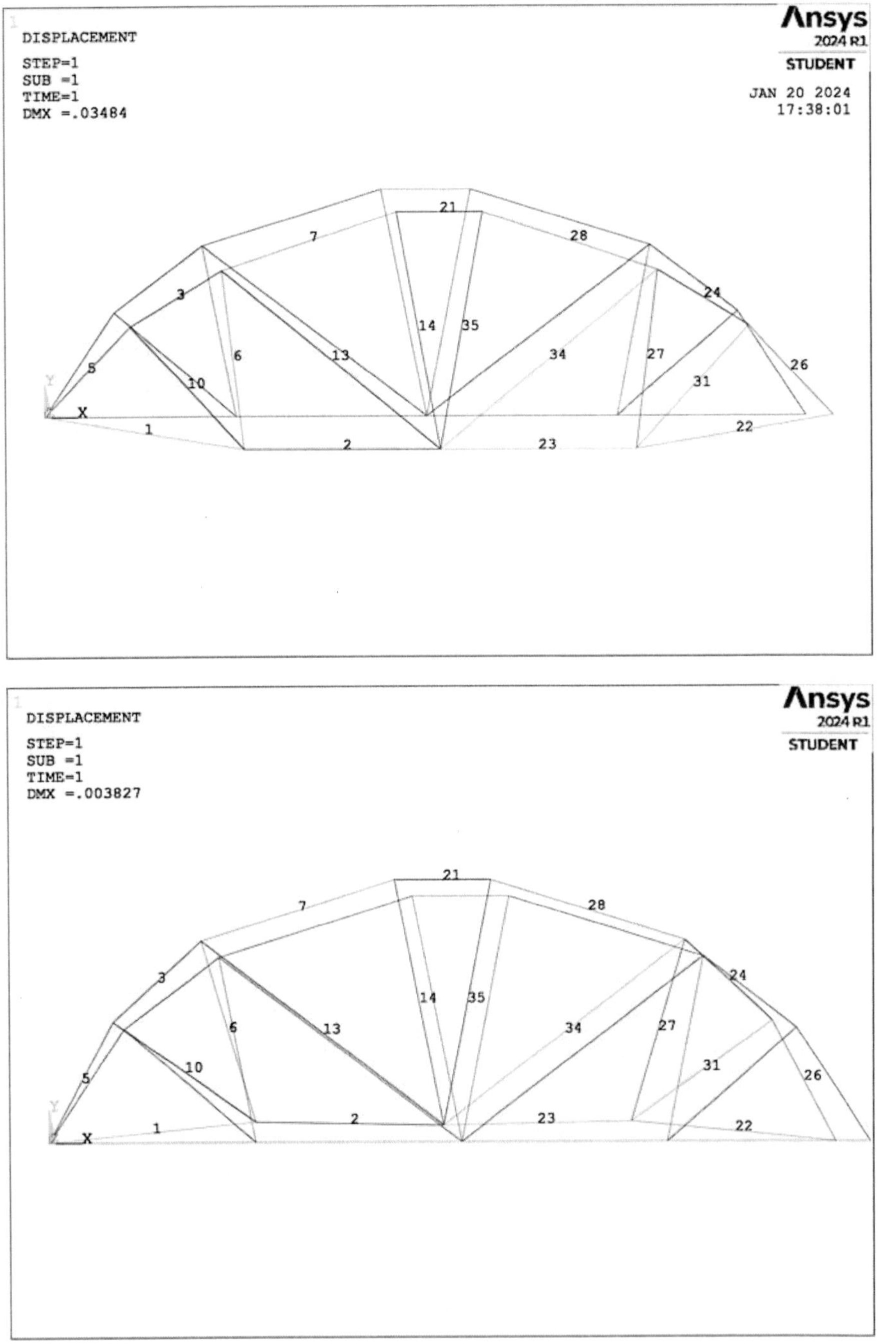

Fig. 7.7 Displacement diagrams of the 39-bar truss under multiple load conditions

Table 7.7 The 39-bar truss

	MATLAB results	ANSYS results	Accuracy (%)	MATLAB results	ANSYS results	Accuracy (%)
	Load case 1			Load case 2		
σ_1 (N/m^2)	1.6617E+08	1.6617E+08	100	0.0000E+00	0.0000E+00	100
σ_2 (N/m^2)	1.2091E+08	1.2091E+08	100	−3.85E+07	−3.85E+07	100
σ_3 (N/m^2)	−1.2169E+08	−1.2169E+08	100	0.0000E+00	0.0000E+00	100
σ_5 (N/m^2)	−1.0595E+08	−1.0595E+08	100	0.0000E+00	0.0000E+00	100
σ_6 (N/m^2)	1.2283E+08	1.2283E+08	100	0.0000E+00	0.0000E+00	100
σ_7 (N/m^2)	−7.1761E+07	−7.1761E+07	100	0.0000E+00	0.0000E+00	100
σ_{10} (N/m^2)	1.5145E+08	1.5145E+08	100	0.0000E+00	0.0000E+00	100
σ_{13} (N/m^2)	1.8821E+06	1.8821E+06	100	0.0000E+00	0.0000E+00	100
σ_{14} (N/m^2)	1.6948E+08	1.6948E+08	100	0.0000E+00	0.0000E+00	100
σ_{21} (N/m^2)	−1.5195E+08	−1.5195E+08	100	0.0000E+00	0.0000E+00	100
σ_{22} (N/m^2)	1.6617E+08	1.6617E+08	100	0.0000E+00	0.0000E+00	100
σ_{23} (N/m^2)	1.2091E+08	1.2091E+08	100	−3.85E+07	−3.85E+07	100
σ_{24} (N/m^2)	−1.2169E+08	−1.2169E+08	100	0.0000E+00	0.0000E+00	100
σ_{26} (N/m^2)	−1.0595E+08	−1.0595E+08	100	0.0000E+00	0.0000E+00	100
σ_{27} (N/m^2)	1.2283E+08	1.2283E+08	100	0.0000E+00	0.0000E+00	100
σ_{28} (N/m^2)	−7.1761E+07	−7.1761E+07	100	0.0000E+00	0.0000E+00	100
σ_{31} (N/m^2)	1.5145E+08	1.5145E+08	100	0.0000E+00	0.0000E+00	100
σ_{34} (N/m^2)	1.8821E+06	1.8821E+06	100	0.0000E+00	0.0000E+00	100
σ_{35} (N/m^2)	1.6948E+08	1.6948E+08	100	0.0000E+00	0.0000E+00	100
δ_{x_2} (m)	7.2249E-03	7.2249E-03	100	0.0000E+00	0.0000E+00	100
δ_{y_2} (m)	−2.9986E-02	−2.9986E-02	100	1.8608E-03	1.8608E-03	100
δ_{x_3} (m)	1.2482E-02	1.2482E-02	100	−1.6722E-03	−1.6722E-03	100
δ_{y_3} (m)	−3.0254E-02	−3.0254E-02	100	1.5353E-03	1.5353E-03	100
δ_{x_4} (m)	1.7739E-02	1.7739E-02	100	−3.3445E-03	−3.3445E-03	100
δ_{y_4} (m)	−2.9986E-02	−2.9986E-02	100	1.8608E-03	1.8608E-03	100
f_1 (Hz)	9.3406	9.3406	100			
f_2 (Hz)	10.7717	10.772	100			
f_3 (Hz)	11.5302	11.53	100			
f_4 (Hz)	17.4925	17.492	100			

7.9 Validation of the 45-Bar Truss

The 45-bar truss of topology, shape, and size optimization yields the minimum truss mass when optimized with the TLBO algorithm, as reported in the study. Figure 7.8 shows that in the finite element model, the two bottom nodes (nodes 1 and 5) of the truss are fixed, and external forces are applied at nodes 7, 8, and 9. The study reports

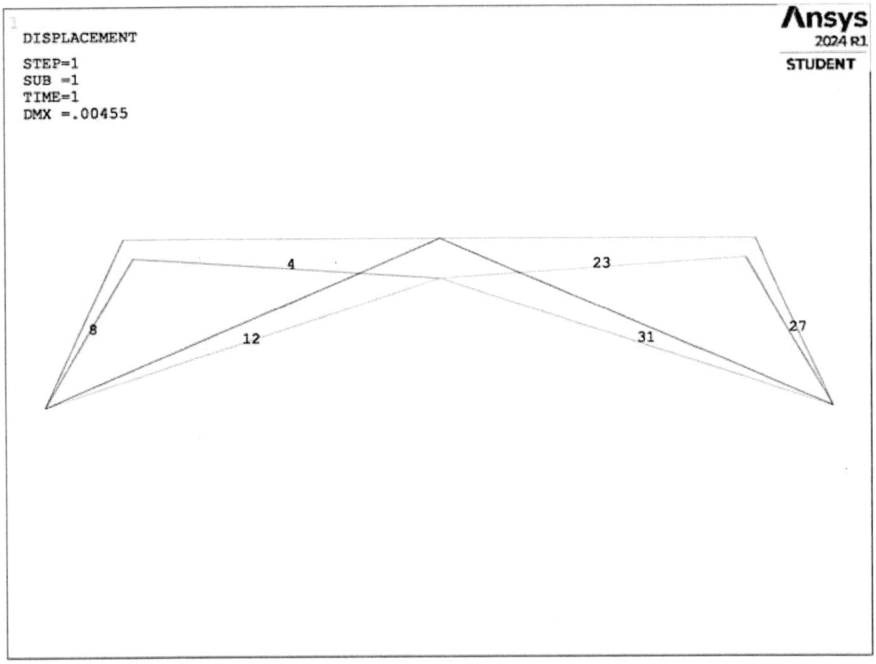

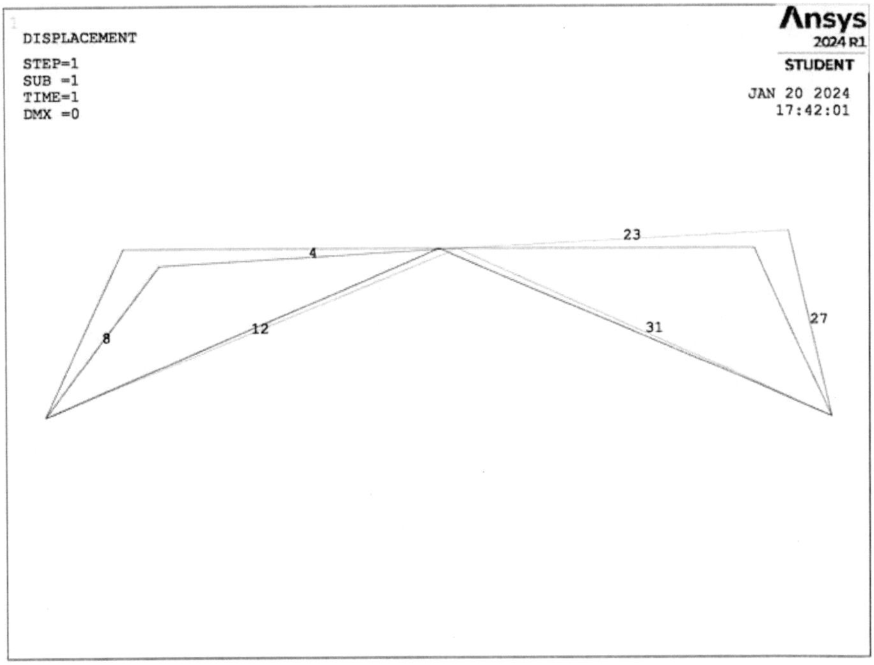

Fig. 7.8 Displacement diagrams of the 45-bar truss under multiple load conditions

Table 7.8 The 45-bar truss

	MATLAB results	ANSYS results	Accuracy (%)	MATLAB results	ANSYS results	Accuracy (%)
	Load case 1			Load case 2		
σ_4 (N/m^2)	−2.9946E+07	−2.9946E+07	100	−6.3624E+06	−6.3624E+06	100
σ_8 (N/m^2)	−7.0319E+07	−7.0319E+07	100	−1.8626E-09	−1.8702E-09	100
σ_{12} (N/m^2)	−3.5588E+07	−3.5588E+07	100	4.5367E+06	4.5367E+06	100
σ_{23} (N/m^2)	−2.9946E+07	−2.9946E+07	100	6.3624E+06	6.3624E+06	100
σ_{27} (N/m^2)	−7.0319E+07	−7.0319E+07	100	3.7253E-09	1.8702E-09	100
σ_{31} (N/m^2)	−3.5588E+07	−3.5588E+07	100	−4.5367E+06	−4.5367E+06	100
δ_{max}	−4.5496E-03	−4.5496E-03	100	4.8066E-04	4.8066E-04	100
f_1 (Hz)	19.9705	19.97	100			
f_2 (Hz)	21.9783	21.978	100			
f_3 (Hz)	25.7898	25.79	100			
f_4 (Hz)	41.2565	41.257	100			

the obtained stress, displacement, and frequency values using MATLAB and ANSYS software in Table 7.8. The results show that all the values obtained using MATLAB software are 100% accurate when compared to those obtained using ANSYS software, thereby validating the results. These results can be used for further analysis. The displacement diagrams of the optimized truss can be seen in Fig. 7.8.

7.10 Validation of the 25-Bar 3D Truss

The TLBO algorithm yields the best result for the 25-bar 3D truss of topology, shape, and size optimization, with the least truss mass obtained. As shown in Fig. 7.9, the finite element model fixes the four bottom nodes (nodes 7, 8, 9, and 10), while external forces are applied at nodes 1, 2, 3, and 6. The MATLAB and ANSYS programs were used to calculate the stress, displacement, and frequency values, which are presented in Table 7.9. All the results obtained using MATLAB software are 100% accurate compared to those obtained using ANSYS software, as indicated in the result table. Therefore, the results are validated and can be utilized for further analysis. The displacement diagrams of this truss are illustrated in Fig. 7.9.

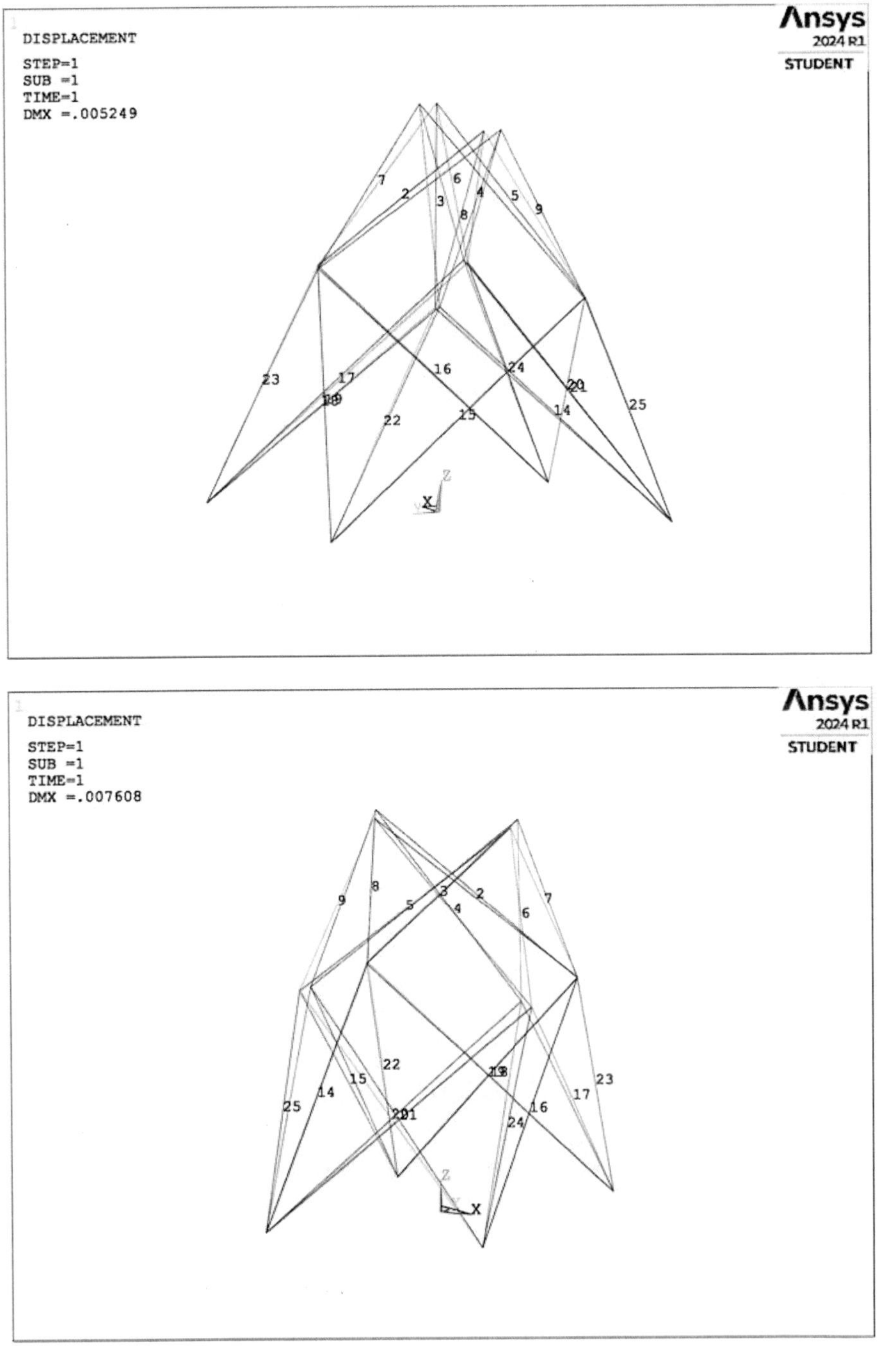

Fig. 7.9 Displacement diagrams of 25-bar truss under multiple load conditions

Table 7.9 The 25-bar 3D truss

	MATLAB results	ANSYS results	Accuracy (%)	MATLAB results	ANSYS results	Accuracy (%)
	Load case 1			Load case 2		
σ_2 (N/m^2)	−3.1432E+07	−3.1432E+07	100	−1.1739E+07	−1.1739E+07	100
σ_3 (N/m^2)	2.7059E+07	2.7059E+07	100	−9.5788E+06	−9.5788E+06	100
σ_4 (N/m^2)	2.7059E+07	2.7059E+07	100	−1.1073E+06	−1.1073E+06	100
σ_5 (N/m^2)	−3.1432E+07	−3.1432E+07	100	8.3267E+05	8.3267E+05	100
σ_6 (N/m^2)	−5.9237E+07	−5.9237E+07	100	−5.4418E+07	−5.4418E+07	100
σ_7 (N/m^2)	4.4021E+07	4.4021E+07	100	2.3985E+07	2.3985E+07	100
σ_8 (N/m^2)	−5.9237E+07	−5.9237E+07	100	2.5444E+07	2.5444E+07	100
σ_9 (N/m^2)	4.4021E+07	4.4021E+07	100	−5.3108E+07	−5.3108E+07	100
σ_{14} (N/m^2)	2.2395E+06	2.2395E+06	100	−1.0373E+06	−1.0373E+06	100
σ_{15} (N/m^2)	−1.2463E+06	−1.2463E+06	100	2.8724E+06	2.8724E+06	100
σ_{16} (N/m^2)	−1.2463E+06	−1.2463E+06	100	−8.7791E+05	−8.7791E+05	100
σ_{17} (N/m^2)	2.2395E+06	2.2395E+06	100	3.0156E+06	3.0156E+06	100
σ_{18} (N/m^2)	−1.4579E+07	−1.4579E+07	100	1.3128E+05	1.3128E+05	100
σ_{19} (N/m^2)	2.6207E+07	2.6207E+07	100	6.9655E+06	6.9655E+06	100
σ_{20} (N/m^2)	2.6207E+07	2.6207E+07	100	1.4707E+07	1.4707E+07	100
σ_{21} (N/m^2)	−1.4579E+07	−1.4579E+07	100	2.1858E+07	2.1858E+07	100
σ_{22} (N/m^2)	−2.3891E+07	−2.3891E+07	100	1.3696E+07	1.3696E+07	100
σ_{23} (N/m^2)	6.1101E+06	6.1101E+06	100	8.8691E+06	8.8691E+06	100
σ_{24} (N/m^2)	−2.3891E+07	−2.3891E+07	100	−4.8648E+07	−4.8648E+07	100
σ_{25} (N/m^2)	6.1101E+06	6.1101E+06	100	−4.3910E+07	−4.3910E+07	100
δ_{x_1} (m)	5.8988E-04	5.8988E-04	100	1.9351E-03	1.9351E-03	100
δ_{y_1} (m)	5.2495E-03	5.2495E-03	100	−7.0542E-03	−7.0542E-03	100
δ_{z_1} (m)	1.1194E-03	1.1194E-03	100	−2.0905E-03	−2.0905E-03	100
f_1 (Hz)	15.0151	14.944	100.47			
f_2 (Hz)	17.4986	17.377	100.70			
f_3 (Hz)	22.4448	22.095	101.56			
f_4 (Hz)	28.8465	28.605	100.84			

7.11 Validation of the 39-Bar 3D Truss

The PVS algorithm yielded the best result with the least truss mass for the 39-bar 3D truss of topology, shape, and size optimization. In the finite element model, the four bottom nodes (nodes 7, 8, 9, and 10) were fixed, and external forces were applied at nodes 1, 2, 3, and 6, as shown in Fig. 7.9. The obtained stress, displacement, and

Table 7.10 The 39-bar 3D truss

	MATLAB results	ANSYS results	Accuracy (%)
σ_2 (N/m^2)	−1.8765E+07	−1.8765E+07	100
σ_3 (N/m^2)	−1.8765E+07	−1.8765E+07	100
σ_4 (N/m^2)	−1.8765E+07	−1.8765E+07	100
σ_5 (N/m^2)	−1.8765E+07	−1.8765E+07	100
σ_6 (N/m^2)	−6.2030E+07	−6.2030E+07	100
σ_7 (N/m^2)	−6.2030E+07	−6.2030E+07	100
σ_8 (N/m^2)	−6.2030E+07	−6.2030E+07	100
σ_9 (N/m^2)	−6.2030E+07	−6.2030E+07	100
σ_{14} (N/m^2)	−1.2714E+06	−1.2714E+06	100
σ_{15} (N/m^2)	−1.2714E+06	−1.2714E+06	100
σ_{16} (N/m^2)	−1.2714E+06	−1.2714E+06	100
σ_{17} (N/m^2)	−1.2714E+06	−1.2714E+06	100
σ_{18} (N/m^2)	−4.8169E+06	−4.8169E+06	100
σ_{19} (N/m^2)	−4.8169E+06	−4.8169E+06	100
σ_{20} (N/m^2)	−4.8169E+06	−4.8169E+06	100
σ_{21} (N/m^2)	−4.8169E+06	−4.8169E+06	100
σ_{22} (N/m^2)	−6.2669E+07	−6.2669E+07	100
σ_{23} (N/m^2)	−6.2669E+07	−6.2669E+07	100
σ_{24} (N/m^2)	−6.2669E+07	−6.2669E+07	100
σ_{25} (N/m^2)	−6.2669E+07	−6.2669E+07	100
δ_{z_1} (m)	5.6793E-03	5.6793E-03	100
f_1 (Hz)	15.0000	14.949	100.34
f_2 (Hz)	16.9233	16.644	101.65
f_3 (Hz)	21.8346	21.632	100.93
f_4 (Hz)	24.2890	23.955	101.38

frequency values using MATLAB and ANSYS programs are listed in Table 7.10. It is noteworthy that all the results obtained using MATLAB software are 100% accurate compared to the results obtained using ANSYS software, validating the accuracy of the results. Therefore, the outcomes can be used for further analysis. The displacement diagram of this truss is illustrated in Fig. 7.10.

All the models are validated by comparing the results obtained using MATLAB software with those achieved by using ANSYS software.

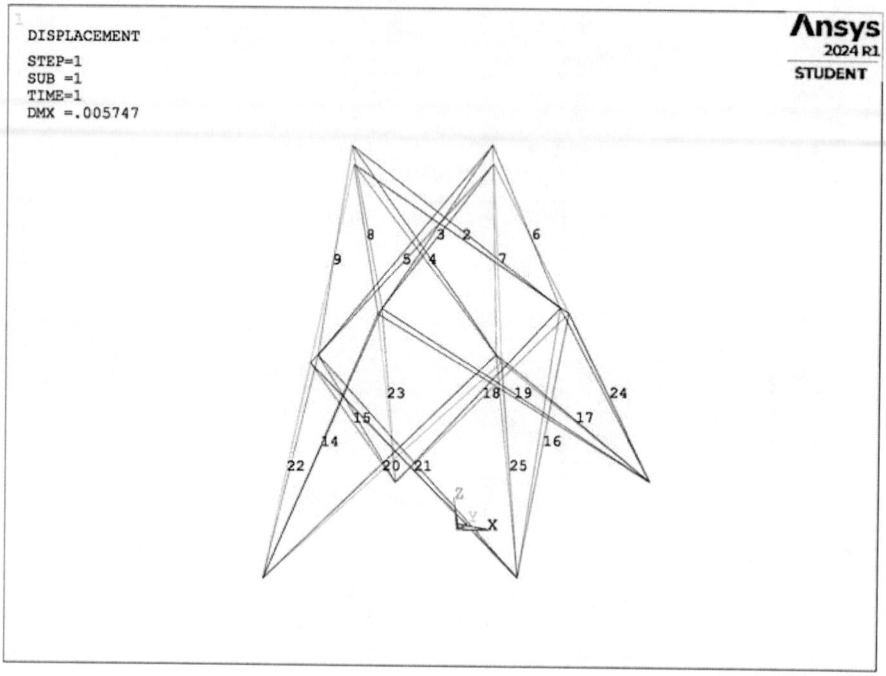

Fig. 7.10 Displacement of the 39-bar 3D truss

Reference

Tejani, G. G. (2017). Investigation of advanced metaheuristic techniques for simultaneous size, shape, and topology optimization of truss structures. Ph.D. Thesis, Pandit Deendayal Petroleum University, India.

Chapter 8
MATLAB Codes of Metaheuristics Methods

8.1 The Dragonfly Algorithm

```
function DA()
run=1;
popsize=50;
gen=400;
fun_no=1;
for i=1:run
mainDA(popsize,gen,fun_no);
end
%%%%%%%%%%%%%%%%%%%%%%%%%%%%%%%%%%%%%%%
function [Best_score,Best_pos]=mainDA(SearchAgents_no,
Max_iteration,fun_no)
tic;
if fun_no==1
  Function_name='F1'; % Sphere function
end
lb,ub,dim,fobj]=Get_Functions_details(Function_name);
[Best_score,Best_pos]=DA(SearchAgents_no,Max_iteration,lb,ub,dim,
fobj);
disp(['Funnum:', num2str(Function_name)])
fprintf('\n%e\n',Best_score);
fprintf('\n%6.15f\n',Best_pos);
toc
%%%%%%%%%%%%%%%%%%%%%%%%%%%%%%%%%%%%%%%
function [lb,ub,dim,fobj] = Get_Functions_details(F)
switch F
  case 'F1'
```


V. Savsani et al., *Truss Optimization*, https://doi.org/10.1007/978-3-031-49295-2_8

```
    fobj = @F1; % Sphere function
    dim=30;
    lb = -100*ones(1,30);
    ub = 100*ones(1,30);
    maxFE=20000;
end
end
function o = F1(x)
o=sum((x).^2);
end
%%%%%%%%%%%%%%%%%%%%%%%%%%%%%%%%%%%%%%%%
function [Best_score,Best_pos]=DA(SearchAgents_no,Max_iteration,
lb,ub,dim,fobj)
display('DA is optimizing your problem');
cg_curve=zeros(1,Max_iteration);
if size(ub,2)==1
  ub=ones(1,dim)*ub;
  lb=ones(1,dim)*lb;
end
%The initial radius of dragonflies' neighbourhoods
r=(ub-lb)/10;
Delta_max=(ub-lb)/10;
Food_fitness=inf;
Food_pos=zeros(dim,1);
Enemy_fitness=-inf;
Enemy_pos=zeros(dim,1);
X=initialization(SearchAgents_no,dim,ub,lb);
X=X';
Fitness=zeros(1,SearchAgents_no);
DeltaX=initialization(SearchAgents_no,dim,ub,lb);
DeltaX=DeltaX';
FE=0;
maxFE=Max_iteration*SearchAgents_no;
iter=1;
while FE<maxFE
  r=(ub-lb)/4+((ub-lb)*(iter/Max_iteration)*2);
  w=0.9-iter*((0.9-0.4)/Max_iteration);
  my_c=0.1-iter*((0.1-0)/(Max_iteration/2));
  if my_c<0
    my_c=0;
  end
  s=2*rand*my_c; % Seperation weight
  a=2*rand*my_c; % Alignment weight
  c=2*rand*my_c; % Cohesion weight
  f=2*rand;    % Food attraction weight
```

```
e=my_c;       % Enemy distraction weight
for i=1:SearchAgents_no % Calculate all the objective values first
  Fitness(1,i)=fobj(X(:,i)');
  FE=FE+1;
  if Fitness(1,i)<Food_fitness
    Food_fitness=Fitness(1,i);
    Food_pos=X(:,i);
  end
  if Fitness(1,i)>Enemy_fitness
    if all(X(:,i)<ub') && all( X(:,i)>lb')
      Enemy_fitness=Fitness(1,i);
      Enemy_pos=X(:,i);
    end
  end
end
for i=1:SearchAgents_no
  index=0;
  neighbours_no=0;
  clear Neighbours_DeltaX
  clear Neighbours_X
  %find the neighbouring solutions
  for j=1:SearchAgents_no
    Dist2Enemy=distance(X(:,i),X(:,j));
    if (all(Dist2Enemy<=r) && all(Dist2Enemy~=0))
      index=index+1;
      neighbours_no=neighbours_no+1;
      Neighbours_DeltaX(:,index)=DeltaX(:,j);
      Neighbours_X(:,index)=X(:,j);
    end
  end
  % Seperation%
  S=zeros(dim,1);
  if neighbours_no>1
    for k=1:neighbours_no
      S=S+(Neighbours_X(:,k)-X(:,i));
    end
    S=-S;
  else
    S=zeros(dim,1);
  end
  % Alignment%
  if neighbours_no>1
    A=(sum(Neighbours_DeltaX')')/neighbours_no;
  else
    A=DeltaX(:,i);
```

```
      end
      % Cohesion%
      if neighbours_no>1
        C_temp=(sum(Neighbours_X')')/neighbours_no;
      else
        C_temp=X(:,i);
      end
      C=C_temp-X(:,i);
      % Attraction to food%
      Dist2Food=distance(X(:,i),Food_pos(:,1));
      if all(Dist2Food<=r)
        F=Food_pos-X(:,i);
      else
        F=0;
      end
      % Distraction from enemy%
      Dist2Enemy=distance(X(:,i),Enemy_pos(:,1));
      if all(Dist2Enemy<=r)
        Enemy=Enemy_pos+X(:,i);
      else
        Enemy=zeros(dim,1);
      end
      for tt=1:dim
        if X(tt,i)>ub(tt)
          X(tt,i)=lb(tt);
          DeltaX(tt,i)=rand;
        end
        if X(tt,i)<lb(tt)
          X(tt,i)=ub(tt);
          DeltaX(tt,i)=rand;
        end
      end
      if any(Dist2Food>r)
        if neighbours_no>1
          for j=1:dim
        DeltaX(j,i)=w*DeltaX(j,i)+rand*A(j,1)+rand*C(j,1)+rand*S
(j,1);
            if DeltaX(j,i)>Delta_max(j)
              DeltaX(j,i)=Delta_max(j);
            end
            if DeltaX(j,i)<-Delta_max(j)
              DeltaX(j,i)=-Delta_max(j);
            end
            X(j,i)=X(j,i)+DeltaX(j,i);
          end
```

```
        else
          X(:,i)=X(:,i)+Levy(dim)'.*X(:,i);
          DeltaX(:,i)=0;
        end
      else
        for j=1:dim
         DeltaX(j,i)=(a*A(j,1)+c*C(j,1)+s*S(j,1)+f*F(j,1)+e*Enemy
(j,1)) + w*DeltaX(j,i);
          if DeltaX(j,i)>Delta_max(j)
            DeltaX(j,i)=Delta_max(j);
          end
          if DeltaX(j,i)<-Delta_max(j)
            DeltaX(j,i)=-Delta_max(j);
          end
          X(j,i)=X(j,i)+DeltaX(j,i);
        end
      end
      Flag4ub=X(:,i)>ub';
      Flag4lb=X(:,i)<lb';
      X(:,i)=(X(:,i).*(~(Flag4ub+Flag4lb)))+ub'.*Flag4ub+lb'.
*Flag4lb;
    end
    Best_score=Food_fitness;
    Best_pos=Food_pos;
    iter=iter+1;
    cg_curve(iter)=Best_score;
    if mod(FE,20)==0
      fprintf('%e\n',Best_score);
    end
end
disp(['FE:',num2str(FE)])
```

8.2 The Multiverse Optimizer

```
function MVO()
run=1;
popsize=50;
gen=400;
fun_no=1;
for i=1:run
mainMVO(popsize,gen,fun_no);
end
```

```
%%%%%%%%%%%%%%%%%%%%%%%%%%%%%%%%%%%%%%%%
function function [Best_score, Best_pos]=mainMVO(Universes_no,
Max_iteration,fun_no)
tic;
if fun_no==1
  Function_name='F1'; % Sphere function
end
lb,ub,dim,fobj]=Get_Functions_details(Function_name);
[Best_score,Best_pos]=MVO(Universes_no,Max_iteration,lb,ub,dim,
fobj);
disp(['Funnum:', num2str(Function_name)])
fprintf('\n%e\n',Best_score);
fprintf('\n%6.15f\n',Best_pos);
toc
%%%%%%%%%%%%%%%%%%%%%%%%%%%%%%%%%%%%%%%%
function [lb,ub,dim,fobj] = Get_Functions_details(F)
switch F
  case 'F1'
    fobj = @F1; % Sphere function
    dim=30;
    lb = -100*ones(1,30);
    ub = 100*ones(1,30);
    maxFE=20000;
end
end
function o = F1(x)
o=sum((x).^2);
end
%%%%%%%%%%%%%%%%%%%%%%%%%%%%%%%%%%%%%%%%
function [Best_universe_Inflation_rate,Best_universe,
Convergence_curve]=MVO(N,Max_time,lb,ub,dim,fobj)
display('MVO is optimizing your problem');
% Two variables for saving the position and inflation rate (fitness) of the
best universe
Best_universe=zeros(1,dim);
Best_universe_Inflation_rate=inf;
%Initialize the positions of universes
Universes=initialization(N,dim,ub,lb);
%Minimum and maximum of Wormhole Existence Probability
WEP_Max=1;
WEP_Min=0.2;
Convergence_curve=zeros(1,Max_time);
%Iteration(time) counter
```

```
Time=1;
FE=0; %
maxFE=Max_time*N;
%Main loop
while FE<maxFE
  WEP=WEP_Min+Time*((WEP_Max-WEP_Min)/Max_time);
  %Travelling Distance Rate (Formula)
  TDR=1-((Time)^(1/6)/(Max_time)^(1/6));
  %Inflation rates (I) (fitness values)
  Inflation_rates=zeros(1,size(Universes,1));
  for i=1:size(Universes,1)
    % Boundary checking (to bring back the universes inside search
    % space if they go beyoud the boundaries
    Flag4ub=Universes(i,:)>ub;
    Flag4lb=Universes(i,:)<lb;
Universes(i,:)=(Universes(i,:).*(~(Flag4ub+Flag4lb)))+ub.*Flag4ub
+lb.*Flag4lb;
    %Calculate the inflation rate (fitness) of universes
    Inflation_rates(1,i)=fobj(Universes(i,:));
    FE=FE+1;
    %Elitism
    if Inflation_rates(1,i)<Best_universe_Inflation_rate
      Best_universe_Inflation_rate=Inflation_rates(1,i);
      Best_universe=Universes(i,:);
    end
  end
  [sorted_Inflation_rates,sorted_indexes]=sort(Inflation_rates);
  for newindex=1:N
    Sorted_universes(newindex,:)=Universes(sorted_indexes
 (newindex),:);
  end
  %Normaized inflation rates
  normalized_sorted_Inflation_rates=normr(sorted_Inflation_rates);
  Universes(1,:)= Sorted_universes(1,:);
  %Update the Position of universes
  for i=2:size(Universes,1) %Starting from 2 since the firt one is the
 elite
    Back_hole_index=i;
    for j=1:size(Universes,2)
      r1=rand();
      if r1<normalized_sorted_Inflation_rates(i)
        White_hole_index=RouletteWheelSelection
 (-sorted_Inflation_rates);% for maximization problem
```

```
-sorted_Inflation_rates should be written as sorted_Inflation_rates
        if White_hole_index==-1
          White_hole_index=1;
        end        Universes(Back_hole_index,j)=Sorted_universes
(White_hole_index,j);
      end
      if (size(lb,2)==1)
        r2=rand();
        if r2<WEP
          r3=rand();
          if r3<0.5
          Universes(i,j)=Best_universe(1,j)+TDR*((ub-lb)*rand+lb);
          end
          if r3>0.5
          Universes(i,j)=Best_universe(1,j)-TDR*((ub-lb)*rand+lb);
          end
        end
      end
      if (size(lb,2)~=1)
        %Eq. (3.2) in the paper if the upper and lower bounds are
        %different for each variables
        r2=rand();
        if r2<WEP
          r3=rand();
          if r3<0.5
         Universes(i,j)=Best_universe(1,j)+TDR*((ub(j)-lb(j))*rand
+lb(j));
          end
          if r3>0.5
         Universes(i,j)=Best_universe(1,j)-TDR*((ub(j)-lb(j))*rand
+lb(j));
          end
        end
      end
    end
  end
  %Update the convergence curve
  Convergence_curve(Time)=Best_universe_Inflation_rate;
  if mod(FE,20)==0
    fprintf('%e\n',Best_universe_Inflation_rate);
  end
  Time=Time+1;
end
disp(['FE:',num2str(FE)])
```

8.3 The Sine Cosine Algorithm

```
function SCA()
run=1;
popsize=50;
gen=400;
fun_no=1;
for i=1:run
mainSCA(popsize,gen,fun_no);
end
%%%%%%%%%%%%%%%%%%%%%%%%%%%%%%%%%%%%%%%%%%%
function [Best_score, Best_pos]=mainSCA(SearchAgents_no,
Max_iteration,fun_no)
tic;
if fun_no==1
  Function_name='F1'; % Sphere function
end
lb,ub,dim,fobj]=Get_Functions_details(Function_name);
[Best_score,Best_pos]=SCA(SearchAgents_no,Max_iteration,lb,ub,
dim,fobj);
disp(['Funnum:', num2str(Function_name)])
fprintf('\n%e\n',Best_score);
fprintf('\n%6.15f\n',Best_pos);
toc
%%%%%%%%%%%%%%%%%%%%%%%%%%%%%%%%%%%%%%%%%%%
function [lb,ub,dim,fobj] = Get_Functions_details(F)
switch F
  case 'F1'
    fobj = @F1; % Sphere function
    dim=30;
    lb = -100*ones(1,30);
    ub = 100*ones(1,30);
    maxFE=20000;
end
end
function o = F1(x)
o=sum((x).^2);
end
%%%%%%%%%%%%%%%%%%%%%%%%%%%%%%%%%%%%%%%%%%%
function [Destination_fitness,Destination_position,
Convergence_curve]=SCA(N,Max_iteration,lb,ub,dim,fobj)
display('SCA is optimizing your problem');
%Initialize the set of random solutions
```

```
X=initialization(N,dim,ub,lb);
Destination_position=zeros(1,dim);
Destination_fitness=inf;
Convergence_curve=zeros(1,Max_iteration);
Objective_values = zeros(1,size(X,1));
% Calculate the fitness of the first set and find the best one
for i=1:size(X,1)
  Objective_values(1,i)=fobj(X(i,:));
  if i==1
    Destination_position=X(i,:);
    Destination_fitness=Objective_values(1,i);
  elseif Objective_values(1,i)<Destination_fitness
    Destination_position=X(i,:);
    Destination_fitness=Objective_values(1,i);
  end
  All_objective_values(1,i)=Objective_values(1,i);
end
%Main loop
t=2; % start from the second iteration since the first iteration was
dedicated to calculating the fitness
FE=0; %
maxFE=Max_iteration*N;
while FE<maxFE
  a = 2;
  Max_iteration = Max_iteration;
  r1=a-t*((a)/Max_iteration); % r1 decreases linearly from a to 0
  % Update the position of solutions with respect to destination
  for i=1:size(X,1) % in i-th solution
    for j=1:size(X,2) % in j-th dimension
      % Update r2, r3, and r4
      r2=(2*pi)*rand();
      r3=2*rand;
      r4=rand();
      if r4<0.5
        X(i,j)= X(i,j)+(r1*sin(r2)*abs(r3*Destination_position(j)-X
(i,j)));
      else
        X(i,j)= X(i,j)+(r1*cos(r2)*abs(r3*Destination_position(j)-X
(i,j)));
      end
    end
  end
  for i=1:size(X,1)
    % Check if solutions go outside the search spaceand bring them back
    Flag4ub=X(i,:)>ub;
```

```
        Flag4lb=X(i,:)<lb;
        X(i,:)=(X(i,:).*(~(Flag4ub+Flag4lb)))+ub.*Flag4ub+lb.*Flag4lb;
        % Calculate the objective values
        Objective_values(1,i)=fobj(X(i,:));
        FE=FE+1;
        % Update the destination if there is a better solution
        if Objective_values(1,i)<Destination_fitness
            Destination_position=X(i,:);
            Destination_fitness=Objective_values(1,i);
        end
    end
    Convergence_curve(t)=Destination_fitness;
    if mod(FE,20)==0
        fprintf('%e\n',Destination_fitness);
    end
    % Increase the iteration counter
    t=t+1;
end
disp(['FE:',num2str(FE)])
```

8.4 The Whale Optimization Algorithm

```
function WOA()
run=1;
popsize=50;
gen=400;
fun_no=1;
for i=1:run
mainWOA(popsize,gen,fun_no);
end
%%%%%%%%%%%%%%%%%%%%%%%%%%%%%%%%%%%%%%%%
function [Best_score, Best_pos]=mainWOA(SearchAgents_no,
Max_iteration,fun_no)
tic;
if fun_no==1
    Function_name='F1'; % Sphere function
end
lb,ub,dim,fobj]=Get_Functions_details(Function_name);
[Best_score,Best_pos,WOA_cg_curve]=WOA(SearchAgents_no,
Max_iteration,lb,ub,dim,fobj);
disp(['Funnum:', num2str(Function_name)])
fprintf('\n%e\n',Best_score);
```

```
fprintf('\n%6.15f\n',Best_pos);
toc
%%%%%%%%%%%%%%%%%%%%%%%%%%%%%%%%%%%%%%%%%%%
function [lb,ub,dim,fobj] = Get_Functions_details(F)
switch F
  case 'F1'
    fobj = @F1; % Sphere function
    dim=30;
    lb = -100*ones(1,30);
    ub = 100*ones(1,30);
    maxFE=20000;
end
end
function o = F1(x)
o=sum((x).^2);
end
%%%%%%%%%%%%%%%%%%%%%%%%%%%%%%%%%%%%%%%%%%%
function [Leader_score,Leader_pos,Convergence_curve]=WOA
(SearchAgents_no,Max_iter,lb,ub,dim,fobj)
display('WOA is optimizing your problem');
% initialize position vector and score for the leader
Leader_pos=zeros(1,dim);
Leader_score=inf; %change this to -inf for maximization problems
%Initialize the positions of search agents
Positions=initialization(SearchAgents_no,dim,ub,lb);
Convergence_curve=zeros(1,Max_iter);
t=0;% Loop counter
FE=0; %
maxFE=Max_iter*SearchAgents_no;
% Main loop
while FE<maxFE
  for i=1:size(Positions,1)
    % Return back the search agents that go beyond the boundaries of the
search space
    Flag4ub=Positions(i,:)>ub;
    Flag4lb=Positions(i,:)<lb;
Positions(i,:)=(Positions(i,:).*(~(Flag4ub+Flag4lb)))+ub.*Flag4ub
+lb.*Flag4lb;
    % Calculate objective function for each search agent
    fitness=fobj(Positions(i,:));
    FE=FE+1;
    % Update the leader
```

```
    if fitness<Leader_score % Change this to > for maximization problem
      Leader_score=fitness; % Update alpha
      Leader_pos=Positions(i,:);
    end
  end
  a=2-t*((2)/Max_iter); % a decreases linearly fron 2 to 0
  % a2 linearly dicreases from -1 to -2 to calculate t
  a2=-1+t*((-1)/Max_iter);
  % Update the Position of search agents
  for i=1:size(Positions,1)
    r1=rand(); % r1 is a random number in [0,1]
    r2=rand(); % r2 is a random number in [0,1]
    A=2*a*r1-a;
    C=2*r2;
    b=1;
    l=(a2-1)*rand+1; p = rand();
    for j=1:size(Positions,2)
      if p<0.5
        if abs(A)>=1
          rand_leader_index = floor(SearchAgents_no*rand()+1);
          X_rand = Positions(rand_leader_index, :);
          D_X_rand=abs(C*X_rand(j)-Positions(i,j));)
          Positions(i,j)=X_rand(j)-A*D_X_rand;
        elseif abs(A)<1
          D_Leader=abs(C*Leader_pos(j)-Positions(i,j));
          Positions(i,j)=Leader_pos(j)-A*D_Leader;
        end
      elseif p>=0.5
        distance2Leader=abs(Leader_pos(j)-Positions(i,j));
Positions(i,j)=distance2Leader*exp(b.*l).*cos(l.*2*pi)+Leader_pos
(j);
      end
    end
  end
  t=t+1;
  Convergence_curve(t)=Leader_score;
  if mod(FE,20)==0
    fprintf('%e\n',Leader_score);
  end
end
disp(['FE:',num2str(FE)])
```

8.5 The Ant Lion Optimizer

```
function ALO()
run=1;
popsize=50;
gen=400;
fun_no=1;
for i=1:run
mainALO(popsize,gen,fun_no);
end
%%%%%%%%%%%%%%%%%%%%%%%%%%%%%%%%%%%%%%%%
function [Best_score, Best_pos]=mainALO(SearchAgents_no,
Max_iteration,fun_no)
tic;
if fun_no==1
  Function_name='F1'; % Sphere function
end
lb,ub,dim,fobj]=Get_Functions_details(Function_name);
[Best_score,Best_pos]=ALO(SearchAgents_no,Max_iteration,lb,ub,
dim,fobj);
disp(['Funnum:', num2str(Function_name)])
fprintf('\n%e\n',Best_score);
fprintf('\n%6.15f\n',Best_pos);
toc
%%%%%%%%%%%%%%%%%%%%%%%%%%%%%%%%%%%%%%%%
function [lb,ub,dim,fobj] = Get_Functions_details(F)
switch F
  case 'F1'
    fobj = @F1; % Sphere function
    dim=30;
    lb = -100*ones(1,30);
    ub = 100*ones(1,30);
    maxFE=20000;
end
end
function o = F1(x)
o=sum((x).^2);
end
%%%%%%%%%%%%%%%%%%%%%%%%%%%%%%%%%%%%%%%%
function [Elite_antlion_fitness,Elite_antlion_position,
Convergence_curve]=ALO(N,Max_iter,lb,ub,dim,fobj)
display('ALO is optimizing your problem');
% Initialize the positions of antlions and ants
```

```
antlion_position=initialization(N,dim,ub,lb);
ant_position=initialization(N,dim,ub,lb);
% Initialize variables to save the position of elite, sorted antlions,
% convergence curve, antlions fitness, and ants fitness
Sorted_antlions=zeros(N,dim);
Elite_antlion_position=zeros(1,dim);
Elite_antlion_fitness=inf;
Convergence_curve=zeros(1,Max_iter);
antlions_fitness=zeros(1,N);
ants_fitness=zeros(1,N);
% Calculate the fitness of initial antlions and sort them
for i=1:size(antlion_position,1)
  antlions_fitness(1,i)=fobj(antlion_position(i,:));
end
 [sorted_antlion_fitness,sorted_indexes]=sort(antlions_fitness);
for newindex=1:N
 Sorted_antlions(newindex,:)=antlion_position(sorted_indexes
(newindex),:);
end
Elite_antlion_position=Sorted_antlions(1,:);
Elite_antlion_fitness=sorted_antlion_fitness(1);
% Main loop start from the second iteration since the first iteration
% was dedicated to calculating the fitness of antlions
FE=0; %
maxFE=Max_iter*N;
Current_iter=1;
while FE<maxFE
  % This for loop simulate random walks
  for i=1:size(ant_position,1)
   % Select ant lions based on their fitness (the better anlion the %higher
chance of catching ant)
     Rolette_index=RouletteWheelSelection(1./
sorted_antlion_fitness);
     if Rolette_index==-1
       Rolette_index=1;
     end
     % RA is the random walk around the selected antlion by rolette %wheel
     RA=Random_walk_around_antlion(dim,Max_iter,lb,ub,
Sorted_antlions(Rolette_index,:),Current_iter);
     % RA is the random walk around the elite (best antlion so far)
     [RE]=Random_walk_around_antlion(dim,Max_iter,lb,ub,
Elite_antlion_position(1,:),Current_iter);
     ant_position(i,:)= (RA(Current_iter,:)+RE(Current_iter,:))/2;
  end
  for i=1:size(ant_position,1)
```

```
    % Boundar checking (bring back the antlions of ants inside search
    % space if they go beyoud the boundaries
    Flag4ub=ant_position(i,:)>ub;
    Flag4lb=ant_position(i,:)<lb;
    ant_position(i,:)=(ant_position(i,:).*(~(Flag4ub+Flag4lb)))
+ub.*Flag4ub+lb.*Flag4lb;
    ants_fitness(1,i)=fobj(ant_position(i,:));
    FE=FE+1;
  end
  % Update antlion positions and fitnesses based of the ants (if an ant
  % becomes fitter than an antlion we assume it was cought by the antlion
  % and the antlion update goes to its position to build the trap)
  double_population=[Sorted_antlions;ant_position];
  double_fitness=[sorted_antlion_fitness ants_fitness];
  [double_fitness_sorted I]=sort(double_fitness);
  double_sorted_population=double_population(I,:);
  antlions_fitness=double_fitness_sorted(1:N);
  Sorted_antlions=double_sorted_population(1:N,:);
  % Update the position of elite if any antlinons becomes fitter than it
  if antlions_fitness(1)<Elite_antlion_fitness
    Elite_antlion_position=Sorted_antlions(1,:);
    Elite_antlion_fitness=antlions_fitness(1);
  end
  % Keep the elite in the population
  Sorted_antlions(1,:)=Elite_antlion_position;
  antlions_fitness(1)=Elite_antlion_fitness;
  % Update the convergence curve
  Convergence_curve(Current_iter)=Elite_antlion_fitness;
  if mod(FE,20)==0
    fprintf('%e\n',Elite_antlion_fitness);
  end
  Current_iter=Current_iter+1;
end
disp(['FE:',num2str(FE)])
```

8.6 The Heat Transfer Search

```
function HTS()
run=1;
for i=1:run
HTS(@Tryvp);
end
```

```
%%%%%%%%%%%%%%%%%%%%%%%%%%%%%%%%%%%%%%%%%%%%%%%%%%
function [InitFunction, CostFunction, FeasibleFunction,
FeasibleFunction_1,CostFunction_1] = Tryvp
InitFunction = @TryvpInit;
CostFunction = @TryvpCost;
CostFunction_1 = @TryvpCost_1;
FeasibleFunction = @TryvpFeasible;
FeasibleFunction_1 = @TryvpFeasible_1;
return;
%%%%%%%%%%%%%%%%%%%%%%%%%%%%%%%%%%%%%%%%%%%%%%%%%%
function [MaxParValue, MinParValue, Population, OPTIONS] = TryvpInit
(OPTIONS)
global MinParValue MaxParValue ll ul
Granularity = 1;
MinParValue = 1;
ll=-100*ones(1,30);
ul=100*ones(1,30);
MaxParValue = ul;
% Initialize population
for popindex = 1 : OPTIONS.popsize
  for k = 1 : OPTIONS.numVar
    chrom(k) =round((ll(k)+ (ul(k) - ll(k)) * rand));
  end
  Population(popindex).chrom = chrom;
end
OPTIONS.OrderDependent = true;
return;
%%%%%%%%%%%%%%%%%%%%%%%%%%%%%%%%%%%%%%%%%%%%%%%%%%
function [Population] = TryvpCost(OPTIONS, Population)
% Compute the cost of each member in Population
global MinParValue MaxParValue
popsize = OPTIONS.popsize;
for popindex = 1 : popsize
  for k = 1 : OPTIONS.numVar
    x(k) = Population(popindex).chrom(k);
  end
% Benchmark objective function
  Population(popindex).cost = objective(x);
end
return
%%%%%%%%%%%%%%%%%%%%%%%%%%%%%%%%%%%%%%%%%%%%%%%%%%
function [Population_1] = TryvpCost_1(OPTIONS, Population)
% Compute the cost of each member in Population
global MinParValue MaxParValue
popsize = size(Population,1);
```

```
Population_1 = [];
for popindex = 1 : popsize
% Benchmark function number from objective
  for k = 1 : OPTIONS.numVar
    x(k) = Population(popindex,k);
  end
   Population_1(popindex) = objective(x);
end
return
%%%%%%%%%%%%%%%%%%%%%%%%%%%%%%%%%%%%%%%%%%%%%%%%%%%%%%%%
function [Population] = TryvpFeasible(OPTIONS, Population)
global MinParValue MaxParValue ll ul
for i = 1 : OPTIONS.popsize
  for k = 1 : OPTIONS.numVar
    Population(i).chrom(k) = max(Population(i).chrom(k), ll(k));
    Population(i).chrom(k) = min(Population(i).chrom(k), MaxParValue
(k));
  end
end
return;
%%%%%%%%%%%%%%%%%%%%%%%%%%%%%%%%%%%%%%%%%%%%%%%%%%%%%%%%%%%%%
function [Population] = TryvpFeasible_1(OPTIONS, Population)
global MinParValue MaxParValue ll ul
for i = 1 : size(Population,1)
  for k = 1 : OPTIONS.numVar
    Population(i,k) = max(Population(i,k), ll(k));
 Population(i,k) = min(Population(i,k), MaxParValue(k));
  end
end
return;
%%%%%%%%%%%%%%%%%%%%%%%%%%%%%%%%%%%%%%%%%%%%%%%%
function [OPTIONS, MinCost, AvgCost, InitFunction, CostFunction,
FeasibleFunction, ...
  MaxParValue, MinParValue, Population,FeasibleFunction_1,
CostFunction_1] = Init(DisplayFlag, ProblemFunction, RandSeed)
% Initialize population.
OPTIONS.popsize = 50; % total population size
OPTIONS.Maxgen = 400; % generation count limit
OPTIONS.numVar = 30; % number of genes in each population member
OPTIONS.maxFE = 20000; % mutation probability
if ~exist('RandSeed', 'var')
  RandSeed = round(sum(100*clock));
end
rand('state', RandSeed); % initialize random number generator
% Get the addresses of the initialization, cost, and feasibility
```

```
functions
[InitFunction, CostFunction, FeasibleFunction,FeasibleFunction_1,
CostFunction_1] = ProblemFunction();
% Initialize the population.
[MaxParValue, MinParValue, Population, OPTIONS] = InitFunction
(OPTIONS);
% Make sure the population does not have duplicates.
Population = ClearDups(Population, MaxParValue, MinParValue);
% Compute cost of each individual
Population = CostFunction(OPTIONS, Population);
% Sort the population from most fit to least fit
Population = PopSort(Population);
% Compute the average cost
AverageCost = ComputeAveCost(Population);
% Display info to screen
MinCost = [Population(1).cost];
AvgCost = [AverageCost];
if DisplayFlag
end
return;
%%%%%%%%%%%%%%%%%%%%%%%%%%%%%%%%%%%%%%%%%%%%%%
function [fit] = objective(c)
fit = sum((x).^2);
end
%%%%%%%%%%%%%%%%%%%%%%%%%%%%%%%%%%%%%%%%%%%%%%
function HTS(ProblemFunction, DisplayFlag, ProbFlag)
display(HTS is optimizing your problem');
format long eng;
global ll
global ul
global ov;
if ~exist('DisplayFlag', 'var')
  DisplayFlag = true;
end
if ~exist('ProbFlag', 'var')
  ProbFlag = true;
end
 [OPTIONS, MinCost, AvgCost, InitFunction, CostFunction,
FeasibleFunction, ...
  MaxParValue, MinParValue, Population,FeasibleFunction_1,
CostFunction_1] = Init(DisplayFlag, ProblemFunction);
Keep=0;
maxFE=OPTIONS.maxFE;
FE=0;
TF=1;
MR=1;
```

```
% Begin the optimization loop
for GenIndex = 1 : OPTIONS.Maxgen
while FE<maxFE
  % Save the best habitats in a temporary array.
  for i = 1 : Keep
    chromKeep(i,:) = Population(i).chrom;
    costKeep(i) = Population(i).cost;
  end
  for i=1:length(Population)
    pp(i,:)=Population(i).chrom;
    pp_cost(i)=Population(i).cost;
  end
  Mean_pop=mean(pp);
  R=rand;
   for i = 1 : 1 : OPTIONS.popsize
      if R<=0.3333 %% main if

      ii=ceil(rand*(OPTIONS.popsize));
      while ii==i
        ii=ceil(rand*(OPTIONS.popsize));
      end
       jj=ceil(rand*(OPTIONS.numVar));
      if Population(i).cost<Population(ii).cost
        if GenIndex < (OPTIONS.Maxgen /2)
        pp(ii,jj) = (1-R^2)*pp(i,jj);
        else
        pp(ii,jj) = (1-rand)*pp(i,jj);
        end
      else
        if GenIndex < (OPTIONS.Maxgen /2)
       pp(i,jj) = (1-R^2)*pp(ii,jj);
        else
         pp(i,jj) = (1-rand)*pp(ii,jj);
        end
      end
      pp_new = FeasibleFunction_1(OPTIONS, pp(i,:));
      pp_new_cost = CostFunction_1(OPTIONS, pp_new);
      % Increase the number of function evaluation counter
      FE=FE+1;
      if pp_new_cost<Population(i).cost
        Population(i).chrom =pp_new;
        Population(i).cost=pp_new_cost;
      end
 elseif 0.3333< R && R <=0.6666 %%main if
     ii=ceil(rand*(OPTIONS.popsize));
     while ii==i
```

```
        ii=ceil(rand*(OPTIONS.popsize));
      end
      j=ceil(rand*(OPTIONS.numVar));
      if Population(i).cost<Population(ii).cost
       for j=1:OPTIONS.numVar
         if GenIndex < (OPTIONS.Maxgen *1/2)
      pp(i,j) =pp(i,j)+ (R)*(pp(i,j)-pp(ii,j));
         else
      pp(i,j) =pp(i,j)+ (rand)*(pp(i,j)-pp(ii,j));
         end
       end
      else
       for j=1:OPTIONS.numVar
         if GenIndex < (OPTIONS.Maxgen *1/2)
       pp(i,j) =pp(i,j)+ (R)*(pp(ii,j)-pp(i,j));
         else
       pp(i,j) =pp(i,j)+ (rand)*(pp(ii,j)-pp(i,j));
         end
       end
      end
      Island_1 = FeasibleFunction_1(OPTIONS, pp(i,:));
      Island_1_cost = CostFunction_1(OPTIONS, Island_1);
      % Increase the number of function evaluation counter
      FE=FE+1;
       if Island_1_cost<Population(i).cost
       Population(i).chrom =Island_1;
       Population(i).cost=Island_1_cost;
      end
  else %%%%% main if
      for j=1:OPTIONS.numVar
        if GenIndex < round(OPTIONS.Maxgen *1/10)
       pp(i,j) =pp(i,j)+ (R)*(pp(1,j)- Mean_pop(j)*abs(R-rand));
        else
       pp(i,j) =pp(i,j)+ (R)*(pp(1,j)- Mean_pop(j)*round(1+rand));
        end
      end
      Island_1 = FeasibleFunction_1(OPTIONS, pp(i,:));
      Island_1_cost = CostFunction_1(OPTIONS, Island_1);
      % Increase the number of function evaluation counter
      FE=FE+1;
       if Island_1_cost<Population(i).cost
       Population(i).chrom =Island_1;
       Population(i).cost=Island_1_cost;
       end
   end %%% main if end
   end
```

```
Population = PopSort(Population);
% Replace the worst with the previous generation's elites.
n = length(Population);
for i = 1 : Keep
  Population(n-i+1).chrom = chromKeep(i,:);
  Population(n-i+1).cost = costKeep(i);
end
% Make sure the population does not have duplicates.
Population = ClearDups(Population, MaxParValue, MinParValue);
% Sort from best to worst
Population = PopSort(Population);
% Compute the average cost
[AverageCost, nLegal] = ComputeAveCost(Population);
% Display info to screen
MinCost = [MinCost Population(1).cost];
AvgCost = [AvgCost AverageCost];
if mod(FE,20)==0
  if DisplayFlag
     disp([num2str(MinCost(end))]);
  end
end
end
end
Chrom = (Population(1).chrom);
fprintf('\n %f',MinCost(end));
fprintf('\n %f',Chrom);
fprintf('\n %f',FE);
```

8.7 The Passing Vehicle Search

```
function PVS()
run=1;
for i=1:run
PVS(@Tryvp);
end
%%%%%%%%%%%%%%%%%%%%%%%%%%%%%%%%%%%%%%%%%%%%%%%%%
function [InitFunction, CostFunction, FeasibleFunction,
FeasibleFunction_1,CostFunction_1] = Tryvp
InitFunction = @TryvpInit;
CostFunction = @TryvpCost;
CostFunction_1 = @TryvpCost_1;
FeasibleFunction = @TryvpFeasible;
FeasibleFunction_1 = @TryvpFeasible_1;
```

```
return;
%%%%%%%%%%%%%%%%%%%%%%%%%%%%%%%%%%%%%%%%%%%%%%%%
function [MaxParValue, MinParValue, Population, OPTIONS] = TryvpInit
(OPTIONS)
global MinParValue MaxParValue ll ul
Granularity = 1;
MinParValue = 1;
ll=-100*ones(1,30);
ul=100*ones(1,30);
MaxParValue = ul;
% Initialize population
for popindex = 1 : OPTIONS.popsize
  for k = 1 : OPTIONS.numVar
    chrom(k) =round((ll(k)+ (ul(k) - ll(k)) * rand));
  end
  Population(popindex).chrom = chrom;
end
OPTIONS.OrderDependent = true;
return;
%%%%%%%%%%%%%%%%%%%%%%%%%%%%%%%%%%%%%%%%%%%%%%%%
function [Population] = TryvpCost(OPTIONS, Population)
% Compute the cost of each member in Population
global MinParValue MaxParValue
popsize = OPTIONS.popsize;
for popindex = 1 : popsize
  for k = 1 : OPTIONS.numVar
    x(k) = Population(popindex).chrom(k);
  end
% Benchmark function number from objective
  Population(popindex).cost = objective(x);
end
return
%%%%%%%%%%%%%%%%%%%%%%%%%%%%%%%%%%%%%%%%%%%%%%%%
function [Population_1] = TryvpCost_1(OPTIONS, Population)
% Compute the cost of each member in Population
global MinParValue MaxParValue
popsize = size(Population,1);
Population_1 =[];
for popindex = 1 : popsize
% Benchmark function number from objective
  for k = 1 : OPTIONS.numVar
    x(k) = Population(popindex,k);
  end
   Population_1(popindex) = objective(x);
end
return
```

```
%%%%%%%%%%%%%%%%%%%%%%%%%%%%%%%%%%%%%%%%%%%%%%%%%%
function [Population] = TryvpFeasible(OPTIONS, Population)
global MinParValue MaxParValue ll ul
for i = 1 : OPTIONS.popsize
  for k = 1 : OPTIONS.numVar
    Population(i).chrom(k) = max(Population(i).chrom(k), ll(k));
    Population(i).chrom(k) = min(Population(i).chrom(k), MaxParValue
(k));
  end
end
return;
%%%%%%%%%%%%%%%%%%%%%%%%%%%%%%%%%%%%%%%%%%%%%%%%%%%%%%%
function [Population] = TryvpFeasible_1(OPTIONS, Population)
global MinParValue MaxParValue ll ul
for i = 1 : size(Population,1)
  for k = 1 : OPTIONS.numVar
    Population(i,k) = max(Population(i,k), ll(k));
    Population(i,k) = min(Population(i,k), MaxParValue(k));
  end
end
return;
%%%%%%%%%%%%%%%%%%%%%%%%%%%%%%%%%%%%%%%%%
function [OPTIONS, MinCost, AvgCost, InitFunction, CostFunction,
FeasibleFunction, ...
  MaxParValue, MinParValue, Population,FeasibleFunction_1,
CostFunction_1] = Init(DisplayFlag, ProblemFunction, RandSeed)
% Initialize population
OPTIONS.popsize = 50; % total population size
OPTIONS.Maxgen = 400; % generation count limit
OPTIONS.numVar = 30; % number of genes in each population member
OPTIONS.maxFE = 20000; % mutation probability
if ~exist('RandSeed', 'var')
  RandSeed = round(sum(100*clock));
end
rand('state', RandSeed); % initialize random number generator
% Get the addresses of the initialization, cost, and feasibility
functions.
[InitFunction, CostFunction, FeasibleFunction,FeasibleFunction_1,
CostFunction_1] = ProblemFunction();
% Initialize the population.
[MaxParValue, MinParValue, Population, OPTIONS] = InitFunction
(OPTIONS);
% Make sure the population does not have duplicates.
Population = ClearDups(Population, MaxParValue, MinParValue);
% Compute cost of each individual
```

```
Population = CostFunction(OPTIONS, Population);
% Sort the population from most fit to least fit
Population = PopSort(Population);
% Compute the average cost
AverageCost = ComputeAveCost(Population);
% Display info to screen
MinCost = [Population(1).cost];
AvgCost = [AverageCost];
if DisplayFlag
end
return;
%%%%%%%%%%%%%%%%%%%%%%%%%%%%%%%%%%%%%%%%%%%%%%%%%%%%%%%
function [fit] = objective(c)
fit = sum((x).^2);
end
%%%%%%%%%%%%%%%%%%%%%%%%%%%%%%%%%%%%%%%%%%%%%%%%%%%%%%%
function PVS(ProblemFunction, DisplayFlag, ProbFlag)
display(PVS is optimizing your problem');
format long
if ~exist('DisplayFlag', 'var');DisplayFlag = true; end
Population.chrom=[];
Population.cost=[];
 [OPTIONS, MinCost, AvgCost, InitFunction, CostFunction,
FeasibleFunction, ...
  MaxParValue, MinParValue, Population,FeasibleFunction_1,
CostFunction_1] = Init(DisplayFlag,ProblemFunction);
FE=0;
maxFE=OPTIONS.maxFE;
Keep=0;
norm_factor=1; %%%constant for normalization
VF=1; %% velocity factor
% Begin the optimization loop
for GenIndex = 1 : OPTIONS.Maxgen
while FE<maxFE
  %Save the solutions in a temporary array.
  for i = 1 : Keep
    chromKeep(i,:) = Population(i).chrom;
    costKeep(i) = Population(i).cost;
  end
  % Save population in variable temp_1
  for i=1:OPTIONS.popsize
    temp_1(i,:)=Population(i).chrom;
    temp_1_cost(i)=Population(i).cost;
  end
  Pop_size=OPTIONS.popsize;
```

```
    for k = 1 : OPTIONS.popsize
      r=[]; v=[];
      r(1) = k;
      while true;
        r(2) = round(OPTIONS.popsize * rand + 0.5);
        if (r(2) ~= r(1)), break, end
      end
      while true
        r(3) = round(OPTIONS.popsize* rand + 0.5);
        if (r(3) ~= r(1)) && (r(3) ~= r(2)), break, end
      end
      r2=r; % temp save r in r2
      r=(norm_factor/Pop_size)*r;%normalization
      r1=r;% save r in r1
      v1=(rand(1,3)).*(VF-r1);% generate random velocity
      %%%% Passing vehicle procedure %%%
      y=abs(r1(3)-r1(2));
      x=abs(r1(3)-r1(1));
      x1=(x*v1(3))/(v1(1)-v1(3));%% dist travelled by 3
      y1=(x*v1(2))/(v1(1)-v1(3));%%%dist travelled by 2
      %%% passing vehicle conditions%%%
      if v1(1)>v1(3)
        if (y-y1)>x1;
          con1=v1(1)/(v1(1)-v1(3));
          temp_1_new= Population(r2(1)).chrom+ con1*rand(1,OPTIONS.
numVar).*(Population(r2(1)).chrom-Population(r2(3)).chrom);
        else
        temp_1_new= Population(r2(1)).chrom+ rand(1,OPTIONS.numVar).*
(Population(r2(1)).chrom-Population(r2(2)).chrom);
        end
      else
       temp_1_new = Population(r2(1)).chrom+ rand(1,OPTIONS.numVar).*
(Population(r2(3)).chrom-Population(r2(1)).chrom);
      end
      temp_1_new = FeasibleFunction_1(OPTIONS, temp_1_new);
      temp_1_new_cost = CostFunction_1(OPTIONS, temp_1_new);
      % Increase the number of function evaluation counter
      FE=FE+1;
      if temp_1_new_cost<Population(k).cost
        Population(k).chrom =temp_1_new;
        Population(k).cost=temp_1_new_cost;
      end
    end %% pop end%
    Population = PopSort(Population);
    %Replace the worst with the previous generation's elites.
```

```
  n = length(Population);
  for i = 1 : Keep
    Population(n-i+1).chrom = chromKeep(i,:);
    Population(n-i+1).cost = costKeep(i);
  end
  Population = ClearDups(Population, MaxParValue, MinParValue);
  Population = PopSort(Population);
  MinCost = [MinCost Population(1).cost];
  %% print each iteration result
  if mod(FE,20)==0
    fprintf('\n %f',MinCost(end));
  end
end
end %%%%% end of generation
Chrom = (Population(1).chrom);
fprintf('\n %f',MinCost(end));% print objective function value
fprintf('\n design variables\n'); % print design variables
fprintf('\n %f',Chrom); % print design variables
fprintf('\n %f',FE);
```

8.8 The Symbiotic Organisms Search

```
function SOS()
run=1;
for i=1:run
SOS(@Tryvp);
end
%%%%%%%%%%%%%%%%%%%%%%%%%%%%%%%%%%%%%%%%%%%%%%%%
function [InitFunction, CostFunction, FeasibleFunction,
FeasibleFunction_1,CostFunction_1] = Tryvp
InitFunction = @TryvpInit;
CostFunction = @TryvpCost;
CostFunction_1 = @TryvpCost_1;
FeasibleFunction = @TryvpFeasible;
FeasibleFunction_1 = @TryvpFeasible_1;
return;
%%%%%%%%%%%%%%%%%%%%%%%%%%%%%%%%%%%%%%%%%%%%%%%%
function [MaxParValue, MinParValue, Population, OPTIONS] = TryvpInit
(OPTIONS)
global MinParValue MaxParValue ll ul
Granularity = 1;
MinParValue = 1;
```

```
ll=-100*ones(1,30);
ul=100*ones(1,30);
MaxParValue = ul;
% Initialize population
for popindex = 1 : OPTIONS.popsize
  for k = 1 : OPTIONS.numVar
    chrom(k) =round((ll(k)+ (ul(k) - ll(k)) * rand));
  end
  Population(popindex).chrom = chrom;
end
OPTIONS.OrderDependent = true;
return;
%%%%%%%%%%%%%%%%%%%%%%%%%%%%%%%%%%%%%%%%%%%%%%%%%%%%%%
function [Population] = TryvpCost(OPTIONS, Population)
% Compute the cost of each member in Population
global MinParValue MaxParValue
popsize = OPTIONS.popsize;
for popindex = 1 : popsize
  for k = 1 : OPTIONS.numVar
    x(k) = Population(popindex).chrom(k);
  end
% Benchmark function number from objective
  Population(popindex).cost = objective(x);
end
return
%%%%%%%%%%%%%%%%%%%%%%%%%%%%%%%%%%%%%%%%%%%%%%%%%%%%%%
function [Population_1] = TryvpCost_1(OPTIONS, Population)
% Compute the cost of each member in Population
global MinParValue MaxParValue
popsize = size(Population,1);
Population_1 = [];
for popindex = 1 : popsize
% Benchmark function number from objective
  for k = 1 : OPTIONS.numVar
    x(k) = Population(popindex,k);
  end
  Population_1(popindex) = objective(x);
end
return
%%%%%%%%%%%%%%%%%%%%%%%%%%%%%%%%%%%%%%%%%%%%%%%%%%%%%%
function [Population] = TryvpFeasible(OPTIONS, Population)
global MinParValue MaxParValue ll ul
for i = 1 : OPTIONS.popsize
  for k = 1 : OPTIONS.numVar
    Population(i).chrom(k) = max(Population(i).chrom(k), ll(k)); %
```

```
always chage this mini value
    Population(i).chrom(k) = min(Population(i).chrom(k), MaxParValue
(k));
  end
end
return;
%%%%%%%%%%%%%%%%%%%%%%%%%%%%%%%%%%%%%%%%%%%%%%%%%%%%%%%%%%%%%%%%%%%%
function [Population] = TryvpFeasible_1(OPTIONS, Population)
global MinParValue MaxParValue ll ul
for i = 1 : size(Population,1)
  for k = 1 : OPTIONS.numVar
    Population(i,k) = max(Population(i,k), ll(k)); % always chage this
mini value
    Population(i,k) = min(Population(i,k), MaxParValue(k));
  end
end
return;
%%%%%%%%%%%%%%%%%%%%%%%%%%%%%%%%%%%%%%%%%%%%%
function [OPTIONS, MinCost, AvgCost, InitFunction, CostFunction,
FeasibleFunction, ...
  MaxParValue, MinParValue, Population,FeasibleFunction_1,
CostFunction_1] = Init(DisplayFlag, ProblemFunction, RandSeed)
% Initialize population-based optimization software.
OPTIONS.popsize = 50; % total population size
OPTIONS.Maxgen = 400; % generation count limit
OPTIONS.numVar = 30; % number of genes in each population member
OPTIONS.maxFE = 20000; % mutation probability
if ~exist('RandSeed', 'var')
  RandSeed = round(sum(100*clock));
end
rand('state', RandSeed); % initialize random number generator
% Get the addresses of the initialization, cost, and feasibility
functions.
[InitFunction, CostFunction, FeasibleFunction,FeasibleFunction_1,
CostFunction_1] = ProblemFunction();
% Initialize the population.
[MaxParValue, MinParValue, Population, OPTIONS] = InitFunction
(OPTIONS);
% Make sure the population does not have duplicates.
Population = ClearDups(Population, MaxParValue, MinParValue);
% Compute cost of each individual
Population = CostFunction(OPTIONS, Population);
% Sort the population from most fit to least fit
Population = PopSort(Population);
% Compute the average cost
```

```
AverageCost = ComputeAveCost(Population);
% Display info to screen
MinCost = [Population(1).cost];
AvgCost = [AverageCost];
if DisplayFlag
end
return;
%%%%%%%%%%%%%%%%%%%%%%%%%%%%%%%%%%%%%%%%%%
function [fit] = objective(c)
fit = sum((x).^2);
end
%%%%%%%%%%%%%%%%%%%%%%%%%%%%%%%%%%%%%%%%%%
function SOS(ProblemFunction, DisplayFlag, ProbFlag)
display(SOS is optimizing your problem');
global ll ul
tic
if ~exist('DisplayFlag', 'var')
  DisplayFlag = true;
end
if ~exist('ProbFlag', 'var')
  ProbFlag = true;
end
 [OPTIONS, MinCost, AvgCost, InitFunction, CostFunction,
FeasibleFunction, ...
  MaxParValue, MinParValue, Population,FeasibleFunction_1,
CostFunction_1] = Init(DisplayFlag, ProblemFunction);
Keep=1;
maxFE=OPTIONS.maxFE;
FE=0;
TF=1;
MR=1;
% Begin the optimization loop
for GenIndex = 1 : OPTIONS.Maxgen
  % Save the best habitats in a temporary array.
while FE<maxFE
  for i = 1 : Keep
    chromKeep(i,:) = Population(i).chrom;
    costKeep(i) = Population(i).cost;
  end
  for i=1:length(Population)
    pp(i,:)=Population(i).chrom;
    pp_cost(i)=Population(i).cost;
  end
  pp_new=pp;
  pp;
  Mean_pop=mean(pp);
```

```
  Stdev_pop=std(pp);
  for i=1:length(Population)
  %%%%%%%%%%%%%%%%%%%%%%%%%%%%%%%%%%%%%%%%%%%%%%%%%%%%%%%%%%%%
%%%
  % Mutualism Phase
  % Choose organism k randomly other than organism i
  k=i;
  while i==k
    seed=randperm(length(Population));
    k=seed(1);
  end
  mutualVector=mean([Population(i).chrom;Population(k).chrom]);
  % Determine Mutual Vector & Beneficial Factor
  BF1=round(1+rand);
  BF2=round(1+rand);
  % Calculate new solution after Mutualism Phase
  pp_new1=Population(i).chrom+rand(1,size(pp,2)).*(Population(1).
chrom-BF1.*mutualVector);
  pp_new2=Population(k).chrom+rand(1,size(pp,2)).*(Population(1).
chrom-BF2.*mutualVector);
  pp_new1 = FeasibleFunction_1(OPTIONS, pp_new1);
  pp_new2 = FeasibleFunction_1(OPTIONS, pp_new2);
  % Evaluate the fitness of the new solution
  pp_new_cost1 = CostFunction_1(OPTIONS, pp_new1);
  pp_new_cost2 = CostFunction_1(OPTIONS, pp_new2);
  % Accept the new solution if the fitness is better
  if pp_new_cost1<Population(i).cost
      Population(i).chrom =pp_new1;
      Population(i).cost=pp_new_cost1;%%%%%%
  end
  if pp_new_cost2<Population(k).cost
      Population(k).chrom =pp_new2;
      Population(k).cost=pp_new_cost2;%%%%%%
  end
  % Increase the number of function evaluation counter
  FE=FE+2;
  % End of Mutualism Phase
  %%%%%%%%%%%%%%%%%%%%%%%%%%%%%%%%%%%%%%%%%%%%%%%%%%%%%%%%%%%%
%%%
  % Commensialism Phase
  % Choose organism k randomly other than organism i
  k=i;
  while i==k
    seed=randperm(length(Population));
    k=seed(1);
  end
```

```
  % Calculate new solution after Commensalism Phase
  pp_new1=Population(i).chrom+(rand(1,size(pp,2))*2-1).*
(Population(1).chrom-Population(k).chrom);
  pp_new1 = FeasibleFunction_1(OPTIONS, pp_new1);
  % Evaluate the fitness of the new solution
  pp_new_cost1 = CostFunction_1(OPTIONS, pp_new1);
  % Accept the new solution if the fitness is better
  if pp_new_cost1<Population(i).cost
      Population(i).chrom =pp_new1;
      Population(i).cost=pp_new_cost1;%%%%%%
  end
  % Increase the number of function evaluation counter
  FE=FE+1;
  % End of Commensalism Phase
  %%%%%%%%%%%%%%%%%%%%%%%%%%%%%%%%%%%%%%%%%%%%%%%%%%%%%%%%%%%%%%%%
%%%
  % Parasitism Phase
  % Choose organism k randomly other than organism i
  k=i;
  while i==k
    seed=randperm(length(Population));
    k=seed(1);
  end
  % Determine Parasite Vector & Calculate the fitness
  parasiteVector=Population(i).chrom;
  seed=randperm(size(pp,2));
  pick=seed(1:ceil(rand*size(pp,2))); % select random dimension
  parasiteVector(:,pick)=rand(1,length(pick)).*(ul(pick)-ll(pick))
+ll(pick);
  % Evaluate the fitness of the new solution
  parasiteVector_cost = CostFunction_1(OPTIONS, parasiteVector);
  % Accept the new solution if the fitness is better
  if pp_new_cost1<Population(k).cost
    Population(k).chrom =parasiteVector;
    Population(k).cost=parasiteVector_cost;%%%%%%
  end
  % Increase the number of function evaluation counter
  FE=FE+1;
  % End of Parasitism Phase
  %%%%%%%%%%%%%%%%%%%%%%%%%%%%%%%%%%%%%%%%%%%%%%%%%%%%%%%%%%%%%%%%
%%%
  end
  % Sort from best to worst
  Population = PopSort(Population);
  % Replace the worst with the previous generation's elites.
```

```
  n = length(Population);
  for i = 1 : Keep
    Population(n-i+1).chrom = chromKeep(i,:);
    Population(n-i+1).cost = costKeep(i);
  end
  if rand<1
  % Make sure the population does not have duplicates.
  Population = ClearDups(Population, MaxParValue, MinParValue);
  end
  % Sort from best to worst
  Population = PopSort(Population);
  % Compute the average cost
  [AverageCost, nLegal] = ComputeAveCost(Population);
  % Display info to screen
  MinCost = [MinCost Population(1).cost];
  AvgCost = [AvgCost AverageCost];
  if mod(FE,50)==0
  if DisplayFlag
       fprintf('\n %f',MinCost(end));
       fprintf('\n %f',MinCost(end));
   end
  end
end
end
Chrom = (Population(1).chrom);
fprintf('\n %f',MinCost(end));
fprintf('\n %f',Chrom);
fprintf('\n %f',FE);
```

8.9 The Grey Wolf Optimizer

```
function GWO()
run=1;
popsize=50;
gen=400;
fun_no=1;
for i=1:run
mainGWO(popsize,gen,fun_no);
end
%%%%%%%%%%%%%%%%%%%%%%%%%%%%%%%%%%%%%%%%%%
function [Best_score, Best_pos]=mainGWO(SearchAgents_no,
Max_iteration,fun_no)
```

```
tic;
if fun_no==1
  Function_name='F1'; % Sphere function
end
lb,ub,dim,fobj]=Get_Functions_details(Function_name);
[Best_score,Best_pos,GWO_cg_curve]=GWO(SearchAgents_no,
Max_iteration,lb,ub,dim,fobj);
disp(['Funnum:', num2str(Function_name)])
fprintf('\n%e\n',Best_score);
fprintf('\n%6.15f\n',Best_pos);
toc
%%%%%%%%%%%%%%%%%%%%%%%%%%%%%%%%%%%%%%%%%
function [lb,ub,dim,fobj] = Get_Functions_details(F)
switch F
  case 'F1'
    fobj = @F1; % Sphere function
    dim=30;
    lb = -100*ones(1,30);
    ub = 100*ones(1,30);
    maxFE=20000;
end
end
function o = F1(x)
o=sum((x).^2);
end
%%%%%%%%%%%%%%%%%%%%%%%%%%%%%%%%%%%%%%%%%
function [Alpha_score,Alpha_pos,Convergence_curve]=GWO
(SearchAgents_no,Max_iter,lb,ub,dim,fobj)
display('GWO is optimizing your problem');
% initialize alpha, beta, and delta_pos
Alpha_pos=zeros(1,dim);
Alpha_score=inf; %change this to -inf for maximization problems
Beta_pos=zeros(1,dim);
Beta_score=inf; %change this to -inf for maximization problems
Delta_pos=zeros(1,dim);
Delta_score=inf; %change this to -inf for maximization problems
%Initialize the positions of search agents
Positions=initialization(SearchAgents_no,dim,ub,lb);
Convergence_curve=zeros(1,Max_iter);
l=0;% Loop counter
FE=0; %
maxFE=Max_iter*SearchAgents_no;
% Main loop
while FE<maxFE
  for i=1:size(Positions,1)
```

```
    % Return back the search agents that go beyond the boundaries of the
search space
    Flag4ub=Positions(i,:)>ub;
    Flag4lb=Positions(i,:)<lb;
Positions(i,:)=(Positions(i,:).*(~(Flag4ub+Flag4lb)))+ub.*Flag4ub
+lb.*Flag4lb;
    % Calculate objective function for each search agent
    fitness=fobj(Positions(i,:));
    FE=FE+1;
    % Update Alpha, Beta, and Delta
    if fitness<Alpha_score
      Alpha_score=fitness; % Update alpha
      Alpha_pos=Positions(i,:);
    end
    if fitness>Alpha_score && fitness<Beta_score
      Beta_score=fitness; % Update beta
      Beta_pos=Positions(i,:);
    end
    if fitness>Alpha_score && fitness>Beta_score && fitness<Delta_score
      Delta_score=fitness; % Update delta
      Delta_pos=Positions(i,:);
    end
  end
  a=2-l*((2)/Max_iter); % a decreases linearly fron 2 to 0
  % Update the Position of search agents including omegas
  for i=1:size(Positions,1)
    for j=1:size(Positions,2)
      r1=rand(); % r1 is a random number in [0,1]
      r2=rand(); % r2 is a random number in [0,1]
      A1=2*a*r1-a;
      C1=2*r2;
      D_alpha=abs(C1*Alpha_pos(j)-Positions(i,j));
      X1=Alpha_pos(j)-A1*D_alpha;
      r1=rand();
      r2=rand();
      A2=2*a*r1-a;
      C2=2*r2;
      D_beta=abs(C2*Beta_pos(j)-Positions(i,j));
   X2=Beta_pos(j)-A2*D_beta;
      r1=rand();
      r2=rand();
      A3=2*a*r1-a;
      C3=2*r2;
      D_delta=abs(C3*Delta_pos(j)-Positions(i,j));
```

```
        X3=Delta_pos(j)-A3*D_delta;
        Positions(i,j)=(X1+X2+X3)/3;
      end
    end
    l=l+1;
    Convergence_curve(l)=Alpha_score;
    % Display the iteration and best optimum obtained so far
    if mod(FE,20)==0
      fprintf('%e\n',Alpha_score);
    end
end
disp(['FE:',num2str(FE)])
```

8.10 The Teaching-Learning-Based Optimization

```
function TLBO()
run=1;
for i=1:run
TLBO(@Tryvp);
end
%%%%%%%%%%%%%%%%%%%%%%%%%%%%%%%%%%%%%%%%%%%%%%%%%%%%
function [InitFunction, CostFunction, FeasibleFunction,
FeasibleFunction_1,CostFunction_1] = Tryvp
InitFunction = @TryvpInit;
CostFunction = @TryvpCost;
CostFunction_1 = @TryvpCost_1;
FeasibleFunction = @TryvpFeasible;
FeasibleFunction_1 = @TryvpFeasible_1;
return;
%%%%%%%%%%%%%%%%%%%%%%%%%%%%%%%%%%%%%%%%%%%%%%%%%%%%
function [MaxParValue, MinParValue, Population, OPTIONS] = TryvpInit
(OPTIONS)
global MinParValue MaxParValue ll ul
Granularity = 1;
MinParValue = 1;
ll=-100*ones(1,30);
ul=100*ones(1,30);
MaxParValue = ul;
% Initialize population
for popindex = 1 : OPTIONS.popsize
```

```
  for k = 1 : OPTIONS.numVar
    chrom(k) =round((ll(k)+ (ul(k) - ll(k)) * rand));
  end
  Population(popindex).chrom = chrom;
end
OPTIONS.OrderDependent = true;
return;
%%%%%%%%%%%%%%%%%%%%%%%%%%%%%%%%%%%%%%%%%%%%%%%%%%%%
function [Population] = TryvpCost(OPTIONS, Population)
% Compute the cost of each member in Population
global MinParValue MaxParValue
popsize = OPTIONS.popsize;
for popindex = 1 : popsize
  for k = 1 : OPTIONS.numVar
    x(k) = Population(popindex).chrom(k);
  end
% Benchmark function number from objective
  Population(popindex).cost = objective(x);
end
return
%%%%%%%%%%%%%%%%%%%%%%%%%%%%%%%%%%%%%%%%%%%%%%%%%%%%
function [Population_1] = TryvpCost_1(OPTIONS, Population)
% Compute the cost of each member in Population
global MinParValue MaxParValue
popsize = size(Population,1);
Population_1 =[];
for popindex = 1 : popsize
% Benchmark function number from objective
  for k = 1 : OPTIONS.numVar
    x(k) = Population(popindex,k);
  end
   Population_1(popindex) = objective(x);
end
return
%%%%%%%%%%%%%%%%%%%%%%%%%%%%%%%%%%%%%%%%%%%%%%%%%%%%
function [Population] = TryvpFeasible(OPTIONS, Population)
global MinParValue MaxParValue ll ul
for i = 1 : OPTIONS.popsize
  for k = 1 : OPTIONS.numVar
    Population(i).chrom(k) = max(Population(i).chrom(k), ll(k));
Population(i).chrom(k) = min(Population(i).chrom(k), MaxParValue
(k));
  end
end
return;
```

```
%%%%%%%%%%%%%%%%%%%%%%%%%%%%%%%%%%%%%%%%%%%%%%%%%%%%%%%%%%%%
function [Population] = TryvpFeasible_1(OPTIONS, Population)
global MinParValue MaxParValue ll ul
for i = 1 : size(Population,1)
  for k = 1 : OPTIONS.numVar
    Population(i,k) = max(Population(i,k), ll(k));
 Population(i,k) = min(Population(i,k), MaxParValue(k));
  end
end
return;
%%%%%%%%%%%%%%%%%%%%%%%%%%%%%%%%%%%%%%%%%%%
function [OPTIONS, MinCost, AvgCost, InitFunction, CostFunction,
FeasibleFunction, ...
  MaxParValue, MinParValue, Population,FeasibleFunction_1,
CostFunction_1] = Init(DisplayFlag, ProblemFunction, RandSeed)
% Initialize population-based optimization software.
OPTIONS.popsize = 50; % total population size
OPTIONS.Maxgen = 400; % generation count limit
OPTIONS.numVar = 30; % number of genes in each population member
OPTIONS.maxFE = 20000; % mutation probability
if ~exist('RandSeed', 'var')
  RandSeed = round(sum(100*clock));
end
rand('state', RandSeed); % initialize random number generator
% Get the addresses of the initialization, cost, and feasibility
functions.
[InitFunction, CostFunction, FeasibleFunction,FeasibleFunction_1,
CostFunction_1] = ProblemFunction();
% Initialize the population.
[MaxParValue, MinParValue, Population, OPTIONS] = InitFunction
(OPTIONS);
% Make sure the population does not have duplicates.
Population = ClearDups(Population, MaxParValue, MinParValue);
% Compute cost of each individual
Population = CostFunction(OPTIONS, Population);
% Sort the population from most fit to least fit
Population = PopSort(Population);
% Compute the average cost
AverageCost = ComputeAveCost(Population);
% Display info to screen
MinCost = [Population(1).cost];
AvgCost = [AverageCost];
if DisplayFlag
end
return;
```

```
%%%%%%%%%%%%%%%%%%%%%%%%%%%%%%%%%%%%%%%%%%%%
function [fit] = objective(c)
fit = sum((x).^2);
end
%%%%%%%%%%%%%%%%%%%%%%%%%%%%%%%%%%%%%%%%%%%%
function TLBO(ProblemFunction, DisplayFlag, ProbFlag)
%function [MinCost, Hamming] = BBO(ProblemFunction, DisplayFlag,
ProbFlag)
global ll ul
tic
if ~exist('DisplayFlag', 'var')
  DisplayFlag = true;
end
if ~exist('ProbFlag', 'var')
  ProbFlag = true;
end
[OPTIONS, MinCost, AvgCost, InitFunction, CostFunction,
FeasibleFunction, ...
  MaxParValue, MinParValue, Population,FeasibleFunction_1,
CostFunction_1] = Init(DisplayFlag, ProblemFunction);
Keep=0;
maxFE=OPTIONS.maxFE;
FE=0;
TF=1;
MR=1;
% Begin the optimization loop
for GenIndex = 1 : OPTIONS.Maxgen
while FE<maxFE
  % Save the best habitats in a temporary array.
  for i = 1 : Keep
    chromKeep(i,:) = Population(i).chrom;
    costKeep(i) = Population(i).cost;
  end
  for i=1:length(Population)
  pp(i,:)=Population(i).chrom;
  pp_cost(i)=Population(i).cost;
  end
  pp_new=pp;
  pp;
  Mean_pop=mean(pp);
  Stdev_pop=std(pp);
 TF=round(1+rand*(1));
  for i=1:size(pp,2)
```

```
    New_Mean(i)=((pp(1,i)));
    for j=1:size(pp,1)
      pp_new(j,i)=pp(j,i)+rand*(New_Mean(i)-TF*Mean_pop(i));
    end
   end
     pp_new = FeasibleFunction_1(OPTIONS, pp_new);
     pp_new_cost = CostFunction_1(OPTIONS, pp_new);
     FE=FE+length(Population);
     % Increase the number of function evaluation counter
     for i = 1 : length(Population)
      if pp_new_cost(i)<Population(i).cost
        Population(i).chrom =pp_new(i,:);
        Population(i).cost=pp_new_cost(i);%%%%%%
      end
     end
  % Compute the average cost
  [AverageCost, nLegal] = ComputeAveCost(Population);
  % Display info to screen
  MinCost = [MinCost Population(1).cost];
  AvgCost = [AvgCost AverageCost];
  if mod(FE,20)==0
  if DisplayFlag
        fprintf('\n %f',MinCost(end));
  end
  end
   %%%student loop
  for i = 1 : 1 : length(Population)
    ii=ceil(rand*(length(Population)));
    while ii==i
        ii=ceil(rand*(length(Population)));
    end
    z=rand;
    z1=z*(1-z);
    if Population(i).cost<Population(ii).cost
      Island_1(i,:) = (Population(i).chrom + rand*(Population(i).
chrom-Population(ii).chrom));
    else
      Island_1(i,:) = (Population(i).chrom+ rand*(Population(ii).
chrom-Population(i).chrom));
    end
 end
 % Make sure the trial vector is feasible
   Island_1 = FeasibleFunction_1(OPTIONS, Island_1);
```

```
        % Compute the cost of the trial vector and the test vector
        Island_1_cost = CostFunction_1(OPTIONS, Island_1);
        % Increase the number of function evaluation counter
        FE=FE+length(Population);
        % Restore the original population size
        for i = 1 : length(Population)
         if Island_1_cost(i)<Population(i).cost
            Population(i).chrom =Island_1(i,:);
            Population(i).cost=Island_1_cost(i);%%%%%%
         end
        end
    % Sort from best to worst
    Population = PopSort(Population);
    % Replace the worst with the previous generation's elites.
    n = length(Population);
    for i = 1 : Keep
        Population(n-i+1).chrom = chromKeep(i,:);
        Population(n-i+1).cost = costKeep(i);
    end
    if rand<1
    % Make sure the population does not have duplicates.
    Population = ClearDups(Population, MaxParValue, MinParValue);
    end
    % Sort from best to worst
    Population = PopSort(Population);
    % Compute the average cost
    [AverageCost, nLegal] = ComputeAveCost(Population);
    % Display info to screen
    MinCost = [MinCost Population(1).cost];
    AvgCost = [AvgCost AverageCost];
    if mod(FE,20)==0
    if DisplayFlag
          fprintf('\n %f',MinCost(end));
    end
    end
end
end
Chrom = (Population(1).chrom);
fprintf('\n %f',MinCost(end));
fprintf('\n %f',Chrom);
fprintf('\n %f',FE);
```

Index

A

ANSYS, 6, 361–364, 367, 369, 371, 373, 375, 378, 380, 382, 383
Ant lion optimizer (ALO), 5, 6, 38, 39, 44–45, 58, 71–80, 82–84, 86–90, 92, 94–96, 98–101, 103–105, 108–111, 113–116, 118, 120, 121, 123–128, 130, 132, 134, 135, 137–141, 148, 149, 156, 159, 171–175, 177, 178, 180–184, 186–189, 191, 192, 194, 196, 198, 199, 201, 202, 204, 205, 207, 208, 210, 211, 213–215, 217, 219, 220, 222, 223, 225, 226, 228, 230, 232, 233, 235, 236, 244, 246, 248, 250, 252, 254, 264, 267, 269, 271–274, 276, 277, 279, 280, 282, 284–287, 289, 292, 293, 295–297, 299, 303, 304, 306–308, 310, 311, 314, 315, 317–319, 321, 322, 326, 327, 330, 332, 333, 335–337, 340, 341, 343, 345, 347–350, 354–356, 398–400

B

Buckling, 1, 3, 11, 20, 24–31, 38, 57–61, 64, 74, 88, 91, 95, 107, 112, 124, 127, 131, 155–158, 160, 163, 164, 176, 186, 190, 193, 209, 219, 224, 227, 242–245, 268, 278, 283, 286, 306, 314, 329, 334, 338

C

Continuous sections, 31, 32, 72, 75, 77, 80, 82, 84, 87, 90, 93, 98–101, 103–105, 132, 172, 175, 178, 180–182, 184, 187, 189, 190, 192, 196–199, 201, 202, 245–304

D

Discrete sections, 32, 33, 108, 111, 113, 116, 119, 123, 126, 128, 129, 134, 137–140, 144, 205, 208, 211, 214, 215, 217, 220, 223, 224, 226, 230, 231, 233, 235, 236, 244, 305–356
Dragonfly algorithm (DA), 5, 6, 38–40, 58, 71–80, 82–84, 86–90, 92, 94–96, 98–101, 103–105, 108–111, 113–116, 118, 120, 121, 123–128, 130, 132, 134, 135, 137–141, 148, 149, 156, 159, 171–175, 177, 178, 180–184, 186–189, 191, 192, 194, 196, 198, 199, 201, 202, 204, 205, 208, 210, 211, 213–215, 217, 219, 220, 222, 223, 225, 226, 228, 230, 232, 233, 235, 236, 244, 246, 248, 250, 252, 254, 264, 267, 269, 271–274, 276, 277, 279, 280, 282, 284–287, 289, 292, 293, 295–299, 303, 304, 306–308, 310, 311, 314, 315, 317–319, 321, 322, 326, 327, 330, 332, 333, 335–337, 340, 341, 343, 345, 347–350, 354–356, 385–389

F

Finite element method (FEM), 11–20

G

Grey wolf optimizer (GWO), 5, 6, 38, 39, 50–51, 58, 71–80, 82–84, 86–92, 94–96, 98–106, 108–111, 113–116, 118, 120, 121, 123–128, 130–132, 134–141, 148, 149, 156, 159, 171–175, 177, 178, 180–184, 186–189, 191, 192, 194, 196,

V. Savsani et al., *Truss Optimization*, https://doi.org/10.1007/978-3-031-49295-2

198–202, 204, 205, 207, 208, 210, 211, 213–215, 217, 219, 220, 222, 223, 225, 226, 228, 230, 232–236, 244, 246, 248, 250, 252, 254, 264, 265, 267, 269, 271–274, 276, 277, 279, 280, 282, 284–287, 289, 292, 293, 295–299, 303, 304, 306–308, 310, 311, 314, 315, 317–319, 321, 322, 326, 327, 330, 332, 333, 335–337, 340, 341, 343, 345, 347–350, 354–356, 417–420
Ground structure approach, 20, 157

H
Heat transfer search (HTS), 5, 6, 38, 39, 45–47, 58, 71–80, 82–84, 86–90, 92, 94–96, 98–105, 108–111, 113–116, 118, 120, 121, 123–128, 130, 132, 134–142, 144, 148, 149, 156, 159, 171–175, 177, 178, 180–184, 186–189, 191, 192, 194, 196, 198, 199, 201, 202, 204, 205, 207, 208, 210, 211, 213–215, 217, 219, 220, 222, 223, 225, 226, 228, 230, 232–236, 244, 246, 248, 250, 252, 254, 264, 267, 269, 271–274, 276, 277, 279, 280, 282, 284, 285, 287, 289, 292, 293, 295–299, 303, 304, 306–308, 310, 311, 314, 315, 317–319, 321, 322, 326, 327, 330, 332, 333, 335–337, 340, 341, 343, 345, 347–350, 354–356, 400–406

K
Kinematic instability, 3, 24, 58, 155, 158

M
MATLAB, 25, 68, 244, 363, 364, 367, 369, 371, 373, 375, 378, 380, 382, 383, 385–425
Metaheuristics, 3–6, 10, 37–52, 58, 60, 78, 79, 83, 86, 89, 91, 100, 106, 109, 110, 114, 115, 118, 121, 124, 125, 128, 131, 132, 135, 142, 156, 157, 159, 174, 180, 182, 186, 188, 198, 203, 204, 207, 210, 213, 215, 219, 222, 225, 228, 232, 242, 244, 245, 264, 267, 269, 271, 273, 282, 284, 287, 292, 298, 305, 308, 310, 319, 321, 326, 333, 337, 340, 343, 350, 355
Multi verse optimizer (MVO), 5, 6, 38–42, 58, 71–80, 82–84, 86–90, 92, 94–96, 98–101, 103–105, 108–111, 113–116, 118, 120, 121, 123–128, 130, 132, 134–141, 147–149, 156, 159, 171–175, 177, 178, 180–184, 186–189, 191, 192, 194, 196, 198, 199, 201, 202, 204, 205, 207, 208, 210, 211, 213–215, 217, 219, 220, 222, 223, 225, 226, 228, 230, 232, 233, 235, 236, 244, 246, 248, 250, 252, 254, 264, 267, 269, 271–274, 276, 277, 279, 280, 282, 284–287, 289, 292–297, 299, 303, 304, 306–308, 310, 311, 314, 315, 317–319, 321, 322, 326, 327, 330, 332, 333, 335–337, 340, 341, 343, 345–350, 354–356, 371, 373, 389, 390

N
Natural frequencies, 3, 10, 11, 13, 14, 20, 22, 23, 25–29, 57, 58, 60, 61, 63, 64, 74, 76, 79, 83, 86, 88, 89, 92, 95, 97, 99, 100, 106, 107, 110, 112, 115, 118, 122, 124, 125, 128, 131–133, 136, 155–157, 160, 163, 164, 174, 176, 177, 179, 183, 185, 186, 188, 190, 191, 193, 195, 197, 198, 204, 207, 209, 210, 213, 216, 218, 221, 222, 224, 225, 227, 229, 231, 232, 242, 244, 245, 247, 249, 251, 253, 268, 269, 275, 279, 283, 284, 286, 288, 290, 291, 306, 310, 313, 314, 317, 321, 326, 330, 334, 338, 339, 343, 344

O
Optimization techniques, 3–5, 31, 32, 79, 82, 190, 312, 320

P
Passing vehicle search (PVS), 5, 6, 38, 39, 47–48, 58, 71–80, 82–84, 86–92, 94–96, 98–106, 108–111, 113–116, 118, 120, 121, 123–128, 130–132, 134–142, 144, 147–149, 156, 159, 171–175, 177, 178, 180–184, 186–194, 196, 198–202, 204–208, 210, 211, 213–217, 219, 220, 222, 223, 225, 226, 228, 230, 232–236, 244, 246, 248, 250, 252, 254, 264, 265, 267, 269, 271–274, 276, 277, 279, 280, 282, 284, 285, 287, 289, 292–299, 303, 304, 306–308, 310, 311, 314, 315, 317–319, 321, 322, 326–328, 330, 332, 333, 335–337, 340, 341, 343–350, 354–356, 367, 369, 371, 375, 382, 406–411

Planar and space trusses, v, 12, 14–20, 60, 63, 106, 159, 163, 203, 245, 256
Problem complexity, 29–31

S

Shape optimization, 1, 9, 11, 60, 156, 157, 204, 241, 305, 309, 312
Sine cosine algorithm (SCA), 5, 6, 38, 39, 42, 58, 71–80, 82–84, 86–90, 92, 94–96, 98–101, 103–105, 108–111, 113–116, 118, 120, 121, 123–128, 130, 132, 134, 135, 137–141, 148, 149, 156, 159, 171–175, 177, 178, 180–184, 186–189, 191, 192, 194, 196, 198, 199, 201, 202, 204, 205, 207, 208, 210, 211, 213–215, 217, 219, 220, 222, 223, 225, 226, 228, 230, 232, 233, 235, 236, 244, 246, 248, 250, 252, 254, 264, 267, 269, 271–274, 276, 277, 279, 280, 282, 284–287, 289, 292, 293, 295–299, 303, 304, 306–308, 310, 311, 314, 315, 317–319, 321, 322, 326, 327, 330, 332, 333, 335–337, 340, 341, 343–345, 347–350, 355, 356, 393–395
Single-stage approach, 11, 14, 39, 157, 244
Size optimization, 1–4, 6, 9, 11, 21–23, 26, 29, 38, 57–150, 155–235, 241–356, 363, 364, 367, 369, 371, 373, 375, 378, 380, 382
Structural optimization (SO), 1, 38, 39, 57, 58, 155, 156
Symbiotic organisms search (SOS), 5, 6, 38, 39, 48–49, 58, 71–80, 82–84, 86–90, 92, 94–96, 98–105, 108–111, 113–116, 118, 120, 121, 123–128, 130–132, 134–142, 147–149, 156, 159, 171–175, 177, 178, 180–184, 186–194, 196, 198–202, 204–208, 210, 211, 213–217, 219, 220, 222, 223, 225, 226, 228, 230, 232–236, 244, 246, 248, 250, 252, 254, 264, 265, 267, 269, 271–274, 276, 277, 279, 280, 282, 284, 285, 287, 289, 292–299, 303, 304, 306–308, 310, 311, 314, 315, 317–319, 321, 322, 326, 327, 330, 332, 333, 335–337, 340, 341, 343, 345, 347–350, 354–356, 411–417

T

Teaching learning based optimization (TLBO), 5, 6, 38, 39, 51–52, 58, 71–80, 82–84, 86–90, 92, 94–96, 98–106, 108–111, 113–116, 118, 120, 121, 123–128, 130, 132, 134, 135, 137–141, 148, 149, 156, 159, 171–175, 177, 178, 180–184, 186–189, 191, 192, 194, 196, 198–202, 204, 205, 207, 208, 210, 211, 213–215, 217, 219, 220, 222, 223, 225, 226, 228, 230, 232–236, 244, 246, 248, 250, 252, 254, 264, 267, 269, 271–274, 276, 277, 279, 280, 282, 284, 285, 287, 289, 292–299, 303, 304, 306–308, 310, 311, 314, 315, 317–319, 321, 322, 326–328, 330, 332, 333, 335–337, 340, 341, 343, 345, 347–350, 354–356, 363, 364, 367, 373, 378, 380, 420, 423
Topology optimization, 1, 9, 38, 57, 156, 241
Truss optimization, 1, 3, 6, 9, 10, 20, 25, 29, 38, 57, 58, 60, 61, 143, 156, 197, 204, 231, 242, 247, 274, 308, 321, 325
Truss topology optimization (TTO), 1, 3, 5, 9, 11, 24–29, 38, 39, 155–236, 297, 298, 303, 349
Two-stage approach, 11, 157

W

Whale optimization algorithm (WOA), 5, 6, 38, 39, 43–44, 58, 71–80, 82–84, 86–90, 92, 94–96, 98–105, 108–111, 113–116, 118, 120, 121, 123–128, 130, 132, 134, 135, 137–141, 148, 149, 156, 159, 171–175, 177, 178, 180–184, 186–189, 191, 192, 194, 196, 198–202, 204, 205, 208, 210, 211, 213–215, 217, 219, 220, 222, 223, 225, 226, 228, 230, 232–236, 244, 246, 248, 250, 252, 254, 264, 267, 269, 271–274, 276, 277, 279, 280, 282, 284, 285, 287, 289, 292–299, 303, 304, 306–308, 310, 311, 314, 315, 317–319, 321, 322, 326, 327, 330, 332, 333, 335–337, 340, 341, 343, 345, 347–350, 354–356, 395–397